ECOLOGY OF
FRESHWATER FISH
PRODUCTION

ECOLOGY OF FRESHWATER FISH PRODUCTION

EDITED BY

SHELBY D. GERKING

PhD

Department of Zoology

Arizona State University

Tempe

A HALSTED PRESS BOOK

JOHN WILEY & SONS

NEW YORK–TORONTO

© 1978 Blackwell Scientific Publications
Osney Mead, Oxford OX2 0EL
8 John Street, London WC1N 2ES
9 Forrest Road, Edinburgh EH1 2QH
P.O. Box 9, North Balwyn, Victoria, Australia

First published 1978

Library of Congress
Cataloging in Publication Data

Main entry under title:
Ecology of freshwater fish production.
  Includes indexes.
    1. Fish populations. 2. Fishes—Growth.
3. Fishes—Behavior. 4. Fish-culture. I. Gerking,
Shelby Delos, 1918-
QL618.3.E27   1978   639'.31   77-92407

ISBN 0-470-99362-6

Published in the U.S.A. by
Halsted Press, a Division of
John Wiley & Sons, Inc., New York.

Printed in Great Britain

Contents

VITAL STATISTICS OF
FISH POPULATIONS

THE FISH POPULATION AND
ITS FOOD SUPPLY

COMPETITION AND SOCIAL BEHAVIOUR
INFLUENCING PRODUCTION

THE CONTRIBUTION OF FISH PRODUCTION
TO HUMAN NUTRITION AND WELL-BEING

Preface

This book is an outgrowth of a symposium on 'The Biological Basis of Freshwater Fish Production' held in 1966 at The University, Reading, England, under the sponsorship of the sectional committee on Productivity of Freshwater Communities of the International Biological Programme. The proceedings of that meeting appeared in 1967 as a volume carrying the same title as the symposium. The response to that volume stimulated this book, now appearing exactly ten years later. The title has been changed because it is no longer a symposium volume; it has some new chapters, some new authors and a somewhat different approach.

The question of publishing this book was raised in a conference with Blackwell Scientific Publications in 1972. The idea germinated for two years and plans were laid in 1974. Correspondence with prospective authors at that time was more than encouraging; it was enthusiastic. As a result, assembling the final author list and blocking out the topics began in earnest. The task was accomplished in late 1976.

My sincere thanks go to the authors of the various chapters. The literature review of a broad topic is no small task in our field. The fisheries literature is scattered, and it takes persistence and a great deal of effort to bring it together. Each author performed his task willingly and with much insight. Additionally, no sharp yelps of pain were heard even though the editor's scalpel was wielded without anaesthesia, and that helped me maintain my equilibrium when the work load became heavy.

Each author submitted a manuscript which was edited and returned for polishing. Several authors had their chapters reviewed additionally by colleagues. The inclusion of authors from several countries was essential to the presentation of the subject matter. Consequently, English language usage varied considerably, and I attempted in the editing to retain the author's meaning but at the same time to present the subject in an easily read style. I have not attempted to eliminate the 'flavour' given to a subject by an author using English as a second language.

Deserved thanks are extended to Mr. Robert Campbell of Blackwell Scientific Publications who encouraged this work from its inception.

Tempe, Arizona, USA   Shelby D. Gerking
1 December 1976

Introduction

Fish have been regarded historically as a staple, dependable, and nutritious food whether from the sea or freshwater. Certain lore has grown up about them which is exemplified by our admiration for the 'men who go down to the sea in ships' to catch them and the rod-and-reel fisherman who proudly brings home a meal for the family. The fish is even a religious symbol for millions of people. This favourable image is based in part on the abundance of fish in the waters of the earth and our ability to catch them whenever we needed them. That part of the image is becoming tarnished, however.

We are now confronted with the fact that many, once dependable, marine fish stocks are overexploited by modern technological hunting methods. We track them down by helicopter; we locate their depth with sonar; we design special ships and nets to surround them; and we even package and freeze them at sea to make them ready for the market immediately upon docking. As for freshwater fish, we have reduced some of the most favorable habitats drastically by our sewage, industrial wastes, acidification, and the slow, accumulative effects of eutrophication. The gradual erosion of our commercial and sport fish stocks and the species composition changes which accompany heavy exploitation and alteration of the habitat is one reason why our science, fishery biology, came into existence.

Science exists to benefit mankind, and it does so by grappling with abstract concepts and with practical application. Fishery biology is a satisfying blend of the theoretical and practical. The mathematician and statistician who catch their fish 'on paper' and the fellow in hip boots on the deep end of a seine net are both contributing to the progress of our science. This book reflects this variety. Mathematical models of estimating abundance and survival are found here alongside laboratory experiments on metabolism and the harvesting of fish from ponds.

We have made much progress in the relatively few years that scientific fishery biology has been practiced, and we are in a good position to deploy our knowledge to help meet the increased demand for human protein food that is almost sure to develop. We make no bones about the fact that we are interested in production studies because we believe that the scientific principles that we espouse can be employed to meet this need for additional protein. In what form these principles will appear is a matter of conjecture. Many feel that pond culture and aquaculture in general will make the greatest contribution, if we pool our human and capital resources for a thrust in this direction. We would also hope that our knowledge can be put to good use in making our inland lakes and streams more suitable for sport and commercial fishing.

At the present time the freshwater fish yield amounts to 23·7 per cent (10·2 million metric tons) of a global aquatic harvest (both freshwater and marine) of 43·0 million metric tons of fish used for human food, not counting the sport and subsistence fishing which do not enter the statistics. If we in freshwater fishery biology are to contribute our share to improving the percentage of protein in the average diet of an increasing population, we must develop some rather idealistic goals. The commercial catch should be increased two-fold, bringing it to about 20 million metric tons, and pond culture must increase by about 15 times, or from 700,000 metric tons to 10·5 million metric tons, by the year 2000. The doubling of the commercial catch is possible if pollution abatement is progressive and lasting and if, through regulation, we are able to harvest a greater share of the catchable population than we do now, without overexploitation. Increasing the harvest of pond culture by 15 times in about 25 years is truly an enormous task; and we have no way of predicting whether it is over-ambitious. The goal was set to meet the need for at least twice the animal protein in the diet that we now supply, if everyone is to have an improved, though not adequate, diet [see S. J. Holt (1967) *The Biological Basis of Freshwater Fish Production*. Blackwell Scientific Publications, Oxford]. The goal is only obtainable under the most favourable conditions where the needed resources are committed to an extensive programme of pond construction and after society concludes that fish must be eaten in increasing amounts to satisfy its basic need for protein.

The common denominator of this demand for fish as sport and fish as food is production, or simply the rate of growth of a population. The basic questions of reproduction, food assimilation, growth, causes of mortality, movements, etc., must all be investigated to understand the production of fish in a body of water, whatever use is made of the product. The scientific approach is, then, to identify these processes, to define their relative importance, and to study the factors which limit them. Viewed in this fashion, fish production becomes a fundamental biological problem. For this reason the book has been titled *Ecology of Freshwater Fish Production* in order to distinguish its subject matter from purely technical aspects such as stocking rates, species combinations, dam construction, egg rearing, feeding, and the like.

Fish production is defined in the technical sense as the total quantity of tissue elaborated over a stated period of time regardless of whether or not all of it survives to the end of that time. If has often been confused with the yield to man because the term production is used in the rearing of domestic animals and plants. In an intensively managed fishery, such as a carp farm, production and yield are synonymous for all practical purposes. Causes of mortality can be controlled and the number of fish in the harvest is the same as the number used to stock the ponds. Since the weight of introduced fry is negligible compared with that of the adults, the weight at the end of the period of growth is the production and, if every fish is captured, it is also the yield.

In a natural lake, river or reservoir a real distinction must be made between production and yield. Under these circumstances neither reproduction and survival of the young nor mortality of the older members of the population is under direct control. The yield, or catch, represents only a portion of that part of the population vulnerable to the fishing gear, and the yield plus the fish that escape the gear represent only a part of the population that was present at the start of a given period of time. In order to measure total production and gain a true picture of the capacity of a body of water to produce fish, some account must be made of those that were not captured and those that died in the time interval under question. Since the fish cannot be individually counted and weighed at stated time periods, indirect methods have been invented to measure the population in terms of numbers and weight, its rate of growth, and its rate of mortality. The estimate of production under these circumstances is subject to considerable error since each of the vital statistics has its own bias.

The principles of fish production are the same in the sea as in freshwater. This explains why the book freely dips into the marine literature, although freshwater production has been stressed in most of the chapters. Fish production can be measured more easily in freshwater than in the oceans, simply because the size of the ecosystem is smaller and the research is more easily carried out. Also, the practical side of the subject, pond culture, is more highly developed than is marine aquaculture as it pertains to fishes.

The inevitable overlap of subject matter in different chapters written by different authors will be recognized. Most of the overlap involves metabolism and growth, concepts which are fundamental to understanding production in natural populations as well as those which are artificially contained. These ideas are presented in various contexts by various authors and in my view act to bind somewhat unrelated material together rather than as disconcerting repetition. A minor distraction is the division of the genus *Tilapia* into the genera *Sarotherodon*, mouth brooders, and *Tilapia*, nest guarders (see E. Trewawas (1973) *Bull. Brit. Mus. Nat. Hist.* **25**: 1–26). The separation was made during the last review of the manuscript. The designation 'tilapia' has been adopted as a vernacular to refer to all species in the two genera.

In the earlier volume I said that bringing this material together should stimulate research in the field. It is gratifying to learn that this has proved to be correct. I also said that a sense of urgency prevails to advance our science because we see the resource dwindling away at a time when it is needed more than ever. I trust that this sense of urgency has been conveyed in an even keener manner in this volume because it expresses the feeling of the authors and fishery biologists in general. Finally, it was stated that the earlier volume might act as a landmark in the progress of establishing sound scientific principles for freshwater fish production. That, I suppose, still has to be proved one way or the other, but at the very least, the earlier book and the present version provide a broad view of the 'state of the art' as of this time.

VITAL STATISTICS OF
FISH POPULATIONS

The measurement of production rate in natural fish populations of lakes and streams depends upon knowing the average biomass between two points in time and the growth rate, the average biomass being determined by the balance between growth rate and mortality rate during the time in question. The average biomass multiplied by the growth rate yields the production rate, or the addition of new tissue to the population.

This bare-bones description of production gives an oversimplified impression of the problems involved in its measurement, however. In actual practice the estimation of each of the three vital statistics—abundance, mortality rate and growth rate—is a test of the biologist's power to sample the population effectively. Abundance measurements require time-consuming, costly procedures because the fish population cannot be observed and counted directly under most conditions. Indirect methods, employing catch-per-unit-effort or mark-and-recapture statistics, have been devised to measure abundance, and these are subject to greater error than we would wish. The same can be said for the assessment of natural mortality, or those fish which die from disease, predation, senescence or some other cause. These methods have most often been applied to the catchable (usually adult) members of a commercially exploited species. On only a few occasions has fish production included all of the age-groups in the population. The larvae and juveniles of many species are omitted because methods have not been devised to catch them quantitatively.

Growth rate measurements are the most accurate of the three vital statistics needed to measure production, because fish are blessed with marks on their scales and other hard parts of the body which identify periods of growth cessation. In temperate climates these periods correlate with winter; in the tropics they often correlate with drastic but regular changes in weather conditions, such as monsoons. By comparing the diameter of the scale and the distance from the scale focus to the various annuli with the known length of the fish at the time of capture, populations of measurements are developed which provide a sensitive measure of the response of the fish throughout its life to its food supply and other environmental variables.

The major objective of the first section of the book is to boil down the

3

individual elements of production to the point where the serious worker can identify the problems he faces and choose among various alternatives for overcoming these problems successfully. The first chapter describes how to measure fish production and presents the range of results acquired in different kinds of aquatic habitats. The second and third chapters deal with abundance, mortality and growth. The material on abundance and mortality is methodological, while that on growth is concerned largely with basic concepts of this fundamental life process and introduces the method of calculating growth as it applies to fishes.

A second objective is to examine the early life history stages, because both the rate of growth and the rate of mortality are greater during the first few weeks and months of life than at any other time. The ultimate production of the whole population often depends upon events occurring in the larval and juvenile stages. For example, an increase in the larval survival rate of 0·001 in a population of 10 million, normally subject to a survival rate of 0·020, yields an additional 10,000 juveniles and a significant increase in the numbers entering the fishery. Thus, the unpredictable shifts in larval abundance are crucial to the immediate production and also that of the future as the younger age-groups move through the population year by year.

In order to achieve the second objective, we examine fecundity and larval survival. A variety of factors affect fecundity, ranging from heredity to nutrition of the female. These influences can be more easily studied than can those impinging on the larvae after they are left to fend for themselves. Larval survival investigations have been hindered by the difficulty of identifying these small fishes to species in field collections. This should not, however, prevent basic studies of nutrition, time of fish feeding, feeding behaviour, escape behaviour, and critical environmental limits of survival, as our chapter on the subject is designed to demonstrate.

Chapter 1: Production in Fish Populations

D. W. Chapman

1.1 Introduction

In reviewing fish production research and concepts we sometimes forget that the pioneer in production work was Boysen-Jensen (1919), who estimated production of benthic animals in Limfjord (Denmark) from 1910 to 1917. This was an era when fish specialists still confined their energies to taxonomic and fish cultural enterprises and long before modern ecologists recognized the significance of production. Most of the pertinent research on fish production is less than 30 years old, and was stimulated by insights into community metabolism (Lindeman 1942, Ivlev 1945), population dynamics (Ricker 1946) and, more recently, environmental interactions.

My aim in this chapter is to summarize certain production concepts, methods, and errors, and to indicate the significance of production research.

1.2 Terminology

Production has been variously defined to mean yield to man, yield of smolt emigrants, products of oviposition, or recruits to a fishery. Clarke *et al.* (1946) used gross and net production as assimilation and production, respectively. Hatch and Webster (1961) and Hayne and Ball (1956) considered gross production in a manner equivalent to 'net production' of Clarke *et al.* (1946). Ricker and Foerster (1948) termed production to mean 'actual net production' of Clarke *et al.* (1946).

Production is used here as the total elaboration of new tissue in a time period of interest by a species-population. It includes the sum of growth increments for all population members alive at any time in the period. Growth increment is the net increase or decrease in amount of tissue contained in bodies of population members, regardless of the fate of the tissue.

With this definition of production, the term 'gross production' becomes superfluous (Winberg 1971), being equivalent to 'assimilation', the energy of which vectors to either growth or maintenance.

A list of meanings and symbols used in this chapter follows.

Δt = time interval t_0 to t_1, or time 0 and a later time.

w = weight of an individual fish.

$\bar{w}$ = mean weight per individual in the population.

N = number of fish in the population.

B = biomass of the population, or total population weight.

$\bar{B}$ = mean biomass in Δt.

G = instantaneous growth rate or coefficient of growth in weight in Δt in the population.

Z = instantaneous mortality coefficient in Δt.

P = production in Δt.

B_d = sum of weights of fish which die in Δt in the population, with weight assessed as each fish dies.

B_r = total biomass respired in Δt by fish in the population.

B_u = total faeces weight in Δt in the population.

B_v = total weight of urine or other excretions in Δt in the population.

B_c = total weight of food consumed in Δt in the population.

B_a = total weight of food digested and absorbed (assimilated) by fish in the population in Δt, including B_r and excluding B_u.

All components of B and P could be expressed in energy or nitrogen quantities, but fresh or wet weight will be used here.

Production in Δt by the population, in summary, can be shown in the following gravimetric terms:

$B_1 - B_0$ = net biomass change

$$P = B_1 - B_0 + B_d = \text{production}$$

$$B_a = P + B_r = \text{assimilation}$$

$$B_c = P + B_r + B_u + B_v = \text{food consumption}$$

The average weight per fish may decrease over Δt, for example in winter (Higley 1963, Hunt 1966, MacKinnon 1973, Kelso & Ward 1972, Alexander & MacCrimmon 1974). In this event 'negative' production occurs, in contrast to tissue elaboration which is the usual connotation of production. A decrease in total biomass because of emigration or mortality of any type is a reduction of capital (rather than interest) and should not be considered as negative production.

If negative production occurs and the population later regains the lost weight, the regained tissue constitutes additional energy transformed from other community trophic components. But this 'replacement' production merely cancels out negative production if we view the phenomenon from the narrower outlook of population dynamics. This difference in viewpoints illustrates a weakness of production data as measures of food consumption and an exception to the idea that the ultimate fate of produced tissue can be ignored in estimating production.

Gonad growth may be considered separately from that of other tissues, partly because it contains higher calorific values per gram, and also because it forms a capital base for the progeny population rather than for the fish which produced the reproductive tissue. Research objectives will govern whether gonad weight should be included in the calculation of production.

The seasonal interrelationships (Fig. 1.1) in energetics terms are illustrated

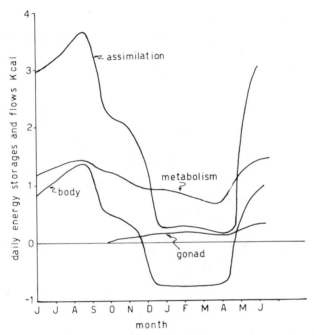

Fig. 1.1. Seasonal variation in daily body and gonad growth rate, metabolism and assimilation of a mature age 11 American plaice (*Hippoglossoides platessoides*) female. Adapted from MacKinnon (1973).

by MacKinnon (1973). He points out that production, including energy stored and used for later maintenance, must equal elimination (mortality and negative production) in a steady state, although of course certain age classes will enjoy positive production over Δt while others will show greater elimination than production (Fig. 1.2).

1.3 Models for production statistics

Production may be estimated from a knowledge of numbers and weight of population members over time; from information on food consumption and conversion efficiency; or from estimates of food consumption by predators, which in the steady state may take up most of the production of the fish population. The first of these methods is the most commonly used.

Chapter 1

Allen (1971) thoroughly explored the combinations of model alternatives available for growth and survivorship. He gives formulae for cohort production, biomass integral, and $P/\bar{B}$ ratios where possible.

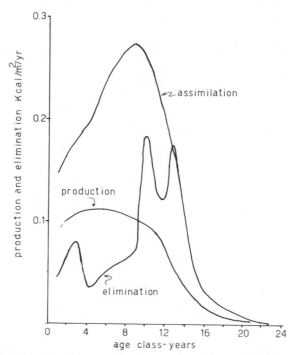

Fig. 1.2. Distribution among age–classes of annual production and elimination for the American plaice (*H. platessoides*) population of St. Margaret's Bay, N.S. Adapted from MacKinnon (1973).

One of the two most common models is that of Ricker (1946), in which $P = G\bar{B}$. If we can assume a constant and known Z, then

$$P = \frac{GB\,(e^{G-Z} - 1)}{G - Z}$$

and can be calculated from one observation of biomass. This formula is developed from exponential rates of growth and mortality which usually serve satisfactorily if calculated over small Δt. For accurate computations, estimates of N and $\bar{w}$ are needed over fairly short intervals, in which case the arithmetic mean of two adjacent biomasses is more useful than a $\bar{B}$ calculated from Ricker's formula.

Allen (1950) and Beverton and Holt (1957) developed formulae for P

where fish are assumed to be growing toward an asymptote (negative exponential curve) or in accord with the von Bertalanffy model, respectively.
These models have rarely been used in published production estimates.

The frequently-used model of Allen (1951) requires no assumed functions
for growth and mortality, relying on a curve of N against $\bar{w}$ over Δt (Fig. 1.3)
and integrating the area under the curve by planimeter.

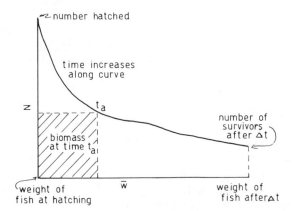

Fig. 1.3. An Allen curve in which mean individual weight is plotted against cohort size
at given times.

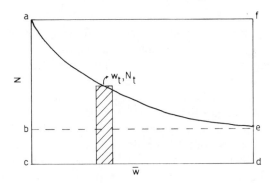

Fig 1.4. An Allen curve. The sum of all small increments $\Delta\bar{w}N_t$ is production between
times a and e. Shaded area is one such increment.

In Fig. 1.4, over any small growth increment $\Delta\bar{w}$, at time t, the amount
of tissue produced is $\Delta\bar{w} N_t$. As $\Delta\bar{w} \to 0$, production $\to N_t d\bar{w}$. Therefore,
integrating the production function or assessing area beneath the curve in
Δt gives the sum of all $N_t d\bar{w}$:

$$P= \int_{\bar{w}_0}^{\bar{w}_t} N_t d\bar{w}$$

Where point estimates of N or $\bar{w}$ are far apart, curve placement may again require model assumptions for growth and survivorship. The method relies for accuracy on frequent point estimates.

Allen (1971) explores the consequences of using different models for production components. For examples of production computation by the Ricker formula ($P = G\bar{B}$) one can examine Coche (1964), Gerking (1962), Hunt (1966), Ricker and Foerster (1948), Waters (1966), Chapman (1965), Alexander and MacCrimmon (1974), Staples (1975), Mathews (1970) and Patriarche (1968). Estimates from Allen curves were used by Allen (1951), LeCren (1962), Chapman (1965), Goodnight and Bjornn (1971), and Egglishaw (1970). Johnson (1966) adopted the Beverton and Holt (1957) equation.

Predictions of weight and survivorship functions over long intervals are rarely satisfactory except for very gross estimates of production. These rates often change seasonally and annually (Ricker & Foerster 1948, and Coche 1964) in response to many biological or physical factors. In temperate areas temperature changes trigger many mortality or growth stanzas; in tropical waters rainfall and water level (Dudley 1974) or even wind (Coulter 1963) more often govern growth, recruitment and mortality. In certain long-lived species (e.g. *Salvelinus namorycush*) mortality and growth rates may be predictable over extended periods.

Ricker (1948, 1958) and Neess and Dugdale (1959) indicate that if the ratio G/Z remains constant over Δt, the growth–survivorship curve will trace out the same path even if absolute values of the two rates change. If G/Z is constant and fish do not immigrate or emigrate, Ricker's (1946) production formula should serve. But one should not expect constancy in G/Z. To account for changes in G/Z immigration or emigration, or even recruitment of new age classes, one must assess growth and numbers over short time intervals, e.g. monthly or each 2 weeks. Short-term growth estimates are feasible and often accurate and precise, but numbers or mortality estimation is time-consuming and often neither accurate nor precise.

Winberg (1971) points out that production is easier to calculate in monocyclic species in which all the population members belong to a single age-class and new young appear once per year at a time when few individuals remain from the previous age-class. He discusses at length some of the means used for studying production of aquatic invertebrates. Many of the techniques are quite ingenious and involved, especially for heterotopic, or multi-aged animals with continuous recruitment.

Cooper *et al.* (1963, 1971) estimated 'surplus production', meaning the summed weight of harvest by anglers and change in biomass over Δt. Surplus production is always less than production because natural mortalities or, in certain cases, migrations are not included. Where natural mortality may be small, as in older age classes of large predators, surplus production approaches production.

Mathews (1970) notes that instantaneous growth rates are calculable for multi-cohort species by assuming that the population is in equilibrium in the sense of Beverton and Holt (1957). Growth rates for each cohort, obtained across cohorts, can then be applied to biomass data. The advantage, of course, is that all data could be obtained in a single intense sampling effort at one time. The method would not apply for annual species or those which migrate before multi-aged cohorts develop (Bjornn 1975, Chapman 1965). Neither is it likely to offer accuracy where cropping of older ages is size-specific. Mathews demonstrated efficacy of the method in cyprinid populations in the River Thames. These fish apparently did not suffer heavy harvest. He also described situations in which the method may offer the best alternative means of production calculations.

1.4 Errors in the methods

The two methods most frequently used for production estimation, the exponential formula (Ricker 1946) and graphic (Allen 1951), give the same results, contrary to Horton's (1961) statement that the former is mathematically more sound. Allen integrates mechanically by measuring the area under a curve of N against $\bar{w}$ by a planimeter, while Ricker integrates growth and mortality computationally, starting from a known biomass. The principal inaccuracy of both methods is large sampling error and possible bias in population and survivorship estimations. Even without bias, standard errors in population estimates are usually large. Often 95% confidence limits exceed $\pm 50\%$ of the point estimate (Gerking 1962, Chapman 1965).

There is little point in being greatly concerned about minor shifts or bends in growth functions when most survivorship curves are based on point estimates widely scattered in time and of dubious accuracy. Such concern risks the 'Fallacy of Misplaced Concreteness'. Allen (1971) demonstrates the considerable effects of different estimates of mortality and mean age upon estimates of production by Antarctic krill.

The sampling errors associated with growth estimates can be quite small if sampling is intensive. Hall (1963) has growth data for which standard error of the mean is less than 5% of the estimated mean length, and Chapman (1965) computed standard errors of mean lengths from 1–10% of the mean, and usually less than 5%. A possible hidden error in growth curves may be that small or large population members suffer selective mortality such as predation or exploitation by man, so that a change in mean size over Δt may not be entirely due to growth. Another source of error might be caused by size-specific emigration or imigration in Δt leading to erroneous apparent growth (Chapman 1965, Waters 1966). Ricker (1969) points out examples of the large errors (up to 30% in his examples) which can result in production calculations on incorrect (biased) growth rates.

Growth could be assessed from individually recognizable fish (branded or tagged) in the population in Δt, with the caution that tagged fish may not grow as rapidly as untagged ones. If scales are used in growth calculations, Ricker (1969) suggests that the best estimate of G comes from comparing the mean size at the last annulus with that at the second-last annulus among randomly sampled fish of each age.

Mean weight of older population members could decline in the growing season because eggs or sperm are shed. Under these circumstances G may be negative when, in fact, somatic tissue has been elaborated. Mann (1965) estimated that germ cell shedding amounted to 10% of annual production in some fishes of the River Thames.

Discrete treatment of each age-group is mandatory in production calculations in order to minimize problems of size-selective mortality, emigration, and bias.

1.5 Variance of estimates

Goodman (1960) suggested an assessment of variance (V) of products that appears to be useful for production estimates. The appropriate formula, which does not require independence of G and $\bar{B}$, is

$$V(G\bar{B}) = G^2 V(\bar{B}) + \bar{B}^2 V(G) + 2\, G\bar{B}\, \text{cov}(G, \bar{B})$$

Chapman (1968) discusses components of the formula, which provides a means of estimating variance of a single estimate of P in Δt. The method requires careful planning and execution to develop the components of variance and covariance.

There is substantial merit in estimating production regionally in waters that are similar (H. Regier, personal communication). Production estimates over several years from a stratified random sample of these waters would permit application of analysis of variance to permit factor analyses among waters and years. Error estimates obtained in this way are preferable to error estimates derived from model properties, and environmental factor analyses would permit the construction of better predictive models useful in different communities (Regier & Henderson 1973).

1.6 Results of production studies

Standing waters in temperate areas have produced up to 15 gm^{-2}yr^{-1} with the exception of Spectacles Lake, New Zealand, in which *Philypnodon breviceps* produced 40 gm^{-2}yr^{-1}. (Table 1.1) This fish was the only species occupying the moderately eutrophic shallow lake, and fed upon distinctly different invertebrate life forms at different fish sizes and seasons.

Other high lentic production figures were reported in two shallow Michigan ponds, 18·1 gm^{-2}yr^{-1} (Hayne & Ball 1956) with carbonate hardness of 160 ppm; in a eutrophic Oregon reservoir, 15·6 gm^{-2}yr^{-1} (Higley 1963); and in shallow Alanconnie Lake, Pennsylvania, 14·6 gm^{-2}yr^{-1} where the value represented surplus production only (Cooper *et al.* 1971).

Tropical and fertilized standing waters should produce at very much higher levels than the more or less natural waters listed in Table 1.1. Some tropical lakes may have maximum equilibrium yields higher than 10 gm^{-2}yr^{-1} (Welcomme 1972). Naiman (1976) reports 155·4 gm^{-2}yr^{-1} in Tecopa Bore, a desert stream in Death Valley, California. This exceedingly high value came from a monotypic population of *Cyprinodon nevadensis*, a pupfish which adopted an herbivorous habit of feeding mainly on diatoms in this unusual habitat. Intense year-round solar radiation contributed to the high production rate.

Most cold water streams, listed in Table 1, produced up to 18 gm^{-2}yr^{-1}. The very notable exceptions include Hinau and Hinaki streams, N.Z., which produced 24 and 73 gm^{-2}yr^{-1}, respectively (Hopkins 1971), Horokiwi Stream, N.Z., 54 gm^{-2}yr^{-1} (Allen 1951), and River Tarrant, U.K., 60 gm^{-2}yr^{-1} (Mann 1971). LeCren (1972) generalized that salmonid production rarely exceeded 12–18 gm^{-2}yr^{-1} but results of 30 gm^{-2}yr^{-1} from Big Spring, Pennsylvania, (Cooper & Scherer 1967), 73 gm^{-2}yr^{-1} from Hinaki Stream (Hopkins 1971) and of 54 gm^{-2}yr^{-1} from the Horokiwi Stream exceed this level by a considerable amount.

There appears to be no *a priori* reason why salmonid production should not exceed an average of 18 gm^{-2}yr^{-1}, except that most streams are not more productive. As cited above the Horokiwi is an obvious exception, as is Big Springs, Pennsylvania. Attempts to explain overestimates in Allen's Horokiwi work reduced the production estimate from 54 to 45 gm^{-2}yr^{-1} (Chapman 1967). There appears no doubt that Big Spring and Horokiwi Stream are, or were, very productive waters.

One can generalize, in these terms of careful allocation of limited water supplies, that much more stream water is required to produce a given quantity of fish flesh than is the case in a standing water environment. Deer Creek (Chapman 1965) and a eutrophic Oregon reservoir (Higley 1963) produced nearly equal fish tissue in gm^{-2}yr^{-1}, but 4300 acre-feet of water in Deer Creek produced 76 kg of tissue while 278 acre-feet produced 1150 kg in the reservoir. It is true, of course, that the same basin outflow from Deer Creek, supplemented by other down-basin flows, produced more tissue downstream. Nevertheless we can state that lotic environments are not efficient tissue producers, from the somewhat narrow viewpoint of water allocation and

Table 1.1. Production estimates in selected waters gm^{-2} yr^{-1}

Water	Species	P	P/$\bar{B}$	Temperature C Winter/Summer/mean	Reference
Wyland Lake, Indiana	*Lepomis macrochirus*	9·1			Gerking 1962
Wyland Lake, Indiana	*L. microlophus*	13·6[1]			Gerking 1962
	L. gibbosus				
	Pomoxis nigromaculatus				
	Chaenobryttus gulosus				
	Ictalurus nebulosus				
	I. natalis				
2 ponds, Michigan	*L. macrochirus*	18·1[4]			Hayne & Ball 1956
	L. gibbosus				
	L. microlophus				
Eutrophic reservoir, Oregon	*Oncorhynchus tshawytscha*	15·6			Higley 1963
5 dystrophic lakes, Wisc., Mich.	*Salmo gairdneri*	1·9–8·4			Johnson & Hasler 1954
4 lakes, New York	*Salvelinus fontinalis*	3·3–6·5			Hatch & Webster 1961
Cultus Lake, B.C.	*O. nerka*	5·9[2]			Ricker & Foerster 1948
Pond, Pennsylvania	*Micropterus salmoides*	5–8[3]			Cooper *et al.* 1963
Ponds, Oregon	*M. salmoides* and mixed with *L. macrochirus*	7·5–12·6			Hansen 1963
Jewett Lake, Michigan	*L. macrochirus*	2·8–3·3 (age V and older)	0·45		Patriarche 1968
Lodge Lake, Michigan	*L. macrochirus*	0·8 (age III and older)	0·34		Patriarche 1968
West Blue Lake, Canada	*Stizostedion vitreum vitreum*	0·21	0·43		Kelso & Ward 1972
Beaver Reservoir, Arkansas	*Dorosoma cepedianum*	4·5			Houser & Netsch 1971
	D. petenense	1·2			
Matamek Lake, Quebec	*S. fontinalis*	0·22			Saunders & Power 1970
Lake Windermere, U.K.	*Esox lucius*	0·24–0·67			Kipling & Frost 1970
Alanconnie Lake, Pennsylvania	*L. macrochirus*	14·6[3]			Cooper *et al.* 1971
	M. salmoides				
	I. nebulosus				
	Catostomus commersoni				
	L. gibbosus				
	P. nigromaculatus				
	Notemigonus crysoleucas				

Water	Species	P	P/B̄	Winter	Summer	mean	Reference
Horokiwi Stream, N.Z.	*Salmo trutta*	54·7	2	7	19	13	Allen 1951
Hinau Stream, N.Z.	*Philypnodon breviceps*	24·2	1·15 (trout)				Hopkins 1971
	S. trutta						
Hinaki Stream, N.Z.	*Anguilla dieffenbachii*	73·5	1·25 (trout)				Hopkins 1971
	A. australis schmidti						
North Branch Hinau Stream	*Galaxias divergens*	14·3					Hopkins 1971
Valley Creek, Minn.	*S. fontinalis*	14·5[6]	1·4	6	14	12	Hanson & Waters 1974
	S. gairdneri						
Valley Creek, Minn.	*S. fontinalis*	4·4–6·1[7]	1·34				Elwood & Waters 1969
Larry's Creek, Pennsylvania	*S. fontinalis*	5·8	1·30				Cooper & Scherer 1967
Big Spring, Pennsylvania	*S. gairdneri*	30·0	1·4				Cooper & Scherer 1967
Big Springs Creek, Idaho	*S. fontinalis*	11·8[8]		1	15	10	Goodnight & Bjornn 1971
Lemhi River, Idaho	*S. gairdneri*	13·6	0·64	1	12·5	8	Goodnight & Bjornn 1971
	O. tshawytscha						
	Prosopium williamsoni		1·47 w/o *Prosopium*				
Bothwell's Creek, Canada	*S. gairdneri*	13·2	2·39		17		Alexander & MacCrimmo 1974
Loucka Creek, Czech.	*S. trutta*	1·7–8·6	0·9	1	12	7	Libosvarsky 1968
Shelligan Brook, Scotland	*S. salar*	6·5–11·1	1·2				Egglishaw 1970
	S. trutta	7·7–12·3					
Lawrence Creek, Wisconsin	*S. fontinalis*	9·3–10·6	1·5–2·9				Hunt 1966
Walla Brook, U.K.	*S. trutta*	10–18	1·5	7	13	9	Horton 1961
Deer Creek, Oregon	*O. kisutch*	16	1·89			10	Chapman 1965
	S. gairdneri						
	S. clarki						
	Cottus perplexus						
Flynn Creek, Oregon	*O. kisutch* / *S. clarki* [8]	11	1·86	7	13	10	Chapman 1965
Needle Branch, Oregon	*C. perplexus*	9	1·96	6	14	10	Chapman 1965
BlackBrows Stream, U.K.	*S. trutta*	10				9	LeCren 1969
Kingswell Stream, U.K.	*S. trutta*	7·4				9	LeCren 1969
Hall Stream, U.K.	*S. trutta*	5·2					LeCren 1969
Appletreeworth Stream, U.K.	*S. trutta*	3·0					LeCren 1969
Nether Hearth Stream, U.K.	*S. trutta*	5·0				6	LeCren 1969
	C. gobio	1·0					

Table 1.1. Production estimates in selected waters gm⁻²yr⁻¹ *continued.*

Water	Species	P	$P/\bar{B}$	Temperature C Winter/Summer/mean	Reference
Bere Stream, U.K.	*S. trutta*	2·6–13	1·5	12	Mann 1971
	S. salar	7·2			
	C. gobio	6·2–30			
	Phoxinus phoxinus	2·0			
	Gasterosteus aculeatus	1·8			
Tarrant Brook, U.K.	*S. trutta*	12	3·0	11	Mann 1971
	C. gobio	43			
	P. phoxinus	3·9			
	G. aculeatus	0·6			
Devil's Brook, U.K.	*S. trutta*	4·8	2·0		Mann 1971
	C. gobio	14			
	G. aculeatus	0·6			
Docken's Stream, U.K.	*S. trutta*	12	1·9	11	Mann 1971
	P. phoxinus	1·9			
Berry Creek, Oregon unenriched sections with controlled fish density	*S. clarki*	1·0			Warren *et al.* 1964
	C. perplexus	6·0			
Berry Creek, Oregon enriched sections with controlled fish density	*S. clarki*	5–6			Warren *et al.* 1964
	C. perplexus	6·0			
Laboratory controlled stream	*C. perplexus*	6·0			Davis & Warren 1965
Manistee River, Michigan	*Ichthyomyzon castaneus*	0·15			Hall 1963
River Thames, U.K.	*Rutilus rutilus*	42·6			Mann 1965
	Alburnus alburnus				
	Perca fluviatilis				
	Leuciscus leuciscus				
	Gobio gobio				
River Thames, U.K.	*Leuciscus leuciscus*	154 (1967)			Mathews 1971
	Gobio gobio	78 (1968)			
	Alburnus alburnus				
	Rutilus rutilus	197 (10 spp) 1967	1·77		

Water	Species	P	P/B	Winter/Summer/mean	Reference
Vistula River, Poland	*E. lucius* *Aspius aspius* *Stizostedion lucioperca* *Siluris glanis* *Leuciscus cephalus* *Perca fluviatilis*	0·43	0·44		Backiel 1970
Tecopa Bore, California	*Cyprinodon neadensis*	155	5·0		Naiman 1976
Clemons Fork, Kentucky					Lotrich 1973
(First order)	*Semotilus atromaculatus*	8·35[9]	1·52	1	22
(Second order)	*S. atromaculatus* *Campostoma anomalum* *Etheostoma sagitta* *E. nigrum* *E. flabellare* *E. caeruleum* *Hypentelium nigricans* *Catostomus commersoni*	10·55[9]	1·66	1	24
(Third order)	All spp. in second order plus *Ericymba buccata* *Notropis ardens* *N. chrysocephalus* *Pimephales notatus* *Ambloplites rupestris* *Lepomis megalotis* *Micropterus dolomieui*	7·72[9]	1·08	1	30

[1] I used 4·5 g/m²/yr for all other species, assuming that since the standing crop of these fish was half that of bluegills, production may have been proportional to crop percentage.
[2] Other species present but not estimated.
[3] Does not include natural mortalities, hence is less than actual production.
[4] 21 April–9 September.
[5] After recovery from flood damage.
[6] After flood damage.
[7] Excluding *Prosopium williamsoni*.
[8] Vertical lines indicate 'all those species in both waters'.
[9] Excludes 0 age class.

economics of protein yield. Further, less of the total production can usually be harvested economically from streams.

A striking 197 gm^{-2}yr^{-1} was produced in 1967 by 10 fish species in the River Thames (Mathews 1971). Densities of fish reached 96 fish/m^2 in August 1967. Production of four species (bleak, roach, gudgeon and dace) equalled 154 and 78 gm^{-2}yr^{-1} in 1967 and 1968, respectively. Of this, age 0+ fish made up 66–73% of the production in 1967 and 39–64% in 1968. Z for the important 4 species in the first year ranged from 6·4 to 8·7. Less than 5% of total production was shed as sex products. One can only conclude that the Thames is an extremely productive environment for cyprinids, and little information is available to permit exhaustive examination of the causes.

Fish species living sympatrically appear to produce more total tissue than those living allopatrically. Big Springs Creek, Idaho, provides an interesting demonstration of this (T. C. Bjornn, personal communication). Production by allopatric steelhead trout (*S. gairdneri*) totaled 8·2 and 13·7 gm^{-2}yr^{-1} in 1969 and 1973, while production by sympatric steelhead and chinook salmon (*O. tshawytscha*) totaled 18·4 and 28·2 gm^{-2} in 1971 and 1972. Because both species migrated out of the stream largely as yearlings, and made up most of the ichthyomass, the examples are limited to production by under-yearlings in all cases.

Calcium content and total fish production are apparently positively related in most streams (LeCren 1969), and it is reasonable to accept the general thesis that increased general stream productivity and fish production are correlated (Warren *et al.* 1964). Useful now would be more demonstrated relationships between primary production or allochthonous input and fish production.

1.7 The $P/\bar{B}$ ratio

For all streams for which data from intensive studies were available (Table 1.1) I calculated the $P/\bar{B}$ ratio which in the commonly-used models gives instantaneous growth rate in Δt. In all cases $\Delta t = 1$ year, so $P/\bar{B} = G_{annual}$.

My main interest was to compare $P/\bar{B}$ for waters with much in common, e.g. streams containing salmonids. For those waters in which intensive studies were conducted on salmonids, $P/\bar{B}$ varied from a low of 0·9–1·5 (e.g. Shelligan Brook, Scotland; Big Springs Creek, Idaho; Lemhi River, Idaho) to a high of 2–2·4 (e.g. Needle Branch, Oregon; Bothwell's Creek, Canada, and Horokiwi Stream). Admittedly, estimates of $\bar{B}$ from published data were necessarily somewhat rough in some cases, but it appears that the lowest $P/\bar{B}$ ratios were measured in streams in which the winter temperature regime is near 1C and mean annual temperature near 8C. The highest $P/\bar{B}$ ratios were in streams in which the winter low temperature did not reach below 6–7C and the yearly mean was 10C or more. A shortage of time for produc-tion may limit the rate of tissue elaboration in the colder streams. Although

Shelligan Brook, Scotland, contains only about 7–8 ppm calcium and 12–34 ppm alkalinity as $CaCO_3$, Big Springs Creek and Lemhi River in Idaho contained 31 and 50 ppm Ca, respectively, and alkalinity (as $CaCO_3$) of 134 and 160 ppm, respectively. With total dissolved solids of 298 and 273 ppm, Big Springs Creek and Lemhi River can be identified as relatively rich waters when compared to many others which had much higher $P/\bar{B}$ ratios. Because the rainbow trout, *Salmo gairdneri*, stocked as embryos made up most of the biomass in Big Springs Creek, and because these fish could move out of the stream, one assumes that they adjusted density by emigration and that overstocking did not cause the low G. In the Lemhi River, low G was partly due to presence of an unexploited, slow-growing stock of whitefish (*Prosopium williamsoni*). The $P/\bar{B}$ ratio for species other than whitefish equalled 1·47; and with whitefish 0·64.

Allen (1971) notes that the $P/\bar{B}$ ratio is not always simply G when the growth model in the production function is other than exponential. Most fish cohorts appear to follow a single or multiple exponential decay model for mortality and an exponential, negative exponential, or von Bertalanffy growth model. In the latter two cases, cohorts with exponential decay rates appear to have $P/\bar{B}$ ratios which equal Z or functions of Z. Survivorship decay may follow a linear model in some periods, and where growth is isometric, fish may follow a linear growth in length model (examples in Chapman 1965).

The key point in considering the $P/\bar{B}$ ratio is that in the majority of environments in which salmonids dominate, $P \simeq 2B$ for annual data, and the multiplier probably can be considered as a secondary productivity function of use in estimating P without measuring more than average biomass. As noted earlier the multiplier varies from water to water. In over 4 years of intensive production research in Deer Creek, Oregon, the $P/\bar{B}$ ratio ranged from 1·69 to 2·04; in Flynn Creek 1·43 to 2·06, and in Needle Branch 1·68 to 2·18. Probably one could use 1·5 as an estimator in cold environments and 1·7 to 2·0 in warmer waters with reasonable confidence.

Waters (1969) thoroughly discusses $P/\bar{B}$ ratios for freshwater invertebrates, terming them 'turnover rates', and examines families of Allen curves which covered many possible combinations of growth and mortality. In available studies of benthic invertebrates the $P/\bar{B}$ ratio varied from 2·5 to 5 with a mode of 3·5. He concluded that an annual $P/\bar{B}$ ratio should be about 3·5 for univoltine species which spend little time in the egg or adult stage; and higher for multivoltine or brooding species.

1.8 Significance of production

Production estimates based on knowledge of the seasonal vectors of fish growth and mortality help explain magnitude of biomass at a given time and may provide clues about factors affecting production and yield. Obviously, a year-class which ends the year at a given mean size and abundance may have reached that biomass by more than one path (Fig. 1.5), and have quite

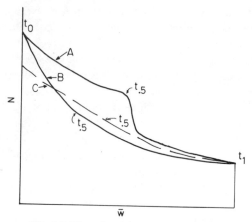

Fig. 1.5. Three hypothetical Allen curves.

different structures at some point in Δt. The similarities of $P/\bar{B}$ ratios among years should not be interpreted to mean that structural differences do not exist within Δt. In growth-survivorship curve A of Fig. 1.5, growth was rapid and mortality low until about mid-way through the year, when mortality markedly increased. In curve B, a more typical early mortality was suffered and growth was better in the latter half of the year. In C, recruitment was lower than in A and B, but low mortality and rapid growth compensated. In all three curves the same biomass was present at year-end. In situation A production was nearly twice that in B or C and the latter two were about equal.

Measures of assimilation, consumption, and production of adjacent trophic levels also permit calculation of efficiency statistics:

Physiological
Growth coefficient (1) = production/consumption
Growth coefficient (2) = production/assimilation
Assimilation efficiency = assimilation/consumption

Ecological
Ecotrophic coefficient or $=$ consumption by predator/
coefficient of hunting $\qquad$ production by prey

Trophic level production $=$ production of consumer/
efficiency $\qquad$ production of prey

These and other possible combinations of production and trophic statistics were discussed by Lindeman (1942), Ivlev (1945), Odum (1959), Gerking (1962), Teal (1957) and Winberg (1971). These estimates can permit general evaluations of ecological relationships and comparisons of different communities.

Production per unit area per year is not an appropriate indication of the importance of fish as a group turning over energy in the aquatic ecosystem. Assimilation would be a more realistic estimator of the turnover, since energy used in respiration is included. In captive animals where natural regulatory factors, such as socially induced emigration or perhaps predation, cannot act with realistic intensity, production may give an unrealistic picture of importance of fish as a consumer of energy. Davis and Warren (1965) studied trophic relationships of simple communities in artificial streams, and found that total production of cottids (*Cottus perplexus*) increased as fish biomass increased up to a point, after which production decreased with further increases of biomass. Due to interactions among individuals at high levels of fish biomass, most of the food available was used for maintenance; at lower biomasses more of the available food was converted to growth. Hunt (1966) reported the same result in Lawrence Creek. In short, the instantaneous growth coefficient was inversely correlated with biomass.

Hall (1963) provides an example of utility of efficiency statistics. He assessed food conversion efficiency for chestnut lampreys feeding on fish in aquaria (grams of fish killed per gram of lamprey growth), then applied this to estimates of lamprey production in the Manistee River, Michigan, to estimate total mortality of trout due to lamprey predation.

Warren *et al.* (1964), Gerking (1962), Allen (1951), and others applied results of laboratory studies of food consumption by fish of prey organisms. Eventually a reasonably accurate, quantified energy flow diagram for the trophic links from fish production backward to incoming radiant energy may be produced in the manner suggested by Ivlev (1945). The two greatest obstacles to definition of energy flow in fish and their principal prey groups are the lack of a valid method of estimating consumption rates for fish and the difficulty of estimating production of prey organisms. Winberg (1971) offers some ideas for attacking the latter problem.

Potential rewards are great if the factors limiting production can be identified. As in many ecological problems, we must develop experimental techniques for testing field correlates as causal or limiting factors. Research based on sampling natural populations, which has occupied the bulk of aquatic ecologists in the past, must of course continue. But mechanistic research, so under-emphasized to the present, must receive much more attention. Kitchell *et al.* (1974) developed a model for fish biomass dynamics which should suggest critical interactions and physiological parameters for mechanistic research.

In the literature there is a large gap between the relatively intensive production studies, such as those in Table 1.1, and the extensive research on populations in oceans or large lakes where surplus production (yield or potential yield) is the statistic often sought. Models of equilibrium yield and surplus production range from the detailed (Beverton & Holt 1957, Ricker 1975) to 'simple' (Schaefer 1954, Pella & Tomlinson 1969, Fox 1970). The simple models are useful in regulation of fisheries even though they are not based on detailed mortality or growth data. Allen (1971) illustrated an application of production calculations to a 'big-environment' problem, and offers many combinations of models useful in calculating production and biomass. This contribution should prove valuable in expanding the application of production estimates to large lakes and river systems. Development of acoustic (echo-sounding) estimates of ichthyomass in lakes (Nunnallee & Mathisen 1972) also offers potential for production estimation. Where species of fish move differently in space, it appears quite feasible to estimate ichthyomass by species. If these estimates can be obtained several times each year and growth samples obtained by any reasonably unbiased technique, production can be estimated by one of the model combinations (Allen 1971). Species continually mixed in space in lakes or large rivers pose very difficult estimation problems, and may force investigators to rely on surplus production and $P/\bar{B}$ ratios for rough production estimates.

Finally, we should note that production is one of the sensitive dynamic properties of populations which can provide measures of response of stocks to a fishery, water pollutants or development projects, stock augmentation, or other environmental disturbance. Models and empirical production data should be pursued to the point where feedback mechanisms can be incorporated and tested (Jensen 1972). These ought to include time-lag functions (Walter 1973) which are usually completely ignored in yield models and certainly in production and trophic efficiency modelling.

References

ALLEN K.R. (1950) The computation of production in fish populations. *N.Z. Sci. Rev.* **8**, 89.

ALLEN K.R. (1951) The Horokiwi Stream: a study of a trout population. *Fish. Bull. N.Z.* **10**, 1–238.

ALLEN K.R. (1971) Relation between production and biomass, *J. Fish. Res. Bd. Canada* **28**, 1573–1581.

ALEXANDER D.R. & MacCRIMMON H.R. (1974) Production and movement of juvenile rainbow trout (*Salmo gairdneri*) in a headwater of Bothwell's Creek, Georgian Bay, Canada. *J. Fish. Res. Bd. Canada* **31**, 117–121.

BACKIEL T. (1971) Production and food consumption of predatory fish in the Vistula River. *J. Fish Biol.* **3**, 369–405.

BEVERTON R.J.H. & HOLT S.J. (1957) On the dynamics of exploited fish populations. *Fishery Invest., Lond., Ser. 2* **19**, 1–533.

BOYSEN-JENSEN P. (1919) Valuation of the Limfjord. I. Studies on the fish food in the Limfjord 1909–1917. *Rep. Dan. Biol. Stn.* **26**, 1–44.

CHAPMAN D.W. (1965) Net production of juvenile coho salmon in three Oregon streams. *Trans. Am. Fish. Soc.* **94**, 40–52.

CHAPMAN D.W. (1967) Production in fish populations. *The Biological Basis of Freshwater Fish Production*, pp. 3–29 (S.D. Gerking, ed.) Blackwell Scientific Publications, Oxford.

CHAPMAN D.W. (1968) Production. *Methods for Assessment of Fish Production in Fresh Waters*, pp. 199–214 (W. E. Ricker, ed.) I.B.P. Handbook No. 3. Blackwell Scientific Publications, Oxford.

CLARKE G.E., EDMONDSON W.T. & RICKER W.E. (1946) Mathematical formulation of biological productivity. *Ecol. Monogr.* **16**, 336–337.

COCHE A.G. (1964) Net production of juvenile steelhead, *Salmo gairdneri* Richardson, in a freshwater impoundment. PhD Dissertation, Oregon State University, Corvallis, 274pp.

COOPER E.L., HIDU H. & ANDERSON J.K. (1963) Growth and production of largemouth bass in a small pond. *Trans. Am. Fish. Soc.* **92**, 391–400.

COOPER E.L. & SCHERER R.C. (1967) Annual production of brook trout (*Salvelinus fontinalis*) in fertile and infertile streams of Pennsylvania. *Proc. Penn. Acad. Sci.* **41**, 65–70.

COOPER E.L., WAGNER C.C. & KRANTZ G.E. (1971) Bluegills dominate production in a mixed population of fishes. *Ecology* **52**, 280–291.

COULTER G.W. (1963) Hydrological changes in relation to biological production in southern Lake Tanganyika. *Limnol. and Oceanog.* **8**, 463–467.

DAVIS G.E. & WARREN C.E. (1965) Trophic relations of a sculpin in laboratory stream communities. *J. Wildl. Mgmt.* **29**, 846–871.

DUDLEY R.G. (1974) Growth of *Tilapia* of the Kafue floodplain, Zambia: Predicted effects of the Kafue Gorge dam. *Trans. Am. Fish. Soc.* **103**, 281–291.

EGGLISHAW H.J. (1970) Production of salmon and trout in a stream in Scotland. *J. Fish Biol.* **2**, 117–136.

ELWOOD J.W. & WATERS T.F. (1969) Effects of floods on food consumption and production rates of a stream brook trout population. *Trans. Am. Fish. Soc.* **98**, 253–262.

FOX W.W., Jr. (1970) An exponential surplus yield model for optimizing exploited fish populations. *Trans. Am. Fish. Soc.* **99**, 80–88.

GERKING S.D. (1962) Production and food utilization in a population of bluegill sunfish. *Ecol. Monogr.* **32**, 31–78.

GOODMAN L.A. (1960) On the exact variance of products. *J. Am. Stat. Ass.* **55**, 708–713.

GOODNIGHT W.H. & BJORNN T.C. (1971) Fish production in two Idaho streams. *Trans. Am. Fish. Soc.* **100**, 769–780.

HALL J.D. (1963) An ecological study of the chestnut lamprey, *Ichthyomyzon castaneus* Girard, in the Manistee River, Michigan., PhD. Dissertation, University of Michigan, Ann Arbor, 101 pp.

HANSEN H.L. (1963) Fish production and fishing success in four experimental warmwater fish ponds. M.S. Thesis, Oregon State University, Corvallis, 61 pp.

HANSON D.L. & WATERS T.F. (1974) Recovery of standing crop and production rate of a brook trout population in a flood-damaged stream. *Trans. Am. Fish. Soc.* **103**, 431–439.

HATCH R.W. & WEBSTER D.A. (1961) Trout production in four central Adirondack mountain lakes. Cornell Univ. *Ag. Exp. Sta., N.Y. St. Coll. Agric., Ithaca*, Memoir **373**, pp. 3–81.

HAYNE D.W. & BALL R.C. (1956) Benthic productivity as influenced by fish production. *Limnol. and Oceanogr.* **1**, 162–175.

HIGLEY D.L. (1963) Food habits, growth, and production of juvenile chinook salmon, *Oncorhynchus tshawytscha* (Walbaum) in an eutrophic reservoir. M.S Thesis, Oregon State University, Corvallis, 55 pp.

HOPKINS C.L. (1971) Production of fish in two small streams in the North Island of New Zealand. *N.Z. J. of Mar. and Freshw. Res.* **5**, 280–290.

HORTON P.A. (1961) The bionomics of brown trout in a Dartmoor stream. *J. Anim. Ecol.* **30**, 311–338.

HOUSER A. & NETSCH N.F. (1971) Estimates of young-of-year shad production in Beaver Reservoir. *Reservoir Fish. and Limn. Spec. Publ. No. 8, Am. Fish. Soc.*, 359–370.

HUNT R.L. (1966) Production and angler harvest of wild brook trout in Lawrence Creek, Wisconsin. *Wis. Conserv. Bull.* **35**, 1–52.

IVLEV V.S. (1945) [The biological productivity of waters] in Russian. *Adv. Mod. Biol.* **19**, 98–120. (Trans. by W.E. Ricker).

JENSEN A.L. (1972) Population biomass, number of individuals, average individual weight, and the linear surplus-production model. *J. Fish. Res. Bd. Canada* **29**, 1651–1655.

JOHNSON L. (1966) Consumption of food by the resident population of pike, *Esox lucius*, in Lake Windermere. *J. Fish. Res. Bd. Canada* **23**, 1523–1535.

JOHNSON W.E. & HASLER A.D. (1954) Rainbow trout production in dystrophic lakes. *J. Wildl. Mgmt.* **18**, 113–134.

KELSO J.R.M. & WARD F.J. (1972) Vital statistics, biomass, and seasonal production of an unexploited walleye (*Stizostedion vitreum vitreum*) population in West Blue Lake, Manitoba. *J. Fish. Res. Bd. Canada* **29**, 1043–1052.

KIPLING C. & FROST W.E. (1970) A study of the mortality, population numbers, year class strengths, production and food consumption of pike, *Esox lucius* L. in Windermere from 1944 to 1962. *J. Anim. Ecol.* **39**, 115–157.

KITCHELL J.F., KOONCE J.F., O'NEILL R.V., SHUGART J.J., Jr., MAGNUSON J.J. & BOOTH R.S. (1974) Model of fish biomass dynamics. *Trans. Am. Fish. Soc.* **103**, 786–798.

LECREN D.E. (1962) The efficiency of reproduction and recruitment in freshwater fish. *The Exploitation of Natural Animal Populations* (eds. E. D. LeCren and M. W. Holdgate) pp. 283–296. Blackwell, Oxford.

LECREN D.E. (1969) Estimates of fish populations and production in small streams in England. *Symposium on Salmon and Trout in Streams* (T. Northcote ed.) pp. 269–280. MacMillan Lect. Univ. of Brit. Col., Vancouver.

LECREN D.E. (1972) Fish production in freshwaters. *Symp. Zool. Soc. Lond.* **29**, 115–133.

LEWIS S.L. (1974) Population dynamics of Kokanee salmon in Odell Lake. Dingell–Johnson Rept. F 71-R-11 Jan. 1, 1973–Dec. 31, 1974. *Oregon Wildl. Comm.*, Portland, Oregon. 32 pp.

LINDEMAN R.L. (1942) The trophic–dynamic aspect of ecology. *Ecology* **23**, 399–418.

LIBOSVARSKY J. (1968) A study of brown trout population (*Salmo trutta* morpha *fario* L.) in Loucka Creek (Czechoslovakia). *Acta Sci. Nat. Brno.* **2**, 1–56.

LOTRICH V.A. (1973) Growth, production, and community composition of fishes inhabiting a first-, second-, and third-order stream of eastern Kentucky. *Ecol. Monogr.* **43**, 377–397.

LOWRY G.R. (1964) Net production, movement, and food of cutthroat trout (*Salmo clarki clarki* Richardson) in three Oregon coastal streams. M.S. Thesis, Oregon State Univ., Corvallis. 72 pp.

MACKINNON J.C. (1973) Analysis of energy flow and production in an unexploited marine flatfish population. *J. Fish. Res. Bd. Canada* **30**, 1717–1728.

MANN K.H. (1965) Energy transformation by a population of fish in the River Thames. *J. Anim. Ecol.* **34**, 253–275.

MANN R.H.K. (1971) The populations, growth and reproduction of fish in four small streams in southern England. *J. Anim. Ecol.* **40**, 155–190.

MATHEWS C.P. (1970) Estimates of production with reference to general surveys. *Oikos* **21**, 129–133.

MATHEWS C.P. (1971) Contribution of young fish to total production of fish in the River Thames near Reading. *J. Fish. Biol.* **3**, 157–180.

NAIMAN R.J. (1976) Productivity of a herbivorous pupfish (*Cyprinodon nevadensis*) in a warm desert stream. *J. Fish. Biol.* **9**, 125–137.

NEESS J. & DUGDALE R.C. (1959) Computation of production for populations of aquatic midge larvae. *Ecology* **40**, 425–430.

NUNNALLEE E.P. & MATHISEN O.A. (1972) Acoustic survey of Shuswap Lake, British Columbia, Canada. *Rept. to Int. Pac. Salmon Comm. by Fish. Res. Inst.*, University Washington, 36 pp.

ODUM E.P. (1959) *Fundamentals of Ecology.* Saunders, Philadelphia,

PATRIARCHE M.H. (1968) Production and theoretical equilibrium yields for the bluegill (*Lepomis macrochirus*) in two Michigan lakes. *Trans. Amer. Fish. Soc.* **97**, 242–251.

PELLA J.J. & TOMLINSON P.K. (1972) A generalized stock production model. *Inter-Amer. Trop. Tuna Comm. Bull.* **13**, 421–496.

PETROSKY C.E. & WATERS T.F. (1975) Annual production by the slimy sculpin population in a small Minnesota front stream. *Trans. Amer. Fish. Soc.* **104**, 237–244.

REGIER H.A. & HENDERSON H.F. (1973) Towards a broad ecological model of fish communities and fisheries. *Trans. Am. Fish. Soc.* **102**, 56–72.

RICKER W.E. (1946) Production and utilization of fish populations. *Ecol. Monogr.* **16**, 374–391.

RICKER W.E. (1948) Methods of estimating vital statistics of fish populations. Indiana Univ. Publs. Sci. Ser. **15**, 1–101.

RICKER W.E. (1975) *Computation and Interpretation of Biological Statistics of Fish Populations. Bull. Fish. Res. Bd. Canada* **191**, 1–382.

RICKER W.E. (1969) Effects of size-selective mortality and sampling bias on estimates of growth, mortality, production, and yield. *J. Fish. Res. Bd. Canada* **26**, 479–541.

RICKER W.E. & FOERSTER R.E. (1948) Computation of fish production. *Bull. Bingham Oceanogr. Coll.* **11**, 173–211.

SAUNDERS L.H. & POWER G. (1970) Population ecology of the brook trout, *Salvelinus fontinalis*, in Matamek Lake, Quebec. *J. Fish. Res. Bd. Canada* **27**, 413–424.

SCHAEFER M.B. (1954) Some aspects of the dynamics of populations important to the management of the commercial marine fisheries. *Bull. Inter-Amer. Trop. Tuna Comm.* **1**, 27–56.

STAPLES D.J. (1975) Production biology of the upland bully *Philypnodon breviceps* Stockall in a small New Zealand lake. III. Production, food consumption and efficiency of food utilization. *J. Fish Biol.* **7**, 57–69.

TEAL J.M. (1957) Community metabolism in a temperate cold spring. *Ecol. Monogr.* **27**, 283–302.

WALTER G.G. (1973) Delay-differential equation models for fisheries. *J. Fish. Res. Bd. Canada* **30**, 939–945.

WARREN C.E., WALES J.H., DAVIS G.E. & DOUDOROFF P. (1964) Trout production in an experimental stream enriched with sucrose. *J. Wildl. Mgmt.* **28**, 617–660.

WATERS T.F. (1966) Production rate, population density, and drift of a stream invertebrate. *Ecology.* **47**, 595–604.

WATERS T.F. (1969) The turnover ratio in production ecology of freshwater invertebrates. *Amer. Nat.* **103**, 173–185.

WELCOMME R.L. (1972) The inland waters of Africa. *CIFA-FAO Tech. Pap.* No. 1, 1–117.

WINBERG G.G. (1971) *The Estimation of Production of Aquatic Animals.* Academic Press, London, New York.

Chapter 2: Estimation of
Population Abundance and Survival

D. S. Robson and
G. R. Spangler

2.1 Introduction

The literature of population estimation methodology has grown largely in response to the need for answers to specific questions related to exploitation or management strategy. Given the broad spectrum of activities in applied ecology, it is inevitable that some methods would be re-discovered independently in a number of guises. In an attempt to avoid redundancy we have organized the subject matter into three areas which encompass most of the special methods applied in aquatic ecology to estimate population number and mortality rates.

The extent to which a quantitative estimate of a population parameter is meaningful depends upon the degree to which the population under study conforms to the requirements of the mathematical model which underlies the estimation procedure. The actual mechanics of estimation all too frequently obscure the need for a critical examination of the realism of the underlying model. For this reason we introduce our topic with a brief summary of the major biological and statistical concepts encountered in the selection of a model for the estimation process.

A freshwater fish population is generally confined to a relatively well-defined water mass which undergoes seasonal change throughout the year. Irrespective of whether the seasonal variability results from the influence of altitude, latitude or biotic factors, it is evident that the vital processes of such a population are unlikely to be adequately described by a stationary model. Thus it may be necessary to choose either an evolutionary model which will allow changes through time or to select short intervals of time within a year over which constant conditions might reasonably obtain. The objective, of course, is to describe the dynamics of the fish population in such a manner that responses to varying environmental stimuli are not obscured. These responses, coupled with measurement of the relevant environmental variables ultimately provide the basis for understanding the functional relationship between the population and its environment.

Quantitative descriptions of the dynamics of a population are developed in three interactive steps: (1) definition of the 'population' under study, (2) identification of the major factors characterizing population change, and

(3) derivation of procedures for extracting quantitative information from the population. The first two elements interact strongly in the initial planning of the study. For example, the method selected for determination of a mortality rate will depend upon whether the population is open or closed with respect to migration. The third element in approaching a population dynamics problem is generally peculiar to the particular time and location of the study. The state-of-the-art in both sampling hardware and analytical models often determines what variables will be measured and what estimates may be obtained.

Assuming that the population definition problem is tractable, the most comprehensive quantitative characteristics of the population are the size distribution and numbers of individuals within the various age groups. The dynamics of the population are described by the manner in which these characteristics change with time.

A change in the number of individuals in a population may be expressed as a rate. The term 'rate' can have a number of mathematically and biologically different connotations, and the specific connotation intended should always be clearly indicated. If $\Delta t = t_1 - t_2$, $\Delta N = N_1 - N_2$, with dt and dN their infinitesimal analogues, then the quantities implied by the following terms are all commonly referred to as 'rates':

$$\frac{\Delta N}{N_1}, \frac{\Delta N}{\frac{1}{2}(N_1 + N_2)}, \frac{\Delta N}{N_1 \Delta t}, \frac{\Delta N}{\Delta t}, \frac{dN}{N}, \frac{dN}{N dt}, \frac{dN}{dt}.$$

These are deterministic definitions, and their stochastic analogues in terms of probabilities are also called 'rates'. In those connotations where time is not explicitly mentioned, it is understood to be of unit length, however the unit happens to be defined.

It is expedient, in the pages that follow, to distinguish between 'interval' and 'continuous' models of rates or probabilities. The interval models are those in which a continuous process is not specified within the interval. The continuous models do specify the mathematical properties of the dynamic processes within such intervals.

Changes in the number of individuals in a population may be measured on a 'per individual' basis; thus, for a specified time period:

$$\text{Rate of change} = \frac{\begin{bmatrix} \text{Number present at the} \\ \text{start of the period} \end{bmatrix} - \begin{bmatrix} \text{Number present at the} \\ \text{end of the period} \end{bmatrix}}{\text{Number present at the start of the period}}.$$

This rate of change will in general represent a compound of at least three specific rates of change due to death, in-migration, and out-migration, though in some freshwater situations the population may be closed to one or both forms of migration. For a population which is closed to both forms of migration this rate of change becomes a mortality rate for the specified period:

$$\text{Mortality rate} = \frac{\text{Number of deaths during the period}}{\text{Number alive at the start of the period}}$$

$$= \text{Proportion dying during the period.}$$

Notice that if the number of deaths during the period is regarded as a chance outcome then the mortality rate estimates the probability that any randomly chosen fish will die during the period or, in other words, estimates the average probability of death for the fish alive at the start of the period.

In an exploited population, the mortality rate is a compound of two specific rates resulting from mortality due to fishing and mortality due to natural causes, and separation of the total mortality rate into these two components then becomes a matter of prime concern to the biologist who wishes to predict the effect of altering the fishing intensity. While such a separation is readily accomplished by the definition,

Total mortality rate = Fishing mortality rate + Natural mortality rate,

$$= \frac{\text{Catch during the period}}{\text{Number alive at start}} + \frac{\text{Natural deaths during the period}}{\text{Number alive at start}}$$

conceptual difficulties would usually arise in seeking a realistic interpretation of these two component rates. The model does apply, i.e. we do not encounter conceptual difficulties, where any particular fish has a finite probability of death, during the specified time interval, either from fishing or from natural causes but not from both. But this contradicts our experience with most fish populations since both types of mortality usually occur concurrently.

Clearly, a fish which is removed by capture early in the period is not among those subjected to the risk of natural death for the remainder of the period, thus reducing the number that could conceivably die from natural causes. Therefore, these two forces of mortality are competing. Decreasing the intensity of fishing, for example, would have the effect of increasing the natural mortality rate as defined above, since more fish would then be left uncaptured to run the risk of death from natural causes during the period in question. The utility of the above definition of fishing and natural mortality rates is distinctly limited by the fact that the rates defined in this manner depend upon each other, and the true functional form of this dependence is generally unknown. But the average form of the functional relationship will likely become discernible in a large body of data from a series of populations in a region.

The interpretational difficulty associated with competing risks operating over the same time period will decrease as the measure of competition increases. In the limiting case where deaths of one kind preclude the possibility of deaths of the other kind then the mortality rates defined above become functionally independent and hence operationally useful. This limiting case can be approached by the simple device of decreasing the length of the time period. By making the time sufficiently short we can be assured of overwhelming odds that at most one kind of death will occur during the period, thus creating a situation where the two kinds of events are mutually exclusive. This device is impracticable to implement, but does suggest a mathematical approach for obtaining a functional form of the dependence between fishing and natural mortality rates over longer time periods.

For a time period of very short duration Δt—say, minutes or seconds—the probability of any death at all is extremely small, and approximately proportional to the length of the period. Thus

probability of *at least one death* in $(t, t+\Delta t) \simeq [F(t)+M(t)]\Delta t$

where the proportionality factors $F(t)$ and $M(t)$ are called, respectively, the (instantaneous) fishing and natural mortality coefficients. If deaths occur singly, as opposed to the simultaneous occurrence of two or more deaths, then for Δt sufficiently small the probability of *more than one death occurring* will be negligible and the

probability of exactly one death in $(t, t+\Delta t) \simeq [F(t)+M(t)]\Delta t$

$$\simeq \frac{N(t)-N(t+\Delta t)}{N(t)}$$

where $N(t)$ is the number alive at time t. That is, the mortality rate estimates the probability that a randomly chosen fish, alive at time t, will die before $t+\Delta t$. If further, each fish alive at t has the same probability of dying and if the events occurring in successive time periods are statistically independent then it follows from the differential calculus that the probability of a particular fish dying in the period $(t, t+\Delta t)$ is exactly

$$\int_t^{t+\Delta t} [F(x)+M(x)] \frac{1}{\exp\left\{\int_t^x [F(y)+M(y)]dy\right\}} dx$$

$$= \int_t^{t+\Delta t} F(x) \frac{1}{\exp\left\{\int_t^x [F(y)+M(y)]dy\right\}} dx$$

$$+ \int_t^{t+\Delta t} M(x) \frac{1}{\exp\left\{\int_t^x [F(y)+M(y)]dy\right\}} dx.$$

We refer to this model as a continuous Poisson model (Bailey 1964) with the functions of time as the parameters.

An expression of the above form shows, under the aforementioned special condition, the functional relationship between fishing mortality rate and natural mortality rate over a specified time period. With the functions $F(t)$ and $M(t)$ completely unspecified, however, this mathematical result has no practical applications. Where this model has been applied, simplifying assumptions have been made concerning the behaviour of F and M. The usual assumptions are that F is proportional to fishing effort f, $F = qf$ (where q is the catchability coefficient), and that M is constant over time. During periods of constant fishing intensity this gives the relation

$$1 - \frac{1}{\exp\{(qf+M)\Delta t\}}$$

$$= \frac{qf}{qf+M}\left\{1 - \frac{1}{\exp\{(qf+M)\Delta t\}}\right\} + \frac{M}{qf+M}\left\{1 - \frac{1}{\exp\{(qf+M)\Delta t\}}\right\}$$

between total per-individual mortality rate, on the left, and per-individual fishing and natural mortality rates on the right, for a period of duration Δt.

In a freshwater fishery the simplified model might conceivably apply within segments of a year, with different values for the constants q, f and M in the different segments. It cannot realistically apply in situations where simultaneous multiple deaths occur; the purse seine, for example, in a single set removes almost an entire school of fish. This model, or modifications thereof, can certainly be tested if adequate data are available on a regional basis.

Our ignorance of the details of the interactions between individual fish in the population does not, of course, absolve the investigator of the responsibility to do everything in his power to eliminate or account for interactions which are quantifiable or known to exist. Thus, for example, where the catchability coefficient (q) is known to differ between identifiable segments of the population due, perhaps, to sexual dimorphism or to size selectivity of the fishing gear, it may be necessary to redefine the 'population' with respect to sex or size in order that the ensuing estimates will be the product of a more realistic model than would otherwise have been the case. The degree to which the estimates might differ on the basis of alternate models can be examined through sensitivity analysis (Kowal 1971).

We have referred to the (interval) mortality rate over a specified period as an estimate of the average probability of death, recognizing that the actual number of fish dying in a specific situation is not predetermined but is, in fact, a chance event. It may be possible to observe this chance event, in order to obtain an exact count of the number of deaths and a direct estimate of the mortality rate. Despite the complete count, this estimate is still subject to the random error generated by chance variation in number of actual deaths.

If deaths occur in batches, as when fishing operates on schools, then chance variation in number of deaths will be relatively large for any given intensity of fishing, i.e. for any given probability of death from fishing. Chance variation will be smaller if deaths occur singly, and is reduced still more if the intraclass correlation is negative; that is, if the death of one fish increases the chances of survival of those remaining. A *stochastic* model specifies the probability associated with each possible outcome of this chance process. The model most commonly assumed for this situation asserts that each fish alive at the start of the period has the same probability of dying during the period and that the death of one in no way influences the chances of death for another. In effect, this binomial model requires that each fish behave independently of all others. The absurdity of such an assumption is self-evident but with the present incomplete knowledge of fish behaviour and the nature of mortality no specific alternative has been offered.

Usually an exact count of the number of deaths cannot be obtained, and we must estimate the proportion actually dying by some sampling procedure, thus introducing an additional source of error into the estimate of the average probability of death. The sample actually obtained is but one out of some very large number of possible samples, each of which had some unknown probability of being drawn. A model which associates a probability to each possible sample will, in general, depend upon the unknown proportion that actually died during the period under consideration. Thus, if the sampling was planned properly, a sample which indicates that there were very few deaths should have very small probability of appearing if there were, in fact, very many deaths, while this same sample should have a relatively high probability of occurrence if there were in fact very few deaths. A fully defined *statistical* model, therefore, states explicitly, in mathematical form, how the probability attaching to any given sample varies with the value of the unknown quantity that is being estimated.

When a statistical model has been specified in such detail, it is then possible to determine the numerical value for this unknown quantity which maximizes the probability (likelihood) of having drawn the sample in hand. This numerical value is called the *maximum likelihood estimate* of the unknown quantity. This method of estimation, widely used in connection with population estimation, is statistically efficient for large samples, provided that the statistical model is valid. Unfortunately, specifying the model in sufficient detail to permit use of the maximum likelihood method requires many assumptions which, in the present context, usually reduce fish to totally senseless particles bumping about at random.

A more robust estimation procedure and more widely used in fishery statistics is the so-called *method of moments*, which equates statistics computed from the sample to their average value expressed in terms of unknown population quantities, and then solves these equations for the unknowns.

When a collection of observations are available, each of which estimates a known function of the same unknown constants, the *principle of least squares* may be applied to determine the values of these unknowns which minimize the sum of squares of deviations of the observations from their predicted values. The least squares solution is most easily calculated if these functions are linear in terms of the unknown constants, reducing the computational problem to ordinary linear regression analysis; in the nonlinear case a transformation of the scale of measurement is, therefore, sought to yield linear functions.

An estimate with a small variance is regarded as more precise than an estimate with a large variance. As noted above, the appropriateness of any estimate is a function of the realism of the underlying model, hence an estimate with a small variance is not necessarily 'right', merely precise! Paulik and Robson (1969) observe that an 'approximate right answer is always preferred to a highly precise but wrong answer'. The need obviously exists for a method of quantifying the trustworthiness of an estimate and the method most frequently chosen is that of calculating variances and constructing a confidence interval about the estimate. We will not go into the methods of calculating individual variances for each of the estimators discussed below except to note that many final estimates of population parameters are, themselves, based upon estimates of more basic parameters. In effect, we are confronted with the problem of combining all of the contributing individual variances into a single expression for the over-all variance of the parameter of interest. One general procedure for accomplishing this is the 'propagation-of-error' technique or 'delta-method' (Deming 1943, 1960, Guest 1961, Keyfitz 1968, Seber 1973). The approximate variances produced by the delta-method become more reliable as sample size increases (Paulik & Robson 1969).

The foregoing considerations are implicit in many of the commonly applied methods of estimating population number and mortality rates. In presenting the three general methods which follow (adapted from Regier & Robson 1967), we have made no attempt to assure that these categories are mutually exclusive. More elaborate taxonomies of estimation methods are available in the comprehensive reviews of Seber (1973) and Ricker (1975). We begin with the most generally applicable methodology and proceed to the more restricted techniques for estimating abundance and mortality rates.

2.2 Change-in-ratio estimators

Methods of this general type have been variously known as change of composition, survey-removal or dichotomy methods (Regier & Robson 1967). The basis for the method is an observed change in the relative abundance of two classes of animals within a population. The classes may be naturally occurring groups such as age, species, or sex classes or they may be artificially constructed classes such as marked or unmarked animals. The nature of the change-in-ratio of the classes allows us to estimate population abundance and survival (Paulik & Robson 1969).

Chapman and Murphy (1965) credit origin of the method to G. H. Kelker in 1940 but Paulik and Robson (1969) subsequently reviewed the method and identified an entire family of estimators (including mark-recapture methods) which they named CIR (acronym for Change-In-Ratio) estimators. Recognition of mark-recapture methods as members of this family extends the recorded history of development back to 1662 when John Graunt applied the principle of the method in a demographic analysis (Ricker 1975).

We follow the notation of Paulik and Robson (1969) in developing the CIR estimators. Assuming that the population consists of 2 types of animals, x and y and that a differential change occurs in the numbers of x- and y-type animals between Time 1 (t_1) and Time 2 (t_2)
let

N_i = total number of animals in the population at t_i, (i = 1, 2),
X_i = number of x-type animals in the population at t_i,
Y_i = number of y-type animals in the population at t_i,

then the following relations hold:

$$N_i = X_i + Y_i,$$

$$p_i = \frac{X_i}{N_i} = \text{the proportion of } x\text{-type animals in the population at } t_i,$$

$R_x = X_2 - X_1$ = net change in numbers of x-type animals in the population
between t_1 and t_2,

$R_y = Y_2 - Y_1$,
$R = R_x + R_y$ = net removal ($-$) or addition ($+$) to the population between
t_1 and t_2,

$$f = \frac{R_x}{R}, \ 0 \leq f \leq 1, \ (f \text{ is defined only when } R_x \text{ and } R_y \text{ have the same sign}).$$

Counts obtained in the sample surveys conducted to provide CIR estimates are designated by lower case letters corresponding to their population counterparts, thus,

$$n_i = x_i + y_i \quad \text{and}$$
$$r = r_x + r_y.$$

The symbols and relationships defined above allow us to state the basic CIR formula from which abundance and survival estimates will be made, that is,

$$p_2 = \frac{p_1 N_1 + R_x}{N_1 + R}$$

A basic word equation for this relationship demonstrates its applicability in situations where males or females might be selectively removed from the population:

$$\left(\begin{array}{c}\text{proportion of males in total} \\ \text{population after removal}\end{array}\right) = \frac{\left(\begin{array}{c}\text{no. of males} \\ \text{before removal}\end{array}\right) - \left(\begin{array}{c}\text{no. of males} \\ \text{removed}\end{array}\right)}{\left(\begin{array}{c}\text{population size} \\ \text{before removal}\end{array}\right) - \left(\begin{array}{c}\text{total no. of} \\ \text{animals removed}\end{array}\right)}.$$

Notice that the signs in the word equation depend upon whether animals are entering or leaving the population (in this case, leaving).

Rearranging terms in the basic formula, Paulik and Robson (1969) obtain expressions for estimating the absolute abundance of the population and the number of x-type animals in the population at t_1:

$$\hat{N}_1 = \frac{R_x - \hat{p}_2 R}{\hat{p}_2 - \hat{p}_1}, \text{ and}$$

$$\hat{X}_1 = \frac{\hat{p}_1 (R_x - \hat{p}_2 R)}{\hat{p}_2 - \hat{p}_1}.$$

Estimates of population parameters are indicated by a caret over the symbol for the parameter, thus, $\hat{p}_i$ is the estimate of $\hat{p}_i$ and $\hat{f}$ is the estimate of f. Variance (V) of $\hat{N}_1$ derived by the delta-method for independent estimates of p_1 and p_2 is,

$$V(\hat{N}_1) \simeq (p_2 - p_1)^{-2} [N_1^2 V(\hat{p}_1) + (N_1 + R)^2 V(\hat{p}_2) + (1 - p_2)^2 V(\hat{R}_x) + p_2^2 V(\hat{R}_y)].$$

Seber (1973) presents the simplified formula for estimating the variance of $\hat{N}_1$ where the removals (or additions) R_x and R_y are known exactly.

Paulik and Robson (1969) suggest application of the CIR method in a mark-recapture context for largemouth bass, *Micropterus salmoides*, and for estimating the abundance of brook trout, *Salvelinus fontinalis*, prior to opening of the fishing season. In the latter example, the x- and y-type animals are different species with differential catchabilities.

CIR methods may also be applied directly to the problem of estimating survival rates. Again, following Paulik and Robson (1969), we define s_x and s_y as the respective proportions of x- and y-type animals surviving from t_1 to t_2. At t_2 the numbers of x- and y-type animals are:

$$X_2 = X_1 s_x = p_1 N_1 s_x, \text{ and}$$
$$Y_2 = Y_1 s_y = (1 - p_1) N_1 s_y.$$

Substituting into the basic CIR formula and rearranging terms yields an expression for the number of surviving x-type animals per surviving y-type animal:

$$\frac{s_x}{s_y} = \left(\frac{p_2}{1-p_2}\right)\left(\frac{1-p_1}{p_1}\right) = \frac{Y_1 X_2}{X_1 Y_2}.$$

By the delta-method, the variance is,

$$V\left(\frac{s_x}{s_y}\right) \simeq \left[p_1(1-p_2)\right]^{-4}\left[(1-p_2)^2 p_2^2 V(p_1) + (1-p_1)^2 p_1^2 V(p_2)\right].$$

Paulik and Robson (1969) note that s_x could be handily estimated from this expression in situations where s_y could be contrived to be equal to 1. Thus, if the Y_1 animals are introduced into the population just before time t_2 so that no removals of y-type animals will occur before the survey is made to estimate p_2, then,

$$Y_2 = Y_1 \qquad \text{and}$$

$$\hat{s}_x = \left(\frac{x_2}{X_1}\right)\bigg/\left(\frac{y_2}{Y_1}\right)$$

The utility of this result and its relationship to multiple mark-recapture studies is demonstrated by Paulik and Robson (1969) with the following data from Chadwick (1968) on tagged striped bass, *Roccus saxatilis*. In the spring of 1958 (t_1), 3891 bass were tagged and released in the Sacramento–San Joaquin delta area of San Francisco Bay. The following year (t_2), an additional 2965 bass were similarly tagged and released. Over the next 8 years, 430 recoveries from the first batch and 1026 from the second were reported in the sport fishery. Survival from t_1 to t_2 is therefore:

$$\hat{s}_x = \left(\frac{430}{3891}\right)\bigg/\left(\frac{1026}{2965}\right) = 0{\cdot}319$$

This example typifies applications where an identifiable experimental group is subjected to a force of mortality which kills an unknown fraction of the group and an unstressed control group provides a base population for CIR estimation of the mortality induced in the stressed group. The example above (i.e. two releases of marks) requires only an identifiable batch-mark, but for cases where numerous releases would quickly exhaust the available batch-marks, say fin-clip combinations, it might be feasible to apply uniquely numbered tags to the fish so that an individual animal could be identified throughout its capture history in the fishery. Tag recoveries by sport and commercial fishermen provide a basis for estimating successive, year specific, annual survival rates in the manner described by Seber (1970) and Youngs and Robson (1975), and extended by Brownie and Robson (1976) to provide estimates of age- as well as year-specific survival rates.

The literature of multiple mark-recapture methods is far too extensive to review here. The reader is referred to the volumes by Seber (1973) and Ricker (1975) for historical and case history development of the subject or to Cormack (1968, 1972) for a penetrating, but concise, discussion of the models, logic and assumptions underlying these methods. We restrict our discussion to the model developed by Jolly (1965), and variously known as the Jolly, Seber–Jolly, and Jolly–Seber method, which allows immigration, emigration, death and losses of tagged animals upon recapture. The model derived by Seber (1965) is almost identical to Jolly's (and nearly simultaneous in development) having sprung from earlier deterministic models by Darroch (1958, 1959), Seber (1962) and Jolly (1963).

The estimation of abundance depends upon estimating both the number of marked animals (M_i) and the proportion (a_i) of marked animals in the total population at time i. Using Jolly's symbols for these estimates:

$$\hat{a}_i = \frac{m_i}{n_i} \ (i = 2, 3, \ldots, l),$$

$$\hat{M}_i = \frac{s_i K_i}{R_i} + m_i \ (i = 2, 3, \ldots, l-1),$$

where the number of samples l is indexed through time by the subscript i (the t_i of CIR estimators discussed above) and,

  m_i = number of *marked* animals in the ith sample,
  n_i = number captured in the ith sample,
  s_i = number released from the ith sample after marking,
  R_i = number of the s_i animals which are subsequently caught,
  K_i = number of animals marked before time i which are not caught in the ith sample, but are caught subsequently. (Jolly used Z_i for this class of animals but we have adopted Ricker's K_i to avoid confusion with the international symbol for total instantaneous mortality rate.)

The CIR nature of the estimator of M_i becomes apparent upon consideration of the groups of animals involved. Jolly (1965) provides a lucid description: 'The *number* of marked animals can be thought of as consisting of the observed number captured, m_i, plus the number not captured, $M_i - m_i$. Immediately after time i there are two groups of marked animals, the $M_i - m_i$ and the s_i animals just released. Of the former [K_i] are subsequently caught; of the latter R_i are subsequently caught.' Intuitively, one would expect that the ratio of K_i to the marked animals alive and free at time i would be equal to the ratio of R_i to the marked animals just released *providing that the probability of recapture is the same for both groups.*

The proportion of marked animals in the sample is taken to represent the proportion of marked animals in the population, hence the estimate of abundance (N):

$$\hat{N}_i = \frac{\hat{M}_i n_i}{m_i} \quad (i=2, 3, \ldots, l-1).$$

Survival (ϕ_i) from immediately after the ith release to time ($i+1$) is estimated by the ratio of marked animals alive at these times,

$$\hat{\phi}_i = \frac{\hat{M}_i + 1}{\hat{M}_i - m_i + R_i} \quad (i=2, 3, \ldots, l-2).$$

The final estimate of special interest to us in Jolly's model is that for immigration into the population. This parameter may be the item of greatest interest to the investigator concerned with the dynamics of a population. It could be useful in determining hatching (or emergence) rates for insects as in Jolly's example of Black-kneed capsids (*Blepharidopterus angulatus*) or for estimating recruitment to a population from an adjacent stock. The estimate takes the form:

$$\hat{B}_i = \hat{N}_{i+1} - \hat{\phi}_i(N_i - n_i + s_i) \quad (i=2, 3, \ldots, l-2),$$

where B_i is the number of new animals joining the population during the interval between t_i and t_{i+1} and alive at t_{i+1}. The following interpretation by Cormack (1968) is appealing: 'Of the $(N_i - n_i + s_i)$ animals alive in the population immediately after t_i, $\phi_i(N_i - n_i + s_i)$ will be alive at t_{i+1}: the difference between this figure and N_{i+1} must be accounted for by immigration.'

The asymptotic variances derived by Jolly (1965) apply only to situations where reasonably large numbers of recaptures might be expected. However, he points out that the variance of $\hat{M}_i$ will be a minimum when the number of recaptures, R_i, is as large as possible. Small sample bias is likely to affect the results of studies where recaptures might be very few (on sampling occasions when $R_i = 0$, no estimate of M_i is possible). Ricker (1975) suggests the use of a modified formula due to Seber (1965) for situations involving low numbers of recaptures and concludes with the rule that each of the R_i or m_i should be at least 3 or 4. Budd *et al.* (1968), in discussing a population estimator similar to Jolly's, make it clear that the K_i necessary to estimate M_i invariably provides a lower boundary for the estimate of the number of marked animals extant.

In spite of the generality of the Jolly method for estimating abundance, an important constraint remains. The model does not allow 'circular migration', i.e. once an animal enters the population it must not depart and subsequently return. None of the models developed thus far, including the general stochastic model of Robson (1969), have overcome this difficulty. Until an alternative becomes available, the wary investigator will simply have to contrive his experiments so that known cyclic migratory behaviour in the population does not interfere with the estimation procedure.

There have not been too many fisheries applications of the Jolly method to date, but Seber (1973) lists four applications in the entomological literature and three for mice. In fisheries investigations the method has been applied to striped bass (Chadwick 1968), dead pink salmon, *Oncorhynchus gorbuscha* (Parker 1968), lake whitefish, *Coregonus clupeaformis* (Spangler 1970), and splake hybrids, *Salvelinus fontinalis* X *S. namaycush* (Berst & Spangler 1970). The investigator interested in planning CIR experiments should consult the papers of Robson and Regier (1964) and Paulik and Robson (1969) for extensive discussion of sample size requirements in relation to bias, accuracy and precision of the estimates.

2.3 Direct enumeration methods

Total population

A direct count of the population is practicable in some circumstances. The draining of a pond, for example, may result in complete recovery of all fish present; fish migrating through a narrow channel may be completely enumerated, as may highly sedentary fish inhabiting clear shallow water. In such circumstances the rate of population change becomes directly measurable if separate counts can be made at two or more points in time.

In order to interpret and to assess the reliability of an estimate of the probability of death obtained from two direct counts N_1 and N_2 separated by an interval Δt, some form of stochastic model must be adopted for the change in population number. First, if the change is to represent only mortality, or mortality plus out-migration, then the population must either be closed to in-migration and other forms of recruitment or the recruits entering during Δt must be recognizable and excluded from the count N_2. The simplest stochastic model then assigns equal probability to every combination of N_2 fish from the initial set of N_1 fish; that is, the N_2 survivors are assumed to be a random sample from the initial population.

Randomness, while a necessary condition, is not the only condition needed to specify a stochastic model of survival. The process of selecting N_2 survivors could be completely random while the process which determined the number N_2 to be selected was completely deterministic. Thus, in a period Δt when no natural mortality occurred, a fishery operating on a quota system might non-selectively remove a precisely predetermined catch $C = N_1 - N_2$. In this slightly unrealistic situation the probability of death was exactly $D = (N_1 - N_2)/N_1$ for each of the N_1 fish, the N_2 survivors were a random sample from the initial stock of N_1 and, despite the randomness, there was no error due to chance in the observed mortality rate $\hat{D}=(N_1-N_2)/N_1$ as an estimate of the true probability of death D.

At the other extreme is the catastrophic type mortality where, say, there is a probability D that the catastrophe will occur and exterminate the population, and a probability $1-D$ that no mortality will occur. In the latter instance $N_2=N_1$ and $\hat{D}=0$, while in the former $N_2=0$ and $\hat{D}=1$; thus, with probability $1-D$ the error of estimate will be $\hat{D}-D=0-D=-D$ and with probability D the error will be $\hat{D}-D=1-D$. The average error is then zero,

$$(1-D)(-D)+D(1-D)=0$$

but the average squared error, or the variance of the estimate, is as large as it can possibly be with direct enumeration,

$$V(\hat{D})=(1-D)(-D)^2+D(1-D)^2=D(1-D).$$

The real situation lies somewhere between these two extremes, $0<V(\hat{D})<D(1-D)$, and a formula to express the variance of $\hat{D}$ in terms of N_1 and D can only be obtained by specifying some stochastic model intended to approximate the true probability distribution of the number surviving. As mentioned earlier, the standard assumption invoked at this point to completely specify the stochastic model is independence; each fish is assumed to follow its own independent course to death. This assumption, together with randomness, is equivalent to assuming that N_2 is a binomial random variable and that the variance of the estimator $\hat{D}$ is $V(D)=D(1-D)/N_1$; thus if N_1 is large this model optimistically predicts that the variance is very near to its lower bound of 0. Since any mortality force such as fishing on schools which operates in the manner of an epidemic serves to increase the variance, we should in fact expect to find that the binomial variance $D(1-D)/N_1$ can be expected to be but a small fraction of the true variance of $\hat{D}$. There is, unfortunately, no possible means of testing the independence assumption of the binomial model in a single, two-count experiment. When such experiments are repeated through time and space within a geographic region, however, unbiased variance estimates are obtainable directly by computing the variance among all estimates, thus not requiring a completely specified stochastic model.

A direct count experiment on an exploited population may include also a count of the number C of fish removed by capture during the period Δt. Then total mortality,

$$\hat{D}=\hat{D}_F+\hat{D}_M$$

or

$$1-\frac{N_2}{N_1}=\frac{C}{N_1}+\frac{N_1-N_2-C}{N_1}$$

where D_F and D_M are, respectively, the probability of death from fishing and the probability of death from natural causes during Δt. If the N_1 fish alive at the start follow independent and identical stochastic death processes then the joint probability distribution of the number of fishing deaths and natural deaths will be trinomial, with variances:

$$V(\hat{D}_F) = \frac{D_F(1 - D_F)}{N_1}, \; V(\hat{D}_M) = \frac{D_M(1 - D_M)}{N_1}.$$

Again, however, consideration of the contagious nature of fishing deaths (due to schooling) and natural deaths suggests that the trinomial model is invalid, grossly underestimating variances.

Imposing on the trinomial model the further assumption that events occurring in successive time intervals within Δt are independent and identically distributed leads to the continuous Poisson model of the death process with

$$D_F = \frac{F}{F+M}\left(1 - \frac{1}{\exp\{(F+M)\Delta t\}}\right) D_M = \frac{M}{F+M}\left(1 - \frac{1}{\exp\{(F+M)\Delta t\}}\right),$$

from which

$$\hat{F} = -\frac{\hat{D}_F \log(1 - \hat{D})}{\hat{D}\Delta t} \; \hat{M} = -\frac{\hat{D}_M \log(1 - \hat{D})}{\hat{D}\Delta t}.$$

Approximate variance formulae for $\hat{F}$ and $\hat{M}$ are given by Rothschild (1966) for the trinomial model.

The negative exponential model of mean population number as a function of time is derived from the continuous Poisson death process. More frequently, in the fisheries literature, it is derived deterministically from the assumption that at any instant the rate of change in population number is directly proportional to the population number. Thus

$$\frac{dN}{dt} = -MN$$

or, with fishing also present,

$$\frac{dN}{dt} = -(F+M)N.$$

Solution of these equations, and setting observations equal to the parameters, yields method of moments estimators identical to those with the maximum likelihood estimator. Since this latter model is *deterministic*, it would be contradictory to seek to estimate a variance of $\hat{F}$ or $\hat{M}$, when the latter is estimated from complete counts. If the study were replicated it would be reasonable to estimate a variance between replicates.

Partial population

There are circumstances where it is not feasible to enumerate a whole population but where it is possible to completely enumerate *randomly selected* segments of that population. If the population is migratory, earlier studies may show that it will very likely pass a vantage point (under a bridge, through a fish ladder, etc.) over a period of, say, 20 days in May. Available resources may permit a census of only 10 complete days at random. Other examples are the capture of all fish in small selected sections of a stream or the capture of all fish in enclosed small areas in lakes. Practical aspects of obtaining estimates of the total population number under such circumstances have been discussed by Rounsefell and Everhart (1953) and by Lambou and Stern (1958). Much of the necessary theory for choosing an efficient design and determining the reliability of estimates was summarized by Lambou (1963); further details may be found in text books on sampling theory, e.g. Cochran (1963). An outline of the simplest design follows.

Suppose the boundary of the total population space is known but the population distribution in this space is not known and hence there is no basis for stratification into high density or low density areas. We arbitrarily divide the population space into A equal unit spaces and select a of these to enumerate completely, the number a being chosen usually on the basis of resources available. The problem is to estimate

$$N. = \sum_{i=1}^{A} N_i = N_1 + N_2 + N_3 + \cdots + N_A;$$

where the 'dot' notation is an alternative to using the Greek letter sigma (Σ) and is interpreted to mean summation for all values of the subscript for the place occupied by the dot (Steel & Torrie 1960). The experiment yields error-free numbers $N_1, N_2, N_3, \ldots, N_a$ corresponding to subspaces 1, 2, 3, ..., a. Then

$$\hat{N}. = \frac{A}{a} \sum_{i=1}^{a} N_i$$

and

$$\hat{V}(\hat{N}.) = \frac{A^2 - aA}{a} \hat{V}(N_i)$$

where

$$\hat{V}(N_i) = \frac{a \sum_{i=1}^{a} N_i^2 - \left(\sum_{i=1}^{a} N_i \right)^2}{a(a-1)}.$$

This is, of course, a non-parametric estimate of N. and its sampling variance, in contrast to the parametric maximum likelihood method used earlier. The estimators are valid whether the population is randomly dispersed in space, or contagiously distributed (sometimes called 'under dispersed'), or approximately uniformly distributed ('over dispersed'). But they are likely inefficient except where the researcher is in fact totally ignorant of the distribution.

Consider the case where two estimates of the above type, $\hat{N}_1$. and $\hat{N}_2$., are available at two points in time, t_1 and t_2. The population has been closed except to death. We seek an estimate of the fraction that died and utilize a deterministic model. Such an estimate (not unbiased) is

$$\hat{D} = 1 - \hat{S}$$

where

$$\hat{S} = \frac{\hat{N}_2.}{\hat{N}_1,}$$

The approximate sampling variance of $\hat{S}$ by the delta method is

$$(\hat{V}\hat{D}) = \hat{V}(\hat{S}) \simeq \left(\frac{1}{\hat{N}_1.}\right)^2 \hat{V}(\hat{N}_2.) + \left(\frac{\hat{N}_2.}{\hat{N}_1.^2}\right)^2 \hat{V}(\hat{N}_1.)$$

with the expressions for $\hat{V}(\hat{N}_2.)$ and $\hat{V}(\hat{N}_1.)$ analogous to $\hat{V}(\hat{N} .)$ above.

The variance estimator given above measures only the sampling variability in the estimates of $N_1.$, $N_2.$ and D due to the random choice of sub-spaces; it does not include a 'stochastic component' as it should if we had chosen to consider $N_2.$ as a chance variable depending on $N_1.$ and D (where D = probability that a particular fish dies during the corresponding Δt). If we were to use the latter stochastic approach then the estimator $\hat{V}(\hat{D})$ would depend partly on the particular stochastic model used and we would be facing again the usually frustrating problem of finding a reasonably realistic stochastic model.

Confidence intervals (C.I.) and tests of hypotheses concerning the actual mortality rate

$$1 - \frac{N_2}{N_1}$$

may be calculated when values of the N_{1i} are large and $\hat{D}$ can be assumed to be distributed approximately normally. Under these conditions

$$100(1-\alpha) \text{ per cent C.I.} = \hat{D} \pm t_\alpha \sqrt{\{\hat{V}(\hat{D})\}}$$

where t_α is Student's t with ∞ degrees of freedom. A confidence interval estimate of D itself cannot be constructed without specifying a stochastic model for N_2.

The above discussion concerned visual counting in precisely defined subspaces of the population space. The echo-sounder has for some years been used to indicate not only the presence of fish but also their relative abundance. Efforts are being made to identify species by reflected sound waves, to count individual fish, and to measure the volume and population space scanned in order to obtain absolute estimates of pelagic populations. The use of acoustic instruments for detection of fish and estimation of their abundance has been reviewed by Forbes and Nakken (1972), and Cushing (1973). Most of the applications to date have been in the marine environment, but Kelso *et al.* (1974) report limited success in estimating abundance of smelt, *Osmerus mordax* and alewives, *Alosa pseudoharengus*, in lakes Erie Erie and Ontario. Estimation of abundance in situations where the fish are very densely schooled or where various co-habiting species are not individually identifiable will require additional research. The system described briefly by Lord (1973) utilizing an array of upward-looking transducers to enumerate migrating sockeye salmon, *Oncorhynchus nerka*, suggests that acoustic sampling equipment may play a conspicuous role in future efforts to estimate abundance of mobile fish stocks.

CATCH-EFFORT METHODS

Methods are included here that utilize primarily data on catch and the effort expended in obtaining various segments of the catch. In general, the method requires that a population be fished intensively enough that the removals result in a decline in the catch obtained per unit of fishing effort. Ricker (1975) traces the origin of these methods to Helland (cited by Hjort 1914) who used hunting statistics to estimate the abundance of Norwegian bears. The reviews of Gulland (1964), Paloheimo and Dickie (1964), Seber (1973) and Ricker (1975) provide extensive discussion of the general method and its variants as well as details of the assumptions underlying catch-effort models.

Two periods of sampling

An analogue of direct enumeration methods will be presented first. A population, open only to death is sampled with f units of standard non-selective gear at randomly chosen locations in the population space. The sampling period is very brief, and it is assumed that no natural mortality occurs during this period, also that the probability of capture of any particular fish is $q'f$, where q' is called the binomial catchability coefficient.

Taking N_1 fixed at t_1, then C_1 (obtained with f_1 units of effort employed during the short sampling interval) is distributed binomially with the maximum likelihood estimator of N_1

$$\hat{N}_1 = \frac{C_1}{q'f_1}.$$

This assumes that we know the value of q'. Note that when C_1 is much less than N_1, the interval Poisson approximates the binomial distribution.

If at t_2, f_2 units of effort are used, with q' unchanged and catch C_2 fish, then considering N_2 fixed

$$\hat{N}_2 = \frac{C_2}{q'f_2}.$$

The estimator of deterministic mortality is

$$\hat{D} = 1 - \hat{S} = 1 - \frac{\hat{N}_2}{\hat{N}_1} = 1 - \frac{C_2}{C_1}.$$

Thus an estimate of q' is not required to obtain estimates of total mortality between t_1 and t_2.

Multiple periods of sampling

The binomial stochastic model can now be extended, retaining all assumptions, to repeated periods of capture. Every member captured is removed from the population, otherwise the population is closed. For simplicity, constant effort in all sampling periods will be assumed. Starting with N_1 members at t_1, we expect to capture $N_1 q'f$ the first period, $(N_1 - N_1 q'f)q'f = N_1 q'f(1-q'f)$ in the second period, $\{N_1 - N_1 q'f - N_1 q'f(1-q'f)\}q'f = N_1 q'f(1-q'f)^2$ in the third period, and so on to $N_1 q'f(1-q'f)^{n-1}$ in the nth period. Now if the observed catches are set equal to their expectations (i.e. using the method of moments)

$$C_i = N_1 q'f(1-q'f)^{i-1}$$

and

$$\log_{10} C_i = \log_{10}(q'fN_1) + (i-1)\log_{10}(1-q'f)$$

This corresponds to the linear equation

$$Y_i = a + bX_i, \text{ where } 1 \leqslant i \leqslant n.$$

Here

$$Y_i = \log_{10} C_i$$

and

$$X_i = i - 1.$$

Thus

$$b = \log_{10}(1-q'f)$$

and

$$a = \log_{10}(q'fN_1)$$

may be estimated by linear regression methods, $q'f$ estimated from $\hat{b}$ and $\hat{q}'f$ used to obtain $\hat{N}_1$ from $\hat{a}$.

This is a simplified version of what is often called the Leslie estimator, after P. H. Leslie who developed it in 1939. Moran (1951) examined the method and noted that conventional regression methods are not fully valid here. He derived maximum likelihood estimators of the parameter and its variance using a binomial model, and compared the results in a practical instance to those obtained from a weighted regression method. He stressed the need for critical examination of the assumptions which would usually require that the study be designed to provide additional information. More recently, Ricker (1973) has reviewed applications of regression methodology in fishery research and has recommended that the geometric mean functional regression be used in place of the conventional predictive regression models when the X and Y regression variates are subject to both natural variability and measurement error. An alternative way of writing the expression

$$E(C_i) = N_1 q'f(1 - q'f)^{i-1}$$

is

$$E(C_i) = q'f\left\{ N_i - \sum_{j=1}^{i-1} E(C_j) \right\},$$

and

$$\frac{E(C_i)}{f} = q'N_1 - q' \sum_{j=1}^{i-1} E(C_j)$$

where $E(C)$ is the expectation of C.

DeLury (1947) independently of Leslie derived this formulation and also discussed methods of estimating N_1. Paloheimo (1963) has extended the method to a practical situation, and has corrected the estimators for certain types of violations of assumptions.

Commercial catch statistics

The stimulus for development of these methods must often have been the fact that large masses of commercial catch statistics (often gathered primarily for other purposes) were available and that analyses of such data would be far less costly than to obtain mortality estimators using alternative approaches.

One of the specialized methods was developed independently by Beverton and Holt (1957) and Paloheimo (1958, 1961), the latter stimulated apparently by Fry's (1949) 'virtual population' method. Beverton and Holt used the mean number in various subpopulations during successive intervals of time as basic data; Paloheimo used catch statistics directly. The rationale of the former approach rests on the fact that with the exponential model:

$$\bar{N} = \frac{C}{F} = \frac{C}{qf}$$

for a particular subpopulation and interval of time.

Individual year classes are designated by i, and individual years by j. Complete commercial catch statistics yield the number of each year class caught in each year, C_{ij}. Restrict j to those years after the year class is fully recruited into the fishery and take the gear to be non-selective beyond the age of recruitment. Set all $q_{ij}=q$, all $f_{ij}=f_j$, and all $M_{ij}=M$. Under these conditions the basic relationship can be transformed to logarithms and rewritten

$$\log_e \left\{ \frac{C_{i,j}f_{j+1}}{C_{i,j+1}f_j} \right\} = qf_j + M - \log_e \left\{ \frac{(qf_j + M)\left(1 - \dfrac{1}{\exp(qf_{j+1} + M)}\right)}{(qf_{j+1} + M)\left(1 - \dfrac{1}{\exp(qf_j + M)}\right)} \right\}.$$

This is an equation with two unknowns, thus if an additional set of data were on hand (e.g. $C_{i+1,j}$ and $C_{i+1,j+1}$ or $C_{i,j+2}$ and f_{j+2}) we could estimate both q and M iteratively. Normally of course far more than two sets of data are available; and there are various iterative ways in which a pooled estimate can be obtained. Paloheimo (1961) and Chapman (1961) independently derived approximate pooled estimators in closed form. Paloheimo noted that the first two terms of an infinite series expansion of

$$\log_e \left\{ \frac{(qf_j + M)\left(1 - \dfrac{1}{\exp(qf_{j+1} + M)}\right)}{(qf_{j+1} + M)\left(1 - \dfrac{1}{\exp(qf_j + M)}\right)} \right\} \simeq \frac{q(f_j - f_{j+1})}{2}.$$

The above equation thus becomes

$$\log_e \left\{ \frac{C_{i,j}f_{j+1}}{C_{i,j+1}f_j} \right\} = M + q\frac{(f_j + f_{j+1})}{2}$$

which corresponds to the linear equation

$$Y = M + qX.$$

Paloheimo suggested the use of simple linear regression estimators of M and q. He tested this and some iterative methods with contrived data and noted that they were all sensitive to the various types of error usually met in catch and effort data.

So far as we know, catch-effort methods of the Silliman–Beverton–Holt–Paloheimo–Chapman type have seldom been used with sport fisheries, though both the catch curve and Leslie methods have been used extensively. The reason the former has not been used is no doubt partly due to the practical and theoretical difficulties in estimating both catch and effort in such situations. Robson (1960, 1961) and Grosslein (1961) have studied sport catch and effort estimation problems. We conclude from their analyses that it will seldom be practical to estimate catch and effort in a *large* sport fishery with sufficient accuracy to yield reliable mortality estimates by this method.

Catch curves

The catch curve method is a special case of catch-effort methods. A single random sample of the population is obtained at some point in time with a gear that is non-selective beyond the age of recruitment. The population is in equilibrium, the number of recruits from annual spawnings is constant from year to year and just balances all deaths of post-recruits. The number in each year class is assumed to follow an exponential function with time, with the total mortality coefficient constant over all year classes and years, in the deterministic model. Alternatively the expected number will be exponential in the analogous simple stochastic time-continuous Poisson mortality process, or geometric in the interval binomial or Poisson processes. The problem is to estimate the mortality parameter from data obtained from such a population.

The catch curve method dates to T. Edser in 1908. Ricker (1975) has discussed some of the implications of various types of catch curves and practical difficulties in obtaining valid estimates of mortality by these means. Chapman and Robson (1960) have developed the statistical theory for this approach and Robson and Chapman (1961) have illustrated the application of the theory to a specific population.

With the deterministic exponential model, but allowing for random sampling error, Chapman and Robson (1960) derived as an estimator of the annual survival rate

$$S = \frac{T}{n+T-1},$$

where

$$T = \sum_{i=1} i\, C_i$$

C_i being the number caught of age group i, and n being the total catch from all age groups. They found identical estimators using the stochastic continuous Poisson model. Theory states that these estimators are unbiased and have smaller variances than any other estimator under conditions where the model is realistic. The authors devised a test of the applicability of the model in a practical situation and methods of adapting the method for certain types of violations of the model.

There are a number of alternative estimators sometimes used under these conditions, e.g. average of ratios and regression estimator. These latter methods are based on less restrictive assumptions than the one outlined above, and as a result have some advantages where the earlier assumptions are clearly unrealistic. But the estimators are not well defined and no clear rules-of-thumb are or will be available as to which is the best specific method in a particular case.

The catch curve method, and particularly the regression estimator, has probably been used more frequently than any other method for estimating survival in freshwater fish populations. It has often been used rather un-critically where little attention was paid to the assumptions on population equilibrium and non-selectivity of gear. Both assumptions are quite un-realistic for most freshwater fish populations.

2.5 Discussion

The principles underlying the three general methods presented above for estimating population number and mortality rate have provided the basis for almost all of the special methods found in the literature of applied ecology. The most obvious commonality among the methods lies in the need for the investigator to carefully define or identify a number of groups of animals and the relationships between them. This process requires some knowledge of the life history and habits of the animal in order that the investigator can take advantage of differences between sub-units of the population. Thus, it may be impossible to estimate the size of an entire population of fish due to differential catchability of younger and older age groups, while at the same time, the spawning segment may be readily estim-able with considerable precision.

Having identified a group of animals or sub-unit of the population which is amenable to random sampling (or of uniform vulnerability with respect to the fishing gear) we recommend a two-stage approach to estimating its vital parameters. The first stage is to select a statistical model which requires the fewest assumptions about the behaviour, distribution, or sampling of the animals. Then, assuming a 'best case' hypothetical situation in which all basic assumptions are fulfilled, calculate the minimum sample sizes required to provide estimates of an acceptable precision. If funding is inadequate to meet the minimum requirements, then we concur with Paulik and Robson (1969) that under real conditions the sampling will be even less adequate and the experiment should be abandoned or refinanced. Often a preliminary experiment will have to be conducted just to allow an estimation of sample size requirements or to examine an assumption in a tentative model.

The second stage in the procedure is to design the sampling in such a way that the resulting data may be input to more than one estimation method. Paulik and Robson (1969) suggest, for example, that a simple pre-season CIR estimate of a sport fish population using relative abundance of the sport fish and a coarse fish species might be augmented by marking some of the sport fish during the pre-season survey in order that a mark-recapture estimate might be made from the sport fish catch sampling. Since the different classification methods (i.e. natural classes compared to marked and un-marked fish) involve different assumptions with their own errors and biases, the two simultaneous estimates provide a check on each other. Cormack (1968) follows Paulik in advising that, wherever possible, the investigator should collect his sample data in such a manner that he can test the assumptions underlying the models upon which his estimates are based.

Our final admonition to the investigator about to select an estimation method is paraphrased from Cormack (1968). Given a choice between estimates derived from a fully efficient and general method or a less refined alternate method we are inclined to prefer the results of the former. If the alternate approach happens to provide estimates similar to those of the general method, this circumstance increases one's faith in the *alternate* method, *not* in the general one.

References

BAILEY N.T.J. (1964) *The Elements of Stochastic Processes with Applications to the Natural Sciences.* John Wiley and Sons, Inc., New York.

BERST A.H. & SPANGLER G.R. (1970) Population dynamics of F_1 splake (*Salvelinus fontinalis* X *S. namaycush*) in Lake Huron. *J. Fish. Res. Bd. Canada* **27**, 1017–1032.

BEVERTON R.J.H. & HOLT S.J. (1957) On the dynamics of exploited fish populations. *Fishery Invest. Lond., Ser.* 2, **19**, 1–533.

BROWNIE C. & ROBSON D.S. (1976) Models allowing for age-dependent survival rates for band return data. *Biometrics* **32**, 305–323.

BUDD J.C., FRY F.E.J. & SMITH J.B. (1968) Survival of marked lake trout in Lake Manitou, Manitoulin Island, Ontario. *J. Fish. Res. Bd. Canada* **25**(11), 2257–2268.

CHADWICK H.K. (1968) Mortality rates in the California striped bass population. *Calif. Fish and Game* **54**, 228–246.

CHAPMAN D.G. (1961) Statistical problems in the dynamics of exploited fish populations. *Proc. 4th Berkeley Symp.* 1960, vol. 4, 153–168.

CHAPMAN D.G. & MURPHY G.I. (1965) Estimates of mortality and population from survey-removal methods. *Biometrics* **21**, 721–935.

CHAPMAN D.G. & ROBSON D.S. (1960) The analysis of a catch curve. *Biometrics* **16**, 355–368.

COCHRAN W.G. (1963) *Sampling Techniques.* John Wiley and Sons, Inc., New York.

CORMACK R.M. (1968) The statistics of capture-recapture methods. *Oceanogr. Mar. Biol. Ann. Rev.* **6**, 455–506.

CORMACK R.M. (1972) The logic of capture-recapture estimates. *Biometrics* **28**, 337–343.

CUSHING D.H. (1973) *Detection of Fish.* Pergamon Press, Oxford, England.

DARROCH J.N. (1958) The multiple-recapture census. I: Estimation of a closed population. *Biometrika* **45**, 343–359.

DARROCH J.N. (1959) The multiple-recapture census. II: Estimation when there is immigration or death. *Biometrika* **46**, 336–351.

DELURY D.B. (1947) On the estimation of biological populations. *Biometrics* **3**, 145–167.

DEMING W.E. (1960) *Statistical Adjustment of Data.* Dover Publications, New York. (Original text published 1943).

FORBES S.T. & NAKKEN O. (1972) Manual of methods for fisheries resource survey and appraisal, Part 2. The use of acoustic instruments for fish detection and abundance estimation. *FAO Manuals in Fisheries Science* No. 5, Rome.

FRY F.E.J. (1949) Statistics of a lake trout fishery. *Biometrics* **5**, 27–67.

GROSSLEIN M.D. (1961) Estimation of angler harvest on Oneida Lake, New York. PhD Thesis, Cornell University, Ithaca.

GUEST P.G. (1961) *Numerical methods of curve fitting.* Cambridge (England) Univ. Press, London.

GULLAND J.A. (1964) Catch per unit effort as a measure of abundance. *Rapp. P.-V. Reun. Cons. perm. int. Explor. Mer* **155**, 8–14.

HJORT J. (1914) Fluctuations in the great fisheries of northern Europe, viewed in the light of biological research. *Rapp. P.-V. Reun. Cons. Perm. Int. Explor. Mer* **20**, 1–228.

JOLLY G.M. (1963) Estimates of population parameters from multiple recapture data with both death and dilution—deterministic model. *Biometrika* **50**, 113–128.

JOLLY G.M. (1965) Explicit estimates from capture-recapture data with both death and immigration—stochastic model. *Biometrika* **52**, 225–248.

KELSO J.R.M., PICKETT E.E. & DOWD R.G. (1974) A digital echo-counting system used in determining abundance of freshwater pelagic fish in relation to depth. *J. Fish. Res. Bd. Board Canada* **31**, 1101–1104.

KEYFITZ N. (1968) *Introduction to the Mathematics of Population.* Addison-Wesley Inc.

KOWAL N.E. (1971) A rationale for modelling dynamic ecological systems. *Systems Analysis and Simulation in Ecology.* (B. C. Patten ed.), pp. 123–194. Academic Press, New York.

LAMBOU V.W. (1963) Application of distribution pattern of fishes in Lake Bistineau to design of sampling programmes. *Progr. Fish-Cult.* **25**, 79–86.

LAMBOU V.W. & STERN H. Jr. (1958) An evaluation of some of the factors affecting the validity of rotenone sampling data. *Proc. 11th Conf. SE. Assoc. Game and Fish Comm.* (1957), pp. 91–98.

LORD G.E. (1973) Population and parameter estimation in the acoustic enumeration of a migrating fish population. *Biometrics* **29**, 713–725.

MORAN P.A.P. (1951) A mathematical theory of animal trapping. *Biometrika* **38**, 307–311.

PALOHEIMO J.E. (1958) A method of estimating natural and fishing mortalities. *J. Fish. Res. Bd. Canada* **15**, 749–758.

PALOHEIMO J.E. (1961) Studies on estimation of mortalities. I. Comparison of method described by Beverton and Holt and a new linear formula. *J. Fish. Res. Bd. Canada* **18**, 645–662.

PALOHEIMO J.E. (1963) Estimation of catchabilities and population sizes of lobsters. *J. Fish. Res. Bd. Canada* **20**, 59–88.

PALOHEIMO J.E. & DICKIE L.M. (1964) Abundance and fishing success. *Rapp. P.-v. Reun. Cons. perm. int. Explor. Mer* **155**, 152–163.

PARKER R.R. (1968) Marine mortality schedules of pink salmon of the Bella Coola River, central British Columbia. *J. Fish. Res. Bd. Canada* **25**(4), 757–794.

PAULIK G.J. & ROBSON D.S. (1969) Statistical calculations for change-in-ratio estimators of population parameters. *J. Wild. Manage.* **33**(1), 1–27.

REGIER H.A. & ROBSON D.S. (1967) Estimating population number and mortality rates. *The Biological Basis of Freshwater Fish Production.* (S. D. Gerking ed.), pp. 31–66. Blackwell Scientific Pub., Oxford.

RICKER W.E. (1973) Linear regressions in fishery research. *J. Fish. Res. Bd. Canada* **30**, 409–434.

RICKER W.E. (1975) Computation and interpretation of biological statistics of fish populations. *Bull. Fish. Res. Bd. Canada* **191**, 382 p.
fishermen. *Biometrics* **16**, 261–277.

ROBSON D.S. (1961) On the statistical theory of a roving creel census of fishermen. *Biometrics* **17**, 415–437.

ROBSON D.S. & CHAPMAN D.G. (1961) Catch curves and mortality rates. *Trans. Am. Fish. Soc.* **90**, 181–189.

ROBSON D.S. & REGIER H.A. (1964) Sample size in Petersen mark-recapture experiments. *Trans. Am. Fish. Soc.* **93**, 215–226.

ROBSON D.S. (1969) Mark-recapture methods of population estimation. *New Developments in Survey Sampling.* (N. L. Johnson and H. Smith, Jr. eds.). John Wiley and Sons, Inc.

ROTHSCHILD B.J. (1966) Competition for gear in a multiple-species fishery. Cornell University Biometrics Unit Mimeo BU-123, 12 pp.

ROUNSEFELL G.A. & EVERHART W.H. (1953) *Fishery Science: Its Methods and Applications.* John Wiley and Sons, Inc., New York.

SEBER G.A.F. (1962) The multi-sample single recapture census. *Biometrika* **49**, 339–349.

SEBER G.A.F. (1965) A note on the multiple-recapture census. *Biometrika* **52**, 249–259.

SEBER G.A.F. (1970) Estimating time-specific survival and reporting rates for adult birds from band returns. *Biometrika* **57**, 313–318.

SEBER G.A.F. (1973) *The Estimation of Animal Abundance and Related Parameters.* Chas. Griffin and Co. Ltd., London.

SPANGLER G.R. (1970) Factors of mortality in an exploited population of whitefish, *Coregonus clupeaformis*, in northern Lake Huron. *Biology of Coregonid Fishes* (C. C. Lindsey and C. S. Woods eds.), pp. 515–529. Univ. Manitoba, Canada, Press.

STEEL R.G.D. & TORRIE J.H. (1960) *Principles and Procedures of Statistics.* McGraw-Hill Book Co., New York.

YOUNGS W.D. & ROBSON D.S. (1975) Estimating survival rate from tag returns: model tests and sample size determination. *J. Fish. Res. Bd. Canada* **32**, 2365–2371.

Chapter 3: Some Aspects of Age and Growth

Alan H. Weatherley and
Stephen C. Rogers

3.1 Introduction

The purpose of this chapter is to provide a survey of a number of rather general problems in somatic growth as they affect the individual fish, and the population as well. The choice of problems is arbitrary in the sense that many are omitted—e.g., vitamins, nutritional plane, most behaviour, hormones—though some of these are dealt with in other chapters of this book (Chapters 8, 14) and elsewhere (e.g., Weatherley 1972, 1976). The main approach here is to provide some scaffolding of general ideas about animal growth, then treat certain of the fish problems as particular facets of the subject.

Growth is a term with different meanings for different people. It covers a broad array of problems and processes in biology (e.g., Thompson 1942, Bertalanffy 1960, Kleiber 1961, Goss 1964, Needham 1964), so it would be pointlessly limiting to restrict its application to particular processes. Its strength lies in its utility as a general operational concept. It nevertheless behoves us in a work of the present type to try to indicate how fish biologists can view and use growth concepts, because of their wide importance in fishery management and their general scientific interest. We can note that fish are among the few groups of wild animals of which many species may be aged with reasonable accuracy by a study of their bony structures (scales, otoliths, operculars, etc.), and whose growth may thereby be charted throughout life. This fortunate fact gives fishery biologists an extremely convenient method (subject to careful validation for each new fish species examined) for accumulating demographic information for use in population dynamics.

3.2 Some Approaches to Growth and its Definition

An organism's somatic growth is frequently equated to increase in bulk (Thompson 1942, Bertalanffy 1960), and this immediately suggests that growth as a phenomenon will interest both the cell biologist and the bioenergeticist. Increase in bulk may be largely accounted for by increase in metabolically active tissues such as muscle, liver, nerve, endocrine, etc., and

52

the histologist and cell biologist will want to consider whether such increase has resulted from sustained cell division or increase in cell size (Weatherley 1972, 1976). There are remarkably few studies to relieve our ignorance of this problem. Two investigations, that of Love (1970) and of Greer-Walker (1970)—both on the cod *Gadus morhua*, appear to be in partial conflict in describing the relationships between cell number, cell size and fish size. Both describe increases in fibre diameter of white muscle, but Love's (1970) study attributes most of the increase in bulk of fish with age to this increase in fibre diameter, whereas Greer-Walker describes increases in cell numbers as well.

In a recent study of 84 marine fish species representing 42 families, Greer-Walker and Pull (1975) obtained the data depicted in Fig. 3.1. This

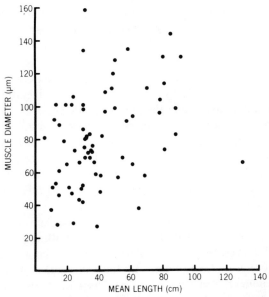

Fig. 3.1. Scattergram of white muscle fibre diameter versus fish length for 84 marine fish species from 42 families. Data from Greer-Walker and Pull (1975).

scattergram suggests a tendency for white muscle fibre diameter to increase with size of fish. The spread of values is considerable, but the correlation coefficient between white muscle fibre diameter and fish length is 0·336 ($P = <0·01$). This degree of significance is rather remarkable, as each point represents a single value on a species-specific regression of fibre diameter against length, the slope of which probably differs considerably between species.

An increase in bulk by cell hypertrophy is quite different from an increase by cell division, and yet we are not able to claim that large or fast-growing fish are so because they retain the power of further cell divisions as they grow, in contrast to small species which do not. However, as a working hypothesis, it does seem reasonable to enquire at what size the cells are incapable of further division—i.e. become limiting to subsequent somatic growth. Beyond a certain size the surface area/volume ratio of cells severely limits their metabolic efficiency (Bertalanffy 1960). We might pose this problem in its most crucial form by proposing the following testable hypothesis:

1   The smallest and slowest-growing fish could be expected to have cells little capable of further division once laid down in major tissues during embryogenesis, and will therefore grow mostly by increase in cell size. Thus their final somatic size will, in at least some cases, be limited by the largest attainable size of their (post-mitotic) cells.

2   Fish having intermediate growth rate and size will have some ability for sustained mitosis and probably also for increase in cell size. The somatic size limit will be imposed by cessation of mitosis, and/or by cell size limitations.

3   Fish attaining great size or manifesting sustained rapid growth may have cells which remain relatively constant in size (and metabolic efficiency) throughout life and which retain the power of mitosis.

Some of these possibilities are currently being investigated in the laboratory of the senior author.

The above observations are meant to clarify the growth problem without an attempt at formal definition, which would put an awkward restriction on the use of the concept. However, problems do arise in distinguishing authentic products of growth.

For instance, some characterize growth as change in amount of living protoplasm, whether expressed in terms of cell bulk, protein, or caloric content, etc. However, we must take into account that protoplasm has a particular carbohydrate:fat:protein ratio (and content) at a given time, and that these components can vary over very short intervals indeed (e.g., the breeding season), seasonally, or with age as measured in years. The fat content of cells may increase after a meal: is this growth? Such fat may subsequently be deposited or may supply energy for work, growth, or gonad elaboration. During their lives animals elaborate and may accrete skin, hair, horn, scales, skeletal structures, and various secretions in addition to increasing the size of their more obvious functional tissues (muscle, liver, nerve, etc.). Should all such products be regarded as manifestations of growth?

Once produced such materials may be biologically inert, but many of them add permanently and substantially to the boundaries, enclosures and structures of the body. Moreover, like such 'permanent' increases in the bulk of the body as the increasing muscle mass of a fish, they have required ingestion and metabolism of food for their production and have exacted a corresponding catabolic price from the organism.

Faced with the problem of deciding what is growth, some biologists have suggested the body protein content as a better index than most other variables, on the premise that protein composition is frequently more stable over time than is carbohydrate or fat content (Gerking 1962, Love 1970). Most organisms are, however, in constant metabolic flux in which proteins are invariably involved (Needham 1964, Shul'man 1974, Weatherley 1976). An animal's total metabolic performance is determined by its fat stores, size (including surface/volume ratio), skeletal bulk, and even its shape. If such attributes are not themselves growth or results of growth, they are certainly its concomitants. It therefore seems too restrictive to view protein accretion as the sole index of growth.

Growth might be holistically viewed as the entire suite of accretions and products of an organism resulting from metabolism, which may increase not only the organism's bulk over time, but also the scope and magnitude of its internal dynamics, plus its structures or its potential for forming new ones.

3.3 Methodologies of Age and Growth Determination

Growth and age structure are essential features in the study of fish populations. As a result, a multitudinous literature has been built up on the subject. This cannot be completely reviewed here; we offer only a general overview of the methodology that has been employed with special reference to certain problems arising in age and growth determinations. For a more comprehensive treatment the reader may refer to relevant reviews of the subject (Van Oosten 1929, 1941, 1957, De Bont 1967, Tesch 1968, Weatherley 1972, Bagenal 1974, Everhart *et al.* 1975).

ANNULI

Characteristic marks which appear at regular intervals, during the growth of bony structures in temperate zone fishes are frequently used as indicators of age. Scales, otoliths and opercular bones are commonly used for age determination, though vertebrae, cleithra, finrays, etc., are employed in special cases (Monon 1950, Bagenal 1974). De Bont (1967) listed five criteria for obtaining reliable determinations of age and growth from bony structures: (1) correspondence of growth rings on individuals of known age with the years of life; (2) a regular increase in body size correlated with an increased number of growth rings; (3) constancy of scale number throughout life; (4) a constant ratio between the annual increment in the radius of the scale and the annual increment in body length, and (5) growth rings formed yearly and at the same time of year.'

Thus, year marks or 'annuli' are the result of resumption of growth following a period of growth cessation or very slow growth. In temperate zones, growth ceases in winter, and an annulus is usually formed in spring when significant growth recommences. False annuli, a source of gross inaccuracy in age determinations, are caused by starvation, which causes growth to cease, or such factors as unfavourable temperatures, injury, spawning checks etc. (Van Oosten 1957). False annuli are seemingly common occurrences in certain temperate zone fishes, so that use of supposed annual rings as indicators of age must consistently be viewed with caution. The problem of accuracy in age determination has been extensively reviewed in recent years (Crichton 1935, Ichikawa 1953, Sych 1970, 1974, Simkiss 1974, Le Cren 1974).

VALIDITY OF ANNULI

In view of this latter fact, the validity of *all* supposed year marks on scales, otoliths, etc., must necessarily be tested by utilizing some additional methodology which involves access to fish populations of *known* age (Le Cren 1947). One such method of validation of the annulus is by use of length-frequency polygons (Petersen method). These may be derived from frequent sampling of a fish population which reproduces seasonally and shows corresponding influxes of new recruits which, in turn, give rise to size modes which represent individual age classes. If the individuals from successive age groups display 'annuli' that correspond with their age as estimated by population size structure, the 'annuli' are probably valid indicators of age. However, this method is unfortunately only valid for younger age groups, because as fish age the size modes become progressively less pronounced. The modes become less recognizable because only a relatively few fish are represented among the older age groups, and they show an increasing tendency to overlap in size.

If, for some reason, such as favourable temperature conditions, or an unusually abundant food supply, mortality of a particular age group is uncommonly low, a larger year class than average may enter a population, indicating its presence by a noticeably large size mode. With adequate sampling, such a notably strong mode may be monitored as it progresses through the population over time and, if followed on a yearly basis, it is possible to test the validity of such marks as age indicators, while the number of supposed annuli laid down on the bony structures of the body is simultaneously checked. This method offers an advantage in that the 'abnormal' age group can sometimes be used by itself as an index of age and growth in the absence of other corroborative data, though this is not desirable.

Perhaps the most satisfactory method of validating annuli is to mark individual fish by staining, fin clipping, tagging, etc., and then release them. Upon recapture of marked specimens, the number of recognizable marks on scales should either coincide with numbers of years since release, or known age of the fish. Thomson (1962), De Bont (1967) and Everhart *et al.* (1975) have extensively described relevant marking techniques. In species that lack hard parts, such as various elasmobranchs, these marking techniques are the only 'certain' ones, as was well demonstrated in the work of Olsen (1954) on the school shark (*Galeorhinus australis*) in Australian waters. Olsen was able to detect age groups among larger school sharks from the very sharp modes of length frequency appearing in the commercially fished population. However, comparatively poor representation of some of the younger age groups made the identification of these modes hinge on the observed growth at sea of a rather small number of tagged sharks of various ages (sizes). Without the use of tags it would have been practically impossible to age these sharks and estimate their growth rate.

The graphical method of Walford (1946) to determine increments of length-for-age involves plotting length at age $t+1$ against length at age t and yields the annual length increments to be expected following any initial length. This method bears the advantage over others that only sparse data from the recapture of marked fish from a range of representative size groups in the population, are needed. The major disadvantage is that the method assumes a steady state population which, over a period of years, exhibits a constant pattern of size-specific growth rate.

A final method for confirming validity of annuli involves holding, or rearing fish in captivity. The number of years a fish is held is compared to the number of supposed annuli laid down during the period. There are obvious disadvantages in that the environment in which fish are held (e.g. hatchery ponds) is artificial, bearing little comparison to natural habitats. Therefore, marks laid down by a natural stock of the same species may differ in number and appearance.

AGE DETERMINATION OF TROPICAL FISH

Tropical species have been studied considerably to find a method of age determination because the constancy of the environment and the non-seasonal breeding habits of many tropical species lead to marks being laid down at any time of the year and in any number. Holden (1955) found that in two species of tilapia in Lake Victoria the scale rings were formed by irregular circuli, a criterion which differs from annulus formation in temperate species. Ring formation in the temperate species is caused by the circuli being closer together at certain times of the year and by the cessation of circuli formation in the winter and the abrupt recommencement of their formation in the succeeding spring. Fagade (1974) pointed out that environmental changes do occur in the tropics which, though differing from those of the temperate zones, have similar effects in initiating ring or annulus-like formations. In his study on *Sarotherodon melanotheron* in Lagos Lagoon, Nigeria, he found that a drastic salinity change (from brackish to fresh) produced by the onset of the rainy season, caused growth rings to appear in the bony parts of these fish. Even with this knowledge, Fagade had difficulty in determining when growth rings were formed, and how many were formed annually.

Jensen (1957) was more conclusive concerning three *Pilapia* species in Lake Marivat (Egypt) in which he found three clear annual zones on the scales, unlike most tropic species investigated to date. He attributed this regular year-mark formation to the considerable difference in the water temperature between summer and winter in that region, which is similar to conditions for temperate zone fishes.

Other causes of growth rings are even less clear cut. Pannella (1974) discussed the use of otolith growth configuration as a means of age determination in both temperate and tropical species. In the latter, he found that the growth rate and patterns of the otoliths do not change greatly over the year 'and that the rings are created by small changes in the ratio of organic matrix and argonite'. It is not annual growth rings, he postulates, that are useful for age determination but the monthly and bimonthly bands that almost all otoliths display. Mathews (1974) suggested that by use of these otolith rings, in conjunction with Cassie's (1954) method, a reliable estimate of age in subtropical and tropical species is certain. Cassie's method is based on the use of probability paper for precise analysis of size frequency distributions and can be used to coordinate age determination by the scale method with length frequency analysis of population samples. For a more substantive view of the problem and new methodologies being developed for age deteminations in these fish the reader may refer to Bagenal (1974) and) De Bont (1967).

BACK CALCULATION

The reading of scales and other structures for back-calculation of past lengths-for-age is an ubiquitous technique, reviewed by many authors (e.g., De Bont 1967, Tesch 1968, Weatherley 1972). As already mentioned, the Petersen method of length-frequency analysis is limited to younger (smaller) groups and as such can render a rough approximation of age and growth of fish in a population. Let us, however, suppose an accurate and reliable method for age determination from marks in the hard parts of the body can be developed for a particular species. Then if the increase in size of such structures bears either a direct linear relationship to the increase in length of the animal or one which can be accounted for in simple mathematical terms, the growth history of most individuals can be computed.

In the case where growth of scale (or other structure) and growth of fish are directly proportional

$$Fx = Fy \frac{Bx}{By}$$

where Bx is the distance from the focus (origin) of the scale to an annulus, By is the scale diameter (distance from focus to periphery), Fx is the length of the fish to be determined, and Fy = the length of the fish at the time of capture. The effect of a heterogonic growth relationship between opercular length and fish length was examined by Le Cren (1947) for a population of perch (*Perca fluviatilis*). He derived an empirical formula for heterogony such that:

Log fish length = $1·357 + 0·9202$ log opercular length

or $F = 22·76B^{0·9202}$

where F = fish length, B = opercular length, $0·9202$ = the 'growth ratio'.

Thus the equation for a directly linear relationship between opercular and fish, given above, becomes:

$$Fx = Fy \frac{Bx^{0·9202}}{By^{0·9202}}$$

From the curve of this equation Le Cren derived corrected lengths-for-age for the perch population.

Nicholls (1957) found for two trout populations in Tasmania that when the regression lines of scale length against fish length were extrapolated to the x axis, one intercepted it at $+1·05$ cm and the other at $-1·24$ cm. This conflicts with the claim of Lee (1912) that all intercepts of the x axis in such plots must have positive values and 'that the position at which the x-axis is cut (by the extrapolated regression line) indicates the fish length at which scales first appear' (Nicholls 1957). It was nevertheless concluded by Nicholls that for 'present purposes it is sufficient to assume that under natural conditions growth of the scale is exactly proportional to growth of the fish throughout life.'

The laboriousness of procedures for age determination from scales and other structures have prompted recent efforts to automate them. Mason (1974) has described his work towards this end, including a machine for counting and measuring circuli in scales. Significant age and growth data were obtained more rapidly than by traditional methods and, if put in common use, such a system could probably revolutionize many aspects of fishery science.

3.4 Simple Biomathematical Expressions of Growth

The so-called exponential

$$\frac{dy}{dt} = ry$$

and logistic

$$\frac{dy}{dt} = ry\,\frac{(k-1)}{k}$$

equations (where dy/dt is the rate of change of size or numbers per unit time, r is an exponent for rate of increase in size or numbers, y is the size of the growing organism or number of increasing individuals, and k is the upper asymptote which y approaches) have long stimulated growth biologists interested in such diverse subjects as yeast cultures, bodily organs, wild animal populations or human demographic problems.

Whereas the exponential portrays growth as a multiplicative increase in numbers (or bulk), the logistic perceives it as initially exponential but undergoing a progressive decay in rate of increase, which is directly related to the numbers (or bulk) already in existence at each moment. The high heuristic value of these equations lies in their continuing ability, and that of certain of their derivatives (Baas Becking 1946), to account for the growth of populations, organisms, or parts of organisms, when these are manifested under defined and limited conditions of space, nutrition, etc.

The concepts underlying these formulae have stimulated considerable debate among ecologists as to whether the numbers of animals in nature are *controlled* by numbers (density-dependent factors) or by climate, or whether, by evolved self-regulatory mechanisms most species have achieved other means of holding their own numbers permanently below values that could seriously deplete such resources as food or space (Nicholson 1933, 1954, Wynne-Edwards 1962).

Whatever view one adopts, three significant points remain:

1 Indefinite maintenance of exponential increase is impossible whether in laboratory culture or wild population.

2 Under natural conditions, many organisms fail to display patterns of increase in size or numbers that bear a close resemblance to laboratory examples of the logistic, though others certainly do so.

3 Most natural populations of animal species do not appear to encounter serious risks from sustained over-exploitation of their food supplies through over-population, except where a prolonged food shortage occurs because of human disturbance of the environment.

In formulae more suitable than the logistics for somatic growth in fishes Bertalanffy (1938, 1957, 1968) visualized growth as the difference between anabolic and catabolic rates, such that:

$$\frac{dw}{dt} = Hw^m - Dw^n$$

H and D are coefficients of rate of anabolism and catabolism respectively and w=body weight (m=2/3, n=1). This basic equation has been used repeatedly to fit curves to somatic growth data for many fish species (Beverton & Holt 1957, Bertalanffy 1960).

Paloheimo and Dickie (1965) preferred the so-called basic energy equation of Winberg (1956) to describe fish growth:

$$pR = T + \frac{\Delta w}{\Delta t}$$

where R=rations (calories) consumed, p=correction factor to convert R to rations assimilated, T=total metabolic rate (calories), Δw=weight change (caloric equivalent), Δt=period of time for weight change. In slightly simpler form

$$\frac{\Delta w}{\Delta t} = R - T$$

Here, 'In place of the abstract concepts of anabolism and catabolism (we) have three quantities which are clearly open to experimental study' (Paloheimo & Dickie 1965).

Then,

$$T = \alpha w^\gamma$$

where α=constant for metabolic rate and γ=metabolic rate of change with change of body weight. From analysis of various fish growth experiments, a value for γ of 0·8 was derived (see also Chapter 8 for more extensive treatment of this point).

Thus,

$$\frac{\Delta w}{\Delta t} = R - \alpha w^\gamma$$

from which $\Delta w/\Delta t$ can, clearly, be positive only when $R > \alpha w^\gamma$. Paloheimo and Dickie (1965) also gave the expressions

$$\frac{\Delta w}{R\Delta t} = K_1 \quad \text{or} \quad \frac{\Delta w}{a'R\Delta t} = K_2$$

where $R\Delta t$=total rations and $a'R\Delta t$='assimilated' rations.

The latter equations are equivalent to somewhat more familiar equations of total and partial growth efficiency used by Kleiber (1961), Warren and Davis (1967) and others.

As Thompson (1942) long ago indicated, both the exponential and logistic growth curves of populations or whole organisms can yield, as first derivatives, reciprocal curves that function as specific growth rate. Such graphs can be more useful than arithmetic plots of growth, because they immediately provide information about the times at which the most significant growth is occurring. Plots of specific growth rate characteristically reveal that the proportional changes in size occurring over the early part of life must exceed those occurring later.

A logarithmic form of the expression for specific growth rate is

$$G = \frac{\text{Log}_e \ \bar{W}_2 - \text{Log}_e \ \bar{W}_1}{\Delta t}$$

where $\bar{W}_2$ is size at latest measurement $\bar{W}$ and $\bar{W}_1$ is the immediately preceding measurement. The time between the two measurements is Δt.

It should be stressed that the practice of obtaining specific growth rates from artificially smoothed curves of growth based on few actual data is hazardous. Growth of fish is frequently discontinuous even during a growing season. Specific growth rate should be used to display ample growth data to better advantage rather than to interpret meagre data.

The problem of a general formulation of growth in fish that will—in terms of energetics—take into consideration the activity of living and the SMR (standard metabolic rate) has recently been partially resolved by Ware (1975) who, basing his studies on data assembled on growth and swimming activity by Ivlev (1960) and the energy equation of Winberg (1956), developed a growth equation of his own.

The version of the Winberg equation, used by Ware was

$$pI' = T' + G' \tag{1}$$

where I'=food intake (cal day^{-1}), T'=total metabolism (cal day^{-1}), G'= growth (cal day^{-1}), p=assimilation factor.

Ware develops from this the equation

$$I = \frac{\gamma v \rho}{1 + \gamma v \rho h} \tag{2}$$

for a foraging herbivorous fish, that swims and feeds, where I=food intake (cal h^{-1}); v=swimming speed (m h^{-1}); ρ=food concentrations (cal m^{-3}); γ=area successfully searched (probability of food capture *times* the cross sectional area of visual field in m^2); h=time required to capture and consume one calorie of food.

The area successfully searched is, in Ware's (1975) model, independent of swimming speed, thus the above equation means that food consumption rate increases asymptotically as swimming speed (v) increases.

Possible energetic advantage derived by increasing the value of v is offset by the increase in energy cost of locomotion (E in cal h^{-1}) which is proportional to some power of v such that:

$$E = z + \alpha v^{\beta} \qquad (3)$$

where SMR $= z$ (cal h^{-1}) is a constant value, and α and β are empirically determined constants related to the additional energy required to swim at various speeds.

Ware (1975) also treats the SDA (specific dynamic action), aptly denoting it as 'an entropic tax paid during food conversion'. A simplified view of the SDA assumes its magnitude as proportional to ration size (only strictly true if the relative proportions of fats, carbohydrates and proteins remain constant; see also Kerr 1971a, b, c, Beamish 1974). Then, total metabolism, T (cal h^{-1}) is expressed according to Ware (1975) as

$$T = sI + z + \alpha v^{\beta} \qquad (4)$$

where I is food intake (cal h^{-1}). Ware combines equations (1), (2) and (4) to give growth (G, cal h^{-1}) as:

$$G = \tau \left[\frac{\gamma v \rho}{1 + \gamma v \rho h} \right] - \left[z + \alpha v^{\beta} \right]$$

where $\tau =$ portion of ration retained as net energy.

This system of equations says nothing about the partitioning of growth between somatic tissues, gonads, etc., and assumes steady state conditions as to swimming rates, SDA, SMR, etc. However, the interested reader is referred to Ware (1975) who describes and critically comments on Ivlev's (1960) data for growing and active bleak (*Alburnus alburnus*) populations. Ware has also made the important observation that we are not as yet aware 'whether swimming speeds have evolved on an ecological time scale and thus are relatively invariant, or if they are controlled to exploit spatial and temporal variations in the food supply.'

3.5 Condition Factor

One major concomitant of somatic growth in fish is change in corpulence during life. The change can be great or small, smoothly progressive, intermittent, or cyclically related to breeding. If the somatic linear proportions remain constant as the fish grows (and average density does not change):

$$K = \frac{W}{L^3}$$

where $K =$ condition, $W =$ body weight (g) and L is some linear dimension (cm), usually body length, and K will not change as length increases.

Innumerable studies have addressed themselves to this relationship, since differences in the value of K between populations can frequently yield insight into the circumstances of their lives—e.g., with regard to food supply, timing and duration of breeding cycle, etc. The value of the condition factor K to fishery science is therefore considerable.

Weatherley (1972) outlined a debate concerning the relative merits of calculating K derived from the above formula as compared to that of calculating a 'relative' condition factor (see especially Le Cren 1951). Where only minor deviations of K from unity occur, as in mature salmonids, K can furnish a sensitive index of somatic differences between populations and age groups and a means of determining the length of the breeding season. However, Le Cren (1951) indicated that for some species or populations a factor c would be required to bring K close to unity, i.e.,

$$K = \frac{W}{cL^3}$$

Since the 'cube law' does not apply very well for many populations, Le Cren proposed that $W \neq L^3$ and that a better weight/length relationship would be:

$$W = aL^n$$

from which a relative condition factor (Kn) can be determined as:

$$Kn = \frac{W}{aL^n}$$

where a is a constant and n is a power empirically derived to bring Kn close to unity.

It is not easy, in our view, to appreciate advantages in the use of such a formula commensurate with the computational effort involved. Whether K has a value greater than unity (as in puffers) or much less than unity (eels) seems immaterial. Each species will have its normal and abnormal range of values which will, in any event, have to be empirically determined, as they would still have to be even if Kn were used.

3.6 Growth Flexibility in Fish and other Vertebrates

A contrast has been drawn many times between the so-called indeterminate growth in fish and the determinate growth of certain other vertebrates. Fish characteristically display a considerable *intraspecific* range of growth rates, manifested according to different conditions of food, space, numbers, competition and temperature. For this reason, a particular species is not associated with a definite adult (final) size.

Mammals and birds, by contrast, have characteristic species-specific adult sizes. Frequently, the size of particular ages (hatchlings, weanlings, juveniles, sub-adults, adults) are constant enough to be used as taxonomic criteria. This practice is less safe for fish.

There is a temptation to infer that fish are basically different as regards their somatic growth. Perhaps they are; but before proceeding to such a conclusion some cautionary notes are in order. In the first place, under certain experimental conditions, pigs, poultry, rats, mice and guinea pigs can all have their growth retarded, or even stopped, for long periods, as a result of dietary insufficiency. Yet they can recover their power to grow again when given adequate food (McCance & Widdowson 1962, Widdowson 1964, Lister, Cowan & McCance 1966). Indeed, Bertalanffy (1968) has proposed that organisms have an innate pattern of maximal somatic growth towards which they tend. If their growth is retarded, for example by sub-optimal diet, it may be restored wholly or partly if they subsequently obtain a better diet. Bertalanffy refers to this tendency of the organism as a condition of equifinality.

It may be concluded, then, that terrestrial vertebrates can, under certain experimental conditions, display considerable variations in somatic growth rate basically similar to those displayed by fish. The reason terrestrial vertebrates do not usually display such variations under natural conditions may be related to population structures and dynamics which are somewhat different to those characteristic of fish, rather than to inherent somatic limitations.

In attempting fuller understanding of this problem it must be remembered that fish do not have to overcome the force of gravity. Within every group of terrestrial animals, increasing size carries with it problems associated with the operation of the 'cube law'—strength (of muscles, skeleton) and general agility tending to increase only according to the square of the body's linear dimensions, whereas volume (and therefore mass) tends to increase according to the cube of the linear dimensions. Fish, because of their buoyancy, have no directly comparable problems.

Many terrestrial vertebrates have evolved elaborate systems of caring for their young, which must either be transported by their parents after birth or be capable of keeping up by themselves with the parents' activities (many ruminants). Clearly, major selection pressures will tend to favour the rapid growth of young of many species to a size at which they can cope for themselves. This does not necessarily mean that the growth rate during this period will be maximal, though juveniles *will* probably be growing exponentially. However, in most mammals there is a tendency for juveniles to grow at a rate characteristic of the species; rates too divergent from the norm can lead to important dislocations in the timing of the life cycle, with subsequent repercussions for the dynamics of the population.

This kind of problem is certainly not as likely to be encountered in such severe form for fish. For many fish there is no after-care of young, but fish are generally much more fecund than terrestrial vertebrates and their populations are thereby buffered against all but the more severe factors causing juvenile mortality.

Whatever its full biological explanation, growth flexibility is a reality in fish populations and we should consider some of its more apparent population functions. Many marine fish populations, even those which are exploited, may remain remarkably stable in age structure, numbers, egg production, etc., over many years. However, if they suffer numerical perturbations, through natural mortality, increase or decrease of fishing pressure, etc., the individuals in the population can and do undergo considerable changes in growth rate (Hile 1936, Swingle & Smith 1942, Alm 1946, Swingle 1951, Le Cren 1958, El-Zarka 1959). Examples of this phenomenon are more frequent among freshwater fish than marine species, probably because their habitats are more seasonally variable and discontinuous, and because their smaller-sized populations are less buffered against numerical perturbation (Weatherley 1972).

Population self-regulation in fish as connected to the relationship between number of young (or recruits) and egg productivity, or to changing population age structure, etc., has been the subject of various monographs (Beverton & Holt 1957, Cushing 1968, Ricker 1975), but the particular importance of somatic growth as a population regulatory mechanism has been comparatively neglected. Weatherley (1972) developed several simple models to demonstrate the population role of growth, always combining it with the other important population factor—age structure.

He pointed out that existing models of egg production and recruitment are usually restricted to the dynamics of the juvenile parts of the population. Even though such models may account adequately for stability of recruitment in the face of numerical change in post-juvenile or adult members, variations in survivorship of older fish may, through changes in their own growth rates, also have extremely important effects on egg numbers. The age-specific somatic growth rates of fish in a population in 'steady-state' will change very little over time unless a perturbation in numbers occurs, when the somatic growth rate response will be reciprocally related to the numerical change.

Now fecundity of female spawners is generally directly related to size. Thus, when the number of females in a population becomes reduced and the remainder grow more rapidly, their age-specific fecundity will rise, and as a consequence the remaining female population's egg production will tend to return to its former level. As long as the perturbation is neither too prolonged nor too severe, recruit numbers should also tend towards a constant value. The reverse arguments apply if, as a result of unusually high survival of young, more females than usual reach breeding age. Then a tendency towards reduced somatic growth rate will eventually lower age-specific fecundity and with it subsequent production of young (Weatherley 1972). We do not in any way underestimate the importance of models of the stock-recruit relationship as proposed by Ricker (1975) and Beverton and Holt (1957), in helping to explain population stability. However, we do suggest that somatic growth responses may also be important.

Another aspect of the role of somatic growth is seen in the decisive influence that the growth of females can have on the rate of build-up of a fish population in a new environment. The pond fish populations of Swingle (1950) and the model outlined by Weatherley (1972) furnish a number of examples of the importance to survival of growth rate through the small stages for prey species in a predator-prey situation. Indeed, Swingle's examples imply that the age/size structures characteristic of many fish communities are a direct result of their predator/prey dynamics, in which time of exposure of a prey to a predator, during the vulnerable stages of the prey species, is decisively important to the subsequent trophic relations and population dynamics of each. The growth rate of all stages of the prey species in relation to the growth and size of the predator is, therefore, of major significance to both.

3.7 Extrinsic Factors that Affect Growth

Internally, growth in fish is a result of dynamics of cell populations, endocrine levels, and the intermediary metabolism which determines the rate of energy flow through the animal or its retention in the form of storage products, growing tissues, etc. However, two major extrinsic factors influence growth—the physical one of temperature and the biotic one of competition for food.

It nowadays seems profitless to attempt to identify *generally* optimal temperatures for growth. Tropical species grow best in the range 25–35C. Cold-adapted species may have optima near 0C, while denizens of temperate regions are likely to have optima somewhere between these limits. It is in this latter group that temperature may have most significance in influencing growth (see Chapter 7). Growth is 'released' in fish at various species-specific threshold temperatures, below which it cannot occur and above which an optimum will be located. The optimum may, for temperate species, be anywhere between the critical temperature and the environmental maximum (Chapter 8) depending on the evolutionary and zoogeographic history of the species.

Backiel and Le Cren (1967) proposed a general form of relationship between growth and population density in fish, implying that changes in population density at low levels of density are associated with greater changes in somatic growth rate than are those when comparable changes occur at high population densities. Though this may apply in special instances—e.g., pond fish populations of simple age structure—it is difficult to perceive how it could operate in populations of more complex age or size structure. If the competitive interaction between fish for food is the mechanism by which somatic growth is affected by changes in population density, it can be inferred that the situation resulting from a change in numbers is not simple for long (Weatherley 1972, 1976). Let it be supposed that a population is exposed to a significant reduction in numbers of several of its age groups. There will, perhaps, then be significantly more food for each remaining fish. But growth flexibility will cause them to respond by more rapid somatic growth. This will result, over a period, in partial or complete restoration of the previous biomass of the age group.

Activity is also important. Clearly, activity requires energy which must be drawn from the net energy. Changes in numbers may also change the average activity level of a group and that will certainly affect the internal (somatic) distribution of energy, and therefore somatic growth. Given a particular level of food supply, a formula which attempts to relate somatic growth rate to numbers with a satisfying general applicability, will need to accommodate the *total metabolic demand* of the actual members of the various age/size groups competing for food in an environment. It will obviously be important to find better ways to compute the energy cost of general activity in populations than are presently available (see also Chapter 8).

3.8 Further Investigations of Fish Growth

More complete understanding of factors affecting maximization of fish growth would contribute importantly to the fundamental biology of one of the more successful animal groups, and give improved insight into various problems of fish culture and fishery management. The role of genetics is one of the more important questions relating to fish growth. To what extent are patterns of growth among fish inherited? As has been already observed, fish can display wide intraspecific somatic growth differences. Yet there are obviously *upper* limits to growth which are species-specific; these are the levels of growth which must be inherited. Much of the genetical investigation of growth in fish has so far tended to ignore the very important effects of intensity of competition for food. In growth experiments, results may therefore be vitiated by failure to observe equality or to allow for inequalities in population density, and by ignorance of the fate of net energy as metabolism distributes it between the various compartments of the energy budget of the fish. Weatherley (1976) proposed a scheme to help geneticists avoid the latter problem in particular (Fig. 3.2). The relative order of magnitude assigned to the various categories of metabolism seem not unrealistic based on general principles, but the scheme's main purpose is to provide a reasonable framework for breeding fish for rapid somatic growth (e.g., of muscle), and values other than those shown could be assigned at will.

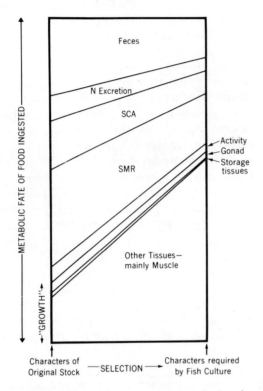

Fig. 3.2. Scheme for attainment of ideal growth characteristics in cultured fish by means of genetical selection of stock. (After Weatherley 1976).

Most of the technical problems of obtaining data for the variables in the energy budget are already soluble. However, activity, which has earlier been distinguished as being in competition with growth for a share of net energy, has been difficult to account for in terms of energy cost, under field conditions. Energy costs of swimming have been estimated from measurements of the average speed of swimming a series of straight lines. This has been made possible by tracking fish fitted with biotelemetric tags. These tags transmit data on point locations (in the horizontal plane) of individual fish over time and as a result, costs of activity must necessarily be underestimated. This may lead to the possible impression that the energetics of activity are of minor significance in the energy budget (Holliday *et al.* 1974), but there seems little present justification for such a view (see also Ware 1975).

Thus, if somatic growth is envisaged as being one fraction of the energy budget interdependent and interchangeable with other fractions, the dynamics of somatic growth will not be fully perceived without a better understanding of activity, which at present remains the most enigmatic value in the energy budget.

Fishery managers and fish culturists have always considered that one of their major tasks is to manage stocks so as to exploit both maximal growth and production. But growth and production of what? The notions expressed in Fig 3.2 are traditional ones to many managers of stocks of terrestrial animals, who breed and manage not only for bulk but to maximise production of preferred tissues (muscle, liver) and to minimize the fraction of net energy channeled to parts less useful to man (skeleton, fat depots, etc.). It is hoped it may become increasingly feasible to apply comparable principles to fish culture (Weatherley 1976). However, the optimal utilization of the products of growth in wild fish stocks is a very challenging problem.

While there are many interspecific differences in fat and protein contents of exploited fish (herring and cod represent two fairly extreme examples of high and low muscle fat contents; see also Love 1970), it is largely from the monograph of Shul'man (1974) that a good deal of information is now available concerning the fat and protein dynamics of certain marine fish. It appears that the major energy source in fish tissues is fat, except in small lean fish, where glycogen replaces it (see also Gerking 1962). Shul'man (1974) stresses the importance of the role of protein in growth, and adds that 'protein growth of young fish is directed wholly to the building up of the body proper, whereas in adult fish protein growth is largely associated with gonad development.'

The dynamics of fats and proteins remain obscure in detail. Optimal feeding may lead to simultaneous increase in fat and protein, but at other times fat may be stored in the muscles (e.g. in migratory species such as salmon) or the liver (cod) and this fat is utilized for activity before protein, though sockeye salmon use protein as fuel for migration (Duncan & Tarr 1958, Idler & Bitners 1959).

Gerking (1962) suggested that the protein content of fish is the least variable and most abundant organic constituent of the fish body, fat being unpredictably variable in many species, and carbohydrate stores slight. Shul'man (1974) and Love (1970) have data to supplement this view. But we should not overlook the established fact that protein is in dynamic equilibrium in the body at all times and that mobilization of various amino acids differs at various times in the life history (Love 1970, Weatherley 1972). This problem of the dynamics of body constituents in fish is not merely of academic interest. It is, after all, not so much fish themselves which should interest the harvester (whether pond culturist or fisherman), but the most useful product of growth: protein. Detailed knowledge of the dynamics of protein during all the exploitable stages of the life cycle could offer future opportunities for exploitation based on maximal protein yields.

Growth is a complex problem and Fig. 3.3 provides an indication of the

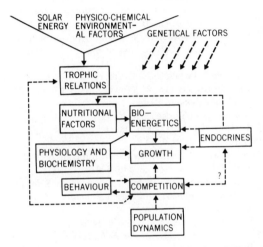

Fig. 3.3. Scheme of factors relating to maximization of growth referred to in this paper. Solid arrows indicate factors influencing growth in a fairly immediate manner. Broken arrows indicate factors that affect, limit or govern growth. (After Weatherley 1976).

factors that influence somatic growth in fish which could be useful as a scheme of reference for investigators. It indicates the probable nature of the relationships between trophic relations in the environment and somatic growth (through competition which in turn is affected by population dynamics), and depicts several major factors that stimulate, depress, or control growth, which growth, in turn, may modify. Genetical factors are indicated as having great potential for influencing many of the factors relating to somatic growth rate.

If one looks to an eventual revolution in the exploitation of growth processes in fishery science, the route will not be a simple one. It will require a sophisticated mixture of physiological, cellular, ecological, and quite possibly ethological knowledge about growth, which we are not yet in a position to exploit in a properly integrated manner. However, time is running out and the problems are consumingly urgent.

References

ALM G. (1946) Reasons for the occurrence of stunted fish populations with special regard to the perch. *Meddn St. Unders. – o. FörsAnst. Sötvatt.* **25**, 1–146.

BAAS BECKING L.G.M. (1946) On the analysis of sigmoid curves. *Acta biotheoret.* **8**, 42–59.

BACKIEL T. & LE CREN E.D. (1967) Some density relationships for fish population parameters. *The Biological Basis of Freshwater Fish Production* (Gerking S. D. ed.). Blackwell Scientific Publications, Oxford.

BAGENAL T.B. (Ed.) (1974) *The Ageing of Fish.* Unwin Brothers, England.

BEAMISH F.W.H. (1974) Apparent specific dynamic action of largemouth bass, *Micropterus salmoides. J. Fish. Res. Bd. Canada* **31**, 1763–1769.

BERTALANFFY L. VON (1938) A quantitative theory of organic growth. *Hum. Biol.* **10**, 181–213.

BERTALANFFY L. VON (1957) Quantitative laws in metabolism and growth. *Q. Rev. Biol.* **32**, 217–231.

BERTALANFFY L. VON (1960) Principles and theory of growth. *Fundamental Aspects of Normal and Malignant Growth* (W. W. Nowinski ed.). Elsevier, Holland.

BERTALANFFY L. VON (1968) *General System Theory: Foundations, Development, Application.* George Braziller, New York.

BEVERTON R.J.H. & HOLT S.J. (1957) On the dynamics of exploited fish populations. *Fishery Invest., Lond. Ser.* 2, **19**, 1–533.

CASSIE R.M. (1954) Some uses of probability paper in the analysis of size frequency distributions. *Aust. J. mar. Freshwat. Res.* **5**, 513–522.

CRICHTON M.I. (1935) Scale resorption in salmon and sea trout. *Salm. Fish., Edinb. No. 4*, 1–8.

CUSHING D.H. (1968) *Fisheries Biology: A Study in Population Dynamics.* Univ. Wisconsin Press.

DE BONT A.F. (1967) Some aspects of age and growth of fish in temperate and tropical waters. *The Biological Basis of Freshwater Fish Production* (S. D. Gerking ed.) Blackwell Scientific Publications, Oxford.

DUNCAN D.W. & TARR H.L.A. (1958) Biochemical studies on sockeye salmon during spawning migration III. Changes in the protein and non-protein nitrogen fractions in muscles of migrating sockeye salmon. *Can. J. Biochem. Physiol.* **36**, 799–803.

EL-ZARKA SALAH EL-DIN (1959) Fluctuations in the population of yellow perch, *Perca flavescens* (Mitchell), in Saginaw Bay Lake Huron, *U.S. Dept. Int. Fish. Bull.* 151.

EVERHART W.H., EIPPER A.W. & YOUNGS W.D. (1975) *Principles of Fishery Science.* Cornell Univ. Press, Ithaca.

FAGADE S.O. (1974) Age determination in *Tilapia malanotheron* (Ruppell) in the Lagos Lagoon, Nigeria, with a discussion of the environmental and physiological basis of growth markings in the tropics. *The Ageing of Fish* (T. B. Bagenal ed.). Unwin Brothers England.

GERKING S.D. (1962) Production and food utilization in a population of bluegill sunfish. *Ecol. Monogr.* **32**, 31–75.

GOSS R.J. (1964) *Adaptive Growth.* Academic Press, New York & London.

GREER-WALKER M. (1970) Growth and development of the skeletal muscle fibres of the cod (*Gadus morhua L.*) *J. Cons. perm. int. Explor. Mer.* **33**, 228–244.

GREER-WALKER M. & PULL G.A. (1975) A survey of red and white muscle in marine fish. *J. Fish. Biol.* **7**, 295–300.

HILE R. (1936) Age and growth of the cisco, *Leucichthys artedi* (Le Sueur), in the Lakes of the north-eastern highlands, Wisconsin. *Bull. Bur. Fish., Wash.* **48**, 211–317.

HOLDEN M.J. (1955) Ring formation in the scales of *Tilapia variabilis* Boulenger and *Tilapia esculenta* Graham from Lake Victoria. *East Afr. Fish Res. Organ. Ann. Rep.* 1954–1955, 36–40.

HOLLIDAY F.G.T., TYTLER T.P. & YOUNG A.H. (1974) Activity levels of trout (*Salmo trutta*) in Airthrey Loch, Stirling, and Loch Leven, Kinross. *Proc. R. Soc. Edinburgh Sect.* **B74**, 315–331.

HUXLEY J.S. (1932) *Problems of Relative Growth.* Methuen, London.

ICHIKAWA R. (1953) Absorption of fish scale caused by starvation. *Rec. oceanogr. Wks. Japan* **1**, 101–104.

IDLER D.R. & BITNERS J. (1959) Biochemical studies on sockeye salmon during spawning migration. V. Cholesterol, fat, protein and water in the body of the standard fish. *J. Fish. Res. Bd. Canada* **16**, 235–241.

IVLEV V.S. (1960) On the utilization of food by planktophage fishes. *Bull. Math. Biophys.* **22**, 371–389.

JENSEN K.W. (1957) Determining age and growth of *Tilapia nilotica, Tilapia galilea*, and *Tilapia zillii*, and *Lates niloticus* by means of their scales. *K. norske Vidensk. Selsk. Forh.* **30**.

KERR S.R. (1971a) Analysis of laboratory experiments on growth efficiency of fishes. *J. Fish. Res. Bd. Canada* **28**, 801–808.

KERR S.R. (1971b) Prediction of fish growth efficiency in nature. *J. Fish. Res. Bd. Canada* **28**, 809–814.

KERR S.R. (1971c) A simulation model of lake trout growth. *J. Fish. Res. Bd. Canada* **28**, 815–819.

KLEIBER M. (1961) *The Fire of Life—an Introduction to Animal Energetics.* Wiley, New York.

LE CREN E.D. (1947) The determination of the age and growth of the perch (*Perca fluviatilis*) from the opercular bone. *J. Anim. Ecol.* **16**, 188–204.

LE CREN E.D. (1951) The length-weight relationship and seasonal cycle in gonad weight and condition in the perch (*Perca fluviatilus*). *J. Anim. Ecol.* **20**, 201–219.

LE CREN E.D. (1958) The production of fish in fresh waters. pp. 67–72. *The Biological Productivity of Britain* (W. B. Yapp & D. J. Watson eds.). Institute of Biology Symposium No. 7.

LE CREN E.D. (1974) The effects of errors in ageing in production studies. *The Ageing of Fish* (T. B. Bagenal ed.). Unwin Brothers, England.

LEE R.M. (1912) An investigation into the methods of growth determination in fishes. *Publ. Circonst. hons. perm. int. Explor. Mer* **63**.

LISTER D., COWAN T. & McCANCE R.A. (1966) Severe undernutrition in growing and adult animals. 16. The ultimate results of rehabilitation: poultry. *Br. J. Nutr.* **20**, 633–648.

LOVE R.M. (1970) *The Chemical Biology of Fishes.* Academic Press, London.

McCANCE R.A. & WIDDOWSON E.M. (1962) Nutrition and growth. *Proc. R. Soc.* **156**, 326–337.

MASON J.E. (1974) A semi-automatic machine for counting and measuring circuli on fish scales. *The Ageing of Fish.* (T. B. Bagenal ed.). Unwin Brothers, England.

MATHEWS C.P. (1974) An account of some methods of overcoming errors in ageing tropical and subtropical fish populations when hard tissue growth markings are unreliable and the data sparse. *The Ageing of Fish* (T. B. Bagenal ed.). Unwin Brothers, England.

MONON M.D. (1950) The use of bones, other than otoliths, in determining the age and growth-rate of fishes. *J. Cons. perm. int. Explor. Mer.* 311–335.

NEEDHAM A.E. (1964) *The Growth Processes in Animals.* Pitman, London.

NICHOLLS A.G. (1957) The Tasmanian trout fishery. I. Sources of information and treatment of data. *Aust. J. mar. Freshwat. Res.* **8**, 451–475.

NICHOLSON A.J. (1933) The balance of animal populations. *J. Anim. Ecol.* **2**, 132–178.

NICHOLSON A.J. (1954) An outline of the dynamics of animal populations. *Aust. J. Zool.* **2**, 9–65.

OLSEN A.M. (1954) The biology, migration and growth rate of the school shark(*Galeorhinus australis* Macleay) (Carcharhanidae) in south-eastern Australian waters. *Aust. J. mar. Freshwat. Res.* **5**, 353–410.

PALOHEIMO J.E. & DICKIE L.M. (1965) Food and growth of fishes. I. A growth curve derived from experimental data. *J. Fish. Res. Bd. Canada* **22**, 521–542.

PANNELLA G. (1974) Otolith growth patterns: an aid in age determination in temperate and tropical fishes. *The Ageing of Fish* (T. B. Bagenal ed.). Unwin Brothers, England.

RICKER W.E. (1975) Computation and Interpretation of Biological Statistics of Fish Populations. *Fish. Res. Bd. Canada Bull.* **191**.

SHUL'MAN G.E. (1974) *Life Cycles of Fish: Physiology and Biochemistry.* John Wiley and Sons, England.

SIMKISS K. (1974) Calcium metabolism of fish in relation to ageing. *The Ageing of Fish* (T. B. Bagenal ed.). Unwin Brothers, England.

SWINGLE H.S. (1950) Relationships and dynamics of balanced and imbalanced fish populations. *Bull. Ala. Agric. Exp. Stn.* 274.

SWINGLE H.S. (1951) Experiments with various rates of stocking bluegills, *Lepomis macrochirus* Rafineseque, and largemouth bass, *Micropterus salmoides* (Lacépède), in ponds. *Trans. Am. Fish. Soc.* **80**, 218–230.

SWINGLE H.S. & SMITH E.V. (1942) Management of farm fish ponds. *Bull. Ala. agric. Exp. Stn.* 254.

SYCH R. (1970) Elements of the theory of age determination of fish according to scales. The problem of validity. *E.I.F.A.C. Technical Paper 70/SC*, 1–3.

SYCH K. (1974) The sources of errors in ageing fish and considerations of the proofs of reliability. *The Ageing of Fish* (T. B. Bagenal ed.). Unwin Brothers, England.

TESCH F.W. (1968) Age and Growth. *Methods of Assessment of Fish Production in Fresh Waters* (W. E. Ricker ed.). IBP Handbook No. 3. Blackwell, London.

THOMPSON D'ARCY W. (1942) *On Growth and Form.* Cambridge University Press.

THOMSON J.M. (1962) The tagging and marking of marine animals in Australia. *C.S.I.R.O. Div. Fish. Oceanogr. tech. pap.* 13.

VAN OOSTEN J. (1929) Life history of the Lake herring (*Leucichthys artedi* Le Sueur) of Lake Huron as revealed by its scales, with a critique of the scale method. *Bull. U.S. Bur. Fish.* **44**, 265–428.

VAN OOSTEN J. (1941) The age and growth of freshwater fishes. *A Symposium of hydrobiology.* Madison, Wis., 196–205.

VAN OOSTEN J. (1957) The skin and scales. *The Physiology of Fishes*, Vol. 1. (Brown M. E. ed.), pp. 207–244. Academic Press, New York.

WALFORD L.A. (1946) A new graphic method of describing the growth of animals. *Biol. Bull. mar. biol. Lab., Woods Hole* **90**, 141–147.

WARE D.M. (1975) Growth, metabolism and optimal swimming speed of a pelagic fish. *J. Fish. Res. Bd. Canada* **32**, 33–41.

WARREN C.E. & DAVIS G.E. (1967) Laboratory studies on the feeding, bioenergetics, and growth of fish. *The Biological Basis of Freshwater Fish Production* (S. D. Gerking ed.). Blackwell, Oxford.

WEATHERLEY A.H. (1972) *Growth and Ecology of Fish Populations.* Academic Press Inc., London.

WEATHERLEY A.H. (1976) Factors affecting maximization of fish growth. *J. Fish. Res. Bd. Can.* **33**, 1046–1058.

WIDDOWSON E.M. (1964) Early maturation and later development. *Diet and Bodily Constitution* (G. E. W. Wolstenholme & M. O'Connor eds.). Ciba Foundtn. Study Conf. No. 17., Churchill, Lond.

WINBERG G.G. (1956) Rate of metabolism and food requirements of fishes. *Fish. Res. Bd. Canada Transl. Ser.* No. 362.

WYNNE-EDWARDS V.C. (1962) *Animal Dispersion in Relation to Social Behaviour.* Oliver and Boyd, London.

Chapter 4: Aspects of Fish Fecundity

T. B. Bagenal

4.1 Introduction

For several centuries the attention of laymen, naturalists and fishery scientists has been drawn to the number of eggs in the roes of female fish. Fish fecundity has been studied not only as one aspect of natural history, but also in association with studies of population dynamics, racial characteristics, production and stock-recruitment problems.

The weight of the ovary is often a large proportion of the total weight; for example, in the annual spawners brown trout, *Salmo trutta*, and perch, *Perca fluviatilis*, the mature ovary may form 20% of the total weight, and over 30% in the European plaice, *Pleuronectes platessa*. However in fish that spawn several times each year the mature gonad weight may be much less, as in *Sartherodon variabilis* of Lake Victoria in which the gonad weight is about 4% of the total (Fryer & Iles 1972). In terms of annual production of new tissue, the proportion that goes to the ovary is even larger; in female perch for example 40% of the annual total production is made up of gonad products (Le Cren 1958).

The way in which the material in the gonad may be divided into few or many eggs is of importance in estimating the number of individuals at the start of a new generation, and this estimate is the starting point of many production and population studies. The eggs laid by the females are a link between one generation and the next, but in general no clear relationship between the number of eggs laid and subsequent recruitment to the parent population can be found. Nevertheless, the number of eggs laid by a fish population does appear to adapt to environmental conditions through food supply 'and so is one of the basic means of adjusting the rate of reproduction to changing conditions' (Nikol'skii 1969). Although this mechanism seems to exist, we shall see later that it does not often appear to be effective.

It is not intended in this chapter to describe the methods of fecundity estimation, since these are dealt with in detail in manuals such as the I.B.P. Handbook No. 3 *Methods for Assessment of Fish Production in Fresh Waters* (Bagenal 1977). It is sufficient here to say that the fecundity must be based on unbiased samples and they must be analysed adequately from the statistical point of view.

4.2 Definitions of fecundity

Fecundity is usually defined as the number of ripening eggs found in the female just prior to spawning. This contrasts with 'fertility' which is the number of eggs shed. However, the range of reproductive habits of fish is very large (Breder & Rosen 1966) and definitions that are acceptable in all circumstances have not yet been devised. Welcomme (1967) uses fecundity for the number of eggs produced by *Sarotherodon leucostictus*, a mouth brooding cichlid, and fertility for the number of young produced. Raitt and Hall (1967) use three definitions with the viviparous redfish *Sebastes marinus*; 'prefertilized fecundity' for the number of eggs in the ovary before fertilization, 'fertilized fecundity' for the number of fertilized eggs in the ovary, and 'larval fecundity' for the number of larvae after hatching but before extrusion. In this case the fertility would be the number of larvae extruded. I believe that fertility should be reserved for the number of eggs or larvae when they leave the female parent, or male as in *Hippocampus*, for the last time. I suggest that in a mouth brooding *Sarotherodon* one should recognize 'ovarian fecundity' and 'brooding fecundity'. In those temperate species which shed their eggs in batches, the fecundity is the sum of all the ripening eggs in all the batches during the season. However, in tropical species where batches follow each other continuously because there are no seasonal changes, the fecundity must only include one batch. Lowe (1955) in her paper on *Tilapia* and *Sarotherodon* fecundity defined it as 'the number of young produced during the lifetime of an individual'. This is probably impossible to determine and is not a satisfactory definition.

The situation in typical multiple or batch spawners, such as the bleak *Alburnus alburnus* and sprat *Sprattus sprattus* is probably only different in degree from the majority of fish which have only one spawning time. Immediately prior to spawning the maturing eggs in most marine fish swell very considerably and become translucent. At first these are scattered through the ovary but very soon large numbers develop, the follicles rupture and the eggs lie in the lumen of the ovary and the fish is 'ripe and running'. The eggs swell to such a size that the ovary clearly cannot contain them all and the eggs must be shed in batches even if the intervals between are short. Perhaps only those fish such as the angler *Lophius piscatorius* and the perch, which shed their eggs joined together in a gelatinous mass, are truly non-multiple spawners.

The ripening ovaries of most species are usually found to contain three classes of eggs. The most numerous are small, white and opaque, and are sometimes termed the 'recruitment stock'. These develop into the 'maturing eggs' which are larger, opaque and laden with yolk, and are often yellow, orange or green. The third class consists of the 'atretic eggs'. These are maturing eggs which for one reason or another may be completely resorbed.

There is evidence that diet restriction is one factor leading to an increase in their number. Different species appear to have different numbers of atretic eggs. Wagner and Cooper (1963) report up to 72% in the creek chubsucker *Erimyzon oblongus*. There are also significant numbers in *Tilapia* spp (Peters 1963), and Macer (1974) found such large numbers in the horse mackerel *Trachurus trachurus* that it appears that there could be a substantial difference between the potential and actual fecundity. Vladykov (1956) inferred large numbers of atretic eggs in the brook trout, *Salvelinus fontinalis*, but Henderson (1963) maintained that this was an error produced by the growth in length. However, Wydoski and Cooper (1966) reported that more than 50% of the eggs were atretic. The fecundity as defined above of an individual fish is usually called the *absolute fecundity*. It is usually found to increase with size and age of the fish so in order to compare fecundities of fish of different size or from different places, many authors calculate the *relative fecundity* which is the number of eggs per unit weight. Zotin (1968) developed a number of refinements of the idea of relative fecundity which include such factors as the volume of the eggs, in order to try to find a measure more ecologically meaningful. These relationships assume that absolute fecundity is a simple and direct function of the body weight. The difficulties that this leads to will be considered later.

A third measure is the *population fecundity*. This is the sum of the absolute fecundities of all the breeding females, and so is the number of eggs laid by the population in one season. In some work such as when making life tables or in production studies, it may be useful to relate all the measurements to age, in which case the *age specific fecundity* is found for each age group.

These various measures of fecundity will be considered in more detail later.

Fecundity as defined above is usually similar to the fertility, i.e. there are few residual eggs left unshed. Staples (1975), who worked on the upland bully *Philypnodon breviceps* of New Zealand, is among the few authors to give the number of residual eggs after spawning and its variability. Macer (1974) gives evidence of some horse mackerel (which is a multiple or repeat spawner) going straight from 'part spent' to 'recovering' with many eggs unshed. As a result of this he believes there could be a substantial difference between the fecundity and fertility. Volodin, Mezhin and Kuz'mina (1974) describe the gross morphology and histology of residual egg resorption in bream *Abramis brama* together with the associated biochemical changes that take place in the fish at the same time. Moore (1975) reports small numbers of residual eggs in the Arctic char *Salvelinus alpinus*, and Kolyushev (1973) describes the development and histology of char egg production and resorption in some detail.

It appears, therefore, that nearly all spent fish have some residual eggs after spawning, but in most species this is not a significant number compared with the total shed.

4.3 The Absolute Fecundity

The absolute fecundity is usually related to the length, weight or age of the fish.

FECUNDITY AND LENGTH

A close relationship is usually found between fecundity and length. With repeated sampling prior to spawning, it must be remembered that the fish may grow, but the fecundity does not increase after a certain stage in maturation, so the fecundity for a given length may appear to decrease, although the fecundity of any individual fish has remained the same.

A great many workers have plotted fecundity and length as a scatter diagram (Fig. 4.1 adapted from Bagenal 1966) and they have concluded that the relationship is of the form

$$F = aL^b$$

where F=fecundity, L=length and a and b are a constant and an exponent derived from the data. A logarithmic transformation gives the straight line regression of log fecundity on log length,

$$\log F = \log a + b \log L$$

and the line may be fitted by the method of least squares, allowing the use of standard statistical procedures for subsequent analysis. The transformation tends to equalize the variance throughout the range of lengths thereby avoiding the problem that fecundity of large fish is more variable than that of small ones. Pope *et al.* (1961) who worked on Atlantic salmon, *Salmo salar*, give a table which shows how the transformation removes the dependence of the standard deviation on the mean.

The exponent b in the above equations has been calculated for a great variety of fish and although it usually ranges from about 2·3 to 5·3, most often the values are a little above 3. In a few cases the confidence limits have been calculated as

$$b \pm t \ 1/S_x^2 \ (\text{Residual } S.S./n-2)^{1/2}$$

where t=Student's t for $n-2$ degrees of freedom, S_x^2=sums of squares for length and residual $S.S.$=the residual sums of squares obtained from the covariance analysis. Baxter and Hall (1960) use the standard error of the coefficient b, for Atlantic herrings *Clupea harengus*, as do Pope *et al.* (1961) with Atlantic salmon.

The antilogarithm of a fecundity estimate derived from the equation given above always tends to give an underestimate. Pope *et al.*, Raitt and Hall (1967) and Raitt (1968) apply a correction of

$$1 \cdot 15 \left(\frac{n-1}{n}\right) S^2$$

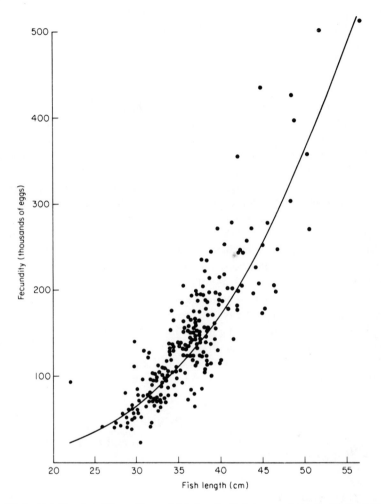

Fig. 4.1. Scatter diagram of fecundity and fish length of plaice caught in the Firth of Clyde Scotland.

where $S^2 =$ the residual variance of the log values. The relation of various parameters following a log transformation are explained by Bagenal (1955).

Some authors have argued that the regression of log fecundity as the dependent variate on log length as the independent variate and measured without error, is not justifiable. Ricker (1973, 1975) proposes a functional regression based on the least products of the points (x·y) from the line, rather than the least squares on the x axis. Mann (1974) has used Ricker's equations in his studies of roach *Rutilus rutilus* and dace *Leuciscus leuciscus* fecundities. Baxter and Pope (1969) have used a far more complex least squares functional regression given by Sprent (1966). I do not recommend this on the grounds of its unnecessary complexity. Ricker's equation and the confidence limits are easy to compute.

FECUNDITY AND WEIGHT

It is often thought that since fish weight is connected with the condition of fish, fecundity is more likely to be more closely correlated with weight than with length. However, where the correlations have been analysed adequately very little advantage has been found in considering weight rather than length. This is shown by Bagenal (1957a), and by Baxter and Hall (1960).

Furthermore, numerous problems arise when correlating fecundity and weight. In a great many fish the somatic weight changes significantly towards spawning. Examples of papers in which these changes have been considered in detail are Le Cren (1951) with perch, Iles (1974) with Atlantic herring and Wootten (1973b) with sticklebacks *Gasterosteus aculeatus*. Where these changes occur, the relation between fecundity and weight will differ as the breeding season approaches. Secondly, if the weight used in the correlations is the total weight (= somatic + gonad) a spurious correlation may be obtained since the greater number of eggs in the more fecund fish will weight more than those in the less fecund. The same kind of problem is present if one tries to associate fecundity with 'condition', if this is measured by a parameter such as $K = 100 \times \text{weight/length}^3$. In addition, relating fecundity and condition assumes either that the power 3 above or b remains constant in the length weight relation

$$W = aL^b$$

It has been shown in some investigations that a and b vary seasonally (Le Cren 1951, Bagenal 1957a, Raitt 1968).

FECUNDITY AND AGE

Various authors have tried unsuccessfully to relate absolute fecundity and age, e.g. Simpson (1951), Bagenal (1957a), Pitt (1964) and Terlecki (1973) with pike, *Esox lucius*. In these cases there was a clearer correlation with length or weight. Raitt and Hall (1967) found with redfish that neither the addition of age nor weight to length produced a significant improvement in the correlation with fecundity. However Berg and Grimaldi (1965) give data for whitefish *Coregonus* spp from Lake Maggiore which suggest an age effect, but this is not supported by Zawisza and Backiel (1970) who in a thorough analysis conclude that age, after eliminating length does not explain the fecundity relations as well as does length alone. Their work supports Gerking's (1959) conclusion that individual fecundity variations are so great that they mask any effects age may have. However Ludwig and Lange (1975) found that a statistical model using age and age-length interaction was more predictive of fecundity than length alone with Northern mottled sculpins, *Cottus bairdi*.

Lear (1970) found in the halibut *Reinhardtius hippoglossoides* that fecundity was proportional to age to the power 1·95. Similarly May (1967) found fecundity to be nearly proportional to the square of the age in cod *Gadus morhua* up to 27 years old. Mackay and Mann (1969) used the semi-logarithmic equation Log $F = n$ (age in years) $+ m$, where m and n are constants. With roach this was as good for predictive purposes as length and with bleak it was better. Winters (1971) also reports a more adequate relation with age than with length with capelin *Mallotus villosus*. This is primarily because the fecundity of repeat spawners is higher than of first time (or recruit spawners) and the proportion of repeat spawners increases with age without, in caplin, a significant increase in length.

The relationship of age and fecundity is most often required in the preparation of life tables. Although it can be obtained directly, in practice (McFadden & Cooper 1964, McFadden 1967) the mean fecundity for each age has been calculated from the fecundity-length relation and an estimate of the mean length for each age group.

4.4 Relative fecundity

Since relative fecundity is based on the weight of the fish, all the problems associated with relating fecundity and weight apply here too. In most fish the number of eggs does not change significantly as the season progresses (Bagenal 1963), but the gonad weight increases due to an increase in water content, or organic matter derived from food, or organic matter transferred from somatic tissue. Only in the latter case will total weight remain constant thus making the calculation of relative fecundity meaningful. In the pike Kipling and Frost (1969) have shown that although the gonads increase fivefold in weight from October to March, this is apparently balanced by a somatic weight loss, and so the relative fecundity remains constant through the season.

The use of relative fecundity assumes that the regression coefficient b, in

$$\log F = \log a + b \log W$$

does not differ significantly from unity. In herrings it has been shown that fecundity increases faster than body weight. I have published (Bagenal 1973) values of b in the above equation with the 95 % confidence limits for plaice, witch *Glyptocephalus cynoglossus*, long rough dab *Hippoglossoides platessoides* and pike.

An example of the problems of using relative fecundity uncritically is shown in Table 4.1, based on Shatunovskii's (1964) data on the flounder,

Table 4.1. The relative fecundity (in thousands of eggs per gram of fish body weight) in seven length groups of flounders from the southeastern Baltic. Taken from Shatunovskii (1964).

Length group	Relative fecundity
16–18	2·1
18–20	2·7
20–22	2·9
22–24	2·9
24–26	3·1
26–28	3·1
28–30	3·6

Platichthys flesus. Here the relative fecundity is clearly related to length, so it cannot be used alone to compare the fecundity of flounders at different places or times. The literature is unfortunately full of similar cases. It cannot be emphasised too strongly that workers use relative fecundity without adequate analysis at their peril. In my view there is only one satisfactory method of comparing fecundities of different populations, and that is to find the equation relating fecundity to length (or weight if this is shown to be reasonable), and from this to calculate the fecundity expected of a fish of a given length (or weight) for the different populations. This method allows statistical tests of significance to be carried out (Bagenal 1966, 1973).

4.5 The Fecundity of Multiple (or Batch) Spawners

The difficulties that are involved with a species that spawns repeatedly over a long period have been considered in most detail by Macer (1974) who worked on the horse mackerel which spawns from May to August. The main problem is to distinguish between the reserve and developing oocytes. A common criterion has been the presence or absence of yolk—a count of the yolky eggs gives the fecundity for the season. This criterion was used by Petrova (1960) for Baltic sprat, Andreu (1954) for Atlantic pilchards, Rao (1971) for Indian mackerel, and other authors. Another method is to estimate the size of a batch and the number of batches. This was attempted by Clark (1934) and MacGregor (1957) for Californian sardines without success since they could not estimate the number of spawnings. Macer's method assumed that the potential annual fecundity for that season is given by the total of those oocytes which are vacuolated or yolked when the ovaries were at the 'late developing' stage. He produces evidence that further oocytes are not recruited to the vacuolated or yolky stock after this stage. Histological examination gave the percentage of vacuolated oocytes in each oocyte size group and the aliquot samples with the stempel pipette gave the total number of oocytes. From these the fecundity, that is the total number of developing eggs, was obtained.

De Silva (1973) worked with the Atlantic sprat and he believed he was successful in estimating the number of batches and the number per batch. He assumed that the oocytes in the most advanced stage of development are shed at one time. The number in this group gives the number shed per batch. The ratio of the total number of yolked oocytes to the number in the most advanced state of maturity is considered to give the number of batches of eggs shed per season.

Among other workers who have considered the problem are Ivanov (1971) who gives evidence for what he believes to be the number of eggs per batch, while Mackay and Mann (1969) provide size frequency diagrams of ovarian egg size which clearly indicate three batches of eggs spawned by the bleak. Moroz (1970) also gives evidence of three egg batches with mean diameters 0·59, 0·81 and 1·17 in the ovaries of *Vimba vimba*. Healey (1971) describes how in the sand goby *Pomatoschistus minutus* the ripe eggs occupy a part of the ovary which becomes constricted, and then separates off, from the posterior end. The eggs contained in this section are shed as one batch, but a second lot is beginning to bud off before the first is shed.

4.6 Viviparous Species and those that show Parental Care

The viviparous species most adequately investigated are the redfish *Sebastes marinus* by Raitt and Hall (1967), the elasmobranch spurdog *Squalus acanthias* by Holden and Meadows (1964) and the guppy *Poecilia reticulata* by Warren (1973).

In the red fish copulation takes place in the autumn and the sperm are stored in the female until March when the eggs are fertilized in the ovary. Raitt and Hall report a significantly lower fertilized fecundity than pre-fertilized fecundity. The relationship between fertilized fecundity and larval fecundity or between fertilized fecundity and fertility (i.e. the number of larvae born) is not known, so no estimate of the spawning potential is possible.

Holden and Meadows (1964) considered fecundity in terms of the large ovarian eggs in the spurdog and the embryos, but had difficulty because some of the latter had already been shed. Holden (1974) shows there is a difference between the fecundity of the heavily exploited Scottish-Norwegian stock and the unexploited North American stock. His hypothesis was that the differences might be a response to the lower population density, but he found no really convincing evidence. In order to fully compensate for the exploitation, the average litter size would have to be 19 which is beyond the capacity of the species.

Warren's (1973) work was primarily a thorough experimental investigation into the relation of population density and reproduction in the guppy. He found that although the situation could be confounded by providing mirrors and by darkness, in general fish kept at high densities, or provided with water from high density populations, produced fewer young.

Of those species which show parental care, most fecundity work has been done on the three-spined stickleback *Gasterosteus aculeatus* and the cichlids *Tilapia* spp. and *Sarotherodon* spp. Much of the work by Wootton (1973a & b) on the stickleback has been of an experimental nature and is mentioned on pages 86 and 91. He makes the point that the high rate of egg production throughout the season is allied to the complex parental care by the male, and it leads to very efficient reproduction. In the course of a breeding season a large female can produce two or three times her weight in eggs. This is not however necessarily associated with a large number of eggs per batch.

Fryer and Iles (1972; other references) discuss the significance of brood and egg size in cichlids. There is a close association between egg number and size; mouth-brooders have fewer and larger eggs than guarders. Thus the mouth-brooders, *Sarotherodon moori* and *S. esculenta*, have eggs 7 mm and 4·5 mm respectively, while the eggs of the guarders, *T. zillii* and *T. mariae*, are 1·8 mm long and 1·5 mm in diameter respectively. The fecundity of *Sarotherodon* has also been considered by Lowe (1955), but there are problems in that the fecundity is greater than the egg number that can be brooded. Welcomme (1967) has considered this problem of the relationship between fecundity (egg production) and fertility (young produced) in *Sarotherodon leucosticta*. The fecundity was approximately equal to the square of the length, whereas fertility was linearly related to parent length. Comparisons of these relationships showed that breeding efficiency decreased with increasing parent length. Welcomme also considered the effects of maturation size, growth rate and spawning frequency.

Fryer and Iles suggest that cichlids 'economize' on gonadal material and that there appears to be a relationship between the risk of being eaten and the size of the egg produced. The young can be brooded to a large size before being released, and at this size they can attempt to collect food in the manner of the adults. In the cichlids the reduction in the number of eggs has been accompanied by the loss of one ovary and very yolky eggs that contain only 50–60% water.

MacArthur and Wilson (1967) coined the terms '*r*-selection' for the evolutionary selection which favours high fecundity and rapid development, and '*K*-selection' for that which is correlated with low fecundity and slow development. No particular organism is completely *r*-selected or *K*-selected, but all can be considered as having reached some compromise on the continuum between the two extremes.

MacArthur and Wilson (see also Southwood 1976) considered that in pioneer communities r-strategists with high fecundity would be at a premium, and it is interesting that Bagenal (1966) found that plaice living at the fringes of their geographical range and apparently suffering greater density independent mortality (an r characteristic: Gadgil and Solbrig 1972) had higher fecundities compared with those living near the population centres, where density dependent mortality (resulting from commercial fishing) would be greater. Similarly in tropical regions, the African Great Lakes appear to have K-selected species compared with the r-selected river populations which have responded to the great density independent wastage during annual flooding (Lowe 1975).

It is characteristic of r-strategists that they tend to be single spawners, favouring early reproduction leading to high intrinsic rates of increase, whereas K-strategists tend to be multiple spawners who allocate a greater proportion of reserves to non-reproductive effort such as redd and nest building, oviparity, brooding and care of the young, and so tend towards efficiency. It is clear that fish span the range of r–K continuum, and equally clearly this is a subject which needs more consideration.

4.7 Freshwater Species that lay Pelagic Eggs

Very little work on the fecundity of species that lay pelagic or semi-bouyant eggs has been reported, or translated, in western languages, though there are a large number of these fishes, spread through a wide range of families (Breder & Rosen 1966). Gorbach (1972) studied the fecundity of 190 specimens of grass carp *Ctenopharyngodon idellus* and although he reports that it varied from 277 thousand in a seven year old fish of 67·5 cm to 1,687 thousand in a fifteen year old specimen of 96 cm, he does not discuss any of the features in the fecundity that might be associated with the spawning habits; perhaps there are none.

4.8 Fecundity of Anadromous Species

Reference has already been made several times to salmonid fecundity. In this section we will consider some of the effects particularly associated with the anadromous habit.

Gritsenko (1971) compared the fecundities of anadromous and resident river chars on Sakalin Island. The author did not compare fish of the same length, but if this is done it appears there is no real difference, but age for age the anadromous race had a higher fecundity. Blackett (1973) found that the resident Dolly Varden *Salvelinus malma* in south-eastern Alaska had larger eggs, fewer eggs per unit length, the same relative fecundity and more eggs per unit weight of ovary than did the anadromous race.

Rounsefell (1957) analysed the fecundity of sockeye salmon of the Karluk River in the years 1938–41. The data refer to 60 cm fish which have spent 2 or 3 summers in fresh water before migrating to the sea, and 2 or 3 years before returning to spawn. Rounsefell concluded that an extra year at sea led to lower fecundity. These fish will have grown more slowly since all the salmon were the same length. The fish that had spent a year longer in fresh water produced only slightly more eggs. The most interesting analysis is of the five-year-old fish which have spent either 2 years in freshwater and 3 in the sea, or 3 in fresh water and 2 in the sea, thus removing any possible age effect. Those that had spent their extra year in the sea were slightly less fecund.

Grachev (1971) maintains that as the weight of the Pacific salmon *Oncorhynchus nerka* gonads increases to between 100 g and 200 g while at sea, the number of oocytes increases at first and then declines. He says the ultimate fecundity depends on the rate of maturation, the growth rate and the length of time at sea.

Wootton (1973b) reanalysed the data on the fecundity characters of the anadromous *trachurus* and freshwater *leiurus* forms of the three-spined stickleback. The slopes of the regression of log fecundity on log length of the two forms were similar, but the intercepts were statistically significantly different. Hagen (1967) reported that the fecundity of the hybrids was midway between the two parent forms.

4.9 Fecundity and Egg Size

Perhaps more uncritical papers have been written about egg size and quality than any other aspect of fish fecundity, and the student is well advised to treat all he reads with caution. To start with, if the sizes of eggs are to be compared, they must be measured at the same developmental stage and this is impossible while they are in the ovary since they are developing fast. They are, therefore, best measured after fertilization, by which time it is impossible to determine the fecundity by the usual methods. This difficulty is less with salmonids, or fish such as the herring or carp which lay adhesive eggs, and it is with these that the best work has been done.

The first point to make is that in general large fish tend to lay larger eggs than do small fish. Pope *et al.* (1961) and Galkina (1970) found this with salmon, Krivobok and Storozhuk (1970) with *Acipenser güldenstädt*, Mackay and Mann (1969) with roach, Hulata *et al.* (1974) with carp, Peters (1963) with tilapia and Templeman (1944) with young of spurdogs, to mention only a few papers on a variety of taxonomic groups.

Pope *et al.* analysed their data thoroughly, but it is clear that there are complications. There were, for example, differences between rivers in the relation of egg size and fish length. Rounsefell (1957) reviewed the fecundity of salmonids and concluded that species with the longest freshwater lives have the largest eggs, that *Oncorhynchus* spp lay larger eggs than other salmonid genera, that river spawners of that genus lay larger eggs than lake spawners and that within the genus, *O. tshawytsha* lay the largest and *O. nerka* the smallest.

Dahl (1918–19) in a classic paper on brown trout showed that large eggs produce larger fry than do small eggs, and he suggested that these larger fry also grow faster. I found (Bagenal 1969b) that under conditions as nearly natural as possible the survival of alevins from large eggs was statistically significantly greater than those from small eggs. These experiments confirmed one of Svärdsen's (1949) postulates, that there must be a balance between egg number and size, otherwise parents with mutations for great fecundity would always leave more progeny and so come to dominate the population. Hence, it is often assumed that fecundity and egg size are negatively correlated. The relationship is complicated by what appears to be a general tendency for fish which spawn later in the season to lay smaller than average eggs (Bagenal 1971). Since egg size and parent length are usually positively correlated, it follows that the larger fish spawn first. This has been found to be true for some species but not for all. Simpson (1959a) showed clearly that older plaice spawn first, but in contrast the older individual perch spawn later.

The relation of egg size and survival of the progeny has been well studied in the herring. The winter-spring spawners have fewer and larger eggs while the summer-autumn spawners lay more and smaller eggs. (Parrish & Saville 1965). The significance of these herring egg size differences has been very adequately studied by Blaxter and Hempel (1963) who showed clearly the survival value of large egg size to an individual. Cushing (1968) suggests that probably young herrings contribute no viable larvae, though Schopka and Hempel (1973) who considered four herring populations in considerable detail, found only a slight increase in egg size with body length.

Detailed investigations have also been carried out on carp *Cyprinus carpio*. Martyshev (cited by Nikol'skii 1969) showed that old carp lay larger and more viable eggs, and Hulata *et al.* (1974) report the results of pond experiments in which 11-month-old females produced eggs less than half the volume of 23-month-old fish regardless of body weight, but the egg size did not increase further with older fish. As they point out, a consideration of the parental age is important in egg size studies, and could explain the variation reported by

Chapter 4

Bagenal (1971) who discussed the seasonal and geographic variations in demersal and planktonic marine and freshwater fish eggs.

A great deal has also been written about the quality of eggs in relation to fecundity. A number of these papers are of little value, since fecundity has been measured as relative fecundity in situations when this is clearly unsatisfactory. For example, Kuznetsov (1973) considered the lipid content of bream eggs in relation to fecundity. He showed first that the egg lipid content decreased with parent length and weight, and over the weight range given, the relative fecundity doubled. Thus for fish of 400–600 g the mean Relative Fecundity $(R.F.) = 90·0$ eggs per gram and for 1400–1600 g fish the mean $R.F. = 183·0$ eggs per g. Also over this range the egg lipid content fell from $5·15\%$ to $4·12\%$. This is clearly unsatisfactory not only because no tests of significance were made, but also because the author did not consider whether in fish of the same weight, length or age the lipid content of the eggs was less in the more fecund fish. The student will find a high proportion of unsatisfactory papers dealing with egg quality.

4.10 Recruit and Repeat Spawners

The effect of previous spawnings has been investigated by some authors to help explain fecundity differences between superficially similar fish. Anokhina (1963), who worked on White Sea herrings, had 38 paired groups; each pair consisted of fish similar in age length and weight, but one group comprised recruit (or first time) spawners and the other repeat spawners. On the whole the recruit spawners contained more eggs but there is evidence that there was very considerable variation between the individuals contributing to each mean. However, there was a definite decrease in fecundity associated with the number of spawning marks on the scales. Bridger (1961), Cushing (1968) and Baxter and Pope (1969) believe that North Sea and Clyde herring recruit spawners have *lower* fecundities. Schopka and Hempel (1973) could find no evidence of this in the herring stocks they studied. The mechanism suggested by Anokhina is that there is a short feeding period for White Sea herrings and she postulates that it is insufficient for the fish to accumulate enough reserves to recover from spawning, produce more eggs and survive the winter.

Hodder (1963) suggests that the number of previous spawnings of Grand Banks haddock is of importance in the fecundity determination. His complicated analysis should be repeated with other species, but attention must be paid to the distinction between individuals which miss a spawning and those that are genuinely immature. Ten per cent of the 7-year-olds on the Grand Banks were classified as immature but they may have spawned previously and be resting through one season. This phenomenon is known for numerous species including the long rough dab (Bagenal 1957a), lake trout *Salvelinus namaycush* (Fry 1949) and whitefish *Coregonus clupeaformis* (Kennedy 1953).

4.11 Annual Variations in Fish Fecundity

I have discussed the annual variations in fecundity in detail (Bagenal 1973) with regard to the North Sea and Clyde plaice, Clyde long rough dabs and witches and Windermere pike. I also reanalysed data on Grand Bank haddock *Melanogrammus aeglefinus* (from Hodder 1965) and reviewed other published work giving evidence of annual variations. Considerable and statistically highly significant annual changes in absolute, relative and population fecundity occur in these data. In plaice the absolute fecundity varied by 48%, for long rough dabs 40%, in witches 28%, in pike 25%, in haddock 56%, in Norway pout *Trisopterus esmarkii* 250% (based on Raitt 1968), and in herring 34% (based on Polder & Zijlstra 1959). The causes of these variations were discussed in relation to population density, temperature, food supply, stress and other environmental effects. Evidence was produced that above average fecundity appears to be associated with low population density and vice versa. This suggests that fecundity variations form a density dependent population regulating mechanism, and so will tend to dampen large fluctuations in the number of young fish produced.

Baxter and Pope (1969) give results of herring fecundity counts over seven years. The analysis by age and year class does not show any annual change, but it is likely from their figures that an analysis by year of capture might have done. Anokhina (1963) found insignificant variations in White Sea herrings.

Unfortunately many authors give their results in the form of relative fecundity without adequate analysis, so it is impossible to tell whether the variations they report are real or spurious.

4.12 Population Fecundity

The population fecundity can be defined as the number of eggs capable of being laid by all the females in one season. This definition applies to the situation in temperate regions where there is one definite spawning season. A crude estimate is obtained from the product of the fecundity expected of an average female in length or weight terms and the number of breeding females. Pitcher and Macdonald (1973) have pointed out that this will give an underestimate if the usual logarithmic relationship between egg numbers and fish length is employed, because fish larger than the mean size will have relatively more eggs. Pitcher and Macdonald say that the error is greatest with large values of the regression coefficient and with a large length range. They give a diagram from which the magnitude of the error may be judged if the regression coefficient, the mean length and its variance are known. They give a computer programme to estimate the population fecundity, but also point out that if the differences in the length—fecundity relation between age groups can be ignored, and all fish sizes are sampled with known efficiency, it will be better to calculate the population fecundity directly from the length frequency distribution. Ivlev (cited by Nikol'skii, 1969) gave an expression for calculating the population fecundity from the age rather than the length distribution. It may indeed be better to consider the mean fecundity of the different age groups and work from these, but this is rarely done. In most fish population work estimates of age group numbers are made separately from the fecundity estimates.

Most often in fishery work it is argued that since fecundity is linearly related to weight, the population biomass may be taken as an index of the total number of eggs laid, (Beverton 1962). I have shown elsewhere (Bagenal 1973) that this is a crude and risky procedure since the fecundity estimated in this way will vary with a and b in the fecundity weight relation, the sex ratio, and the number of mature females. The number of eggs per gram of plaice can vary by over 100% from one locality to another and the absolute fecundity of fish of the same length can also vary by a similar amount.

Few authors have given confidence limits to the calculated population fecundity. Mathisen and Gunnerød (1969) estimated the total egg deposition of sockeye salmon from the expression: Total egg deposition = Number of fish in escapement × the fraction of females in the population × mean fecundity. They say that a large variance is associated with the estimate of the escapement number than the sex ratio estimate and that the least variance is associated with the estimate of the mean fecundity.

4.13 Population Estimation

One great stimulus to fecundity studies has been their use in total population estimation. Simpson (1959a) used his extensive fecundity estimates to calculate the plaice population in the Southern Bight of the North Sea. This varied from 52·5 million spawning females in 1949 to 28·0 million in 1950. Essentially the method involves estimating the number of eggs laid (with marine planktonic eggs from working a grid of quantitative plankton stations) and dividing this by the average fecundity. It has been used by Thompson and van Cleve (1936) for north Pacific halibut, Sette (1943) for eastern American mackerel, Macer (1974) for horse mackerel, and by other authors. These species lay planktonic eggs, but Nagasaki (1958) and Gjøsaeter and Saetre (1974) have used the method with demersal eggs of Pacific herring and

American smelt respectively. The method is discussed by Saville (1964) and English (1964). The error in the population estimate is inevitably very high.

4.14 The relation of fecundity, food supply and population density

A great many of the papers reporting a relationship between fecundity and food supply rely on circumstantial evidence. For example, McFadden *et al.* (1965) found that trout from infertile streams had a lower egg production. Differences in age at first spawning and growth rate, together with an actual lower fecundity, all led to a lower reproductive rate than in fertile streams. Leggett and Power (1969) correlated fecundity and food supply with land-locked salmon. The fecundity variations reported by Bagenal (1957b), Raitt (1968) and Hodder (1965) were associated with food through population density, but Kipling and Frost (1969) rejected food as a cause of pike fecundity fluctuations. Zawisza and Backiel (1970) give a graph which suggests that below a critical level of food quantity the fecundity is reduced, but also above a high level it is again reduced. Nikol'skii (1961 & 1969) reports a number of Russian papers associating food supply and fecundity, and the relationship of nutrition and fecundity was reviewed by Woodhead (1960).

However some experimental work has been done. Scott (1962) fed rainbow trout *Salmo gairdneri* different rations in a complex experimental design and showed that food reduction led to lower fecundity, probably through atresia. I found (Bagenal 1969a) in an experiment with brown trout that food reduction led to fewer but larger eggs by dry weight, and I also found that more of the fish on higher rations became mature.

Wootton (1973a) who studied the effects of food level on the egg production in the stickleback, confirmed that low food intake led to fewer eggs, and he also found that fish weight at maturity was greater and the inter-spawning interval was shorter. With his sticklebacks, however, the food level had no effect on egg size. Hester (1964) gave results of feeding experiments with the guppy in which the low diet led not only to a reduction in the number of embryos, but also to fewer succeeding batches of eggs. Warren (1973a & b) studied the effects of crowding without any food shortage. He found that increased density resulted in fewer ovarian stages and fewer young. This result was also produced by supplying a 'low density' population with water from a 'high density' one.

4.15 Fecundity and stock and recruitment

Ricker (1954, see also 1975) considered the problem of the relation of stock and recruitment to the parent stock, and he formulated an equation which generates a series of curves from one which is slightly convex to one which is dome shaped. If recruitment were determined by fecundity (which may very roughly be taken as a function of stock weight, and so is potentially density dependent) the fluctuations in stock biomass would have to be more variable to produce the observed recruitment variations. By a tortuous argument Cushing (1971, 1975) and Cushing and Harris (1973) associate the shapes of the stock recruitment curves with increasing density dependent mortality during larval life, which in turn they believe has an inverse linear relation with the cube root of fecundity. They conclude that less fecund species are less able to withstand environmental shocks, and that the more fecund fish stocks recover from overfishing more quickly. As a consequence, growth may be regarded teleologically as an agent of fecundity, i.e. that fish grow big in order to be fecund in order to stabilize their stocks under adverse conditions. It is impossible here to do justice to Cushing's hypotheses; suffice it to say that because Cushing arrives at conclusions that seem intuitively obvious, the arguments he used to get there are not necessarily correct.

Nikol'skii, Bogdanov and Lapin (1973) state that for fecundity to be a population regulator there must be evidence of a direct relationship between conditions of existence, variations in population fecundity and corresponding changes in the abundance of the broods produced. My paper (Bagenal 1973) at the same symposium starts with a different view; for fecundity to be a population regulator the fecundity corrected to a given length or weight must be shown to vary inversely with the stock density. To relate fecundity and recruitment may only be relating stock and recruitment. Nikol'skii *et al.* say that there must be evidence of (1) that the fecundity of a population varies with the conditions of existence, and (2) that a change in fecundity leads to a corresponding change in the abundance of the mature progeny. The authors believe that the first is agreed by all fishery workers, but that diametrically opposed views are held concerning the second. They also believe that if the second relationship does not exist, fluctuations in abundance must be regarded as unregulated and fortuitous. Unfortunately, the cases they quote to show the relationship, do not do so. A relationship between the stock and the resulting progeny is clear, since more adults produce more young, but the fecundity of each adult has not necessarily changed. Therefore, there is no evidence that fecundity has been a regulator.

My paper (Bagenal 1973) showed that the fecundity adjusted to a given length could vary considerably. However I could not convincingly demonstrate that they were inversely related to the population density, though there was some circumstantial evidence with plaice, haddock and pike.

Schopka and Hempel (1973) discuss the concept of total spawning potential in herrings which they equate with the egg mass (=fecundity × egg size) and show that it is more affected by fishing than is biomass because of the non-linear relation of fecundity with body weight. The higher value of b

in $F \propto W^b$, the larger the effect will be, and an increase in the growth rate will strengthen the effect as long as there is no growth effect on the fecundity—weight relation. Furthermore the positive relationship of female size on egg size, and hence survival, results in greater effects of fishing on recruitment.

4.16 Other environmental factors

Several environmental effects on fecundity are believed to act through the food supply. For example Hodder (1965) suggested that the fecundity differences of Grand Banks haddock were associated with water temperature. In 1957 the haddock concentrated in the cold water along the edge of the Banks and the resulting overcrowding led to a food shortage and a lower fecundity two years later. In 1958 the water was warmer and the haddock dispersed over the Banks a month earlier; the resulting larger food supply led to higher fecundity in 1960. That temperature itself does not directly affect fecundity is suggested by the three flatfish which I studied in the same area and found that their fecundities did not vary synchronously (Bagenal 1966). Neither did Kipling and Frost (1969) find a temperature effect on pike fecundity.

Brylinska, Platt and Skinsmont (1975) carried out complex, simple and multiple regression analyses on four fecundity indices of bream and eleven environmental factors. All the lake morphology factors were negatively correlated with fecundity, food resources and growth were positively correlated, and population density was negatively correlated with the index of individual fecundity. The individual relative fecundity of small females was influenced by the bream density, the density of all fish and by the mean lake depth.

Macer (1974) suggests a mechanism whereby the environment affects horse mackerel fecundity. With this type of serial spawner which may also resorb a large number of eggs at the end of the season, the number of shed eggs appears to be flexible since the process of asynchronous oocyte development and oocyte resorption make possible a control of the egg numbers during the season. More eggs may be released in a favourable than unfavourable season.

Zawisza and Backiel (1970) showed from a study of 94 Polish lakes, that lake type did not effect the fecundity of *Coregonus albula*. They concluded that depth, temperature and oxygen were not limiting factors in Poland. While metacercaria in the vitreum of the eyes, and *Proteocephalus longicollis* in the gut, did not appear to have an effect on fecundity, heavy infections of *Ergasilus sieboldi* on the gills seemed to depress the fecundity, but not growth, unless the fish growth in the lake was exceptional, in which case growth was affected as well.

The effect of chronic oxygen depletion on the fathead minnow *Pimephales promelas* was studied by Brungs (1971). At $1\cdot06$ mg/$1O_2$ no fish produced eggs, but at $1\cdot99$ mg/1 the number of eggs/spawning was normal but the number of spawnings was lower than in controls.

4.17 Fecundity and racial investigations

The fecundities of different stocks of the same species have been used for racial discrimination for many years particularly with Atlantic herrings by Jenkins (1902), Farran (1938) and Kändler and Dutt (1958) and Pacific herrings by Katz (1948) and Nagasaki (1958). This approach has continued to be used very successfully by European herring workers, for example Anokhina (1959, 1960, 1963), Baxter (1959, 1963), Bridger (1961), Burd and Howlett (1969), Lyamin (1956), and Oyaveyer (1974). Different races have characteristic fecundities and egg size so that the racial origins of herrings caught on a common feeding ground can be determined from egg counts, and conversely, a population can be identified as a homogenous unit and not a mixture of different stocks.

Racial fecundity differences have been reported for flatfish for many years, e.g. Reibisch (1899), Franz (1910) and Kändler and Pirwitz (1957). I investigated (Fig. 4.2) the fecundity of plaice over the whole of its geographic range (Bagenal 1966). The fecundity is expressed as that expected of a 37 cm fish $(\hat{F}_{37})$ and the lowest was found in those localities where plaice are most plentiful—the southern North Sea, Faxa Bay and the Barentz Sea. The fecundity tended to increase radially out from the southern North Sea and contour lines at $\pm20,000$ eggs (based on the confidence limits of $\hat{F}_{37}$) could be drawn. The various possible factors that might explain this pattern are discussed at length in my paper. A racial explanation is only needed for the populations in Trondeim Fjord and the Baltic where the differences are of a different order of magnitude.

Very marked fecundity differences have been found between rivers in salmonids e.g. the Atlantic salmon by Pope *et al.* (1961) and the sockeye salmon by Aro and Broadbent (1950) and by Rounsefell (1957) who re-analysed the data of Eguchi *et al.* (1954). These and other data on sockeye racial differences are well presented by Foerster (1968).

Other possible racial differences have been reported in a variety of species (Määr 1949, Toots 1951, Smyly 1957, Wolfert 1969).

Marshall (1953) discussed at length the variations in egg size and fecundity of fish from different latitudes. He concludes that arctic, antarctic and deep sea fish produce fewer and larger eggs than related boreal species. This feature is also discussed by Permitin (1973) and other examples are given by Nikol'skii (1969).

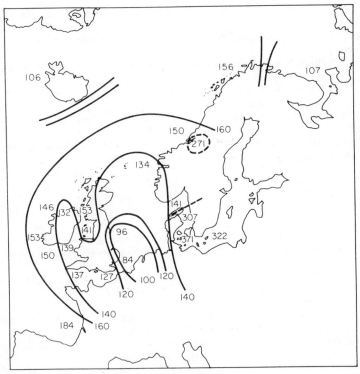

Fig. 4.2. The expected fecundity of 37 cm plaice determined from the regression of log fecundity on log length at each locality throughout the geographical range, divided into areas of similar fecundity by contours at 20,000 egg intervals. (Reproduced by permission from *Journal of the Marine Biological Association* **46**, p. 169).

References

ANDREU B. (1954) The sexuality of sardines. *Proc. tech. Pap. gen. Fish. Coun. Mediterr.* **3**, 45–60.

ANOKHINA L.E. (1959) On the relation between the fertility and fatness in the Baltic herring *Clupea harengus membras* L. *Dokl. Akad. Nauk SSSR* **129**, 1417–1420.

ANOKHINA L.E. (1960) On the relation between fertility, variation in size of eggs and fat content in the White Sea herring *Clupea harengus pallasi maris-albi* Berg. *Dokl. Akad. Nauk SSSR* **133**, 960–963.

ANOKHINA L.E. (1963) Some aspects of the fecundity of the herring in the White Sea. *Rapp. P.-v. Réun. Cons. perm. int. Explor. Mer* **154**, 123–127.

ARO K.V. & BROADBENT G.C. (1950) Differences between egg counts of sockeye salmon at Lakelse and Babine Lakes. *Prog. Rep. Pacif. Cst Stns*, No. 82, pp. 17–19.

BAGENAL M. (1955) A note on the relations of certain parameters following a logarithmic transformation. *J. mar. biol. Ass. U.K.* **34**, 289–296.

BAGENAL T.B. (1957a) The breeding and fecundity of the long rough dab *Hippoglossoides platessoides* (Fabr.) and the associated cycle in condition. *J. mar. biol. Ass. U.K.* **36**, 339–373.

BAGENAL T.B. (1957b) Annual variations in fish fecundity. *J. mar. biol. Ass. U.K.* **36**, 377–382.

BAGENAL T.B. (1963) Variations in plaice fecundity in the Clyde area. *J. mar. biol. Ass. U.K.* **43**, 391–399.

BAGENAL T.B. (1966) The ecological and geographical aspects of the fecundity of the plaice. *J. mar. biol. Ass. U.K.* **46**, 161–186.

BAGENAL T.B. (1969a) The relationship between food supply and fecundity in brown trout *Salmo trutta* L. *J. Fish Biol.* **1**, 167–182.

BAGENAL T.B. (1969b) Relationship between egg size and fry survival in brown trout *Salmo trutta* L. *J. Fish Biol.* **1**, 349–353.

BAGENAL T.B. (1971) The inter-relation of the size of fish eggs, the date of spawning and the production cycle. *J. Fish Biol.* **3**, 207–219.

BAGENAL T.B. (1973) Fish fecundity and its relations with stock and recruitment. *Rapp. P.-v. Réun. Cons. perm. int. Explor. Mer* **164**, 186–198.

BAGENAL T.B. (1977) *Methods in Fish Fecundity Estimation.* In press.

BAXTER I.G. (1959) Fecundities of winter-spring and summer-autumn herring spawners. *J. Cons. perm. int. Explor. Mer* **25**, 73–80.

BAXTER I.G. (1963) A comparison of the fecundities of early and late maturity stages of herring in the north eastern North Sea. *Rapp. P.-v. Réun. Cons. perm. int. Explor. Mer.* **154**, 170–174.

BAXTER I.G. & HALL W.B. (1960) The fecundity of Manx herring and a comparison of the fecundities of autumn spawning groups. Unpublished *I.C.E.S. Herring Committee document* No. 55.

BAXTER I.G. & POPE J.A. (1969) Annual variations in fecundity of Clyde spring spawning herring. Unpublished *I.C.E.S. Pelagic Fish (Northern) Committee* No. H: 33.

BERG A. & GRIMALDI E. (1965) Biologia delle due forme di coregone (*Coregonus sp.*) del lago Maggiore. *Memorie 1st ital. Idrobiol.* **18**, 25–196.

BEVERTON R.J.H. (1962) Long term dynamics of certain North Sea populations, pp. 242–259. (E.D. Le Cren & M.W. Holdgate, Eds.) In *The Exploitation of Natural Animal Populations.* Blackwell Scientific Publications, Oxford.

BLACKETT R.F. (1973) Fecundity of resident and anadromous dolly varden (*Salvelinus malma*) in southeastern Alaska. *J. Fish. Res. Bd. Canada* **30**, 543–548.

BLAXTER J.H.S. & HEMPEL G. (1963) The influence of egg size on herring larvae (*Clupea harengus* L.) *J. Cons. perm. int. Explor. Mer* **28**, 211–240.

BREDER C.M. & ROSEN D.E. (1966) *Modes of Reproduction in Fishes.* Natural History Press, New York.

BRIDGER J.P. (1961) On the fecundity and larval abundance of Downs herring. *Fishery Invest. Lond., Ser.* 2, **23**, 1–30.

BRUNGS W.A. (1971) Chronic effects of low disolved oxygen concentrations on the fat head minnow (*Pimephales promales*). *J. Fish. Res. Bd. Canada* **28**, 1119–1123.

BRYLINSKA M., PLATT C. & SKLINSMONT W. (1975) Analiza zaleznosa miedzy wybranymi wskaznikami plodnosci leszcza, a wskaznikami warunkow srodowiska. *Rocza Nauk. roln.* **96**, 155–170.

BURD A.C. & HOWLETT G.J. (1969) English herring fisheries, fecundity studies. Unpublished *I.C.E.S. Pelagic Fish (N) Committee document* No. H: 28.

CLARK F.N. (1934) Maturity of the Californian sardine (*Sardina caerulea*), determined by ova diameter measurements. *Calif. Fish & Game Fish. Bull.* 42. 49 pp.

CUSHING D.H. (1968) *Fisheries Biology, a study in population dynamics.* The University of Wisconsin Press, Madison.

CUSHING D.H. (1971) The dependence of recruitment on parent stock in different groups of fishes. *J. Cons. int. Explor. Mer* **33**, 340–362.

CUSHING D.H. (1975) *Marine Ecology and Fisheries.* Cambridge, England.

CUSHING D.H. & HARRIS J.G.K. (1973) Stock and recruitment and the problem of density dependence. *Rapp. P.-v. Réun. Cons. Perm. int. Explor. Mer* **164**, 142–155.

DAHL K. (1918–19) Studies of trout and trout waters in Norway. *Salm. Trout Mag.* **17**, 58–79; **18**, 16–33.

DE SILVA S.S. (1973) Aspects of the reproductive biology of the sprat, *Sprattus sprattus* (L.) in inshore waters of the west coast of Scotland. *J. Fish Biol.* **5**, 689–705.

EGUCHI H., HIKITA T. & NISHIDA H. (1954) A comparison between Hokkaido and South Kurile Islands on the salmon egg number. *Scient. Rep. Hokkaido Fish Hatch.* **9**, 151–159.

ENGLISH T.S. (1964) Estimation of the abundance of a fish stock from egg and larval surveys. *Rapp. P.-v. Cons. Perm. int. Explor. Mer* **155**, 174–182.

FARRAN G.P. (1938) On the size and number of ova of Irish herrings. *J. Cons. perm. int. Explor. Mer* **13**, 91–100.

FOERSTER R.E. (1968) The sockeye salmon. *Bull. Fish. Res. Bd. Canada* **162**.

FRANZ V. (1910) Die Eiproduktion der Scholle (*Pleuronectes platessa* L.) *Wiss Meeresunters. Helgoland, N.F.* **9**, 59–141.

FRY F.E.J. (1949) Statistics of a lake trout fishery. *Biometrics.* **5**, 27–67.

FRYER G. & ISLES T.D. (1972) *The Cichlid Fishes of the Great Lakes of Africa.* Oliver & Boyd, London.

GADGILL M. & SOLBRIG O.T. (1972) The concept of *r*- and *K*- selection: Evidence from wild flowers and some theoretical considerations. *Am. Nat.* **106**, 14–31.

GALKINA Z. (1970) Dependence of egg size on the size and age of the female salmon (*Salmo salar* (L.)) and rainbow trout (*Salmo irideus* (Gibb.)). *J. Ichtyol.* **10**, 625–633.

GERKING S.D. (1959) Physiological changes accompanying ageing in fishes. *Ciba Found. Colloq. on Ageing.* 5. *The life span of animals.* J. & A. Churchill, Ltd., London.

GJØSAETER J. & SAETRE R. (1974) The use of data on eggs and larvae for estimating spawning stock of fish populations with demersal eggs. 139–149 in Blaxter J.H.S. (Ed.) *The Early Life History of Fish.* Springer Verlag Berlin.

GORBACH E.I. (1972) Fecundity of the grass carp (*Ctenopharyngodon idella* (Val.)) in the Amur basin. *J. Ichthyol.* **12**, 616–624.

GRACHEV L. YE (1971) Alteration of the number of oocytes in the sockeye (*Oncorhynchus nerka* (Walb.)) during the marine period of its existence. *J. Ichthyol.* **11**, 897–906.

GRITSENKO O.F. (1971) Growth, maturation and fecundity of the char (*Salvelinus alpinus* (L.)) of rivers on Sakalin Island. *J. Ichthyol.* **11**, 555–568.

HAGEN D.W. (1967) Isolating mechanisms in three-spine sticklebacks (*Gasterosteus*). *J. Fish. Res. Bd. Canada* **24**, 1637–1692.

HEALEY M.C. (1971) Gonad development and fecundity of the sand goby, *Gobius minutus* Pallas. *Trans. Am. Fish. Soc.* **100**, 520–526.

HENDERSON N.E. (1963) Extent of atresia in maturing ovaries of the eastern brook trout *Salvelinus fontinalis* (Mitchill). *J. Fish. Res. Bd. Canada* **20**, 899–908.

HESTER F.J. (1964) Effects of food supply on fecundity in the female guppy *Lebistes reticulatus* (Peters). *J. Fish. Res. Bd. Canada* **21**, 757–764.

HODDER V.M. (1963) Fecundity of Grand Bank haddock. *J. Fish. Res. Bd. Canada* **20**, 1465–1487.

HODDER V.M. (1965) The possible effects of temperature on the fecundity of Grand Bank haddock. *ICNAF Spec. Publ.* **6**, 515–522.

HOLDEN M.J. (1974) Problems in the rational exploitation of elasmobranch populations and some suggested solutions. *Sea Fisheries Research.* Elek Science. (F.R. Harden Jones, Ed.), pp. 117–137.

HOLDEN M.J. & MEADOWS P.S. (1964) The fecundity of the spurdog (*Squalus acanthias* L.). *J. Cons. perm. int. Explor. Mer* **28**, 418–424.

HULATA G., MOAV R. & WOHLFARTH G. (1974) The relationship of gonad and egg size to weight and age in the European and Chinese races of common carp *Cyprinus carpio* L. *J. Fish Biol.* **6**, 745–758.

ILES T.D. (1974) The tactics and strategy of growth in fishes. *Sea Fisheries Research.* Elek Science. (F.R. Harden Jones, Ed.), pp. 331–345.

IVANOV S.N. (1971) An analysis of the fecundity and intermittent spawning of Balkhash wild carp (*Cyprinus carpio* (L.)). *J. Ichthyol.* **11**, 666–672.

JENKINS J.T. (1902) Altersbestimmung durch Otolithen bei den Clupeiden. *Wiss. Meeresunters., Abt. Kiel.* **6**, 81–121.

KÄNDLER R. & DUTT S. (1958) Fecundity of Baltic herring. *Rapp. P.-v. Réun. Cons. perm. int. Explor. Mer* **143**, 99–108.

KÄNDLER R. & PIRWITZ W. (1957) Über die Fruchtbarkeit der Plattfische im Nordsee-Ostsee-Raum. *Kieler Meeresforsch.* **13**, 11–34.

KATZ M. (1948) The fecundity of herring from various parts of the North Pacific. *Trans. Am. Fish. Soc.* **75**, 72–76.

KENNEDY W.A. (1953) Growth, maturity, fecundity and mortality in the relatively unexploited whitefish *Coregonus clupeaformis* of Great Slave Lake. *J. Fish. Res. Bd. Canada* **10**, 413–441.

KIPLING C. & FROST W.E. (1969) Variations in the fecundity of pike *Esox lucius* L. in Windermere. *J. Fish Biol.* **3**, 221–237.

KOLYUSHEV A.I. (1973) Maturation and fecundity of chars (Genus *Salvelinus*) of Lakes Imandra and Umbozero. *J. Ichthyol.* **13**, 524–536.

KRIVOBOK M.N. & STOROZHUK A.YE. (1970) The effect of egg size and age of Volga sturgeon on the weight and chemical composition of the mature eggs. *J. Ichthyol.* **10**, 761–771.

KUZNETSOV V.A. (1973) Fecundity of the bream (*Abramis brama* (L.)) and the quality of its eggs. *J. Ichthiol.* **13**, 669–679.

LEAR W.H. (1970) Fecundity of Greenland halibut (*Reinhardtius hippoglossoides*) in the Newfoundland–Labrador area. *J. Fish. Res. Canada* **10**, 1880–1882.

LE CREN E.D. (1951) The length–weight relationship and seasonal cycle in gonad weight and condition in the perch (*Perca fluviatilis*). *J. Anim. Ecol.* **20**, 201–219.

LE CREN E.D. (1958) The production of fish in fresh waters. *The Biological Productivity of Britain.* (W.B. Yapp & D.J. Watson, Eds.), pp. 67–72. *Institute of Biology Symposium No. 7.*

LEGGETT, W.C. & POWER G. (1969) Differences between two populations of land locked Atlantic salmon (*Salmo salar*) in Newfoundland. *J. Fish. Res. Bd. Canada* **26**, 1585–1596.

LOWE R.H. (1955) The fecundity of *Tilapia* species. *E. Afr. agric. J.* **21**, 45–52.

LOWE-MCCONNELL R.H. (1975) *Fish Communities in Tropical Freshwaters.* Longman, London, England.

LUDWIG G.M. & LANGE E.L. (1975) The relationship of length, age, and age–length interaction to the fecundity of the northern mottled sculpin *Cottus b. bairdi. Trans. Am. Fish. Soc.* **104**, 64–67.

LYAMIN K.A. (1956) Investigation into the life-cycle of summer spawning herring of Iceland. *Spec. scient. Rep. U.S. Fish Wildl. Serv. Fisheries*, 1959, No. 327, 166–202. (Translation from *Trudy polyar nauchno-issled Inst. morsk. ryb. Khoz. Okeanogr.*, No. 9.)

MÄÄR A. (1949) Fertility of char (*Salmo alpinus* L.) in the Faxälven water system, Sweden. *Rep. Inst. Freshwater. Res. Drottningholm* **29**, 57–70.

MACARTHUR R.C. & WILSON E.O. (1967) *The Theory of Island Biogeography.* Princeton University Press, Princeton, New Jersey.

MACER C.T. (1974) The reproductive biology of the horse mackerel (*Trachurus trachurus* (L.) in the North Sea and English Channel. *J. Fish Biol.* **6**, 415–438.

MACGREGOR J.S. (1957) Fecundity of the Pacific sardine (*Sardinops caerulea*) *Fishery Bull. Fish Wildl. Serv. U.S.* **121**, 427–449.

MACKAY I. & MANN K.H. (1969) Fecundity of two cyprinid fishes in the River Thames, Reading, England. *J. Fish. Res. Bd. Canada* **26**, 2795–2805.

MANN R.H.K. (1974) Observations on the age, growth, reproduction and food of the dace, *Leucicisus leuciscus* (L.) in two rivers in Southern England. *J. Fish Biol.* **6**, 237–253.

MARSHALL N.B. (1953) Egg size in arctic, antarctic and deep sea fishes. *Evolution.* **7**, 328–341.

MATHISEN O.A. & GUNNERØD T. (1969) Variance components in the estimation of potential egg deposition of sockeye salmon escapements. *J. Fish. Res. Bd. Canada* **26**, 655–670.

MAY A.W. (1967) Fecundity of Atlantic cod. *J. Fish. Res. Bd. Canada* **24**, 1531–1551.

McFADDEN J.T. (1961) A population study of the brook trout, *Salvelinus fontinalis. Wildl. Monogr., Chestertown* **7**, 1–73.

McFADDEN J.T. & COOPER E.L. (1964) Population dynamics of brown trout in different environments. *Physiol. Zööl.* **37**, 355–363.

McFADDEN J.T., COOPER E.L. & ANDERSEN J.K. (1965) Some effects of environment on egg production in brown trout (*Salmo trutta*). *Limnol. Oceanogr.* **10**, 88–95.

MOORE J.W. (1975) Reproductive biology of anadromous arctic char *Salvelinus alpinus* (L.) in the Cumberland Sound of Baffin Land. *J. Fish Biol.* **7**, 143–151.

MOROZ V.M. (1970) Biological description of the vimba from the lower reaches of the Danube. *J. Ichthyol.* **10**, 29–39.

NAGASAKI F. (1958) The fecundity of Pacific herring (*Clupea pallasi*) in British Columbia coastal waters. *J. Fish. Res. Bd. Canada* **15**, 313–330.

NIKOL'SKII G.V. (1961) On some adaptations to the regulation of population density in fish species with different types of stock structure. *The Exploitation of Natural Animal Populations.* (E.D. Le Cren & M.W. Holdgate, Eds.), Blackwell Scientific Publications, Oxford.

NIKOL'SKII G.V. (1969) *Theory of Fish Population Dynamics as the Biological Background for Rational Exploitation and Management of Fishery Resources.* (Translated by Bradley, J.E.S.) Oliver & Boyd, London.

NIKOL'SKII G.·[V.], BOGDANOV A. & LAPIN YU (1973) On fecundity as a regulatory mechanism in fish population dynamics. *Rapp. P.-v. Réun. Cons. perm. int. Explor. Mer.* **164**, 174–177.

OYAVEYER E.A. (1974) The fecundity of autumn Baltic herring (*Clupea harengus membras*) populations in the north eastern Baltic. *J. Ichthyol.* **14**, 552–560.

PARRISH B.B. & SAVILLE A. (1965) The biology of the north-east Atlantic herring populations. *Oceanogr. mar. biol. A. Rev.* **3**, 323–373.

PERMITIN I.E. (1973) Fecundity and reproductive biology of icefish (*Chaenichthidae*), fish of the family Muraenolophidae and dragonfish (*Bathydraconidae*) of the Scotia Sea (Antarctica). *J. Ichthyol.* **13**, 204.

PETERS H.M. (1963) Eizahl, Eigewicht, und Gelegeentwicklung in der Gattung *Tilapia* (Cichlidae, Teleostei). *Int. Revue ges. Hydrobiol. Hydrogr.* **48**, 547–576.

PETROVA E.G. (1960) On the fecundity and maturation of the Baltic sprat. *Trudȳ vses. nauchno-issled. Inst. morsk. rȳb. Khoz. Okeanogr.* **42**, 99–108.

PITCHER T.J. & MACDONALD P.D.M. (1973) A numerical integration method for fish population fecundity. *J. Fish Biol.* **5**, 549–553.

PITT T.K. (1964) Fecundity of the American plaice, *Hippoglossoides platessoides* (Fabr.) from the Grand Bank and Newfoundland areas. *J. Fish. Res. Bd. Canada* **21**, 597–612.

POLDER J. & ZIJLSTRA J.J. (1959) Fecundity of North Sea herring. Unpublished *ICES Herring Committee document 1959*: 84.

POPE J.A., MILLS D.H. & SHEARER W.M. (1961) The fecundity of the Atlantic salmon (*Salmo salar* Linn.). *Freshwat. Salm. Fish. Res.* **26**, 1–12.

RAITT D.F.S. (1968) The population dynamics of the Norway pout in the North Sea. *Mar. Res.* 1968. **(5)**, 1–24.

RAITT D.F.S. & HALL W.B. (1967) On fecundity of the redfish *Sebastes marinus* (L.). *J. Cons. Perm. int. Explor. Mer* **31**, 237–245.

RAO V.R. (1971) Spawning behaviour and fecundity of the Indian mackerel, *Rastrelliger kanagurta* (Cuvier), at Mangalore. *Indian J. Fish.* **14**, 171–186.

REIBISCH J. (1899) Ueber die Eizahl bei *Pleuronectes platessa* und die Altersbestimmung dieser Form aus den Otolithen. *Wiss. Meeresunters. Abt. Keil* **4**, 231–248.

RICKER W.E. (1954) Stock and recruitment. *J. Fish. Res. Bd. Canada* **11**, 559–623.

RICKER W.E. (1973) Linear regressions in fishery research. *J. Fish. Res. Bd. Canada* **30**, 409–434.

RICKER W.E. (1975) Computation and interpretation of biological statistics of fish populations. *Bull. Fish. Res. Bd. Canada* 191.

ROUNSEFELL G.A. (1957) Fecundity of the North American salmonidae. *U.S. Fish and Wildlife Serv. Fish Bull.* **57**, 451–468.

SAVILLE A. (1964) Estimation of the abundance of a fish stock from egg and larval surveys. *Rapp. P.-v. Réun. Cons. Perm. int. Explor. Mer* **155**, 164–170.

SCHOPKA S.A. & HEMPEL G. (1973) The spawning potential of populations of herring (*Clupea harengus* L.) and cod (*Gadus morhua* L.) in relation to the rate of exploitation. *Rapp. P.-v. Réun. Cons. Perm. int. Explor. Mer* **164**, 178–185.

SCOTT D.P. (1962) Effect of food quantity of fecundity of rainbow trout *Salmo gairdneri*. *J. Fish. Res. Bd. Canada* **19**, 715–731.

SETTE O.E. (1943) Biology of the Atlantic mackerel (*Scomber scrombrus*) of North America. I. Early life history including the growth, drift and mortality of egg and larval populations. *U.S. Fish and Wildlife Serv. Fish Bull.* **50**, 149–234.

SHATUNOVSKII M.I. (1964) Some factors in the fecundity dynamics of two populations of flounder (*Pleuronectes flesus* L.) *Nauch. Dokl. vyssh. Shk.* **1**, 27–30.

SIMPSON A.C. (1951a) The fecundity of the plaice. *Fishery Invest., Lond., Ser.* 2, **17**, 1–27.

SIMPSON A.C. (1959a) The spawning of the plaice in the North Sea. *Fishery Invest., Lond., Ser.* 2, **22**, 1–111.

SIMPSON A.C. (1959b) Method used in separating and counting the eggs in fecundity studies on the plaice (*Pleuronectes platessa*) and herring (*Clupea harengus*). *F.A.O. Indo-Pacific Fisheries Council. Occasional paper* 59/12.

SMYLY W.J.P. (1957) The life history of the bullhead or Miller's thumb (*Cottus gobio* L.). *Proc. zool. Soc. Lond.* **128**, 431–453.

SOUTHWOOD T.R.E. (1976) Bionornic strategies and population parameters. *Theoretical Ecology: principles and applications.* (R. M. May, Ed.) Blackwell Scientific Publications, Oxford.

SPRENT P. (1966) A generalised least-squares approach to linear functional relationships. *Jl. R. statist. Soc., Ser.* B, **28**, 278–297.

STAPLES D.J. (1975) Production biology of the upland bully (*Philypnodon breviceps*) Stokell in a small New Zealand lake. *J. Fish Biol.* **7**, 1–24.

SVÄRDSON G. (1949) Natural selection and egg number in fish. *Rep. Inst. Freshwat. Res. Drottningholm* **29**, 115–122.

TEMPLEMAN W. (1944) The life history of the spiny dogfish (*Squalus acanthias*) and the vitamin A values of dogfish liver oil. *Res. Bull. Div. Fish Resour. Newfoundld.* **15**.

TERLECKIE J. (1973) Plodnosc Szczupaka—*Esox lucius* (Linnaeus 1758) z jeziora sniardwy. *Roczn Nauk roln.* **95**, 161–176.

THOMPSON W.F. & VAN CLEVE R. (1936) Life history of the Pacific halibut 2. Distribution and early life history. *Rep. int. Fish. Comm.* No. 9.

Toots H. (1951) Number of eggs in different populations of whitefish (*Coregonus*). *Rep. Inst. Freshwat. Res. Drottningholm* **32**, 133–138.

Vladykov V.D. (1956) Fecundity of the speckled trout (*Salvelinus fontinalis*) in Quebec lakes. *J. Fish. Res. Bd. Canada* **13**, 799–841.

Volodin V.M., Mezhin F.I. & Kuz'mina V.V. (1974) An experimental study of egg resorption in bream (*Abramis brama*). *J. Ichthyol.* **14**, 219.

Wagner C.C. & Cooper E.L. (1963) Population density, growth, and fecundity of the creek chubsucker *Erimyzon oblongus*. *Copeia*, *1963*, 350–357.

Warren E.W. (1973) Modification of the response to high density conditions in the guppy *Poecilia reticulata* (Peters). *J. Fish Biol.* **5**, 737–757.

Welcomme R.L. (1967) The relationship between fucundity and fertility in the mouth-brooding cichlid fish *Tilapia leucosticta*. *J. Zool. Lond.* **151**, 453–468.

Winters G.H. (1971) Fecundity of the left and right ovaries of the Grand Bank capelin (*Mallotus villosus*). *J. Fish. Res. Bd. Canada* **28**, 1029–1033.

Wolfert D.R. (1969) Maturity and fecundity of walleyes from the eastern and western basins of Lake Erie. *J. Fish. Res. Bd. Canada* **26**, 1877–1888.

Woodhead A.D. (1960) Nutrition and reproductive capacity in fish. *Proc. Nutr. Soc.* **19**, 23–28.

Wootton R.J. (1973a) The effect of size of food ration on egg production in the female three spined-stickleback *Gasterosteus aculeatus* L. *J. Fish. Biol.* **5**, 89–96.

Wooton R.J. (1973b) Fecundity of the three-spined stickleback *Gasterosteus aculeatus* L. *J. Fish Biol.* **5**, 683–688.

Wydoski R.S. & Cooper E.L. (1966) Maturation and fecundity of brook trout from infertile streams. *J. Fish. Res. Bd. Canada* **23**, 623–649.

Zawisza J. & Backiel T. (1970) Gonad development, fecundity and egg survival in *Coregonus albula* L. *Biology of Coregonid Fishes.* (C.C. Lindsey & C.S. Woods, Eds.), pp. 363–397. University of Manitoba, Press, Winnipeg, Canada.

Zotin A.I. (1968) Relative fecundity (definition and adaptive variation) *J. Ichthyol.* **8**, 143–146.

Chapter 5: Ecological Aspects of the Survival of Fish Eggs, Embryos and Larvae

Erich Braum

5.1 Introduction

Investigations of the dynamics of fish populations have revealed a great variability in survival from egg to mature fish in various species as well as a considerable variability in survival from year to year for the same species. For example, mortality during the development from egg to adult fish is between 99·87 and more than 99·99% for some freshwater species (Elster 1944, Nikol'skii 1963, Nümann 1961, Wagler 1933). However, the enormous mortality is correlated with high fecundity which provides a sufficient but fluctuating number of survivors. Fluctuations of year class strength are well known from freshwater and marine environments and fisheries biologists assume that a particularly high mortality occurs in the pre-recruit phase.

As rearing experiments show, eggs, embryos and larvae have very sensitive stages where mortality increases considerably if conditions are not optimum. Nevertheless, population parameters of these early phases are scarce. Le Cren (1962) has surveyed the mortality rates of various age groups of freshwater fish using data from Allen (1951), Frost and Smyly (1952), Ricker and Foerster (1948) and his own investigations. He summarizes the difficulties of estimating survival rates during the first year of life when mortality is greater than at any other time. Recently Bannister *et al.* (1974) investigated abundance and mortality of planktonic plaice (*Pleuronectes platessa*) eggs and larvae in the Southern Bight of the North Sea. From the earliest egg stage just prior to metamorphosis, mortality was in excess of 99% in 'normal' years and occurred within the first 130 days of life.

The quantitative analysis of mortality is difficult, but assigning the causes of death is an even greater problem. Eggs and larvae are ontogenetic stages which exhibit rapid changes in morphological development as well as in physiological demand. Synergistic interrelationships of numerous external factors influence the developing organism. The experimental approach to elucidate the relative importance of these factors in a natural ecosystem is extraordinarily complicated because neither the development of the eggs, embryos and larvae nor the environmental factors, which normally show irregular or rhythmic variations, are constant. Though experiments can only solve the problem step by step, they are needed badly in future research on the early life history of fish. Assessment of the chances of survival could be used to great advantage for rearing and stocking, as is practiced in fisheries management and aquaculture.

5.2 Ecological aspects of egg development of oviparous fishes

Fishes can be classified on the basis of their tolerance to environmental factors (Nikol'skii 1963). Apart from this physiological point of view, Kryzhanovsky (1948) and Breder and Rosen (1966) used modes of reproduction as a system for an ecological classification. Recently Balon (1975) surveyed this possibility, based on the literature and his own investigations, and revised earlier classifications.

Most fish are oviparous; eggs and sperm are expelled into the surrounding water, where fertilization occurs immediately and development begins promptly. Hoar (1969) has summarized the variations of fish reproduction and found that 'there is an amazing array of curious modifications so that the fishes as a group exemplify almost every device known among sexually reproducing animals; indeed, they display some variations which may be unique in the animal kingdom'. However, hermaphroditism and gynogenesis are very unusual, and only a minority of teleost species has achieved ovoviviparity or true viviparity.

Reproduction cycles are adapted to the seasons, which guarantees the most favorable conditions for the survival of eggs and larvae. Growth of gonads and timing of spawning are regulated by endocrine processes dependent on annual cycles of temperature, photoperiod, rainy seasons and others. According to their spawning time, fishes of temperature latitudes can be grouped into (1) spring spawners, (2) summer spawners and (3) fall and winter spawners. This classification is only valid for a limited geographic area, as demonstrated by Nikol'skii (1963) for the capelin (*Mallotus villosus*), which spawns from March to May at western Murman and in August in the eastern region of the Barents Sea. Some tropical species have no restricted spawning times, spawning the year round.

Among oviparous fishes there are species which display a more or less intensive parental care for their eggs, and species which leave the eggs to the fate of the waves and currents. Fishes of the first group have developed a number of specializations to guard the eggs. Some of them stick their eggs to a substratum and defend them against predators. This behaviour is typical of cichlids (*Cichlidae*). Others build a nest of plants into which eggs are spawned and protected by the male, as is the case for sticklebacks (*Gasterosteidae*), *Gymnarchus* (Gymnarchidae) or *Crenelabrus* (Labridae). However, nests can also be simple bottom pits built by the fish, e.g. sunfishes (Centrarchidae). Others carry their developing eggs fastened on the skin (*Solenosto-*

mus, Aspredo, Kurtus) where various anatomical differentiations exist for this purpose. Some parents keep the eggs in their buccal cavity, as do the mouth-breeding cichlids. All of these modes of carrying the eggs or guarding the nest obviously improve the chances for survival and are correlated with well defined behaviour patterns. The diversity of adaptations is amazing. The most curious example is perhaps that of *Copella arnoldi* (Lebeasinidae), living in the Amazon area. Males and females jump simultaneously out of the water and spawn sticky eggs on plants roughly 10 cm above the surface. At intervals the guarding male splashes water on the eggs to keep them wet. In general, species which display distinct parental care are characterized by low egg numbers but the low fecundity is compensated by higher survival rates.

Though the breeding habits of only about 300 species are well-described from 20,000 known fish species (Breder & Rosen 1966), it is clear that the majority restricts parental care to the choice of a suitable spawning site which favours egg development and high survival rates. Kryzhanovsky (1948) cate-gorized 5 ecological groups based on the preferred spawning substrate: (1) lithophilous species, which spawn on stoney ground or gravel, as many salmonids do, (2) psammophiles, typically represented by mugilids, which spawn on sandy bottom, (3) phytophilous fishes, which spawn among plants to which the eggs adhere. Most cyprinid species and the pike belong to this guild. (4) Pelagophiles, which release their nonadhesive eggs in the water column, have a high fecundity to compensate for the tremendous mortality during early development. Among freshwater fishes, for example, species of shad, *Alosa sp.* and whitefish, *Coregonus sp.*, spawn near the surface without reference to the type of bottom. This group includes nearly all marine fish families of economic importance, such as Gadidae, Pleuronectidae and Clupeidae (except *Clupea*). The last group (5) is represented by the ostra-cophilous bitterling (*Rhodeus*), which spawns its eggs into the mantle cavity of mussels by means of an ovipositor.

Fertilization occurs externally for the majority of oviparous fish. A more or less distinct courtship behaviour synchronizes spawning readiness, and gametes are extruded simultaneously. Fertilizability of eggs and sperm is time-limited. The fertilizable period of salmon eggs, for example, is very short; the proportion fertilized drops sharply with time and the maximum time period ranges between 15 and 30 minutes (Yamamoto 1951). The short viability of sperms, also restricts the fertilizable period, for example this period is only about 30 seconds in rainbow trout, *Salmo gairdneri*, (Billard *et al.* 1974, Ginsburg 1963).

The physiology of fertilization has been reviewed by Yamamoto (1961) and by Blaxter (1969b). In spite of the restricted time of fertilizability of salmonid eggs, excavations of redds revealed a high degree of natural fertili-zation. Hobbs (1937, 1940) found an average fertilization of 98·2% in New Zealand Chinook salmon (*Oncorhynchus tshawytscha*) redds and of 98·9% in

brown (*Salmo trutta*) and rainbow trout redds. Briggs (1953) found 94·0%
on the average from redds of king salmon, silver salmon (*O. kisutch*) and
steelhead trout (*Salmo gairdneri*) in a California stream, and Roth and
Geiger (1961) found 92,5 to 98,4% fertilized eggs of brown trout in Swiss
streams. Similarly high proportions of fertilized eggs have also been obseved
in other fish groups; the conclusion is clear that low fertilization is not
responsible for the high mortality of deposited eggs.

Nikol'skii (1962) mentioned the importance of different egg and sperm
quality at the intraspecific level. For example, yolk reserve and fat content
of crucian carp (*Carassius auratus*) and roach (*Rutilus rutilus*) eggs vary
with the age of spawning females. Sufficient yolk reserve improves the chances
of larval survival. Blaxter (1969b) pointed out that correlation between
fecundity and dry weight of eggs in Atlantic herring (*Clupea harengus*) races.
He believes that large eggs and low fecundity of winter–spring spawners are
adaptations to poor supply for the young and low predator populations.
In summer–autumn spawning races, fecundity is high and eggs are smaller.
The food supply is better at this time but predators are numerous.

The problem of the interrelationship of the egg volume, which represents
the yolk reserve of hatching larvae, the date of spawning and the seasonal
production cycle was recently surveyed by Bagenal (1971). A comparison of
smallest and largest volumes of planktonic eggs of many marine species
revealed that differences of more than 100% are very common. Similar
differences are found in freshwater fishes. In spawning groups of several
marine fishes, egg volume declined from early spring through summer,
presumably due to older females with larger eggs spawning early in the
season. A correlation in time between the production cycle of an area, which
guarantees adequate food abundance for only a restricted time, and the
onset of external feeding is most probable for herring and plaice larvae.

With respect to specific gravity, two egg types exist among Osteichthyes:
the buoyant of planktonic egg, which is very common in marine fish families,
and the non-buoyant type, which is common in freshwater fishes. Pelagic
eggs have some morphological characteristics which reduce their specific
gravity. Floating qualities are improved by (1) one or more oil globules in
the yolk, (2) an enlarged perivitelline space or a high water content in the
yolk, or (3) an increase in the egg diameter owing to a gelatinous outer layer
of the chorion. The well known marine fish families Gadidae, Pleuronectidae,
Soleidae, Scombridae, Mullidae and most Clupeidae have buoyant eggs.
Pommeranz (1973) investigated the abundance of fish eggs in the uppermost
water layers of the southern North Sea. He found the vertical distribution
to be a function of wind and developmental stage. In waters with a calm
surface, recently spawned eggs were more abundant near the surface and
older developmental stages of plaice floated deeper. This indicates a change
in specific gravity during development. Pelagic eggs in freshwater are scarce.

Examples are known from tropical fish, e.g. giant gourami (*Helostoma temmincki*), climbing perch (*Anabas testudineus*), gourami (*Colisa fasciata, C. lalia*) and from the serpenthead fish (*Channa spp.*). Davis (1959a) described the pelagic egg of the sheepshead (*Aplodinotus grunniens*) from Lake Erie, U.S.A. From rivers of Eastern Asia Nikol'skii (1963) listed the following species of cyprinidae with buoyant eggs: grass carp (*Ctenopharyngodon idellus*, the black Chinese roach (*Mylopharyngodon piceus*), the bighead (*Hypophthalmichthys molitrix*), *Aristichthys nobilis* and the razor fish (*Pelecus cultratus*) and the genera *Saurogobio* and *Rostrogobio*. Normally the eggs float downstream during the season of rising water level and are carried into flood plain lakes, creeks and channels with calm currents where the larvae find a favourable food supply.

Most freshwater fishes have demersal eggs with a specific gravity greater than freshwater. Many of them are temporarily adhesive but the period of adhesiveness is short and restricted to the time immediately after expulsion. Cyprinid species which lay their eggs on aquatic plants are typical of this group. The eggs are prevented from floating away and are located above the bottom mud, thereby insuring a sufficient water circulation over the surface. Spawning beds of herring contain layers or clumps of eggs which stick together and which cover large areas. The Pacific herring (*Clupea harengus pallasi*) has extensive spawning areas along the British Columbia coast. Outram *et al.* (1974) observed an egg density of 1000 eggs per square inch, with the eggs stuck together in many layers. The proportion of larvae hatching from these eggs was lower from heavy depositions up to 16 layers thick than from moderate ones of 2–4 layers. Insufficient water exchange and oxygen supply to deeper layers may well be a reason for increased mortality during development in Pacific and Atlantic herring (Bowers 1969).

5.3 Effect of ecological factors on egg development and hatching

In the majority of teleostean fishes development of the meroblastic egg starts immediately after insemination (Fig. 5.1). Detailed descriptions of cleavage, epibolic overgrowth and morphogenesis are given in standard textbooks such as New (1966), where culturing methods of *Salmo, Oryzias* and *Fundulus* eggs for purposes of experimental embryology are described.

HATCHING

A softening of the chorion by larval enzymes in combination with turning movements of the larva within the egg produces a rupture of the chorion and the larva emerges. The literature on biochemical aspects of hatching

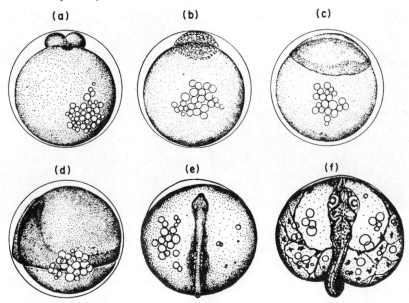

Fig. 5.1. Characteristic stages of fish egg development (*Fundulus heteroclitus*). (From New 1966.) (**a**) After fertilization a cytoplasmatic condensation occurs in the micropylar region forming a single cell which after division forms the 2-cell stage 2·5 hours after fertilization (20C). (**b**) With an increase in cell numbers their size is reduced (10 hours). (**c**) The blastoderm expands over the yolk and the thickening of the rim is forming the embryonic shield at a region where grastrulation begins and the future embryo is formed (27 hours). (**d**) The yolk surface is nearly covered by the ectoderm and the keel of the central nervous system becomes marked (37 hours). (**e**) The embryo is well defined by the eye vesicles and the brain divisions, four pairs of somites are visible, the yolk is completely covered by the blastoderm (52 hours). (**f**) The heart beats and the circulation has begun; a lens is characteristic for the eye and melanophores on the yolk sac surface are expanding; 50% of the incubation time from fertilization to hatch is over, the age of the embryo is 102 hours.

enzymes and the composition of the chorion was reviewed by Blaxter (1969b). Poy (1970) compared the hatching behaviour of ten freshwater fish species and found four dominant types of hatching movements. The last of these is a quick strong movement which breaks the brittled chorion and the larvae may hatch tail or head first. In some cichlid species parent fish open the chorion of the eggs they guard. Scalare (*Pterophyllum scalare*) takes the eggs into its mouth, causing the eggs to burst by gentle chewing movements, and then carefully places the larvae upon plants.

The egg surface is a suitable substratum for protozoa and microorganisms. For example, considerable accumulations of bacteria were observed on herring eggs which were obtained from breeding experiments in a bight of the Baltic Sea (Fig. 5.2). It is not quite clear to what extent the eggs are threatened by bacteria in the outer layers of the egg shell, but one can assume a decreasing stability of the chorion owing to bacterial activities and an

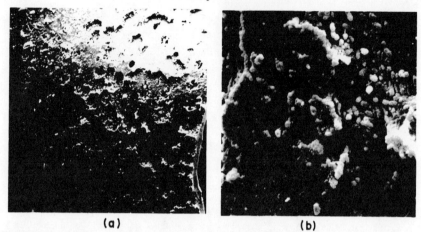

(a)   **(b)**

Fig. 5.2. Surface of an egg of Baltic herring (*Clupea harengus*). Eggs were artificially inseminated, attached to plates and exposed in a spawning site near Travemünde. After seven days the outer layer had (a) flat areas of deteriation obviously caused by bacteria (b) Magnification: (a) × 324, (b) × 3240. (SEM foto K. Hoffmann, Zool. Inst. Hamburg, Braum unpublished).

increased oxygen demand of the egg. According to recent investigations of Hamor and Garside (1973), an intramembranal oxygen utilization in the egg shell of Atlantic salmon (*Salmo salar* L.) does exist. It is the first reference to such dynamic properties of the chorion. Undoubtedly these results will be very important for further considerations of respiratory exchange and osmoregulatory properties of the chorion.

The incubation time from fertilization to hatching varies considerably among teleostean fishes from roughly two days (*Roccus saxatilis*) to one year (*Agonus cataphractus*). Table 5.1 gives some examples of incubation times.

Table 5.1 Duration of egg development from fertilization to hatch, in various species

Species	Days	°C	Author
Roccus saxatilis	2	19·5	Bigelow & Schroeder (1953)*
Cyprinus carpio	10	15·0	Schäperclaus (1961)
Osmerus eperlanus	20	8·5	Lillelund (1961)
Salmo gairdneri	75	4·5	Embody (1934)
Coregonus wartmanni	100	2·1	Braum (1964)
Agonus cataphractus	340–360	4–17†	Lange (1973)

* Reference from Breder and Rosen (1966)
† Incubation time and temperature are based on data from spawning site.

EFFECT OF TEMPERATURE

The egg development of a species tolerates a certain temperature range within the limits of which viable larvae hatch. This range for many species is roughly 10–15C. Incubation time is prolonged by low temperatures and accelerated by high temperatures. Hatching does not occur at a fixed onto-genetic stage, therefore thermic acceleration or prolongation affects some morphological features of the hatching larvae, such as length, yolk mass, meristic patterns and differentiation of the jaws, which have an important bearing on survival. As will be shown below, larval survival chances de-pend to a high degree on frequent encounters with adequate food or-ganisms.

The product of time in days needed to reach a certain developmental stage and the temperature in degrees Celsius, called degree-days, was thought to be constant (Apstein 1909). This is only true for a rather narrow range of suitable temperatures, and the application of degree-days for a wide range of temperatures represents only a rough relation. For many practical purposes this approximation between temperature and egg development is sufficient, but from the more physiological aspect the exponential relation between temperature and development should be examined closely (see Fig. 5.3 for

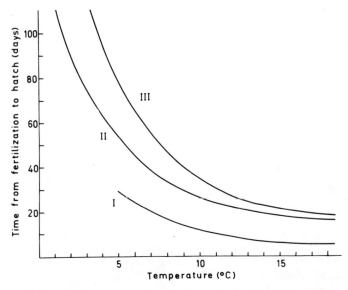

Fig. 5.3 Duration of development from fertilization to hatch of 3 species at different temperatures. I *Esox lucius* $D = 4 + 1 \cdot 29^{19-t}$ (Lindroth 1946), II *Coregonus wartmanni* $D = 15 + 1 \cdot 26^{20.7-t}$ (Braum 1964), III *Salmo gairdneri* $D = 15 + 1 \cdot 26^{23-t}$ (Lindroth 1946).

examples of correlation between temperature and incubation time). The theoretical background was reviewed by Blaxter (1969b).

Short term fluctuations of temperature can be severe under natural conditions. Spawning sites in shallow waters, e.g. overflow areas in river systems, can have day/night variations of 10C and more. Embryonic development is more accelerated by a temperature increase of 5C than retarded by a decrease of 5C, due to the exponential relationship between temperature and development. This explains the differences between calculated and observed incubation time when temperature fluctuations exist and the calculation is based on a mean value of temperature.

EFFECTS OF OXYGEN CONCENTRATION

In addition to temperature the oxygen supply is a factor of great ecological relevance. The developing egg has an increasing oxygen demand which reaches its maximum several days before hatching (Fig. 5.4). The manometric

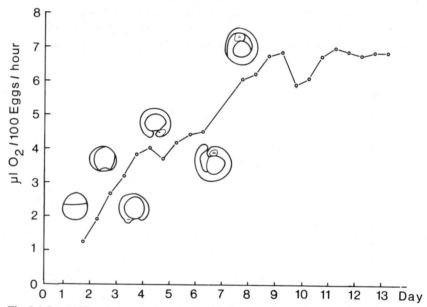

Fig. 5.4. Increasing oxygen consumption by herring eggs during development. Data were obtained by the Warburg-method. Eggs hatched 15 days after fertilization at 8C and a salinity of 15‰ (Braum 1973).

measurements of oxygen consumption of herring eggs revealed a rapid increase during the first four days (Braum 1973). At 8C and a salinity of $15°/_{oo}$, oxygen uptake by eggs from the Baltic spring spawn reached $2·7\mu$ O_2/hour per 100 eggs during the following days. Roughly 30% of total oxygen uptake was needed during the first seven days of incubation and the remaining 70% during the following seven days before hatching, when embryos twist and turn in the egg shell.

Most of the fundamental investigations about oxygen supply and respiration of fish eggs were done with salmonid species (Alderdice, Wickett & Brett 1958, Garside 1959, Hamdorf 1961, Gottwald 1965) or cyprinodontids (Kinne 1962). The 'critical O_2 partial pressure' is the lowest concentration needed for meeting the normal oxygen requirements. If it falls below this critical level, the egg attains a certain chance of survival by reducing its respiration and growth rate, but the chance decreases with the increasing oxygen demand at the advanced developmental stage.

A fundamental condition for oxygenation of eggs is the water circulation at the egg surface. It is generally assumed that the oxygen demand cannot be supplied by diffusion only. A minimum water flow is needed, depending on the oxygen gradient between the egg surface and the surrounding water. Theoretical considerations of the problem are given by Daykin (1965), who calculated dissolved oxygen requirements for eggs of chinook salmon (*Oncorhynchus tshawytscha*) and steelhead trout at various water velocities and compared these results with experimentally determined dissolved oxygen requirements. Critical conditions for salmon eggs in spawning beds were observed by Wickett (1954). He found high mortalities of chum salmon (*Oncorhynchus keta*) eggs in the pre-eyed stage in certain areas of extremely low oxygen content, e.g. 0·2 ppm, or very low subsurface flow of only a few millimeters per hour. The tolerance of steelhead trout and chinook salmon embryos at different oxygen supply and water velocities was investigated in the laboratory by Silver *et al.* (1963). They observed a decreasing hatching size of larvae when water velocity was very slow during incubation.

Field investigations revealed that herring eggs stick together forming clumps of about 10 egg layers (Bowers 1969, Hempel & Schubert 1969, Outram & Humphreys 1974). Water circulation among eggs is obviously reduced in the deeper layers, which can lead to an oxygen shortage for the innermost eggs. The hypothesis agrees with the retardation in development of the lower egg layers. It is well known from experiments (Garside 1959, Devillers 1965, Hamdorf 1961, Braum 1973) that sublethal oxygen shortage retards morphogenesis and can cause developmental failures.

Some pelagic species of the whitefish subfamily, Coregoninae (e.g. *Coregonus wartmanni*), of the prealpine European lakes spawn in the pelagic region of lakes. The eggs of *C. wartmanni* in Lake Constance are demersal and sink to the bottom at 100 m to 250 m depth. The eggs have an average diameter of 2·3 mm and develop at a rather constant temperature of 4C, in total darkness and within the mud-water boundary of the lake, where the oxygen content is usually low. Preliminary experiments suggest that these conditions extend the incubation time and increase mortality. At 100 m depth, where the temperature was 4C, eggs hatched in 105 days instead of 65 days required in a harchery at the same temperature (Braum 1967). Further investigations are necessary to clarify whether considerable delays in this range are a general phenomenon.

EFFECT OF LIGHT

Physiological investigations on the influence of light on fish eggs indicate a changing sensitivity during different egg stages and a reduced growth of rainbow trout embryos (Hamdorf 1960). Under natural conditions, harmful effects of light have not been observed.

5.4 Morphological and behavioral adaptations of embryos and larvae to their environment

THE ELEUTHEROEMBRYO PHASE

I follow the terminology of Balon (1975) here and adopt the term eleuthero-embryo for the phase just after hatching and before exogenous feeding. The word 'embryo' will be used throughout this section in preference to 'pre-larva', which is used in the older literature. The term 'larva' is reserved for the period after external feeding begins. The degree of morphological differentiation varies among newly hatched embryos of different species. The bream (*Abramis ballerus*) and the European pike (*Esox lucius*) are examples of free-living embryos hatching at a relatively early developmental stage. Those of oral incubation species may hatch at a still earlier stage, e.g. embryos of *Labeotropheus* have an 'enormous yolk sac in relation to the size of the developing larva which sits helpless on the top of a mountain of food' (Fryer & Iles 1972) (Fig. 5.5).

Embryos of mouthbrooders continue their development well-protected in the parental mouth. Free living embryos often go through a period of passivity with only slight locomotion. Various species (Ilg 1952) have special dermal glands in the head which produce mucous for attachment to the substratum. This prevents them from drifting away and, compared to the egg stage respiration conditions are improved by direct contact of water with the body surface. Pelagic embryos of marine fishes which hatch at a very early developmental stage, e.g. *Caranx*, keep their buoyancy by oil droplets or a high water content of the yolk. The embryos of the whitefish genus *Coregonus* and the herring start swimming soon after hatching and migrate from the hatching site at the bottom into the water column by undulatory movements of their relatively long trunk.

Other than the yolk sac, typical features of fish embryos are the embryonic fin membrane, which usually surrounds the whole trunk, relatively big eyes

Fig. 5.5. Opposite. Newly hatched larvae of different species with a varying degree of development (a) mouthbreeding cichlid fish (*Labeotropheus*) (Fryer 1959); (b) pike larvae (*Esox lucius*) (Braum 1964); (c) arctic char (*Salvelinus alpinus*) (Braum unpublished); (d) Pink salmon (*Oncorhynchus gorbuscha*) (Nikol'skii 1963); (e) armed bull-head (*Agonus cataphractus*) (Lange 1973). Drawings not to same scale.

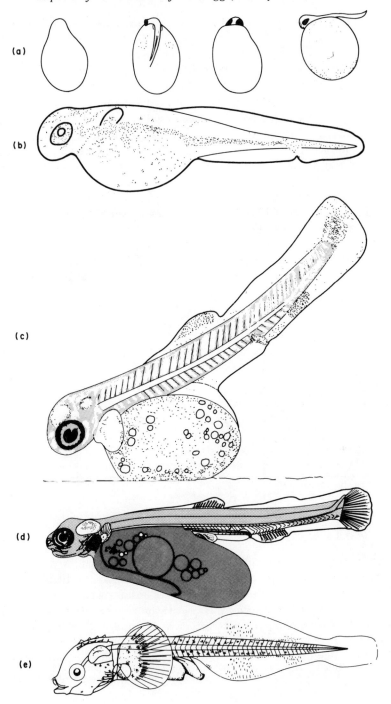

and pectorals. Respiratory blood vessels form a netlike distribution on the yolk of many species and some others, such as *Misgurnus fossilis, Gymnarchus niloticus, Polypterus* and the lung fishes *Protopterus* and *Lepidosiren*, have external larval gills forming filaments or a pseudobranch, which function as gills until the definitive respiratory organs are developed. Respiratory systems of larvae are adapted to the special oxygen conditions of the environment. Nikol'skii (1963) describes the difference in yolk sac blood vessels of closely related species. The better-developed capillary system of the embryos of the chum salmon as compared with the pink salmon (*Oncorhynchus gorbuscha*) is correlated with the lower oxygen content of the water in which they live.

An example of a high degree of development at the hatching stage is the embryo of the armed bull-head (*Agonus cataphractus*) (Fig. 5.6), whose yolk is nearly completely absorbed.

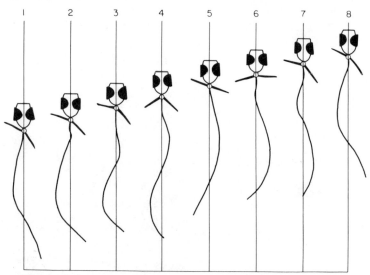

Fig. 5.6. Locomotion pattern of *Coregonus wartmanni* larvae during one undulation period (Braum 1964). From moving pictures.

Temperature during incubation influences body length, yolk sac size, pigmentation and differentiation of the mouth. Most investigations have shown that the body length of embryos increases as eggs are incubated at progressively lower temperatures. But the maximum length of sardine (*Sardinops caerula*) embryos was reached at intermediate temperatures (Lasker 1964), which could be explained by a temperature optimum for yolk utilization. Length variations induced by incubation temperature can reach as much as 4·2 mm for *Coregonus clupeaformis*, or 48% of minimum size (Table 5.2).

Table 5.2. The influence of incubation temperature on the hatching size of some fish larvae (based on Blaxter 1969)

Species	Temp. range °C	Corresponding size range (mm)	Difference in percent from minimum length	Author
Coregonus clupeaformis	0·5–10	13 –8·8	47·7	Price (1940)
Coregonus wartmanni	1–7	11 –9	22·2	Braum (1964)
Osmerus eperlanus	12 –18	5·2–4·6	13·0	Lillelund (1961)
Cyprinodon macularius	28 –35	5·3–3·7	43·2	Kinne & Kinne (1962)
Gadus macrocephalus	2–10	4·1–3·5	17·1	Forrester and Alderdice (1966)

Apparently the advantage of a longer body in embryos and larvae is primarily the improvement of locomotion. Many swim by trunk undulations in the anguilliform technique, and increasing body length means increasing swimming speed and more economical movement. The swimming behaviour in whitefish and herring embryos and its ecological significance will be discussed here as representative of those which swim in both the embryo and larval stages. Whitefish embryos (*Coregonus wartmanni*) incubated at 1·6C start swimming immediately after hatching, while those developed at 6C begin swimming one or two days after hatching. Because they fill their swim bladder roughly three months after hatching by gulping air at the water surface and because the oil globule of the yolk does not compensate for their weight, embryos and larvae sink when they cease to swim. The whole undulation cycle has been followed in frames of moving pictures (Fig 5.6). The trunk undulations are correlated with alternating movements of the pectorals and a larva swims roughly 40% of its own length during one undulation period.

Swimming speed is of ecological importance with regard to the time needed to reach the surface waters after hatching, the frequency of meeting prey at the end of the embryo stage and the ability to escape predators. It is influenced by the temperature, and it may reach values between 50 and more than 100 m per hour (Table 5.3).

Locomotion after hatching is combined with orientation to light and gravity. Embryos leave the egg shell at the bottom of the lake where light is weak or absent. The swimming position, i.e. the angle between the longitudinal axis and a horizontal line, is directed 55° upward in total darkness (Fig. 5.7). The diagram is based on flash pictures during experiments, and each circle integrates the swimming position of one to three embryos. When swimming toward a horizontal light source and guided by positive phototaxis this angle decreases to 30°. The position is also about 30° when light comes from above, which means swimming position is at an angle of 30° when the

Table 5.3. Swimming speed of some fish larvae (Blaxter 1969, 1974)

Species	Length (mm)	Temp. (C)	speed (mm/sec)	speed (m/hour)	Author
Clupea harengus	6,5– 8	?	6	22	Bishai (1960)
Clupea harengus	12–14	9–10	10	36	Blaxter (1960)
Coregonus wartmanni	10–12	4	16	58	Braum (1964)
Coregonus wartmanni	10–12	16	29	104	Braum (1964)
Coregonus clupeaformis	12,8	14	18	65	Hoagman (1974)
Solea Solea	4	15	6–9	17–19	Rosenthal (1966)

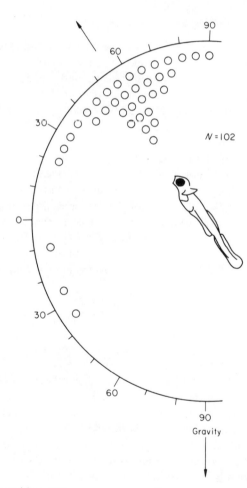

Fig. 5.7. Swimming position of *Coregonus wartmanni* larvae at total darkness (Braum 1964).

embryos move horizontally. The orientation with regard to gravity and light, especially the orientation in total darkness, combined with uninterrupted swimming movements guarantees that embryos reach the upper water layers long before they are forced into first external feeding. A similar swimming pattern was observed by Rosenthal and Hempel (1970) in herring embryos. Swimming was 'serpentine (eel-like) in form with a meandering path', which differs from *Coregonus* larvae which swim more in a straight line. As already mentioned, an obvious relation between body length and swimming velocity exists, as Rosenthal and Hempel have shown (Table 5.4).

Table 5.4. Length of herring larvae and swimming speed (Rosenthal & Hempel 1970) (Data are calculated on a straight line distance between start and end of a single swimming phase)

Total length (mm)	Velocity (cm/sec)
8–11	0·7–0·8
11–15	1·0–1·1
19–24	2·1–2·5
32–40	4·0–5·0

Many fish embryos show a positive response to light, but the reaction may change with age or with temperature. For example, a sudden increase in temperature changed a positive phototaxis response to a negative one in embryos of *Coregonus* (Braum 1967). Blaxter (1968) found positive phototaxis in herring embryos at higher light intensities and a negative phototaxis at lower light intensities.

An example of changing photo- and geotaxis in the embryo stage was investigated by Roth and Geiger (1963). After hatching, brown trout embryos migrate downward in the spaces between the gravel. At this stage and during the following three weeks they are positively geotaxic and negatively phototactic. After 26 to 43 days, depending on temperature, the embryos begin to ascend. The phototactic and geotactic reactions change signs and the larvae migrate upward, reaching the gravel surface. Response to the direction of the water flow is positively rheotactic during the whole stay within the gravel.

Besides the actively swimming type of embryo and the demersal type which spends this phase hidden in the gravel, a third type spends the embryo period fixed on plants or other substrates with a minimum of movement. Embryos of this type have characteristic organs on their head which produce a sticky secretion. After hatching, for example, the embryos of pike swim until they came into contact with a substrate. They then immediately stop their swimming movements. The secretion of the head glands forms a flexible attachment which allows the embryos to hang with the head up and tail down, swinging passively in the water flow (Braum 1964).

The commencement of external feeding marks the transition from the embryo to larval phase. The yolk sac is nearly completely used up and a change in behaviour characterizes the larvae at this time.

The chances of survival at the commencement of feeding depend on encountering enough suitable food just after the yolk is used up. If no food is available after final resorption of yolk, swimming activity of larvae decreases and they become too weak to feed. Blaxter and Hempel (1963) called this the 'point of no return' for herring larvae. From the time of fertilization this point may vary from 15 days for summer spawners in warmer water, to 45 days for winter spawners, depending on the yolk content of eggs and the temperature during development.

Blaxter stressed the relationship between the gape of the jaw, which may be about 0·3–0·4 mm at first feeding in herring larvae, and the size of suitable food items. Gape size increases partly with body length and is very small in species with small eggs and larvae, e.g. the marine species *Microstomus kitt* and *Sardina pilchardus*. These larvae require very small food organisms for survival.

Most fishes have well-developed movable eyes for perceiving food over a short distance. *Coregonus* larvae have independently movable eyes. When they meet a prey object, such as a copepod, both eyes are turned forward symmetrically towards the prey, the trunk is longitudinally aligned with the recognized object, and finally the larva pushes forward from an S-shaped position and tries to suck it in (Braum 1964). The prey is viewed stereoscopically before the larva darts forward. A marked fovea centralis, as found by Kahmann (1934) in the retina of different fish species, is absent in the eyes of *Coregonus* larvae. The visual field of one eye is 145° horizontally and vertically. This was established by histological serial sections and by the analysis of visual reaction to food items taken from moving pictures. A binocular visual field is achieved by turning both eyes forward at an angle of 45° (Fig. 5.8). A distance of 10 mm from eye to prey will stimulate the typical eye movements as a first step of preying. Thus, the prey-sensitive visual field of each eye is a spherical segment 10 mm in diameter. Any object which enters this space can, but may not, release the prey reaction (Braum 1964, 1967).

Knowing the average swimming speed at 4C and the size of the spherical visual field sensitive to prey permits us to estimate that a *Coregonus wartmanni*

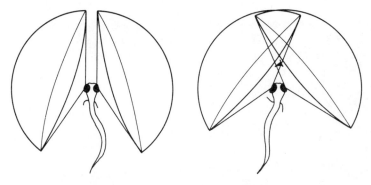

Fig. 5.8. Visual space of a *Coregonus wartmanni* larva at normal eye position (left) and during fixation (right). Visual space is drawn proportional to larval length.

larva in an aquarium can scan a volume of 14·6 l/hour. Further information about prey species which can be captured by larvae and their distribution in a lake will eventually lead to conclusions about the frequency of contacts between the larvae and their prey. Not only is the frequency of contacts important but also the behaviour of larvae and prey during such a meeting.

Recently Hoagman (1974) reported similar data for postlarvae of *Coregonus clupeaformis* from the Great Lakes, which scan 10·8–18 l/h. Blaxter and Staines (1971) give scanning volumes from 6–8 l/h for herring of 13–14 mm.

Larvae of *Coregonus wartmanni* invariably approach their prey from below so that the angle between the water surface and body is roughly 50°. The fixed prey is then seen against the luminous water surface and stationary swimming is less difficult when the body is directed upward. Moreover, the trunk must be in an S-shaped position from which it is possible to push forward by suddenly extending the body. This is accomplished by limiting the undulations to one side (Fig. 5.9). This S-shaped position seems to be common in many species. It was described by Schach (1939), Bückmann *et al.* (1953) and Rosenthal (1969) for herring, and Rosenthal (1966) for sole (*Solea solea*) and Braum (1964) for pike. S-shaped darting movements were observed in *Agonus*, *Pleuronectes* (Rosenthal, personal communication) and in *Lepomis* fry (Werner, personal communication).

In *Coregonus* larvae prey reactions are triggered by both moving and stationary objects up to a maximum size of approximately 1 mm². When the larvae approach the critical distance of 2 or 3 mm, the prey often swims or jumps away before the fish snaps. The prey's sensitivity to the approaching larva is apparently variable; for example *Diaptomus* may flee when a larva approaches within 5 mm or it may only try to jump away when the larva

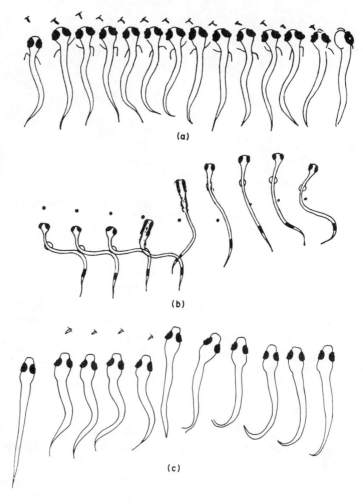

Fig. 5.9. Darting movements of three fish larvae drawn from moving pictures (a) *Coregonus wartmanni* (Braum 1964); (b) *Clupea harengus* (Rosenthal & Hempel 1970); (c) *Esox lucius* (Braum 1964).

pushes forward and snaps (Fig. 5.10). By a remarkable innate behaviour *Coregonus* larvae are adapted to this flight reaction of their prey in the following three ways (Braum 1967):

1.   The main prey organisms are species of the genus *Cyclops* and *Diaptomus*. *Cyclops* has a flight jump between 2 and 10 mm in a straight line away from the larva's head. *Diaptomus* simultaneously jumps upward and backward so that it often comes close to the larva's head. Usually prey that escapes remains in the larva's visual field and optical fixation is repeated immediately.

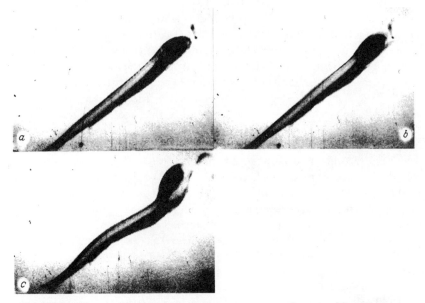

Fig. 5.10. Side view of approach (a) of a *Coregonus wartmanni* larva catching *Diaptomus* b). The prey is missed and jumps away (c). Frames taken from moving picture.

2. The prey may jump out of the larva's visual field. Then the larva looks for it by an innate reaction: it jerkily turns round its vertical axis about 180° without leaving the location where the prey has been lost. In nearly all cases the lost specimen is found again.

3. The disappearance of the prey and especially the jumping movements stimulates and intensifies the larva's tracing instincts.

The observations show clearly the adaptation of the innate behaviour pattern of the larvae to a full exploitation of each meeting with a prey object.

The shift to exogenous feeding in the pike is very different from that of whitefish and herring. Embryos change to larvae within one day by a considerable change of behaviour. As already mentioned, embryos are inactive during the yolk-sac stage. During the transition from embryo to larva, they fill their swim bladder at the water surface, change their locomotion by limiting the trunk undulations to vibrations of pectorals and tail and start prey capture with all its complicated mechanisms. They usually remain at one location waiting for the prey. Only moving organisms such as *Cyclops*, *Diaptomus*, *Daphnia* or *Bosmina* release the binocular eye reaction correlated with a careful approach. The trunk seems to be motionless; only tail tip and pectorals vibrate with a high frequency. Slowly the trunk takes an S-shaped position and the body pushes forward (Braum 1963).

The effectiveness of prey capture is vital for the chances of larval survival.

It seems to be common in fish larvae that not each response to food items is completed by snapping at the fixed object. Without any evident reason, the preying reaction is interrupted. Tugendhat (1960) observed the phenomenon sticklebacks and distinguished between 'initiated and completed feeding responses'. Rosenthal and Hempel (1970) made the same observation in herring larvae, and moreover it seems to be a characteristic of many animals which have a marked predatory behaviour, e.g. toads (Eikmanns 1955) or *Mantis* (Rilling *et al.* 1959). The probability that the feeding response of *Coregonus wartmanni* larvae will be completed increases with the prolongation of fixation time. Ninety percent of all feeding responses lasted between 0·5 and 3 seconds, and between 20 and 55% of these responses were completed. As already mentioned, the duration of prey reactions is influenced by the movement behaviour of the prey. Its flight reaction stimulates the larva to follow, thereby lengthening the feeding response and increasing the probability that the response will be completed by snapping.

The effectiveness of feeding can be measured as the percentage of successful snapping reactions in relation to all completed feeding responses. Braum (1964) compared two groups of larvae (Table 5.5), one of which (group A)

Table 5.5 Feeding responses and effectiveness of capturing prey by *Coregonus wartmanni* (Braum 1964)

Group	Days after first feeding	Initiated feeding responses	Completed feeding responses	Successfully completed feeding responses	Percentage of successful responses
A	0–8	850	454	15	3·2
A	9–16	746	323	70	21·6
B	0–8	221	316	10	3·1
B	9–16	780	380	12	3·1

always had an abundant food supply and the other (group B) encountered prey for the first time. The results of more than 4000 feeding responses were recorded during the first 16 days of the post-larval period. During the first 8 days the effectiveness was approximately 3% in both advanced feeders (group A) and beginners (group B), but the successful responses rose to 21% from 9 to 16 days for group A while it remained at the 3% level for group B (Table 5.5). Thus, the experienced feeders acquired a greater skill at capturing food than did the inexperienced feeders in the time period devoted to the experiments. Because it was impossible to rear larvae with nonliving food, larvae of group B must be regarded as starving and observations of their first feeding responses could therefore only be compared for a limited time. The 'point of no return' (Blaxter & Hempel 1963) when larvae were still alive but too weak for food uptake was dependent on temperature. At 11C unfed larvae were unable to catch prey 11 days after first feeding of controls and at 5·1C after 27 days.

In Downs herring larvae Rosenthal and Hempel (1970) found that the successful snapping rate increased from about 1% for the first attempt to 25% within the first month of life and up to 50% in the second month if optimum food is provided. In pike larvae the effectiveness of first feeding larvae is quite high. It is about 30% during the first 7 days and increases rapidly to 80% within 15 days after first feeding (Braum 1964). Young pike begin predation at a length between 2 and 3 cm, preferring fish larvae to plankton. Mortality due to cannibalism between pike of the same age and nearly the same length increases at greater densities such as occur in cultural conditions. No data exist about the rate of cannibalism of postlarval pike in natural spawning grounds.

By their behaviour the larvae of *Coregonus* are best adapted to rare meetings with suitable food organisms. The flight reaction of prey stimulates the intensity of feeding response, and a sudden loss of optical contact with prey releases a stationary searching activity. Herring larvae do not search when prey disappears from their vision, so they must depend much more on the frequency of meeting than do *Coregonus* larvae. Therefore, the abundance of plankton must be very important for this type of larvae. In contrast to the constantly swimming postlarvae of herring and whitefish with a low feeding effectiveness, post-larvae of the pike hover at one location and lurk for prey. Their very high effectiveness in prey capture is based on the swim bladder, which enables them to dart more precisely, and on their rather large mouth, which produces a strong suction when snapping.

The food ration of *Coregonus wartmanni* obtained by direct observations was 13 to 44 items in 10 hours at 11C. After 31 days a ration of 265 items was found at 15·6C. Einsele (1941, 1963) assumed that 1 g of plankton-feeding whitefish in Austrian lakes needs roughly 1 cm³ of plankton for normal growth at 10C and calculated that *Coregonus* larvae would need about 150 plankton organisms per day 14 days after first feeding. Undoubtedly, larvae can survive with lower food rations. Investigating Swedish *Coregonus lavaretus* larvae, Berzins (1958) found that they could survive over a short period of limited food supply with a ration of three nauplii per day.

Rosenthal and Hempel (1970) calculated a daily ration for 10 mm Downs herring larvae of 40 nauplii. Assuming that herring larvae tend to keep their gut filled during the day, the authors calculate a required number of nauplii of 4–8 per litre if each visually perceived planker would stimulate snapping in 11 mm larvae. This is a minimum estimate based on complete feeding response at each encounter. It varies with the volume of water scanned per day, and if it is assumed that only a part of the meetings with prey will cause snapping reactions, the number of nauplii required per litre will reach 21–42

for optimum growth. No data are available for *Coregonus* or herring larvae
on the nutritive value and the rate of digestion of different food organisms at
commencement of feeding.

The marked eye movements of the investigated larvae indicate that light
is required while feeding, and the minimum light intensity for prey capture
is of interest to estimate the water depth and day length when larvae can
feed. The most detailed studies on vision in fish larvae have been published
by Blaxter (1968, 1969a) on the threshold and spectral sensitivity in herring,
plaice and sole larvae. He found that much more light ($10–10^{-2}$ metre
candles) is required for a feeding than for a phototactic response ($10^{-4}–10^{-6}$
m.c.) in plaice larvae. He found similar data for herring larvae, and he
assumes that more light is required for image formation, movement per-
ception and acuity during feeding behaviour than for phototactic response.

In *Coregonus* larvae Braum (1964) found a minimum limit of white light
intensity of about 1·5 lux at which larvae still show feeding responses.
However, temperature also directly influences preying behaviour. A considerable increase of effectiveness was recorded in *Coregonus* larvae at a
temperature of 12·0C as compared to 5·3C. At 12·0C, 31·2% of the reactions
were successful, while the control group at 5·3C showed only 4·1%. Not
enough data are available to explain whether this correlation is only based
on an improved darting speed of the larvae, or whether the prey reactions
become more favourable for the larvae. Whatever the reason may be, an
optimum in feeding temperature obviously exists, and for rearing purposes
in aquaculture it may be worthwhile to consider this possibility for other
fish larvae as well.

Though the availability of proper food has a strong influence on the
survival rate and year class strength, predation on eggs and larval fish must
be considered as important causes of mortality. One group of predators are
carnivorous copepods. In freshwater rearing experiments with high plankton
density, Davis (1959b) observed increased mortality if copepods were present.
Nikol'skii (1963) observed *Mesocyclops leucarti* and *Acanthocyclops vernalis*
attacking larval shad (*Alosa sp.*). Lillelund (1967) found *Cyclops abyssorum*,
C. vicinus, *C. stenuus* and *C. kolensis* able to fasten themselves onto the trunk
of swimming smelt larvae (*Osmerus eperlanus*) and 'after a few minutes' such
larvae are swallowed. Behavioural and quantitative interrelationships between
predatory copepods of the family Pontelidae and the co-occurring anchovy
larvae (*Engraulis mordax*) were investigated in a detailed study by Lillelund
and Lasker (1971). Westernhagen and Rosenthal (1976) described the predator–prey relationship between the hyperiid amphipod, *Hyperoche medusarum*, and newly hatched Pacific herring (*Clupea harengus pallasi*). Field
and laboratory investigations indicate that the amphipod preys on herring
larvae.

5.5 The critical period concept

The term, critical period, was introduced by Fabre-Domergue and Biétrix (1897) with regard to the high mortality suffered by cultured fish larvae when they shifted from yolk to exogenous food. The term came into common use among fisheries biologists after Hjort (1914) published his well known paper on the fluctuations of the North European marine fisheries. He assumed that the absence of suitable food at the end of yolk nutrition in marine fish larvae would cause catastrophic mortality, determining, in turn, the strength of a year-class. Hjort's hypothesis was based on observations of fish culturists and on the correlation of the spawning of the Atlanto-Scandic herring in spring with a rapid development of plankton. Marr (1956) tried to collect data on a critical period in nature by sampling pelagic eggs and larvae. For larvae of Atlantic mackeral (*Scomber scombrus*) and the Pacific sardine (*Sardinops sagax*) a rather continuous decrease was found, i.e. no evidence for a catastrophic mortality at a certain stage of development was observed. Recently May (1974) has surveyed the available survival curves for natural populations of marine fish larvae. His detailed study reveals that curves 'either show no increased mortality at yolk absorption or permit varying interpretations.' May (1974) concludes that a great number of sources of error prevent a definite answer to the question whether higher mortality at the end of the yolk sac stage exists or not. Those sources of error are the 'patchiness' of plankton, technical sampling errors, poor egg quality or losses by predation.

There are few fundamental differences between the transition from yolk nutrition to external food uptake in the larvae of freshwater and marine species. Elster (1937, 1944) first applied the critical period concept to freshwater in his population studies of pelagic whitefish species in Lake Constance. As Hjort (1914) assumed for the spring-spawning Norwegian herring, Elster estimated that the first feeding of *Coregonus wartmanni* would be accompanied by high mortality because it occurs when plankton density is still low, in February or early March. An attempt was made to compare experimental findings with seasonal plankton distribution and temperature in the lake (Braum 1963, 1964, 1967). The essential data for such a comparison are: (1) the duration of development from egg stage to the time when the first feeding responses occur, (2) the preferred food species, (3) the behavioural relations between the larva and its prey, (4) the sucess of feeding responses, (5) the efficiency of eyes as the essential sense organs for preying, (6) the swimming speed which makes it possible to calculate the water volume scanned by a larva for a time unit, and (7) the required food ration.

Interpretations drawn from the laboratory data outlined above and applied to early life history stages in the lake are not completely satisfactory as the following example demonstrates. To determine first exogenous feeding in the lake, data about plankton distribution, spawning time and temperature are required. Spawning varies in its duration and time of occurrance from year to year. In normal years *Coregonus wartmanni* spawns about 15 December. In this case larvae at the lake bottom will probably hatch about the end of March. At 4·4C the yolk-sac period lasts 20 days under hatchery conditions. The duration of this period in nature is still unknown, but it is probably longer since the stage of inactivity after hatching might be extended and yolk storage would be sufficient for a longer period. First prey capture probably begins in the last third of April.

As mentioned above, larvae can scan 14·6 l/h of water for prey at 4C. In April 915 suitable food items will be encountered in 10 hours. This calculation, of course, is only rough, although it is based on the average quantity of plankton over a period of 10 years. An even distribution of food items has been assumed. The later first exogenous feeding begins the better the chances for survival. In March, probably the earliest date for commencement of feeding, the supply of suitable food is only a quarter of that in April. In this case a larva can encounter only 230 suitable food items in 10 hours.

The percentage of completed feeding responses varies from larva to larva and is influenced by the flight of the prey. It is generally about 50%. Considering that during the first 8 days feeding success is only 3%, a larva in April living at 11C theoretically will eat 14 food items per 10 hours which corresponds to its maximum food requirement based on rearing at optimum supply. In March it would obtain only 3–4 items—only one fourth of the requirement. These examples are minimum calculations which do not consider the ability to endure starvation or to overcome periods of scarce food supply by reducing growth, as Berzins (1958) has found. The variable behavioural interrelationships between larvae and prey are not considered. Though the question of a critical period in the lake is still open, it appears that larvae will have a good chance for survival considering the food supply in the lake and their own ability to exploit each encounter with the prey.

Similar laboratory investigations of marine fish larvae such as herring of different geographic groups (Blaxter & Staines 1971, Rosenthal & Hempel 1970), anchovy (Hunter 1972), needlefish, *Belone belone* (Rosenthal 1970) and of other freshwater larvae such as Great Lakes whitefish (*Coregonus clupeaformis*) (Hoagman 1974) and pike (Braum 1964) reveal that there is a great variability in the ability of different species to survive under conditions of limited food availability. Though it seems to be difficult to prove the existence of a 'critical period' in nature, larval development proceeds irreversibly toward the exhaustion of the yolk. Because of the limited time to meet suitable food, this period contains a high risk during early life history.

5.6 Conclusions

Since the critical period concept was first discussed, it has stimulated field and laboratory research. Knowledge about population dynamics of eggs and larvae and of ecological factors that influence their fluctuations forms the basis for experiments to learn the causes of mortality of eggs and larvae. Survival of post-larvae was primarily seen as a question of food supply and the larval adaptation to exploit it.

Among other factors which influence survival, predation is widely considered to be one of the most significant, particularly in prelarvae (May 1974). Laboratory observations exist on predation of invertebrates, such as amphipods, on fish larvae (v. Westernhager & Rosenthal 1976), carnivorous copepods (Nikol'skii 1963, Lillelund 1967, Lillelund & Lasker 1971) and euphausiid shrimps (Theilacker & Lasker 1974). Fish larvae and predators encounter each other randomly, as experiments revealed. The correlation of experimental results with conditions in the natural environment suggests that predation is an important cause of mortality. The need for further study of this relationship opens an attractive field for further research.

The different motivations for research about 'ecological aspects of the survival of fish eggs and larvae' have produced a great number of investigations extending from population studies to physiological ecology. Eggs and larvae are a part of the ecosystem and they react to any change of the system or the abiotic environment. A combination of field and laboratory work could permit a better integration of results.

References

ALDERDICE D.F., WICKETT W.P. & BRETT J.R. (1958) Some effects of temporary low dissolved oxygen levels on pacific salmon eggs. *J. Fish. Res. Bd. Canada* **15**, 229–249.

ALLEN K.R. (1951) The Horokiwi Stream: a study of a trout population. *Fish. Bull. N.Z.*, **10**, 1–238.

APSTEIN C. (1909) Die Bestimmung des Alters pelagisch lebender Fischeier. *Mitt. dt. Seefisch Ver.* **25**, 364–373.

BAGENAL T.B. (1971) The interrelation of the size of fish eggs the date of spawning and the reproduction cycle. *J. Fish Biol.* **3**, 207–219.

BALON E.K. (1975) Reproductive guilds in fishes: a proposal and definition. *J. Fish. Res. Bd. Canada* **32**, 821–864.

BANNISTER R.C.A., HARDING D. & LOCKWOOD S.J. (1974) Larval mortality and subsequent year-class strength in the plaice (*Pleuronectes platessa*). *The Early Life History of Fish* (J.H.S. Blaxter Ed.), pp. 21–37. Springer Verlag, Berlin, Heidelberg, New York.

BERZINS B. (1958) Södra Sveriges Fiskeriförening. Arsskrift.

BILLARD R., PETIT J., JALABERT B. & SZOLLOSI D. (1974) Artificial insemination in trout using a sperm dilutant. *The Early Life History of Fish* (J.H.S. Blaxter Ed.), pp. 716–723. Springer Verlag, Berlin, Heidelberg, New York.

BISHAI H.M. (1960) The effect of water currents on the survival and distribution of fish larvae. *J. Conseil, Conseil Perm. Intern. Exploration Mer* **25**, 134–146.

BLAXTER J.H.S. & HEMPEL G. (1963) The influence of egg size on herring larvae (*Clupea harengus* L.) *J. Cons. Perm. Intern. Explor. Mer* **28**, 211–240.

BLAXTER J.H.S. (1968) Visual thresholds and spectral sensitivity of herring larvae. I. *J. Exptl. Biol.* **48**, 39–53.

BLAXTER J.H.S. (1969a) Visual threshold and spectral sensitivity of flatfish larvae. I. *J. Exptl. Biol.* **51**, 221–230.

BLAXTER J.H.S. (1969b) Development: Eggs and larvae. *Fish Physiology*, Vol. III, (W.S. Hoar and D.J. Randall Ed.), pp. 177–252. *Reproduction and Growth.* Academic Press, New York and London.

BLAXTER J.H.S. & STAINES M.E. (1971) Food searching potential in marine fish larvae. *4th European Mar. Biol. Symp.* (D.J. Crisp Ed.), pp. 467–489. Cambridge University Press, Cambridge, England.

BOWERS, A.B. (1969) Spawning beds of Manx autumn herring. *J. Fish Biol.* **1**, 355–359.

BRAUM E. (1963) Die ersten Beutefanghandlungen junger Blaufelchen (*Coregonus wartmanni* Bloch) und Hechte (*Esox lucius* L.). *Zeitschr. f. Tierpsychol.* **20**, 257–266.

BRAUM E. (1964) Experimentelle Untersuchungen zur ersten Nahrungsaufnahme und Biologie an Jungfischen von Blaufelchen (*Coregonus wartmanni* Bloch), Weißfelchen (*C. fera* Jurine und Hechten (*Esox lucius* L.). *Arch. Hydrobiol.* **28**, Supple. **5**, 183–244.

BRAUM E. (1967) The survival of fish larvae with reference to their feeding behaviour and the food supply. *The Biological Basis of Freshwater Fish Production.* (S. Gerking Ed.), pp. 113–134. Blackwell Scientific Publications, Oxford.

BRAUM E. (1973) Einflüsse chronischen Sauerstoffmangels auf die Embryogenese des Herings (*Clupea harengus*). *Netherlands Journ. of Sea Res.* **7**, 363–375.

BREDER C.M., Jr. & ROSEN D.E. (1966) *Modes of Reproduction in Fishes.* Natural History Press, Garden City, N.Y.

BRIGGS J.C. (1953) The behaviour and reproduction of salmonid fishes in a small coastal stream. *Dept. Fish and Game of California, Marine Fisheries Branch, Fish Bulletin* **94**, 62 pp.

BÜCKMANN A., HARDER W. & HEMPEL G. (1953) Unsere Beobachtungen am Hering (*Clupea harengus* L.) *Kurze Mitt. d. Fischereibiol. Abt. Max Plank-Inst. f. Meeresbiol. Wilhelmshafen* **3**, 22–42.

DAVIS C.C. (1959a) A planktonic fish egg from fresh water. *Limnol. and Oceanogr.* **4**, 352–355.

DAVIS C.C. (1959b) Damage to fish fry by cyclopid copepods. *Ohio J. Sci.* **59**, 101–102.

DAYKIN P.N. (1965) Application of mass transfer theory to the problem of respiration of fish eggs. *J. Fish. Res. Bd. Canada* **22**, 159–171.

DEVILLERS CH. (1965) Respiration et morphogenèse dans l'oe uf de téléostéens. *Année biol.* **4**, 1965, 157–186.

EIKMANN K.H. (1955) Verhaltensphysiologische Untersuchungen über den Beutefang und das Bewegungssehen der Erdkröte (*Bufo bufo* L.) *Zeitschr. f. Tierpsychol.* **12**, 229–253.

EINSELE W.G. (1941) Fischereiwissenschaftliche Probleme an den deutschen Alpenseen. *Fischere-Zeitung* **44**, 1–16.

EINSELE W.G. (1963) Problems of fish-larvae survival in nature and in the rearing of economically important middle European freshwater fishes. *Calif-Coop. Oceanic Fish. Invest.* **10**, 24–30.

ELSTER H.J. (1937) Versuche zur Hebung der Coregonenfischerei im Bodensee. *Fischerzeitung* **40**, 1–15.

ELSTER H.J. (1944) Über das Verhältnis von Produktion, Bestand, Befischung und Ertrag sowie über die Möglichkeiten einer Steigerung der Erträge untersucht am Beispiel der Felchen des Bodensees. *Zeitschr. f. Fisch.* **42**, 169–357.

EMBODY G.C. (1934) Relation of temperature to the incubation period of eggs of four species of trout. *Trans. Am. Fish. Soc.* **64**, 281–292.

FABRE-DOMERGUE P. & BIÉTRIX, E. (1897) Développement de la sole (*Solea vulgaris*). *Bull. Mus. Nat. Hist.*, Paris **3**, 57–58.

FORRESTER C.R. & ALDERDICE D.F. (1966) Effects of salinity and temperature on embryonic development of the Pacific cod (*Gadus macrocephalus*). *J. Fish. Res. Bd. Canada* **23**, 319–340.

FROST W.E. & SMYLY W.J.P. (1952) The brown trout of a moorland fishpond. *J. Anim. Ecol.* **21**, 62–86.

FRYER G. (1959) The trophic interrelationships and ecology of some littoral communities of Lake Nyassa with especial reference to the fishes and a discussion of the evolution of a group of rock frequenting Cichlidae. *Proc. Zool. Soc. Lond.* **132**, 153–281.

FRYER G. & ILES T.D. (1972) *The Cichlid Fishes of the Great Lake of Africa. The biology and evolution*, Oliver & Boyd, Edinburgh.

GARSIDE E.T. (1959) Some effects of oxygen in relation to temperature on the development of lake trout embryos. *Can. J. Zool.* **37**, 689–698.

GERKING S.D. (Ed.) (1967) *Biological Basis of Freshwater Fish Production*. Blackwell Scientific Publications, Oxford.

GINSBURG A.S. (1963) Sperm egg association and its relationship to the activation of the egg in salmonid fishes. *J. Embryol. exp. Morph.* **11**, 13–33.

GOTTWALD S. (1965) Der Einfluß zeitweiligen Sauerstoffmangels in verschiedenen Stadien auf der Embryonalentwicklung v. d. Regenbogenforelle (*Salmo gairdneri* Rich.) 2. *Z. Fisch. N.F.* **23**, 63–84.

HAMDORF K. (1960) Die Beeinflussing der Embryonal- und Larval entwicklung der Regenbogenforelle (*Salmo irideus* Gibb.) durch Strahlen im sichtbaren Bereich. *Z. vergl. Phys.* **42**, 525–565.

HAMDORF K. (1961) Die Beeinflussung der Embryoral- und Larval entwicklung der Regenbogenforelle (*Salmo irideus* Gibb.) durch die Umweltfaktoren O_2-Partialdruck und Temperatur. *Z. vergl. Phys.* **44**, 523–549.

HAMOR T. & GARSIDE E.T. (1973) Peroxisome-like vesicles and oxidative activity in the zona radiata and yolk of the ovum of the Atlantic salmon (*Salmo salar* L.). *Comp. Biochem. Physiol.* **45**, 147–157.

HEMPEL G. & SCHUBERT K. (1969) Sterblichkeitsbestimmungen an einem Eiklumpen des Nordsee-Herings (*Clupea harengus* L.) *Ber. Dt. wiss. Komm. Meeresforsch.* **20**, 79–83.

HJORT J. (1914) Fluctuations in the great fisheries of northern Europe viewed in the light of biological research. *Rapp. P.-v. Réunions Con. Perm. Intern. Explor. Mer* **20**, 1–228. *Mer* **20**, 1–228.

HOAGMAN W.J. (1974) Vital activity parameters as related to the early life history of larval and postlarval lake whitefish (*Coregonus clupeaformis*) *The Early Life History of Fish*. (J.H.S. Blaxter Ed.), pp. 547–558. Springer Verlag, Berlin, Heidelberg, New York.

HOAR W.S. (1969) Reproduction. *Fish physiology Vol. III, Reproduction and Growth*. (W.S. Hoar & D.J. Randall Eds.), pp. 1–72. Academic Press, New York and London.

HOBBS D.F. (1937) Natural reproduction of Quinnat salmon, brown and rainbow trout in certain New Zealand waters. *New Zealand Marine Dept., Fish. Bull.* **6**, 104 pp.

HOBBS D.F. (1940) Natural reproduction of trout in New Zealand and its relation to density of populations. *New Zealand Marine Dept., Fish. Bull.* **8**, 93 pp.

ILG L. (1952) Über Larvale Haftorgane bei Teleasteem. *Zort. Jahrb.* **72**, 577–600.

HUNTER J.R. (1972) Swimming and feeding behaviour of larval anchovy *Engraulis mordax*. *Fish. Bull., U.S.* **70**, 821–838.

KAHMANN H. (1934) Über das Vorkommen einer Fovea centralis im Knochenfischauge. *Zool. Anz.* **106**, 49–55.

KINNE O. (1962) Rates of development in embryos of a cyprinodont fish exposed to different temperature-salinity-oxygen conditions. *Can. J. Zool.* **40**, 231–253.

KINNE O. & KINNE E.M. (1962) Rates of development in embryos of a cyprinodont fish exposed to different temperature-salinity-oxygen combinations. *Can. J. Zool.* **40**, 231–253.

KRYZHANOVSKY S.G. (1948) Ecological groupings of fishes and regular features in their formation. *Trans. Pacific inst. Fish. USSR* **27** (Reference from Nikol'skii 1963).

LANGE O. (1973) Ein Beitrag zur Embroyonalentwicklung des Steinpickers *Agonus cataphractus* L. (Agonidae, Pisces). *Diplomarbeit, Fachbereich Biologie, Universität Hamburg.*

LASKER R. (1964) An experimental study of the effect of temperature on the incubation time, development and growth of Pacific sardine embryos and larvae. *Copeia* **3**, 399–405.

LE CREN E.D. (1962) The efficiency of reproduction and recruitment in freshwater fish, *The Exploitation of Natural Animal Populations.* (E.D. Le Cren & M.W. Holdgate Eds.), pp. 283–302. Blackwell Scientific Publications, Oxford.

LILLELUND K. (1961) Untersuchungen über die Biologie und Populationsdynamik des Stintes (*Osmerus eperlanus eperlanus* L.) der Elbe. *Arch Fish. Wiss.* **12**, 1–128.

LILLELUND K. (1967) Experimentelle Untersuchungen über den Einfluß carnivorer Cyclopiden auf die Sterblichkeit der Fischbrut. *Zeit. f. Fischerei* (N.F.) **15**, 29–43.

LILLELUND K. & LASKER R. (1971) Laboratory studies of predation by marine copepods on fish larvae. *Fish. Bull., U.S. Fish & Will. Serv.* **69**, 655–667.

LINDROTH A. (1946) Zur Biologie der Befruchtung und Entwicklung beim Hecht. *Mitt. Anst. Binnenfisch. Drottningholm* **24**, 1–173.

MARR J.C. (1956) The critical period in the early life history of marine fishes. *J. Cons. Perm. Intern. Explor. Mer* **21**, 160–170.

MAY R.C. (1974) Larval mortality in marine fishes and the critical period concept. *The Early Life History of Fish* (J.H.S. Blaxter Ed.), pp. 3–19. Springer-Verlag, Berlin, Heidelberg, New York.

NEW D.A.T. (1966) *The culture of vertebrate embryos.* Logos Press & Elek Books, London.

NIKOL'SKII G.V. (1963) *The Ecology of Fishes.* Academic Press, London and New York.

NIKOL'SKII G.V. (1962) On some adaptions to the regulation of population density of fish species with different types of stock structure. *The Exploitation of Natural Animal Populations.* (E.D. Le Cren & M.W. Holdgate Eds.), pp. 267–282. Blackwell Scientific Publications, Oxford.

NÜMANN W. 1961) Das Problem der Ertragssteigerung und Überfischung in der Blaufelchenfischerei am Bodensee. *Zeit. f. Fischerei* **10**, 241–251.

OUTRAM D.N. & HUMPHREYS R.D. (1974) The Pacific herring in British Columbia waters. *Fisheries and Marine Service Pacific Biological Station, Nanaimo, B.C.* Circular 100, 26pp.

POMMERANZ T. (1973) Das Vorkommen von Fischeiern insbesondere von Eiern der Scholle (*Pleuronectes platessa* L.) in den oberflächlichen Wasserschichten der südlichen Nordsee. *Ber. dt. wiss. Komm. Meeresforsch.* **22**, 427–444.

POY A. (1970) Über das Verhalten der Larven von Knochenfischen beim Ausschlüpfen aus dem Ei. *Ber. dt. wiss. Komm. Meeresforsch.* **21**, 377–392.

PRICE J.W. (1940) Time-temperature relations in the incubation of the whitefish *Coregonus clupeaformis* (Mitchill). *J. Gen. Physiol.* **23**, 449–458.

RICKER W.E. & FOERSTER R.E. (1948) Computation of fish production. *Bull. Bingham Oceanogr. Coll.* **11**, 173–211.

RILLING S., MITTELSTAEDT H. & ROEDER K.P. (1959) Prey recognition in the preying mantis. *Behaviour* **14**, 164–184.

ROSENTHAL H. (1966) Beobachtungen über das Verhalten der Seezungenbrut. *Helgol. Wiss. Meeresunters.* **13**, 213–228.

ROSENTHAL H. (1969) Untersuchungen über das Beutefangverhalten bei Larven des Herings (*Clupea harengus* L.) *Mar. Biol.* **3**, 208–221.

ROSENTHAL H. (1970) Anfütterung und Wachstum der Larven und Jungfische des Hornhechtes *Belone belone. Helgol. Wiss. Meeresunters.* **21**, 320–332.

ROSENTHAL H. & HEMPEL G. (1970) Experimental studies in feeding and food requirements of herring larvae (*Clupea harengus* L.) *Marine Food Chains.* (J.H. Steel Ed.), pp. 344–364. Oliver & Boyd, Edinburgh.

ROTH H & GEIGER W (1963) Experimentelle Unter suchungen Über das Verhalten der Bachforellenbrut in der Laichgrube. *Schweiz Zeit. Hydrol.* **25**, 202–218.

SCHACH H. (1939) Die künstlich Aufzucht von *Clupea harengus* L. *Helgol. Wiss. Meeresunters.* **1**, 359–372.

SCHÄPERCLAUS W. (1961) *Lehrbuch der Teichwirtschaft.* P. Parey, Berlin.

SILVER S.J., WARREN C.E. & DOUDOROFF P. (1963) Dissolved oxygen requirements of developing steelhead trout and chinook salmon embryos at different water velocities. *Trans. Am. Fish. Soc.* **92**, 327–343.

THEILACKER G.H. & LASKER R. (1974) Laboratory studies of predation by euphausiid shrimps on fish larvae. *The Early Life History of Fish*, (J.H.S. Blaxter Ed.), pp. 287–289. Springer Verlag, Berlin, Heidelberg, New York.

TUGENDHAT B. (1960) The normal feeding behaviour of three-spined stickleback (*Gasterosteus aculeatus* L.). *Behaviour* **15**, 284–318.

WAGLER E. (1933) Über Eier und Brut der Bodensee-Coregonen. *Arch. Hydrobiol.* **25**, 1–21.

WESTERNHAGEN H.V. & ROSENTHAL H. (1976) Predator-prey relationship between Pacific herring, *Clupea harengus pallasi*, larvae and a predatory hyperiid amphipod, *Hyperoche medusarum. Fish. Bull. U.S. Nat. Fish. Mar. Serv., NOAA.* **74**, 669–674.

WICKETT W.P. (1954) The oxygen supply to salmon eggs in spawning beds. *J. Fish. Res. Bd. Canada* **11**, 933–953.

YAMAMOTO T. (1961) Physiology of fertilization in fish eggs. *Intern. Rev. Cytol.* **12**, 361–405.

YAMAMOTO K. (1951) Activation of the egg of the Dog-salmon by water and the associated phenomena. *J. Fac. Sci. Hokkaido Univ. Ser. VI Zool.* **10**, 303–318.

THE FISH POPULATION AND
ITS FOOD SUPPLY

The book has acquired the flavour of ecosystem ecology, using the concepts and vocabulary associated with that part of our science. Only rarely, however, has the production of all of the fish species in a lake or stream ecosystem been examined or has it been related to the production of other trophic levels. At this stage of our development we have enough trouble obtaining a reasonably accurate picture of the production of, say, the dominant species of fish in an ecosystem without tackling all the rest and gathering data on other parts of the system at the same time. Such large projects must be accomplished with teams of scientists, and such teams have been brought together only a few times thus far. The need for more of those broad-based studies is obvious, and the few that have been done have contributed greatly to our conceptual view of aquatic productivity.

The first chapter of this section reviews three 'total ecosystem' studies—one river, one temperate lake and one tropical lake. By relating fish production to primary production, we see the severity of energy losses as it passes from one trophic level to another. No broad generalizations emerge from this review, primarily because our sample of ecosystems is too small as yet and, as a corollary, we do not know whether or not these ecosystems are typical of those that they are intended to represent.

The remainder of the chapter is devoted to food consumption and growth. Digestion and the partitioning of energy into metabolism and growth are measured on individual animals under controlled conditions in the laboratory. This same kind of approach was adopted in domestic animal research at the turn of the century and has been responsible, in large part, for the great strides that have been made in poultry, beef, hog and mutton production. This same approach will become just as valuable a tool in intensive aquaculture when world food demand forces a shift to new and additional sources of protein. These intensive studies of nutrition and metabolism are the only way to achieve the greatest utilization of food for growth in highly controlled situations, such as ponds and raceways.

These same studies of nutrition and metabolism provide the basic information for understanding food utilization in nature where we are dealing with populations in uncontrolled environments. The last two chapters use these ideas in various ways to relate food consumption by a fish population to its food supply. The efficiency of food utilization for growth is naturally lower in nature than it is in artificial ponds where, for example, high quality food may be fed directly to the fish. Competition is keener, activity is greater, and quality of the food may be lower where the fish must hunt and capture living food organisms. We have learned, however, that hunting and capturing is performed quite efficiently. Whenever cropping rates of fish populations have been estimated, they have turned out to be high. We are rapidly learning that foraging by fish has a tremendous impact on the structure of the ecosystem. Even with these advances, we are not yet at the point of estimating energy utilization efficiency between trophic levels, or the ratio of production of food organisms to the consumption of these organisms by fish. Only after we acquire this knowledge will we be able to follow energy flow accurately at the higher trophic levels in aquatic ecosystems.

Direct your attention to the contrast between the two chapters on food consumption of fish in nature. In Dr. Mann's chapter, the food of the fish is invertebrates, zooplankton and benthos; in Dr. Popova's chapter the food is other fishes. Dr. Popova concentrated her review on the Russian literature which is not well known in the Western world. From both of these chapters has emerged the possibility of separating total mortality of an organism in question into a predatory component and a remaining component which can be obtained by subtraction. Among fish which are caught for human consumption, we have learned how to divide total mortality into fishing and natural mortality, but we have not been able to compartmentalize natural mortality. If a species in question is preyed upon by a piscivore, Dr. Popova's methods of computing food consumption should allow us to subdivide natural mortality into mortality from predation and that from other causes.

Chapter 6: Production in Three Contrasting Ecosystems

Mary J. Burgis
and
I. G. Dunn

6.1 The whole ecosystem approach

The overwhelming practical importance to man of commercial fisheries has led to studies of fish populations in isolation from the other components of the aquatic community. If we are to increase our understanding of fish production dynamics and, hopefully, our ability to predict the effects of exploitation by man, it is essential that we have adequate knowledge of the role of fish within ecosystems. This neglect is due to the difficulty of providing the finance and manpower necessary for coordinated investigations of whole ecosystems. Since Mann (1969) wrote his review, 'Dynamics of aquatic ecosystems', a number of studies of a variety of freshwater ecosystems have been undertaken. Projects such as those on Vorderer Finstertaler See in Austria (Pechlaner et al. 1973), Lake Marion (Efford 1969, 1972) and Char Lake (Rigler 1972, Schindler et al. 1974) in Canada, Tjeukemeer (Beattie et al. 1972) in Netherlands, Lake George (Greenwood & Lund 1973) in Uganda, Loch Leven (Morgan & McLusky 1974) and the River Thames (Mann et al. 1972, Berrie 1972a) in Great Britain and many projects in the USSR (see Winberg 1972), have all included studies of the production processes at many trophic levels and their results enable us to begin a 'whole-ecosystem' approach to the understanding of the dynamics of a production which man may harvest. In this chapter we shall discuss three examples of this approach. Despite the enormous effort put into all these studies and the wealth of data obtained on the biomass and productivity of various organisms, it is still difficult to give a complete quantitative picture of the main energy pathways through any one system. The difficulties of measuring all the relevant parameters with accuracy and the complexity of the systems leave, in every case, gaps in our knowledge which prevent us from showing exactly the efficiency of energy transfer from the primary producers to fish.

6.2 A temperate and a tropical lake compared

Lake George, in Uganda on the equator, has supported a commercial fishery for over twenty years and Loch Leven, in Scotland, a commercial sport

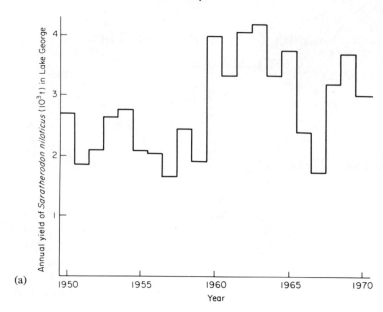

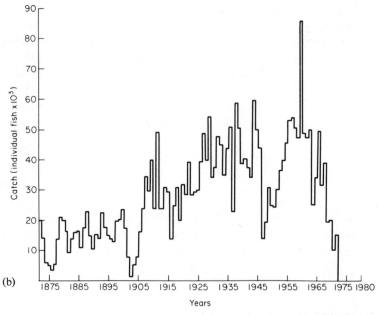

Fig. 6.1. (a) The annual yield of tilapia (*Sarotherodon niloticus*) from Lake George, Uganda from 1950–1970 (from Dunn I.G. (1973). *Afr. J. Trop. Hydrobiol. Fish.* **2**, 109–120) (b) Total annual catch of brown trout (*Salmo trutta*) from Loch Leven, Scotland from 1872–1972 (from Morgan N.G. (1974) *Proc. R. Soc Edinb* (B) **74**, 45–55.)

fishery for more than a hundred years. In both cases reliable yield data is available for a considerable period (Fig. 6.1). The basic physical characteristics of both lakes are set out in Table 6.1. Despite the difference in latitude which accounts for Loch Leven receiving only half the total annual radiation received by Lake George, there are short periods in the summer when the energy available for primary production *per day* is the same.

Table 6.1. The principal physical characteristics of Loch Leven (Scotland) and Lake George (Uganda). (kJ=kilojoules; μmhos=micromhos.)

Description of lake	Loch Leven	Lake George
Latitude	56° 10'N	0° 00'
Longitude	3° 30'W	30° 12'E
Altitude	107 m	913 m
Area of Lake	13·3 km²	250 km²
Depth—mean	3·9 m	2·4±4%
—maximum	25·5 m	*c.* 3 m
Climatic regime	Temperate maritine	Equatorial, two wet seasons
Temperature range	0–20C	23–35C
Rainfall on lake	*c.* 100 cm y⁻¹	82 cm y⁻¹
Area of catchment	145 km²	9955 km²
Run off from catchment	91×10^6 m³	1948×10^6 m³
Renewal time—mean	5·2 months	4·3 months
Incident solar radiation		
—annual total	388×10^4 kJ m⁻²y⁻¹	719×10^4 kJ m⁻²y⁻¹
—daily	0·2–2·0 kJ cm⁻²d⁻¹	1·97 kJ cm⁻²d⁻¹
Day length—annual variation	6·9–17·6 h	12·1 h±0·02 h
Conductivity	200–250 μmhos	*c.* 200 μmhos

Net primary production, which represents the initial fixation of incoming energy, is a function of nutrient supply, usable radiation and temperature. In both the studies of Loch Leven and of Lake George it has proved impossible to give accurate evaluations of net primary production due to the difficulties of distinguishing between algal and heterotrophic respiration in situations where the total community oxygen consumption is very close to the photosynthetic oxygen evolution (Bindloss 1974, Ganf 1974). Both gross photosynthesis and the standing crop of phytoplankton follow a seasonal pattern similar to that of the incident solar radiation at the two lakes (Table 6.2); hence, they are almost constant throughout the year in Lake George (Ganf 1972) and show a wide annual range in Loch Leven (Bindloss 1974, Bailey-Watts 1974). The maximum values for Loch Leven are, in fact, very close to the means for Lake George.

SIMPLIFIED FOOD CHAINS

The energy pool available in the biomass of the primary producers is ultimately the only food available for fish production unless it is supplemented

Table 6.2. Comparison of the annual means of factors involved in planktonic primary production, for Lake George, Uganda with the maximal values obtained for Loch Leven, Scotland. (kJ = kilojoules; chl a = chlorophyll a.)

	Units	Loch Leven (annual range) min	Loch Leven (annual range) max	Lake George (daily mean)
Solar radiation	kJ cm^{-2}d^{-1}	0·2	2·0	1·97 ± 13%
Temperature	°C	0	20	26·5
Phytoplankton standing crop	mg chl a m^{-2}	15	456	450
Gross photosynthesis	g O$_2$ m^{-2}d^{-1}	0·4	15 (21)*	12

* one exceptional day

by a supply of allochthonous material. We shall discuss later a situation (River Thames) where an allochthonous energy input is important. Figure 6.2 illustrates, in simplified, diagrammatic form the main pathways by which the energy fixed by primary producers is transferred to harvested fish in the ecosystems of Loch Leven and Lake George. The energy entering Lake George is fixed by a dense population of planktonic algae dominated by *Microcystis* spp. (Burgis *et al.* 1973). The diet of *Sarotherodon niloticus* (at sizes greater than 6 cm total length) is exclusively phytoplankton, 80% of which is blue-greens. These algae have generally been considered indigestible by fish but Moriarty (1973) has shown that the physiology of *S. niloticus* is adapted to allow this species to assimilate about 43% of the algal carbon it ingests (Moriarty & Moriarty 1973). *Haplochromis nigripinnis*, another herbivorous cichlid, exceeds *S. niloticus* in both numbers and biomass, and it too feeds exclusively on the phytoplankton. In Lake George, therefore, we

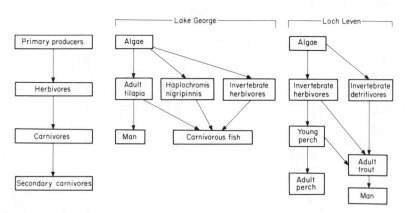

Fig. 6. 2. A generalised food chain applied to simplified diagrams of the main energy pathways leading to the fish exploited by man in Lake George, Uganda and Loch Leven, Scotland.

have the shortest possible food chain by which man can obtain food from an aquatic ecosystem, unless he harvests and consumes algae. Since energy is lost at every transfer between trophic levels, this short pathway should, in theory, allow man to exploit the greatest possible proportion of the energy entering the system. *S. niloticus* forms 80% of the commercial catch (20 year mean 2790 tons y^{-1}: Dunn 1973). Although, at present, the much smaller *H. nigripinnis* is not exploited, there is no theoretical reason why it could not be, as equally small fish are already exploited in Lake Victoria (Kudhongania 1972).

In Loch Leven the dense phytoplankton also forms the basis of the simplified food chain (Fig. 6.2) with energy passing, via chironomid larvae, to brown trout (*Salmo trutta*). As in Lake George, aquatic macrophytes contribute an insignificant fraction of the total primary production (Britton 1974) and the benthic algae probably contribute a considerable, but as yet unmeasured, amount. Besides chironomid larvae the diet of the trout includes fish fry and detritivore invertebrates such as *Asellus*. The large population of the perch (*Perca fluviatilis*) is not harvested. Thus, in contrast to Lake George, where the fish harvest consists mainly of a herbivorous species, the fish harvested from Loch Leven is a carnivore. The exploitation of a carnivore adds at least one additional stage to the food chain and increases the number of transfers at which energy is lost. In Loch Leven the energy fixed by the primary producers passes through a complex web of herbivores, detritivores and carnivores before reaching the fish.

To identify the various factors which determine the level of fish production, we have to look in more detail at the other components of these two systems. To quantify even such simplified diagrams as those in Fig. 6.2 involves several uncertainties, conversion factors and 'order of magnitude estimates'. The authors have been understandably reluctant to apply these conversions to the data of others and do so with due apologies; nevertheless, it is quite an illuminating exercise.

DATA AVAILABLE FROM THE TWO ECOSYSTEMS

Some data obtained from Loch Leven and Lake George (Tables 6.3 and 6.4) have been converted into energy units using the crude conversion factors suggested by Winberg (1971). Gwahaba (1973b) estimated the production of *Sarotherodon niloticus* at 1440 t y^{-1} from his measurements of fish biomass and the analysis of length classes caught over a period of months. This estimate, as he points out, falls below the mean annual yield to the commercial fishery (2790 t y^{-1}). This may imply a decreasing standing crop over a number of years, but we have no means of knowing if this has occurred since these are the first estimates of ichthyomass ever made for the lake. An alternative to this approach is to consider the average yield (2790 t y^{-1})

Table 6.3. Lake George, Uganda: biomass, production and yield of the principal components of the community. (t=metric tonnes=10^3 kg; f.w.=fresh weight; C=carbon; kJ=kilojoules; $\bar{Y}$=mean yield; lake area=250 km²; 1 g C=41·8 kJ; 1 g (f.w.) fish= 4·18 kJ.)

Community component	Data	Source	Energy equivalents
Fish			
YIELD			
S. niloticus (herbivore)	1678–4158 t y⁻¹ (1950–1970)	Dunn 1973	28·1–69·5 kJ m⁻²y⁻¹
	2790 t y⁻¹ (20 year mean)	Dunn 1973	46·6 kJ m⁻²y⁻¹
Carnivore spp.	231–1153 t y⁻¹ (1950–1970)	Dunn 1973	3·9–19·3 kJ m⁻²y⁻¹
	671 t y⁻¹ (20 year mean)	Dunn 1973	11·2 kJ m⁻²y⁻¹
BIOMASS			
S. niloticus (herbivore)	3·7⁹ (f.w.) m⁻²	Gwahaba 1975	15·5 kJ m⁻²
H. nigripinnis (herbivore)	6·7⁹ (f.w.) m⁻²	Gwahaba 1975	28·0 kJ m⁻²
Commercially exploited carnivores	3·5⁹ (f.w.) m⁻²	Gwahaba 1975	14·6 kJ m⁻²
Other fish		See p. 145	6·4 kJ m⁻²
Total fish	29±5 g (f.w.) m⁻²	See p. 145	121·2±20·9 kJ m⁻²
	15·4 g (f.w.) m⁻²	See p. 145	64·4 kJ m⁻²
PRODUCTION			
S. niloticus (adult)	54±8 t (f.w.) month⁻¹	Gwahaba 1973 (a)	10·8 kJ m⁻²y⁻¹
S. niloticus (total)	120±20 t (f.w.) month⁻¹	Gwahaba 1973 (a)	24·1 kJ m⁻²y⁻¹
S. niloticus (estimated total)	5500 t (f.w.) y⁻¹	2$\bar{Y}$ (see p. 145)	92·0 kJ m⁻²y⁻¹
Invertebrate herbivores			
BIOMASS			
Thermocyclops hyalinus	0·25 g C m⁻²	Burgis 1974	10·45 kJ m⁻²
Other herbivorous crustaceae	0·05 g C m⁻²	Burgis 1974	2·09 kJ m⁻²
Benthic herbivores & detritivores	0·33 g C m⁻²	Darlington (pers. comm.)	13·97 kJ m⁻²
Total	0·60 g C m⁻²		25·08 kJ m⁻²

Table 6.3. (*continued*)

Community component	Data	Source	Energy equivalents
PRODUCTION			
T. hyalinus	0·02 g C m^{-2}d^{-1}	Burgis 1974	305 kJ m^{-2}y^{-1}
Invertebrate carnivores			
BIOMASS			
Mesocyclops leukarti	0·065 g C m^{-2}	Burgis 1974	2·72 kJ m^{-2}
Chaoborus spp.	0·45 g C m^{-2}	McGowan 1974	18·8 kJ m^{-2}
	c 0·15 g C m^{-2}	Darlington (pers. comm.)	6·27 kJ m^{-2}
PRODUCTION			
Chaoborus spp.	c 0·03 g C m^{-2}d^{-1}	Darlington (pers. comm.)	457 kJ m^{-2}y^{-1}
Primary producers (Phytoplankton)			
BIOMASS	30 g C m^{-2}	Ganf & Viner 1973	1254 kJ m^{-2}
PRODUCTION			
Gross	5·4 g C m^{-2}d^{-1}	Ganf 1972	82388 kJ m^{-2}y^{-1}
Net	0·6 ± 0·8 g C m^{-2}d^{-1}	Ganf 1972	9154 ± 12206 kJ m^{-2}y^{-1}
	1·52 g C m^{-2}d^{-1}	Estimated from 'back calculation' (see p. 146	23191 kJ m^{-2}y^{-1}

Table 6.4. Loch Leven (Scotland): biomass, production and yield of principal components of the community. (t = metric tonnes = 10^3 kg; d.w. = dry weight; f.w. = fresh weight; C = carbon; kJ = kilojoules; $\bar{Y}$ = mean yield; lake area = 13·3 km²; 1 g C ≡ 41·8 kJ; 1 g (f.w.) fish = 4·18 kJ; 1 g (d.w.) = 21 kJ; assumed C : chl a = 1 : 25 (Bailey-Watts 1974).

Community component	Data	Source	Energy equivalents
Fish			
YIELD			
S. trutta (trout)	12 kg (f.w.) ha⁻¹y⁻¹ (1922–1972)	Thorpe 1974	5·0 kJ m⁻²y⁻¹
BIOMASS (1970–1971)			
S. trutta	4·58 g (f.w.) m⁻²	R. Morgan 1974	19·1 kJ m⁻²
P. fluviatilis (perch)	12·3 g (f.w.) m⁻²	R. Morgan 1974	51·4 kJ m⁻²
Total			70·5 kJ m⁻²
PRODUCTION (1970–1971)			
S. trutta	2·2–3·9 g (f.w.) m⁻²y⁻¹	Thorpe 1974	9·2–16·3 kJ m⁻²y⁻¹
P. fluviatilis	5·1–67·4 g (f.w.) m⁻²y⁻¹	Thorpe 1974	21·3–281·7 kJ m⁻²y⁻¹
Total			30·5–298·0 kJ m⁻²y⁻¹
Benthic invertebrates			
BIOMASS			
in sand	11·4 g (d.w.) m⁻² (1970)	Maitland & Hudspith 1974	239·4 kJ m⁻²
in mud	7·5 g (d.w.) m⁻² (1971–1972)	Charles *et al.* 1974	157·5 kJ m⁻²
Herbivorous chironomids	17 g (d.w.) m⁻²		357 kJ m⁻²
Benthic detritivores (the balance)		See p. 146	40 kJ m⁻²y⁻¹
PRODUCTION			
Chironomid larvae in mud (1971–1972)	34 g (d.w.) m⁻²y	see text p. 146	714 kJ m⁻²y⁻¹
Total benthos in sand (1970)	46·5 g (d.w.) m⁻²y⁻¹	Maitland & Hudspith 1974	976·5 kJ m⁻²y⁻¹
Herbivorous chironomid larvae in sand (1970)	42 g (d.w.) m⁻²y⁻¹	Maitland & Hudspith 1974	882·0 kJ m⁻²y⁻¹

Table 6.4. (*continued*)

Community component	Data	Source	Energy equivalents
in mud (1971–1972)	34 g (d.w.) m^{-2}y^{-1}	Charles *et al.* 1974	714·0 kJ m^{-2}y^{-1}
Total	76 g (d.w.) m^{-2}y^{-1}	see p. 146	1596·0 kJ m^{-2}y^{-1}

Primary producers

BIOMASS

Algae	15–456 mg chl *a* m^{-2}	Bindloss 1974 see p. 146	15·5–476·5 kJ m^{-2}
		R. Morgan 1974	850 kJ m^{-2} (March 1971)

PRODUCTION

Gross (1968–1971)	597–971 g C m^{-2}y^{-1}	Bindloss 1974	25–40 × 10^3 kJ m^{-2}y^{-1}
Net	Estimated	Bindloss 1974	14–25 × 10^3 kJ m^{-2}y^{-1}

of *S. niloticus* and assume total production of this species to be twice this (Chapman 1967, Balon 1973) i.e. 5500 t y^{-1} (92 kJ m^{-2}y^{-1}). This latter estimate is supported by the work of Moriarty *et al.* (1973) who calculated that *S. niloticus* assimilates 17·1 mg C m^{-2} day^{-1} (based on measured assimilation rates and measured biomass figures for the species). This is equivalent to 267·5 kJ m^{-2}y^{-1} (Table 6.3. & above), giving a conversion efficiency of 34%. This is the same efficiency as that obtained by Moriarty and Moriarty (1973) from an estimate of growth related to assimilation by individual fish. Moreover, this value is of the same órder as that obtained by Gerking (1954) for a population of bluegills (*Lepomis macrochirus*), and as the 30–40% suggested by Steele (1974) as a generally reasonable figure for 'young actively growing fish'.

Gwahaba (1975) obtained two figures for the total biomass of fish in the lake as a whole: 29 g m^{-2} fresh weight, based on his survey fishing and 15·4 g m^{-2} derived from the biomasses of individual species obtained from his regular sampling programme. We have decided to use the lower one in order to be conservative and in the hope that it is representative of the condition in the whole lake most of the time.

Despite the difficulties of obtaining an accurate value for net primary production, Ganf (1972) produced an estimate of 600 ± 800 mg C m^{-2} d^{-1}. (9154 kJ m^{-2} y^{-1}). The wide variation is due to the fact that on some days the total carbon lost exceeds the total carbon fixed, and he suggested (1974) that anything from 10–50% of the oxygen uptake by the plankton community may not be due to respiration of the algae. In Lake George, where the standing crop shows little variation, it is a realistic proposition to work 'backwards' from knowledge of consumer requirements, sedimentation and outflow, to an estimate of the minimum net primary production necessary to support the food web and maintain the standing crop observed. Using assimilation data available for the dominant herbivores, and estimating the energy requirements of those benthic animals which are indirectly dependent on the phytoplankton, Moriarty *et al.* (1973) arrived at a figure of 1240 mg C m^{-2} d^{-1} for net primary production. Viner and Smith (1973) estimated that 0·28 g C $m^{-2}d^{-1}$ must be fixed (net) to compensate for the outflow from the lake. Nothing is known about sedimentation. If just these two figures are summed a value of 1·52 g C $m^{-2}d^{-1}$ is obtained, and this is the origin of the second value (23191 kJ $m^{-2}y^{-1}$) quoted for net primary production in Table 6.3.

For Loch Leven (Table 6.4) the figure given for algal biomass in terms of chlorophyll *a* by Bindloss (1974) has been converted to joules using a carbon-to-chlorophyll ratio of 25 (Bailey-Watts 1974) and the conversion factor 1 g C$=41·8$ kJ. Morgan and McClusky (1974) give a figure of 850 kJ m^{-2} (March 1971) in their energy flow diagram for Loch Leven. Since this estimate appears to be at a maximum level of phytoplankton standing crop (Bindloss 1974), we have used it in the following discussion for the purpose of pursuing the idea that levels of biomass and production can be, at times, similar in both Loch Leven and Lake George.

It is now feasible to discuss possible relationships between the trophic levels of these two ecosystems.

INTERRELATIONSHIPS BETWEEN TROPHIC LEVELS WITHIN THE
ECOSYSTEMS OF LOCH LEVEN AND LAKE GEORGE

It is relatively easy to construct a pyramid of biomass for Lake George (Fig. 6.3b) because mean levels are applicable throughout the year. The shape of the pyramid clearly illustrates the preponderance of the phytoplankton within the community. The herbivores and carnivores are only about 5·5% and 2%, respectively, of the phytoplankton biomass. For Loch Leven (Fig. 6.3a) due to the seasonality of the cycles of biomass and production, it is necessary to decide which figures to use. In Fig. 6.3a the primary producers are represented by the planktonic algae which are at their maximum in early spring (see above). The inclusion of benthic algae and macrophytes would increase this value. The value for the herbivores is taken to be that of those species of chironomid larvae which feed primarily on algae (Charles *et al.* 1974). This would be increased by the addition of the zooplankton, particularly the *Daphnia* (see p. 155), for which biomass figures are not yet available Other invertebrates for which biomass has been estimated are detritivores

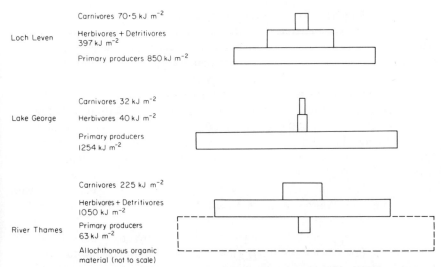

Fig. 6.3. Biomass pyramids for Loch Leven, Scotland, Lake George, Uganda and the the River Thames, England, on the basis of the data presented in Tables 6.3, 6.4 and 6.7 respectively.

(Morgan & McClusky 1974). Brown trout and perch represent the carnivores, although there are, of course, other species of fish for which no estimates have been made and also large populations of *Aythya fuligula* (tufted duck) which also feed on the benthic invertebrates. Within the limited accuracy of these totals the consumers in Loch Leven are shown to form a greater proportion of the total biomass than they do in Lake George.

Biomass levels are only descriptive of an instantaneous situation and they do not indicate the rates at which energy is accumulated in each trophic level and the transfer of this energy from one level to the next. In this respect it is more useful to consider production figures. In Table 6.5 the figure for net primary production for Lake George is that derived as described on p. 146; for Loch Leven it is the mid-point of the range given by Bindloss (Table 6.4). Secondary production for Lake George is the sum of the production of

Table 6.5. Estimates of production in different trophic levels in Loch Leven, Scotland and Lake George, Uganda.

	Loch Leven		Lake George	
	kJ m^{-2}y^{-1}	%	kJ m^{-2}y^{-1}	%
Net primary production	20000	100	23200	100
Secondary production	1614	8	650	3
Tertiary production	150	0·75	150	0·65
Yield	5	0·025	50	0·21

Sarotherodon niloticus (based on known yield, p. 145), production of *Haplochromis nigripinnis*, (derived from its estimated biomass (Gwahaba 1975) on the assumption that its production to biomass ratio is the same as *S. niloticus*), the production of the herbivorous zooplankton (Table 6.3), and production of benthos (assuming, almost arbitarily, that it is about half that of the zooplankton since much of it is feeding on detritus).

In Loch Leven the figure for secondary production is that of the herbivorous chironomid, and the tertiary production is the mid-point of the range estimated for the total of the trout and perch (Table 6.4). Total tertiary production for Lake George has been taken as twice the mean yield of carnivorous fish plus an estimate based on the known biomass of *Mesocyclops leuckarti* and Darlington's figure (Table 6.3) for *Chaoborus* larvae. Comparing these figures, in Loch Leven the production of each tropic level is only slightly less than 10% of the one below. In Lake George, however, the production pyramid is similar to the biomass pyramid. Secondary production (herbivores) is considerably less than 10% of primary production, and it would appear that much of the net primary production is not being converted into secondary production.

In Lake George the ratio of the production of *S. niloticus* to net primary production is less than 1% which may be compared with the $1.02–1.79\%$ obtained by Prowse (1972) from tropical fish ponds containing mainly algal feeding tilapia and some grass carp (*Ctenopharyngodon idellus*). However, the production of *S. niloticus* represents only a part of the total production of herbivorous fish in Lake George.

Although the yield to man, as a proportion of the net primary production, is about ten times higher in Lake George than in Loch Leven we might wonder why it is not even higher in an equatorial situation with the added advantage of harvesting a herbivore. If we take the only two factors from both lakes about which there need be little speculation, we can compare the yield of fish in relation to the solar energy available to the system. In both cases the biomass of the exploited species is about 25% of the total fish biomass and, if we assume that 46% (Talling 1957) of the incident radiation is available for photosynthesis, we get a ratio of solar radiation to yield of 0.0014% in Lake George and 0.00028% in Loch Leven. This is a ratio similar to that obtained using estimates for net primary production and yield. Henderson and Welcomme (1974) in their discussion of possible correlations between a morpho-edaphic index, fish yield and the number of fishermen on African lakes, found a relationship which, when compared with that derived from Ryder (1965), indicated that, in general, the yield from north-temperate lakes is about one-tenth that of similar tropical African waters. That is similar to the results discussed here.

Why is not more of the energy accumulated by primary production not channelled into secondary production in Lake George? Golterman (1975) has suggested that the blue-green algae do not provide an optimal diet for the herbivorous fish but we have already seen (p. 000) that these fish convert a similar proportion of assimilated carbon into growth as other fish. A more probable explanation lies not in the energetics of the system but in its physical limitations. The size of the tilapia population may well be limited by the lack of suitable nursery and breeding areas in Lake George. This fish makes nests on the bottom in shallow water and its young feed, initially, close inshore. Lake George has a simple shoreline, with few bays or indentations, the transition from land to water is generally abrupt with little littoral zone. Thus there may also be competition for space in the inshore areas as only eleven of the thirty-two species of fish recorded from Lake George are found all over the area of the lake and densities are much higher inshore (Gwahaba 1975).

It is also possible that the juveniles suffer heavy predation losses and/or may be limited by the supply of zooplankton food since both species of the herbivorous fish but we have already seen (p. 145) that these fish convert The large size of the dominant algal particles probably determines the structure of the zooplankton community, which is dominated by small cyclopoid copepods able to grasp the clumps of blue-green algae. The fish, however, may not find it easy to select the copepods from among the clumps.

The figures for Loch Leven fish production do not include that of the juveniles, although the major proportion of total production by a fish population may occur in the first two years of life (e.g. Mann *et al.* 1972). The production of young trout does not, however, take place within the Loch itself but in the inflowing streams. Here the young trout, like the juveniles of tilapia in Lake George, may be food limited, since such streams are not highly productive (e.g. Lotrich 1973) and their production may be controlled by quite different factors from those considered in a study confined only to the lake basin. As they migrate into the Loch after one or two growing seasons the young trout can be considered as an input of allochthonous energy to the system.

6.3 The River Thames, England

In both Loch Leven and Lake George the dominant food chains are based primarily on production by planktonic algae. In a river, however, the retention time is normally too short to allow the development of a substantial planktonic community. Considerable development of phytoplankton occurs at some times of the year, in the River Thames, because the time of flow from source to sea (i.e. 'retention time') is increased by natural backwaters, weirs and other man-made modifications which hold back much of the water and reduce the flow. Other possible sources of primary energy are the macro-phytic plants and detritus derived from several sources including leaf fall from terrestrial plants and, above all, man's input of sewage effluent.

The stretch of the River Thames studied by Mann and his co-workers (Mann *et al.* 1972, Berrie 1972a, 1972b) is below the town of Reading at which point sewage effluent constitutes from 2 to 15% of the total river flow, depending on the season. Further effluent, from Reading itself, is added to the main river by the River Kennet within the area studied. On average the river contains 16.5 g dw m^{-3} of suspended material of which 40% is organic. The proportion of this organic material attributable to phytoplankton is highest in summer when it reaches 75% but may fall as low as 2% in the winter (Berrie 1972b). Table 6.6 lists some of the physical parameters of the river at Reading and Table 6.7 data available on the biomass and production of various parts of the community. Where measurements have been taken above and below the mouth of the River Kennet these are given separately, otherwise figures in the first column apply to the general condition of the river at Reading. The material suspended in the water column of the river may occasionally settle on the gravel bed but does not normally remain there for long. Net primary production, as measured in bottles held at one place, does not indicate the total amount of material available to the consumers at any one point in the river, since this also depends on the total load of sus-pended material and the rate of flow. Berrie (1972b) has estimated that the total organic seston carried to the Reading stretch of the river is 8262 t dw y^{-1}, of which 773 t is carbon. If we assume, as before, that 1g C is equivalent to 41.8 kJ we can estimate the calorific value of this as 32311×10^6 kJ y^{-1}, or if we take a mean river width of 60 m, approximately 538×10^6 kJ $m^{-2}y^{-1}$. This may be compared with the 0.018×10^6 kJ $m^{-2}y^{-1}$ of net primary production above the Kennet mouth and 0.008 kJ $m^{-2}y^{-1}$ below Kennet mouth (Mann *et al.* 1972). Contributions from macrophytes, benthic algae and leaf litter are similarly insignificant.

Flowing water with a high load of suspended organic material would appear to be an ideal habitat for filter feeders such as the lamellibranch molluscs (e.g. *Anodonta anatina*), Porifera and Bryozoa which are abundant in the benthic fauna of the Thames. The lamellibranchs dominate in both production and biomass, but they are too large to be exploited directly by the fish, although they may provide a food source when they die. Other components of the benthic fauna such as gastropods, tubificids, insect larvae and *Asellus*, are available as food for the fish but have about one-tenth the production of the filter feeders (Table 6.7). The filter feeding molluscs thus form a sink of energy on the bottom of the river, converting the primary food supply into a form which is not immediately available to fish.

The primary food is more readily made available to fish when channelled to them via planktonic and epiphytic filter feeders. Only small forms, with short life cycles, such as rotifers are found in the zooplankton of the open river but populations of large Cladocera, such as *Sida crystallina* and *Eurycerus lamellatus*, develop in the relatively still water among macrophytes such as *Nuphar lutea* (Bottrell 1975). The macrophytes thus play a very important structural role in this ecosystem, although they contribute relatively little to the total primary production. These Cladocera are filter feeders and not only utilise the suspended organic detritus directly but also contribute to it in the form of cast skins and, as do all the other consumers, as faeces.

The five dominant species of fish in the river are bleak (*Alburnus alburnus*), roach (*Rutilus rutilus*), dace (*Leuciscus leuciscus*), perch and gudgeon (*Gobio gobio*). The juveniles of roach and perch, and to a lesser extent bleak and grudgeon, rely heavily upon the cladocera among the macrophytes for food. Adult roach, dace and gudgeon feed extensively on detritus. Bleak and perch do not feed on detritus at any stage of their life cycle. Bleak feeds on insects at the surface while perch become piscivorous as they grow larger and feed upon fish stocks which themselves may have fed on detritus or detritus feeding invertebrates.

In any fast flowing aquatic system the energy flow through consumers in any one area must be, at least partially, dependent on energy from outside the area rather than that which can be fixed by primary producers within the area. The importance of benthic and epiphytic invertebrates in the food of fish is their ability to fix spatially the mobile organic material and thus present it eventually to the fish in a more readily available form. Whether the fish feed on it directly or indirectly, organic detritus makes an important contribution to the food chains in this river. The majority of this detritus comes, as we have seen, not from within the system itself but from allochthonous sources.

It seems likely that fish production is more efficient when based on animal food sources compared with detritus food sources, despite the energy lost in the transfers. This is partly due to the low energy value of the detritus and the low efficiency with which is is assimilated (Mann *et al.* 1972). Despite the heavy dependence of some species of fish on detritus, we can calculate that their total production is much less than that of the totally carnivorous fish. It we sum the production of bleak, perch and 70% of the roach production (that proportion due to roach juveniles which feed on Cladocera) we get a total of 561 kJ $m^{-2}y^{-1}$ compared with the total fish production of approximately 800 kJ $m^{-2}y^{-1}$.

A biomass pyramid of the River Thames at Reading compared with those of Loch Leven and Lake George (Fig 6.3) has a base that is too narrow; it is an inverted pyramid. When the base is expanded to include all the organic detritus and algae present under a unti area in one day, the pyramidal shape is restored and we have a situation more like that in Lake George where the

Table 6.6 The physical and environmental characteristics of the River Thames at Reading, England (kJ = kilojoules).

Latitude	51° 30′N
Longitude	1° 00′W
Depth—mean	3·0 m
—maximum	4·5 m
Width	40–80 m
Total length of river	338 km
Catchment area	*c.* 10^4 km^2
Distance of Reading from source	150 km
Annual discharge (1967–1968)	1268×10^6 m^3
Water temperature (1967–1968)	1·9–19·9°C
Solar radiation	134–26192 kJ m^{-2}d^{-1}
annual total	305×10^4 kJ m^{-2}y^{-1}
Mean total seston (1967–1968)	19·28 g (d.w.) m^{-3}

Table 6.7. Biomass and production of the various components of the ecosystem in the River Thames, England. Where data are available for the river above and below its confluence with the River Kennet they are given in the appropriate column. Other data, in the first column, may be taken as generally applicable to the river at Reading. All data from Mann (1965) Mann *et al.* (1972), Berrie (1972a and b) and Kowalczewski & Lack (1971). Abbreviations as for Tables 6.3 and 6.4.

Community component	River Thames at Reading	Above Kennet	Below Kennet

Fish

BIOMASS

Carnivore-detritivores (winter)
i.e. Dace, Gudgeon & Adult Roach   57·2 kJ m^{-2}
Carnivores (winter)
i.e. Bleak, Perch & Juvenile Roach   218·2 kJ m^{-2}

PRODUCTION

Carnivore-detritivores	
Dace (*Leuciscus leuciscus*)	33·4 kJ m^{-2}y^{-1}
Gudgeon (*Gobio gobio*)	142·1 kJ m^{-2}y^{-1}
Adult Roach (*Rutilus rutilus*)	36·3 kJ m^{-2}y^{-1}
Carnivores	
Bleak (*Alburnus alburnus*)	447·3 kJ m^{-2}y^{-1}
Perch (*Perca fluviatilis*)	29·2 kJ m^{-2}y^{-1}
Juvenile Roach	84·8 kJ m^{-2}y^{-1}

Table 6.7. (*continued*)

Community component	River Thames at Reading	Above Kennet	Below Kennet
Invertebrate herbivores/detritivores			
BIOMASS			
Bivalves		894 kJ m^{-2}	334 kJ m^{-2}
Other filter feeders		40 kJ m^{-2}	96 kJ m^{-2}
Browsers & grazers		70·6 kJ m^{-2}	54·7 kJ m^{-2}
Total		1004·6 kJ m^{-2}	484·7 kJ m^{-2}
PRODUCTION			
Bivalves		256 ky m^{-2}y^{-1}	107 kJ m^{-2}y^{-1}
Other benthic invertebrates		339 kJ m^{-2}t^{-1}	458 kJ m^{-2}y^{-1}
Invertebrate carnivores			
BIOMASS			
Total		5 kJ m^{-2}	4 kJ m^{-2}
PRODUCTION			
Total		8 kJ m^{-2}y^{-1}	7 kJ m^{-2}y^{-1}
Primary Producers			
BIOMASS			
Algae	1·56 g (d.w.) m^{-3}	1·68 g (d.w.) m^{-3}	0·36 g (d.w.) m^{-3}
PRODUCTION			
Net primary production			
Planktonic algae		18341 kJ m^{-2}y^{-1}	7971 kJ m^{-2}y^{-1}
Macrophytes		99 kJ m^{-2}y^{-1}	184 kJ m^{-2}y^{-1}
Benthic algae	8824 kJ m^{-2} (8 months)		
Allochthonous material			
Total suspended organic matter (1967–1968 instantaneous mean)	0·609 g C m^{-3} ≡ 25·5 kJ m^{-3}		
(1967–1968 annual total)	773 t C ≡ 32311 × 10^6 kJ y^{-1}		
PRODUCTION			
Allochthonous leaf litter	330 kJ m^{-2}y^{-1}		

potential primary food supply is present in excess. In flowing water, however, most of this is carried downstream, and some of it is utilised elsewhere.

The River Thames no longer supports a commercial fishery, although historically anadromous runs of fish were exploited.

Highly productive, commercially exploited river fisheries tend to exist in those rivers which seasonally flood to create shallow, lacustrine conditions. Since these conditions are temporary, the highest production tends to come from fast-growing species. Hence the importance of tropical, warm-water flood-plains (Welcomme 1975) and the relative unimportance of cold-water seasonal floods, which in any case tend to coincide with the cold, unproductive season of the year. Nevertheless, fish biomass in the Thames is high compared with other flowing waters which have been studied and production would probably be increased by exploitation (Mann 1965, Berrie 1972a).

6.4 Discussion

We have examined two lakes from which man exploits fish, one on the equator and one in the north-temperate zone. In the former a herbivore is exploited and in the latter a carnivore. In theory we should expect that in the former, regardless of its latitude, a greater proportion of the energy available to the ecosystem should be channelled into the fish. In our comparisons of the Lake George and Loch Leven ecosystems it has proved impossible to discover whether the greater fish production of the former is due to the trophic status of the species exploited or to the equatorial climatic regime to which it is subjected. Data are lacking that would make it possible to compare the Lake George ecosystem with a similar equatorial ecosystem in which a carnivore is exploited. The very fact that herbivores are a major component of the fish harvest from so many warm-water lakes and fish ponds, suggests a link between a tropical climate and the existence of large populations of herbivorous fish. There are no strictly herbivorous fish in the north-temperate zone. Is this because the primary producers on which herbivores would feed are not able to sustain the necessary biomass for more than a short period of the year? Many fish of the temperate zones feed upon those invertebrates which are available throughout the year, and even the grass carp (*Ctenopharyngodon idellus*), which is able to digest macrophytes, is more omnivorous than is generally supposed (e.g. Fischer 1973).

We have also examined a river ecosystem, from the same latitude as Loch Leven, where the same level of net primary production is achieved but where it is doubtful if it would do so but for man's interference in the flow of the river. Normally suspended material would be swept beyond the reach of all but the few consumers able to trap it. The primary energy supply is, however, supplemented by allochthonous material whose relatively low energy content is partially offset by its vast quantity. Although some adult fish do feed directly on this allochthonous organic material, their growth is slow and they contribute only a small percentage to the total fish production. The greatest proportion of fish production is due to those species and age groups which feed on invertebrates or other fish, which have themselves fed on detritus or are carnivores. This is also true in Loch Leven. It therefore seems possible that detritus is more efficiently converted to fish biomass when channelled through consumers, despite the losses of energy involved in the transfers.

In their study of a trout stream Warren *et al.* (1964) found that enrichment with moderate quantities of sucrose caused an increase in the growth of micro-organisms on which invertebrates could feed. The resulting increase in their food supply led to a seven-fold increase in fish production. The River Thames has been 'enriched' by man for a very long time, and this has no doubt had profound effects on the structure of the ecosystem. Loch Leven has also been affected by man's activities. Sewage effluents, plus the run-off containing increasing amounts of agricultural fertilizers, have almost certainly been prime causes in the increase in phytoplankton, the decrease in macrophytes and the associated changes in the fauna (Morgan 1970). A significant aspect of these changes may have been the disappearance of the once abundant *Daphnia hyalina*. *Cyclops strenuus*, which became the dominant zooplankton after *D. hyalina* disappeared, is usually considered to be a carnivore and is a raptorial rather than a filter feeder. The population of algae increased with the reduction of grazing, small species became dominant, light penetration was reduced, and the macrophytes declined. It is possible that with the return of *Daphnia hyalina* this process may be reversed (R. Morgan 1974). Moreover, although juvenile trout live and feed in the inflowing streams, juvenile perch feed in the lake and probably utilise the *Daphnia*. Young perch fall prey not only to adults of their own species but also trout.

In Lake George the two most abundant species of fish feed on phytoplankton throughout the year. In the temperate zone, where the plankton community does provide food for fish, particularly juveniles, it does so in quantity for only part of the year. As the life-cycles of benthic invertebrates tend to be longer than those of zooplankton, food is available from this source for a more extended period.

The shortness of the growing season in the temperate zone may in part be compensated for by the accumulation of biomass in each trophic level before the build-up of the population of organisms which will exploit it as the season progresses. At the beginning of each growing season, in temperate zones, the primary producers start to build up, often from a very low level, and it is possible for biomass to accumulate in the absence of the consumer population. When consumer populations start to increase in their turn, they are enabled to do so at a faster rate by the presence of an abundant food supply.

A marked contrast is seen in situations with little seasonal change. In Lake Tanganyika despite the removal of considerable quantities of fish which feed on the zooplankton, which in turn must graze on the phytoplankton, the waters of the lake are renowned for their clarity and this has been taken to indicate a lack of productivity (Beadle 1974). It seems more likely that all production is consumed as fast as it is produced. On the other hand in Lake George, which also experiences little seasonal change, there is a large accumulation of primary producers only a small proportion of which is transferred to consumers. It is not yet possible to explain these differences which probably result from the very different morphometry of these two tropical lakes.

Thus we can reach no hard and fast conclusions from our analysis of only three whole-ecosystem studies but their data have provided a basis for some interesting speculations. The kind of questions which have arisen from such studies of pattern in ecosystem structure and function should be the springboard for the next stage of investigations in the field.

The comparisons made in this chapter have shown that although, in theory, short-energy pathways should lead to higher total conversion efficiencies, in natural waters the advantages of short food chains are offset by a multitude of other variables in the total ecosystem complex. To realise the full potential of theoretical efficiencies, the communities must be simple and controlled, as in fish ponds or other artificial ecosystems.

References

BAILEY-WATTS A.E. (1974) The algal plankton of Loch Leven, Kinross. *Proc. R. Soc. Edinb.* (B) **74**, 135–156.

BALON E.K. (1973) Results of fish population size assessments in Lake Kariba coves (Zambia), a decade after their creation. *Geophysical Monographs* **17** (W.C. Ackermann, G.F. White and E.B. Worthington, eds.).

BEADLE L.C. (1974) *The Inland Waters of Tropical Africa.* Longman, London.

BEATTIE M., BROMLEY H.J., CHAMBERS M., GOLDSPINK R., VIJVERBERG J., ZALINGE N.P. VAN & GOLTERMAN H.L. (1972) Limnological studies on Tjeukemeer—a typical Dutch 'polder reservoir'. *Productivity Problems of Freshwaters* (Z. Kajak and A. Hillbricht-Ilkowska, eds.), pp. 421–446. PWN Warsaw.

BERRIE A.D. (1972a) Productivity of the River Thames at Reading. *Symp. zool. Soc. Lond.* **29**, 69–86.

BERRIE A.D. (1972b) The occurrence and composition of seston in the River Thames and the role of detritus as an energy source for secondary production in the river. *Mem. Ist. Ital. Idrobiol.* **29** suppl., 473–483.

BINDLOSS M.E. (1974) Primary productivity of phytoplankton in Loch Leven, Kinross. *Proc. R. Soc. Edinb.* (B) **74**, 157–181.

BOTTRELL H.H. (1975) The relationship between temperature and duration of egg development in some epiphytic Cladocera and Copepoda from the River Thames, Reading, with a discussion of temperature functions. *Oecologia* (Berl.) **18**, 63–84.

BRITTON R.H. (1974) Factors affecting the distribution and productivity of emergent vegetation at Loch Leven, Kinross. *Proc. R. Soc. Edinb.* (B) **74**, 209–218.

BURGIS M.J., DARLINGTON J.E.P.C., DUNN I.G., GANF G.G., GWAHABA J.J. & McGOWAN L.M. (1973) The biomass and distribution of organisms in Lake George, Uganda. *Proc. R. Soc.* (B) **184**, 271–298.

BURGIS M.J. (1974) Revised estimates for the biomass and production of zooplankton in Lake George, Uganda. *Freshwat. Biol.* **4**, 535–541.

CHAPMAN D.W. (1967) Production in fish populations. *The Biological Basis of Freshwater Fish Production* (S.D. Gerking, ed.). Blackwell Scientific Publications, Oxford.

CHARLES W.N., EAST K., BROWN D., GRAY M.C. & MURRAY T.D. (1974) The production of larval Chironomidae in the mud at Loch Leven, Kinross. *Proc. R. Soc. Edinb.* (B) **74**, 241–258.

DUNN I.G. (1973) The commercial fishery of Lake George, Uganda (East Africa). *Afr. J. Trop. Hydrobiol. Fish.* **2**, 109–120.

EFFORD I.E. (1969) Energy transfer in Marion Lake, British Columbia, with particular reference to fish feeding. *Verh. Internat. Verein. Limnol.* **17**, 104–108.

EFFORD I.E. (1972) An interim review of the Marion Lake project. *Productivity Problems of Freshwaters* (Z. Kajak and A. Hillbricht-Ilkowska, eds.), pp. 89–109. PWN Warsaw.

FISCHER Z. (1973) The elements of energy balance in grass carp (*Ctenopharyngodon idellus* Val.) Part IV. Consumption rate of grass carp fed on different types of food. *Polskie. Archwm. Hydrobiol.* **20**, 309–318.

GANF G.G. (1972) The regulation of net primary production in Lake George, Uganda, East Africa. *Productivity Problems of Freshwaters* (Z. Kajak and A. Hillbricht-Ilkowska, eds.), pp. 693–708. PWN Warsaw.

GANF G.G. (1974) Rates of oxygen uptake by the planktonic community of a shallow, equatorial lake (Lake George, Uganda). *Oecologia* (Berl.) **15**, 17–32.

GANF G.G. & VINER A.B. (1973) Ecological stability in a shallow equatorial lake (Lake George, Uganda). *Proc. R. Soc.* (B) **184**, 321–346.

GERKING S.D. (1954) The food turnover of a bluegill population. *Ecology* **35**, 490–498.

GOLTERMAN H.L. (1975) *Physiological Limnology.* Elsevier, Amsterdam.

GREENWOOD P.H. & LUND J.W.G. (1973) A discussion on the biology of an equatorial lake: Lake George, Uganda. *Proc. R. Soc.* (B) **184**, 227–346.

GWAHABA J.J. (1973a) Population studies of the more abundant fish species in Lake George, Uganda. M.Sc. Thesis, Makerere University, Kampala.

GWAHABA J.J. (1973b) Effects of fishing on the *Tilapia nilotica* (Linne 1757) population in Lake George, Uganda over the past 20 years. *E. Afr. Wildl. J.* **11**, 317–328.

GWAHABA J.J. (1975) The distribution, population density and biomass of fish in an equatorial lake, Lake George, Uganda. *Proc. R. Soc.* (B) **190**, 393–414.

HENDERSON H.F. & WELCOMME R.L. (1974) The relationship of yield to morpho-edaphic index and numbers of fishermen in African inland fisheries. FAO CIFA Occasional paper 1. FAO Rome.

KUDHONGANIA W.A. (1972) Past trends and recent research on the fisheries of Lake Victoria in relation to possible future developments. *Afr. J. Trop. Hydrobiol. Fish.* Special Issue No. 2, 93–106.

KOWALCZESKI A. & LACK T. (1971) Primary production and respiration of the phytoplankton of the Rivers Thames and Kennet at Reading. *Freshwat. Biol.* **1**, 197–212.

LOTRICH V.A. (1973) Growth, production and community composition of fishes inhabiting a first, second and third order stream of eastern Kentucky. *Ecol. Monogr.* **43**, 377–397.

MAITLAND P.S. & HUDSPITH P.M.G. (1974) The zoobenthos of Loch Leven, Kinross, and estimates of its production in the sandy littoral area during 1970 and 1971. *Proc. R. Soc. Edinb.* (B) **74**, 219–239.

MANN K.H. (1965) Energy transformations by a population of fish in the River Thames. *J. Anim. Ecol.* **34**, 253–275.

MANN K.H. (1969) The dynamics of aquatic ecosystems. *Advances in Ecological Research* 6. Academic Press, New York.

MANN K.H., BRITTON R.H., KOWALCZEWSKI A., LACK T.J., MATTEWS C.P. & McDONALD I. (1972) Productivity and energy flow at all levels in the River Thames, England. *Productivity Problems of Freshwaters* (Z. Kajak and A. Hillbricht-Ilkowska, eds.), pp. 579–596. PWN Warsaw.

McGOWAN L.M. (1974) Ecological studies on *Chaoborus* (Diptera, Chaoboridae) in Lake George, Uganda. *Freshwat. Biol.* **4**, 483–505.

MORGAN N.C. (1970) Changes in the flora and fauna of a nutrient enriched lake. *Hydrobiologia* **35**, 545–553.

MORGAN N.C. (1974) Historical background to the International Biological Programme project at Loch Leven, Kinross. *Proc. R. Soc. Edinb.* (B) **74**, 45–55.

MORGAN N.C. & McCLUSKY D.S. (1974) A summary of the Loch Leven IBP results in relation to lake management and future research. *Proc. R. Soc. Edinb.* (B) **74**, 407–416.

MORGAN R.I.G. (1974) The energy requirements of trout and perch populations in Loch Leven, Kinross. *Proc. R. Soc. Edinb.* (B) **74**, 333–345.

MORIARTY C.M. & MORIARTY D.J.W. (1973) Quantitative estimation of the daily ingestion of phytoplankton by *Tilapia nilotica* and *Haplochromis nigripinnis* in Lake George, Uganda. *J. Zool. Lond.* **171**, 15–23.

MORIARTY D.J.W. (1973) The physiology of digestion of blue-green algae in the cichlid fish *Tilapia nilotica. J. Zool. Lond.* **171**, 25–39.

MORIARTY D.J.W., DARLINGTON J.E.P.C., DUNN I.G., GANF G.G., GWAHABA J.J., McGOWAN L.M. & TEVLIN M.P. (1973) Feeding and grazing in Lake George, Uganda. *Proc. R. Soc.* (B) **184**, 299–319.

PECHLANER R., BRETSCHKO G., GOLLMAN P., PFEIFER H., TILZER M. & WEISSENBACH H.P. (1973) Das Ökosystem Vorderer Finstertaler See. *Ökosystemforschung.* Springer-Verlag.

PROWSE G.A. (1972) Some observations on primary and fish production in experimental fish ponds in Malacca, Malaya. *Productivity Problems of Freshwaters* (Z. Kajak and A. Hillbricht-Ilkowska, eds.), pp. 555–561. PWN Warsaw.

RIGLER F.H. (1972) The Char Lake project: A study of energy flow in a high Arctic lake. *Productivity Problems of Freshwaters* (Z. Kajak and A. Hillbricht-Ilkowska, eds.), pp. 287–300. PWN Warsaw.

RYDER R.A. (1965) A method for estimating the potential fish production of north-temperate lakes. *Trans. Am. Fish. Soc.* **94**, 214–218.

SCHINDLER D.W., WELCH H.E., KALFF J., BRUNSKILL G.J. & KRITSCH N. (1974) Physical and chemical limnology of Char Lake, Cornwallis Island (75°N Lat.). *J. Fish. Res. Bd. Canada* **31**, 585–607.

STEELE J. (1974) *The Structure of Marine Ecosystems.* Blackwell Scientific Publications, Oxford.

TALLING J.F. (1957) Photosynthetic characteristics of some freshwater plankton diatoms in relation to underwater radiation. *New Phytol.* **56**, 1–132.

THORPE J.E. (1974) Trout and perch populations at Loch Leven, Kinross. *Proc. R. Soc. Edinb.* (B) **74**, 295–313.

VINER A.B. & SMITH I.R. (1973) Geographical, historical and physical aspects of Lake George. *Proc. R. Soc.* (B) **184**, 235–270.

WARREN C.E., WALES J.W., DAVIS G.E. & DOUDOROF P. (1964) Trout production in an experimental stream enriched with sucrose. *J. Wildl. Manage.* **28**(4), 617–660.

WELCOMME R.L. (1975) The fisheries ecology of African floodplains. FAO CIFA Technical Paper No. 3. FAO Rome.

WINBERG G.G. (1971) *Symbols, Units and Conversion Factors in Studies of Freshwater Productivity.* International Biological Programme, Central Office, London.

WINBERG G.G. (1972) Some interim results of Soviet IBP investigations of lakes. *Productivity Problems of Freshwaters* (Z. Kajak and A. Hillbricht-Ilkowska, eds.), pp. 363–404. PWN Warsaw.

Chapter 7: Digestion and the Daily Ration of Fishes

John T. Windell

7.1 Introduction

Knowledge of fish population trophic dynamics is required to interpret the influence of a variety of parameters on fish production (Warren *et al.* 1964, Brett & Higgs 1970, Tyler 1970, Edwards 1971, Swenson & Smith 1973, Elliott 1972, 1975a, 1975b). Appetite, food kind, amount consumed, frequency of consumption, digestibility, rate of movement through the gut, absorption of nutrients and conversion efficiency are important steps in the transformation of fish food into animal tissue. During the growing season these processes interrelate to determine growth and production rates of fish populations.

Rate of digestion for fish is defined as the rate at which food passes from the stomach. Digestion is considered complete when the stomach becomes empty of all measurable remains. The terms 'rate of digestion', and 'gastric digestion' have been used throughout most of the fish literature. However, most workers have measured and reported the amount of food removed from the stomach by peristalsis over a defined time period. Therefore, terms such as 'gastric evacuation', 'gastric decrease', 'gastric depletion' and 'gastric removal' would be more correct to avoid any implication of physiological digestion, intestinal digestion, absorption or passage of undigested material. These terms are used synonomously in this chapter.

Rate of digestion was recognized by Ricker (1946) as having an important bearing on fish production in terms of estimating the daily ration. Since that time, however, digestion rate in combination with diel observations of stomach contents, quantity and quality of food consumed has led to the estimation of daily food consumption in nature in only a few studies (Darnell & Meierotto 1962, Seaburg & Moyle 1964, Kitchell 1970, Swenson & Smith 1973) and current research is intermittent. The literature earlier than 1960 offers only meager information and most often neglects the many variables that affect rates of gastric depletion.

Factors found to be important in assessment of gastric evacuation rates include water temperature, fish size, meal size, meal succession, food particle size, food digestibility and fat concentration. Although not thoroughly studied, the present evidence indicates that season of the year, species

159

differences and/or population differences do not influence gastric removal rates. However, length of the starvation period and force-feeding have a pronounced effect and caution should be exercised when designing and conducting experiments to avoid errors that these procedures produce.

Digestion rate data also provide information for a number of related problems. Values can be used to determine rates of energy passage within certain pathways of marine food webs and underlying relations controlling growth (Tyler 1970); the relations between feeding motivation, appetite, feeding activity, and quantity of food in the stomach (Magnuson 1969); maximal stomach capacity for food consumed at different temperatures (Elliott 1975a, 1975b, Brett & Higgs 1970); different feeding frequencies (Sarokon 1975) ration size for fish cultural practices (Brett & Higgs 1970, Elliott, 1975c, 1975d) and conversion efficiency (Pandian 1967a, 1967b).

7.2 Food consumption and digestion rates

FIELD METHODS

The 'daily meal', was first used by Surber (1930) and soon after, Bajkov (1935) proposed a limited field method for rough estimates of the total daily, seasonal, and annual meal of fish populations. Establishment of the size of the daily meal enables the daily ration to be computed. Ricker (1946) defined the 'daily ration' as the size of the daily meal expressed as a percentage of the body weight.

Early estimates of consumption based on digestion methods were considered to be inaccurate (Mann 1967). Inaccuracy can be traced to variables which include rate of digestion at different temperatures, for different kinds of food organisms and diel variations in feeding activity. Procedures which do not describe feeding rate for all periods of a day result in bias if feeding is periodic (Darnell & Meierotto 1962). By making estimates of consumption for all periods of the day, feeding periodicity can be estimated, thereby eliminating possible bias associated with periodic feeding (Kitchell 1970, Swenson & Smith 1973). Therefore, the methods proposed by Bajkov (1935), Darnell & Meierotto (1962), Windell (1966), Seaburg & Moyle (1964) and Kitchell (1970) although of intrinsic interest, should not be used without substantial modification and refinement.

Reliability of food consumption estimates in wild fish stocks rests ultimately upon an assumption that laboratory results are representative of field conditions. A digestion rate method may be best suited to meet this assumption because other methods depend on laboratory measurements of more complex physiological processes such as metabolism, growth, nitrogen assimilation and efficiency of assimilation (Swenson & Smith 1973). These methods have their own inherent errors and require more assumptions for extrapolation to field conditions that does the digestion method (Mann 1967).

COMBINED FIELD-LABORATORY METHOD

Swenson and Smith (1973) developed a field-laboratory method of estimating daily consumption rate for an average individual from a field population utilizing information from laboratory-based digestion experiments. The technique appears to be suitable for general use and can describe consumption for a given day of individual prey species. When used in conjunction with information of food availability, the method permits the assessment of the effect of grazing on the food supply. Consumption rate combined with information on growth rate also permits estimating conversion efficiency.

All sizes of fish and sizes of meals encountered in the field are included in the laboratory work. The method used to estimate food consumption is summarized in the equation:

$$C = \Sigma_t \Sigma_s \frac{(\Sigma f \times SC)}{F}$$

where C = daily food consumption in grams for the average fish; SC = undigested weight of stomach contents for foods of a given size, not more than 90% digested, consumed during a time period; f = summation corrected weight of stomach content for all fish in the sample, having consumed food of a given size; F = number of fish in the sample which could have contained food of a given size, in a state of digestion not exceeding 90% during a time period; s = summation of food sizes; t = summation of time. This method is the only approach so far developed which describes consumption for a given day. By making estimates of consumption for all periods of a day the problem of periodic feeding is avoided.

LABORATORY METHODS

X-ray method

Molnár and Tölg (1960, 1962a, 1962b) designed and utilized an ingenious X-ray method for determining gastric digestion in several piscivorous species. The fish remain alive and can be used repeatedly. The skeleton, swim bladder and otoliths of the prey fish in the stomach produce X-ray images that are easy to observe on film. A similar X-ray method was used by Edwards (1971). Mean times for test meals to reach the rectum were obtained when pieces of food were injected with barium sulphate paste, force-fed to fish and later exposed to X-ray film. However, viewing the X-ray image on a fluorescent screen and noting the position of the food mass proved to be more rapid and just as accurate.

Radioisotopic method

Peters and Hoss (1974) tagged small fish and shrimp either individually by injecting 25 uCi $^{144}CeCl_3$ dissolved in HCl or as a group by holding the organisms for 12 hr in sea water containing radioactive cerium (200 uCi ^{144}Ce/liter). Fish were fed to satiation after starvation. The gut of live fish placed in glass jars could be repeatedly measured for radioactivity in a whole animal scintillation counter. Comparison of this method with the dried digestible organic weight technique (described later) gave similar estimates of the time to reach 99% gastric evacuation.

Radioactive cesium has been used as a tracer (Kevern 1966) but since it is readily assimilated it is undesirable for evacuation studies. Pappas *et al.* (1973) used chromium oxide (Cr_2O_3) in estimating digestion and absorption of protein, fat, and carbohydrate by neutron activation analysis. Radioactive cerium (^{144}Ce) has been suggested as a substitute for digestion and digestibility studies because it is easily analyzed in samples, very poorly assimilated by fish, permits determinations on the same fish and requires less effort (Cowey & Sargent 1972, Peters & Hoss 1974).

Water displacement method

Hunt (1960) reported rates of digestion by measuring meals and the partly digested food recovered from the stomachs upon autopsy as volume by water displacement. The relationship between volume and weight was established by comparing these two measurements of the prey fish, the mosquitofish, *Gambusia affinis*. Volume was considered to be the equivalent of weight.

Dried digestible organic weight method

A dried digestible organic weight method was described and utilized in several laboratory experiments by Windell (1966). Each food type was analysed for its percentage of organic matter and chitin. Digestible organic matter was defined as dry weight minus ash weight and chitin weight. Rate of digestion was measured as the difference between the weight of food intake and the stomach remains at timed intervals in terms of dried digestible organic matter.

Dry weight method

Windell *et al.* (1969) reported rates of gastric evacuation measured by utilizing a dry weight method. All computations were based on percentage dry weight of the food. Subtraction of dry matter remaining in the stomach at autopsy from dry matter consumed yielded the amount of food evacuated from the stomach per unit time. It is assumed that indigestible materials such as ash, fiber, chitin and debris pose no problem with reference to gastric emptying. Comparison of this method with the dried digestible organic weight method gave equivalent results.

LABORATORY FEEDING TECHNIQUES

The individual method

Difficulty in getting groups of wild fish held in individual tanks or trough compartments to simultaneously consume a measured meal requires special handling techniques (Windell 1966, 1967, 1968, 1971). A dominance and subordinance behaviour has been reported for bluegill sunfish (Gerking 1955) and many other species (Jenkins 1969). When held in groups the more dominant, aggressive fish when fasted for a short period will feed readily when food is offered. Subordinate fish observe the feeding dominants and feed after some time has elapsed. However, when individually placed in separate aquaria or trough compartments, dominant and subordinate fish often require a substantial conditioning period before feeding regularly. A 'teasing' technique developed by Windell (1966) requires that considerable caution be taken to prevent startling the fish during the feeding period. Individually housed fish are offered food 3 or 4 times per day. If not consumed in a short period of time the food items are gently removed in 10–15 minutes. Consequently, most of the subordinate fish are teased into feeding on the desired food item for each experiment.

Group method

Some investigators have preferred to feed groups of fish to satiation over a period of 15 minutes, perform serial slaughter at intervals, freeze until time of dissection, remove stomach contents and apply the digestible organic weight method (Brett & Higgs 1970, Elliott 1972, Peters & Hoss 1974). Subsamples are sacrificed initially (time 0) and at pre-determined time intervals. Data are compiled as the percentage of organic matter in the fish's stomach as milligrams organic content/100 mg dry body weight. The geometric means of these percentages is determined for each time interval and regressed as the logarithm of the percent of stomach contents against time. Because of variation in the amount of food consumed by individual fish, individual measurements are deleted from subsamples on the following basis: (1) no fresh food present in the very early stages of digestion; (2) notable premature empty stomach; (3) highly divergent contents at intermediate stages of digestion; and (4) when chronic poor or nonfeeders are present.

7.3 Water temperature

The successive steps from feeding to the transformation of food by fishes are influenced by numerous physical, chemical and biological factors, but none is more important than water temperature. Whereas growth rate has been shown repeatedly to be characterized by an optimum temperature, digestion rate does not display a distinct optimum. The rate tends to increase with rising temperature, reaching a maximum near the upper limit temperature tolerance for the species (Molnár & Tölg 1962a, 1962b, Smit 1967, Molnár *et al.* 1967, Shrable *et al.* 1969, Brett & Higgs 1970, Tyler 1970). Beyond the maximum, rate drops precipitously (Tyler 1970), when the fish lose their appetite, cease feeding and become extremely lethargic. Data collected near the limit of temperature tolerance is subject to high variability.

The available literature on fish gastric evacuation rates and temperature can be separated into two groups. A number of studies have been performed under controlled temperature conditions and the results expressed as the number of hours for 50% of the food to pass from the stomach (Table 7.1). Other studies express the data as the number of hours required for 100% of the food to pass from the stomach (Table 7.2). Still other studies have been conducted over a broader range of water temperature. (Table 7.3, time to 50% empty; Table 7.4, time to 100% empty). For brevity of presentation three temperature ranges are considered separately: low, intermediate and high.

Low temperature: 1–9°C

Consumption (meal size) and hence growth at low temperatures, depends heavily on the rate of gastric evacuation. For most species the rate of stomach emptying is greatly depressed at 0C (Brett & Higgs 1970). The possibility of temperature and metabolic compensation for seasonal temperatures has been offered by Smit (1968), Molnár *et al.* (1967), and Brett and Higgs (1970). Data for sockeye salmon, *Oncorhynchus nerka*, suggests a compensatory increase in the rate of digestion in winter above the rate expected from experiments conducted at low temperatures in summer. The expected digestion time at 1°C for the sockeye on the basis of experiments done in the summer was 796 hours or approximately 33 days, whereas winter fish demonstrated compensatory increase by growing and consuming 1·5% dry body wt/day.

Intermediate temperature: 10–18°C

As water temperature rises from low to intermediate, gastric evacuation rate increases accordingly. Related processes such as appetite, food intake, absorption, and conversion efficiency increase in roughly the same proportion. Even slight temperature changes within this range result in significant changes in rate.

High temperature: 19–30°C

Data from most studies clearly indicate that the most rapid gastric evacuation rate occurs at the highest tested temperature. Once food is taken into the stomach at high temperature an orderly sequence of events proceeds through which various enzymes are liberated from their site of production. Peristalsis is initiated and the mixture of partially digested food and digestive juices— chyme—is readied for passage into the intestine. Enzyme kinetic studies indicate that the temperature coefficient (Q_{10}) increases as temperature increases. Beyond a critical temperature, however, enzyme denaturation apparently occurs and digestion fails near the thermal death point.

7.4 Body size

Although numerous gastric evacuation rate studies have been conducted, few experiments have been designed to determine the relationship between evacuation time and size. Inasmuch as stomach capacity increases proportionally with fish size, effects of size of fish on evacuation rate are confounded with effects of meal size. Consequently, a slower evacuation rate of larger fish may also reflect effects of greater amounts of food in their stomachs (Noble 1973). However, Pandian (1967b) found that digestion rate varied inversely with size of *Megalops cyprinoides* and Tsunikova (1969, 1970) also found inverse relationships for young roach (*Rutilus rutilus*) and for young pikeperch (*Stizostedium lucioperca*). Larval largemouth bass (*Micropterus salmoides*) fed with zooplankton evacuated more rapidly than adults (Laurence 1971).

Noble (1973) reported the effects of fish size on evacuation rate for young perch (*Perca flavescens*) fed zooplankton. Median gastric evacuation time increased steadily with size of fish at 22°C. Zooplankton passed through the stomachs of 23 mm perch in less than one hour, whereas 45 mm fish required over 3 hours to evacuate the experimental meal under conditions of continual *ad libitum* feeding.

7.5 Meal size

Meal size and rate of gastric emptying has received considerable attention (Hunt 1960, Windell 1966, Kitchell & Windell 1968, Magnuson 1969,

Table 7.1. Number of hours for stomach to empty for various fish species ordered by the number of hours for 50% of the food to pass from the stomach

Species	Food Digested	Hours for 50% of food to pass from stomach	Weight in g or length in cm	Water Temperature °C	Source
Pacific mackerel (*Scomber japonicus*)	—	3	15	20	Kariya (1956)
Black bullhead (*Ictalurus melas*)	*Chironomous riporius*	3	16·4	21–24	Baur (1970)
Pumpkinseed sunfish (*Lepomis gibbosus*)	Damselfly naiads *Ischnura* sp.	4–6	21	20·5	Kitchell & Windell (1968)
Skipjack tuna (*Katsuwonus pelamis*)	Whole white bait (*Osmerdiae*)	5	(39–50 cm)	23·3–25·7	Magnuson (1969)
Pumpkinseed (*Lepomis gibbosus*)	Natural food mixture	5	(12–18 cm)	18–23	Seaburg & Moyle (1964)
Bluegill (*Lepomis macrochirus*)	Crayfish (*Cambarus* sp.)	5	62	22	Windell (1967)
Bluegill (*Lepomis macrochirus*)	Darters (*Etheostoma* sp.)	5	64	21	Windell (1967)
Jack mackerel (*Trachurus japonicus*)	Natural food mixture	6	(10–22 cm)	18–23	Seaburg & Moyle (1964)
	Natural food mixture	6	20–22	ca. 24	Aoyama (1958)

Table 7.1. (*continued*)

Species	Food				Reference
Black crappie (*Pomoxis nigromaculatus*)	Natural food mixture	7	(15–28 cm)	18–23	Seaburg & Moyle (1964)
Largemouth bass (*Micropterus salmoides*)	*Notropis atherinoides*	6·6	91	25	Beamish (1972)
Megalops (*Megalops cyprinoides*)	*Gambusia affinis*	8–9	52	28	Pandian (1967b)
Largemouth bass (*Micropterus salmoides*)	*Gambusia affinis*	9	89	23–26	Hunt (1960)
Rainbow trout (*Salmo gairdneri*)	Encapsulated pellets	10	84–90	15	Windell *et al.* (1969)
Warmouth (*Chaenobryttus gulosus*)	*Gambusia .affinis*	14	93	23–26	Hunt (1960)
Florida gar (*Lepisosteus platyrhincus*)	*Gambusia affinis*	20	110	23–26	Hunt (1960)
Northern pike (*Esox lucius*)	perch	20	(40 cm)	18–23	Seaburg & Moyle (1964)
Rainbow trout (*Salmo gairdneri*)	Stoneflies *Pteronarcys*	22	129	10	Windell, Horak & Reynolds (unpublished)

Table 7.2. Number of hours for stomach to empty for various fish species ordered by the number of hours for 100% of the food to pass from the stomach

Species	Food digested	Hours for 100% of food to pass from stomach	Weight in g or length in cm	Water temperature °C	Source
Megalops (*Megalops cyprinoides*)	Prawn *Metapenaeus monoceros*	6·5	5	28	Pandian (1967a)
Largemouth bass (*Micropterus salmoides*)	Bleak *Alburnus alburnus* & *Acerina cernua*	19	(25–27 cm)	25	Molnár & Tölg (1962a)
Megalops (*Megalops cyprinoides*)	Prawn *Metapenaeus monoceros*	20·5	91	28	Pandian (1967a)
Mackerel (*Scomber japonicus*)	*Anchoviella hepsetus*	21–24	29	20	Kariya & Takahashi (1969)
Rainbow trout (*Salmo gairdneri*)	Caddis fly larvae	24	57	11	Sarokon (1975)
White grunt (*Haemulon plumieri*)	*Anchoviella hepsetus*	25	(19 cm)	24	Pierce (1936)
Pike perch (*Stizostedion lacioperca*)	Bleak *Alburnus alburnus* & *Acerina cernua*	28	400	25	Molnár & Tölg (1962b)
Bowfin (*Amia calva*)	*Gambusia affinis*	32	18	21	Herting & Witt (1968)
Yellowtail (*Ocyurus chrysurus*)	*Anchoviella hepsetus*	33	(20 cm)	24	Pierce (1936)

Table 7.3. Number of hours for stomach to empty by 50% for various fish species at various water temperatures. Numbers in parenthesis represent determinations made at temperatures other than that identified

FAMILY Species	Hours for 50% of Food to Pass from Stomach							Food	Source
	Degrees Centigade								
	<5	5	10	15	20	25	30		
CENTRARCHIDAE									
Bluegill sunfish (*Lepomis macrochirus*)	—	31	11·5	7·5	5·0	4·5	—	Isopods	Kitchell (1970)
GADIDAE									
Cod (*Gadus morhua*)	13 (2)	11	5	4	5 (19)	—	—	shrimp	Tyler (1970)
ICTALURIDAE									
Channel catfish (*Ictalurus punctatus*)	—	—	15·5	13·5 (15·5)	9 (21·1)	4 (26·6) 6 (23·9)	7 (29·4)	pellet	Shrable *et al.* (1969)
PERCIDAE									
Walleye (*Stizostedion vitreum vitreum*)	—	—	—	7 (14·5)	4	—	—	minnows	Swenson & Smith (1973)
Sauger (*Stizostedion canadense*)	—	—	—	7 (14·5)	4	—	—	minnows	Swenson & Smith (1973)

Table 7.3. (*continued*)

Yellow perch (*Perca flavescens*)	—	—	—	10·5	4·0 (22)	—	—	Daphnids	Noble (1973)*
SALMONIDAE									
Sockeye salmon (*Oncorhynchus nerka*)	25·6 (3·1)	12 (5·5)	6·0 (9·9)	3·4 (14·9)	2·7 (20·1)	2·6 (23·0)	—	pellet	Brett & Higgs (1970)
Rainbow trout (*Salmo gairdneri*)	—	25·0	15·1	9·2	5·6	—	—	pellet	Windell *et. al.* (1976)
Rainbow trout (*Salmo gairdneri*)	—	20·0	11·3	7·1	4·5	—	—	oligochaetes	Windell *et al.* (1976)
Brown trout (*Salmo trutta*)	—	9·9 (5·2)	5·9 (9·8)	3·3	—	—	—	*Hydropsyche*	Elliott (1972)
Brown trout (*Salmo trutta*)	—	7·3 (5·2)	4·4 (9·8)	2·4	—	—	—	mixed meal group I	Elliott (1972)
Brown trout (*Salmo trutta*)	—	16·5 (5·2)	9·6 (9·8)	5·5	—	—	—	*Tenebrio molitor*	Elliott (1972)
UMBRIDAE									
Mudminnow (*Umbra limi*)	—	8·9	—	4·4	—	2·7	—	oligochaetes	Holzer (unpublished)

* Used very small fish.

Table 7.4. Number of hours for stomach to empty by 100% for various fish species at various water temperatures. Numbers in parenthesis represent determination made at temperatures other than that identified

FAMILY Species	\<5	5	10	15	20	25	30	Food	Source
				Hours for 100% of Food to Pass from Stomach — Degrees Centigrade					
CENTRARCHIDAE									
Bluegill sunfish (*Lepomis macrochirus*)	—	69	37	27	15	13	—	isopods	Kitchell (1970)
Largemouth bass (*Micropterus salmoides*)	—	110	50	37	24	19	—	bleak	Molnár & Tölg (1962b)
Largemouth bass (*Micropterus salmoides*)	—	110	46	34	22	17	—	bleak and Acerina	Fabian *et al.* (1963)
Largemouth bass (*Micropterus salmoides*) Large size	360 (4)	—	192	32 (16)	16 (22)	—	12 (28)	minnows	Markus (1932)
Small size	360 (4)	—	168	84 (16)	22 (22)	—	16 (28)		
CYPRINIDAE									
Bream larvae (*Abramis brama*)	—	—	—	3·5 (14) / 2·5 (16)	1·75 (18) / 1·75 (20)	—	—	*Cyclops*, copepods *Bosmina*, roiers	Panov & Sorokin (1962)
CYPRINODONTIADAE									
Topminnow (*Fundulus heteroclitus*)	—	27 (5·8)	12·5 (9·8)	8·5 (15·1)	7·0	5·2 (29·5)	3·0 (29·5)	clam mantle	Nicholls (1931)

Table 7.4. (continued)

								Food	Reference
GADIDAE									
Cod (*Gadus morhua*)	72 (2)	58	25	20	25 (19)	—	—	shrimp	Tyler (1970)
ICTALURIDAE									
Channel catfish (*Ictalurus punctatus*)	—	—	—	—	24 (21·1)	24 (26·4) + (23·9)	24 (29·4)	pellet	Shrable et al. (1969)
PERCIDAE									
Walleye (*Stizostedion vitreum vitreum*)	—	147·2 (5·5)	86·3 (11·1) 114·7 (8·3)	63·3 (13·9) 46·6 (16·7)	43·9 (19·4) 31·1 (22·2)	—		fish	Hofmann (1969)
Walleye (*Stizostedion vitreum vitreum*)	—	—	—	16+ (14·5)	12+	—		minnows	Swenson & Smith (1973)
Sauger (*Stizostedion canadense*)	—	—	—	16+ (14·5)	12+	24·9 (24·7)		minnows	Swenson & Smith (1973)
Yellow perch (*Perca flavescens*)	—	—	—	12	6·5	—		daphnids	Noble (1973)
Perch (*Perca fluviatilis*)	—	115	55	43	24 (22)	18		bleak and *Acerina*	Fabian et al. (1963)
Pikeperch (*Stizostedion lucioperca*)	—	257	61	32	18	11		bleak and *Acerina*	Fabian et al. (1963)
PLEURONECTIDAE									
Plaice (*Pleuronectes platessa*)	36 (1)	22	15	11·5	10	—		*Arenicola*	Edwards (1971)

Table 7.4. (*continued*)

									Reference
SALMONIDAE									
Sockeye salmon (*Oncorhynchus nerka*)	147·0 (3·1)	79·4 (5·5)	37·8 (9·9)	22·6 (14·9)	17·7 (20·1)	17·8 (23·0)	—	pellet	Brett & Higgs (1970)
Rainbow trout (*Salmo gairdneri*)	—	72·4	44·2	26·9	16·4	—	—	pellet	Windell *et al.* (1976)
Rainbow trout (*Salmo gairdneri*)	—	58·5	38·3	25·1	16·4	—	—	oligochaetes	Windell *et al.* (1976)
Brown trout (*Salmo trutta*)	—	65·8 (5·2)	39·1 (9·8)	22·1	—	—	—	*Hydropsyche*	Elliott (1972)
Brown trout (*Salmo trutta*)	—	48·5 (5·2)	29·0 (9·8)	16·2	—	—	—	mixed meal	Elliott (1972)
Brown trout *Salmo trutta*)	—	109·6 (5·2)	64·0 (9·8)	36·6	—	—	—	*Tenebrio molitor*	Elliott (1972)
Mountain whitefish (*Prosopium williamsoni*)	—	8·5 (6)	6 (11)	—	—	—	—	salmon alevins	McKone (1971)
SCORPAENIDAE									
Meborn (*Sebastes inermis*)	—	140 (7)	120 (10)	70 (14)	—	—	—	—	Kariya (1969)
SILURIDAE									
Sheatfish (*Silurus glanis*)	—	206	42	24	14	10	—	bleak and *Acernia*	Fabian *et al.* (1963)

Windell *et al.* 1969, Brett & Higgs 1970, Tyler 1970, Beamish 1971, McKone 1971, Swenson & Smith 1973, Elliott 1972, Steigenberger & Larkin 1974). Barrington (1957) suggested that fish digest small meals more rapidly than large meals and small prey more quickly than large prey. However, the results from many studies do not support this statement and show conclusively that the amount of food evacuated from the stomach per unit time is increased as the size of the meal is increased. Under a normal feeding regime for a meal of normal size most data indicate relatively little effect of meal size on the times to reach 50% and 100% stomach depletion.

In rainbow trout, *Salmo gairdneri*, the evacuation of stomach contents was independent of the amount consumed at a single meal except at ration levels below 0·7% body weight. A direct relationship existed between the amount of dry matter consumed and that evacuated from the stomach per hour (Windell *et al.* 1969). For example, a 90 g fish that consumed 900 mg of dry matter or approximately 1·0% of its body weight evacuated approximately 45 mg/hr. A fish that consumed 1800 mg (2% of its body weight) evacuated about 90 mg/hr. In both examples the original meal would have decreased by about 60% after 12 h of digestion at 15°C.

These types of data suggest that gastric motility, once initiated, remains relatively constant as long as food remains in the stomach. Increased amounts of food in the stomach are processed and evacuated per unit time. This only can be accomplished by changes in the volume of gastric contents pumped per peristaltic stroke or an increase in the number of strokes per unit time. Hence, the phenomenon of regulation of gastric emptying in fishes is much like that of higher vertebrates and depends upon the volume of gastric contents.

When small amounts of food are consumed, stomachs may empty at a rate near maximal. Rainbow trout fed 0·24% of their body weight evacuated 90% at the end of 12 hr (Windell *et al.* 1969). Small amounts of dry matter may become saturated with gastric juice more quickly forming a dilute mixture, thereby facilitating gastric removal.

Data on the consumption of large meals are less clear and confounded by satiation and frequency of feeding effects (see below).

7.6 Food particle size

Although closely related to the effect of meal size on digestion rate, few data are available on the effect of food particle size. Tyler (1970) argues that the disintegration of a food particle probably begins at the surface and proposed models for estimating digestion rate based on particle surface area and particle weight (volume). Unfortunately, data are not adequate to eliminate one or the other of the hypothetical surface-dependent or volume-dependent models. It is most likely that both volume and surface effects influence rate of stomach emptying. Accordingly, distension of the stomach is the only natural stimulus known to facilitate gastric evacuation (Hunt & MacDonald 1954). Digestion probably beings at the surface of a particle but food volume might influence peristalsis, which thereby facilitates mechanical and chemical breakdown.

The influence of differences in physical form of foods was observed when minnows of different size were fed to walleye (*Stizostedion vitreum vitreum*) and sauger *Stizostedion canadense* (Swensen & Smith 1973). It was assumed that increased surface area and/or decreased scalation associated with smaller minnows enhanced their rate of gastric removal.

7.7 Meal succession

Increased feeding frequency has been shown to accelerate food progression is several fish species. Early workers reported that rate of food passage in plaice, *Pleuronectes platessa*, is related to the amount of food recently consumed (Dawes 1930, Karzinkin 1935). Frequent feeding may result in either a reduction in the amount or size of food ingested at any one time, or defecation of material before being digested fully (Barrington 1957). In goldfish, *Carassius auratus*, a second meal seemed to stimulate egestion of the first meal contents (Rozin & Mayer 1964). Actively feeding larval largemouth bass have been reported to have a gastric evacuation rate twice as fast as larva fed only a single meal of zooplankton (Laurence 1971). Young yellow perch evacuated zooplankters faster when an initial meal was followed immediately by an excess of food (Noble 1973).

In contrast to these studies, rate of evacuation of three meals per day by brown trout, *Salmo trutta*, was similar to single meal experiments (Elliott 1972). Sarokon (1975) determined the effect of feeding frequency by rainbow trout on gastric evacuation of caddis fly larvae (*Limnophilus* sp.) removed from their case. No significant difference in gastric rate was observed between groups consuming two or eight meals per day. However, fish fed 1431 mg wet weight (324 mg dry wt.) of caddis fly larvae in two meals 8 hours apart on alternate days evacuated 8% faster than daily feeding groups.

As long as the total amount of food consumed remains constant, gastric evacuation is not affected by the number of daily feedings. Sarokon (1975) fed individual rainbow trout (averaging 58 g and 177 mm fork-length) 700 mg (1·2% of the live body weight) of dried pellets in two, four, six or eight portions spaced at eight, two, one-and-a-half and one hour intervals, respectively. The nine fish in each group held in 11°C water were fasted 16 hours before consuming weighed meals and were autopsied 24 hours after consuming their first meal. Average percent evacuation for all groups was

67·1 % with no significant difference between groups ($F = 0·86$, $df = 3$ and 32, $p < 0·05$).

Feeding of pellets on alternate days resulted in a significantly faster gastric evacuation rate than did daily feeding ($F = 7·34$, $df = 1$ and 23, $p < 0·05$). Fish fed 1·2 g on alternate days evacuated 0·84 g dry weight in 24 hours whereas fish fed 0·6 g daily evacuated 0·49 g in 24 hours. On the alternate-day schedule fish evacuated 74 % of their ration during the first 24 hours. Therefore 58·3 % more food was pushed through the stomach in 24 hours by fish fed a double ration on alternate days then fish fed a single ration daily. This suggests a strong 'second meal effect' when a double ration is fed in two feedings eight hours apart on alternate days.

7.8 Fat concentration in food

Lipid concentrations in excess of 15 percent of the dry weight probably have an inhibitory effect on gastric motility. Pelleted diets adjusted to show marked differences in lipid content of 6·5, 10·5 and 14·5 % moved through the stomachs of rainbow trout at the same rate (Windell *et al.* 1972). However, meals of several natural food items fed to bluegill sunfish (*Lepomis macrochirus*) were evacuated more rapidly than meals of mealworm larvae containing 35 % lipid content of the dry weight (Gerking 1955, Windell 1967). This is not to imply that elevated lipid concentrations may not have beneficial effects. Recently, growth, nitrogen metabolism, and fat metabolism were studied in trout along with the effects of temperature and diet change (Atherton & Aitken 1970). Feeding a high fat content diet to rainbow trout at 16C resulted in improved growth rate over fish kept at 12C. A greater percentage of fat was absorbed at the higher temperature which suggested that fat replaces nitrogenous compounds as an energy source.

A considerable amount of the marine and freshwater biomass consists of fats—and thus of diets. Fat is ingested, absorbed and then either metabolized or deposited for future use. Virtually nothing is known about the mechanism of this process in fish. It is assumed that fat metabolism in fish is not very different from that in mammals (Brockerhoff 1966), but further study is required. Most interesting will be the search for the source of the lipase of teleosts. It is possible that the diffuse pancreatic tissue does contain the enzyme but that only the right conditions have not been found to demonstrate its activity.

7.9 Movement of digestible and indigestible food fractions

Little attention has been given to the potential differential movement through the stomach of separate food fractions such as digestible organic matter and indigestible chitin, debris, pebbles, and plant material. Several workers have observed a lingering of indigestible chitinous exoskeletons in the gut of fish (Mann 1948, Gerking 1952, Pandian 1967b, Windell 1966). In some cases at least, chitin was retained in the stomach after the digestible material disappeared. Significant amounts of chitin from aquatic invertebrates were observed in the stomach of bluegill sunfish well after the digestible material was evacuated. Similar results have been observed in black bullheads, *Ictalurus melas*, (Darnell & Meierotto 1962). Total gastric evacuation was affected by the presence of chitin in the food fed to brook trout, *Salvelinus fontinalis*, (Hess & Rainwater 1939) and *Megalops cyprinoides* (Pandian 1967b). Kionka and Windell (1972) reported evidence for retention of large pieces of indigestible chitin in fish stomachs after all other visible organic matter was evacuated. Delay in gastric removal of chitin was attributed to the size of the pieces. Pieces larger than the diameter of the pyloric valve require softening and grinding before movement past the valve by peristalsis. Small pieces of chitin e.g., small appendages, were not delayed in the stomach, but moved freely at essentially the same rate as the digestible organic matter.

7.10 Seasonal effects

No seasonal effects has been reported for tropical, arctic, or temperature zone fishes.

7.11 Species and population Effects

Comparison of results from walleye and sauger from three lakes showed that species and population differences do not influence gastric digestion rate (Swenson & Smith 1973). Means from eight duplicated experiments based on 99 compatible observations with reference to fish size, time, temperature and meal size, showed no significant differences between populations.

7.12 Fasting and starvation

Fasting assumes considerable experimental and ecological significance for studies related to evacuation, digestibility, absorption, efficiency and growth. Evacuation experiments usually consist of fasting prior to feeding and then holding fish without food until the meal is totally digested (Windell 1966). Unfortunately many workers have neglected to report the fasting period and in some cases it is not clear if this variable was considered.

Long term fasting effect on Digestion rate

Fasting periods of 7, 14 and 25 days substantially decreased rate of gastric evacuation of bluegill sunfish (Windell 1966). A seven-day fast decreased gastric evacuation by as much as 22% and after a 25-day period by 51% when compared to 2-day fasting controls.

Short term fasting effect on digestion rate

Fasting periods of less than six days did not affect the rate of evacuation in brown trout fed *Gammarus* or *Tenebrio* and autopsied at six and 15 hours later (Elliott 1972). However, Sarokon (1975) investigated the short term effect of three- and six-day fasting periods on gastric evacuation of a pelleted diet by rainbow trout. Conditions of the experiment included utilization of the dry weight method, 58 g fish held in water at 15°C and fed 1·4% (833 mg) of live body weight.

Trout fasted for three and six days had significantly lower gastric evacuation rates than did fish fasted for 18 hours when compared after 24 hours of digestion. About four days of daily feeding was required for the evacuation process to return to the control level following a three-day fast, whereas with a six-day fast the process approached control values only after six days of resumed feeding (Fig. 7.1).

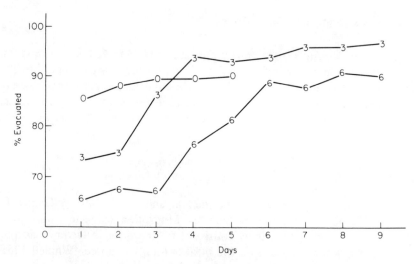

Fig. 7.1. Average percent gastric evacuation of 3 mm pellets by rainbow trout, *Salmo gairdneri* at 15C. Numbers refer to length (days) of pre-experimental fast (Sarokon 1975).

Morphological effects caused by long term fasting

Careful observation of digestive tracts of starved fish showed striking morphological changes which were especially prominent in the pyloric caeca of bluegill sunfish (Windell 1966). A noticeable shrunken condition was observed after 7 days. The condition became progressively advanced with time. Elliott (1972) reported brown trout pyloric caeca to be shrunken after 10 days of fasting.

The pyloric caeca have a dual function in fish, being a site of enzyme production (Lawrence 1950) and an active area of absorption (Greene 1913, Dawes 1930). Reported observations suggests that starvation and the associated morphological changes of the pyloric caeca interfere with normal functioning of these organs.

It has been proposed that rate of gastric evacuation following periods of starvation may be an adaptive mechanism to maximise digestive efficiency of subsequent meals (Sarokon 1975). Food remaining in the alimentary tract for a longer period of time may allow for more complete enzymatic degradation resulting in more efficient digestion and absorption.

When pre-experimental fasting is used in digestion or evacuation experiments one should first determine the normal feeding periodicity of the species under consideration and then structure the fasting period accordingly. Fasts of longer duration will result in underestimating rate of gastric evacuation and size of daily rations.

7.13 Force feeding

A major problem in performing laboratory digestion experiments has been the difficulty of getting a large number of fish to eat a measured meal at the same time. For this reason many workers have resorted to placing food items directly into the stomach (Markus 1932, Hess & Rainwater 1939, Hunt 1960, Molnár & Tölg 1962a and b, Seaburg & Moyle 1964, Windell 1966, Armstrong & Blackett 1966, Herting & Witt 1968, Shrable *et al.* 1969, Baur 1970, Edwards 1971, McKone 1971, Swenson & Smith 1973, Steigenberger & Larkin 1974). However, Windell (1966) and Swenson and Smith (1973) provide convincing evidence that force-feeding may cause psychological disturbance which in turn strongly affects certain physiological body processes. Fish sensitivity to even slight handling is well known and one of the initial effects of the manual manipulation of force-feeding is either immediate or delayed regurgitation (Markus 1932, Molnár & Tölg 1962, Seaburg & Moyle 1964, Windell 1966). Swenson and Smith (1973) reported an approximate two-fold difference in evacuation rate when comparing voluntary with force-feeding fish. Therefore, force-feeding data were considered inappropriate for estimating food consumption rates.

7.14 Profitable lines of future research

One of the most outstanding needs in future research on rate of digestion as related to production dynamics is estimation of food consumption in terms of the 'daily meal' and the 'daily ration'. Appropriate laboratory and field studies will allow the daily ration concept to become even more useful. 'Maximum' and 'minimum' daily rations for each species over the normal range of environmental temperatures would be most useful predictive values. The minimum daily ration is the amount that would just balance weight loss. The maximum daily ration may be a hypothetical value or that amount of food consumed when present in an unlimited supply. Metabolically there also is an 'optimum daily ration' which would lead to optimum growth, efficiency and well being. Once laboratory data on maximum, optimum and minimum rations become available, field work could supply data for the 'actual daily ration' in nature. The actual daily ration probably lies between the minimum and maximum, but seldom approaches optimum (Kitchell 1970). Such comparisons will lead to ecological predictions concerning fish food consumption in relation to growth and production in aquatic ecosystems. The model provided by Swenson and Smith (1973) although quite ambitious, deserves additional trials with different fish species, at different environmental temperatures, and in a variety of contrasting ecosystems to test its validity.

References

AOYAMA T. (1958) On the discharge of stomach contents concerning jack-mackerel *Trachurus japonicus. Bull. Seikai Reg. Fish. Res. Lab.* **15**, 29–32. [In Japanese with English summary.]

ARMSTRONG R.H. & BLACKETT F. (1966) Digestion rate of the dolly varden. *Trans. Am. Fish. Soc.* **95**(4), 429–430.

ATHERTON W.D. & AITKEN A. (1970) Growth, nitrogen metabolism and fat metabolism in *Salmo gairdneri. Comp. Biochem. & Physiol.* **36**(4), 719–747.

BAJKOV A.D. (1935) How to estimate the daily food consumption of fish under natural conditions. *Trans. Am. Fish. Soc.* **65**, 288–289.

BARRINGTON C.J.W. (1957) The alimentary canal and digestion. *The Physiology of Fishes.* Vol. 1. (M.E. Brown, ed.), pp. 109–161. Academic Press, New York.

BAUR R.J. (1970) Digestion rate of the Clear Lake black bullhead. *Proc. Iowa Acad. Sci.* **77**, 112–121.

BEAMISH F.W.H. (1972) Ration size and digestion in largemouth bass, *Micropterus salmoides* Lacépède. *Can. J. Zool.* **50**, 153–164.

BRETT J.R. & HIGGS D.A. (1970) Effect of temperature on the rate of gastric digestion in fingerling sockeye salmon, *Oncorhynchus nerka. J. Fish. Res. Bd Canada* **27**, 1767–1779.

BROCKERHOFF H. (1966) Digestion of fat by cod. *J. Fish. Res. Bd Canada* **23** (12), 1835–1839.

COWEY C.B. & SARGENT J.R. (1972) Fish nutrition. *Advances in Marine Biology*, Vol. 10. (F.S. Russell and M. Yongel, eds.), pp. 383–492. Academic Press, New York.

DARNELL R.M. & MEIEROTTO R.M. (1962) Determination of feeding chronology in fishes. *Trans. Am. Fish. Soc.* **91**(3), 313–320.

DAWES B. (1930) The adsorption of fats and lipids in the plaice. *J. Mar. Biol. Ass. U.K.* **17**, 75–102.

EDWARDS D.J. (1971) Effect of temperature on rate of passage of food through the alimentary canal of plaice *Pleuronectes platssa* L. *J. Fish Biol.* **3**, 433–439.

ELLIOTT J.M. (1972) Rates of gastric evacuation in brown trout, *Salmo trutta* L. *Freshwat. Biol.* **2**, 1–18.

ELLIOTT J.M. (1975a) Number of meals in a day, maximum weight of food consumed in a day and maximum rate of feeding for brown trout, *Salmo trutta* L. *Freshwat. Biol.* **5**, 287–303.

ELLIOTT J.M. (1975b) Weight of food and time required to satiate brown trout, *Salmo trutta* L. *Freshwat. Biol.* **5**, 51–64.

ELLIOTT J.M. (1975c) The growth rate of brown trout (*Salmo trutta* L.) fed on maximum rations. *J. Anim. Ecol.* **44**(3), 805–821.

ELLIOTT J.M. (1975d) The growth rate of brown trout (*Salmo trutta* L.) fed on reduced rations. *J. Anim. Ecol.* **44**(3), 823–842.

FABIAN G., MOLNÁR G. & TÖLG I. (1963) Comparative data and enzyme kinetic calculations on changes caused by temperature in the duration of gastric digestion of some predatory fish. *Acta Biol. Acad. Sci. Hung.* **14**, 123–129.

GERKING S.D. (1952) The protein metabolism of sunfishes of different ages. *Physiol. Zool.* **25**, 358–372.

GERKING S.D. (1952) The protein metabolism of sunfishes of different ages. *Physiol. Zoöl.* abolism of bluegill sunfish. *Physiol. Zoöl.* **28**, 267–282.

GREENE C.W. (1913) The fat-absorbing function of the alimentary tract of the king-salmon. *Bull. Bur. Fish., Wash.* **33**, 149–175.

HERTING G.E. & WITT A. (1968) Rate of digestion in the bowfin. *Prog. Fish-Cult.* **30** (1), 26–28.

HESS A.D. & RAINWATER J.H. (1939) A method for measuring the food preference of trout. *Copeia* 1939, 154–157.

HOFMANN P. (1969) Growth of walleyes in Oneida Lake and its relation to food supply. *Job Progress Report, U.S. Fish and Wildlife Service*, New York State Conservation Department and Cornell University, pp. 1–19.

HUNT B.P. (1960) Digestion rate and food consumption of Florida gar, warmouth, and largemouth bass. *Trans. Am. Fish. Soc.* **89**(2), 206–210.

HUNT J.N. & MACDONALD I. (1954) The influence of volume on gastric emptying. *J. Physiol.* (London) **126**, 459–474.

JENKINS T.M. (1969) Social structure, position choice and microdistribution of two trout species (*Salmo trutta* and *Salmo gairdneri*) resident in mountain streams. *Anim. Behav. Monogr.* **2**, 57–123.

KARIYA T. (1956) Problems concerning the biting of fishes. *Suisan Zashoku*, **4**(2), 1–8. [Translated from Japanese by W.G. VanCampen, Bureau of Commercial Fisheries, Honolulu.]

KARIYA T. (1969) The relationship of food intake to the amount of stomach contents in mebaru, *Sebastes inermis. Bull. Jap. Soc. Sci. Fish.* **35**(6), 533–536. [In Japanese with English summary.]

KARIYA T. & TAKAHASHI M. (1969) The relationship of food intake to the stomach contents in the Mackerel, *Scomber japonicus. Bull. Jap. Soc. Sci. Fish.* **35**(4), 386–390.

KARZINKIN G.S. (1935) K Poznaniyu Rybnoy Produktiviosti Vodoyemov. Soodshchenie. II. Izuchenie Fiziologii Pitaniya Segoletok Zerkal'nogo Karpa. *Trudy limnol. Sta. v Kosine* **19**, 21–66. (Russian, German summary.)

KEVERN N.R. (1966) Feeding rate of carp estimated by a radioisotopic method. *Trans. Am. Fish. Soc.* **95**, 363–371.

KIONKA B.C. & WINDELL J.T. (1972) Differential movement of digestible and indigestible food fractions in rainbow trout, *Salmo gairdneri*. *Trans. Am. Fish. Soc.* **101**(1), 112–115.

KITCHELL J.F. & WINDELL J.T. (1968) Rate of gastric digestion in pumpkinseed sunfish, *Lepomis gibbosus*. *Trans. Am. Fish. Soc.* **97**(4), 489–492.

KITCHELL J.F. (1970) The daily ration for a population of bluegill sunfish (*Lepomis macrochirus* Raf.). Ph.D. Thesis, University of Colorado, Boulder.

LAURENCE G.C. (1971) Digestion rate of larval largemouth bass. *N.Y. Fish. Game J.* **18**, 52–56.

LAWRENCE F.B. (1950) The digestive enzymes of the bluegill bream (*Lepomis macrochirus*). M.A. Thesis, Auburn University, Alabama.

MANN V.H. (1948) Über die Rohfaser-Verdauung des Karfens. *Arch. F. Fisch.* **1**, 1–11.

MANN K.H. (1967) The cropping of the food supply. *The Biological Basis of Freshwater Fish Production* (S.D. Gerking, ed.), pp. 243–257. Blackwell Scientific Publications, Oxford.

MAGNUSON J.J. (1969) Digestion and food consumption by skipjack tuna *Katsuwonus pelamis*. *Trans. Am. Fish. Soc.* **98**, 99–113.

MARKUS H.C. (1932) The extent to which temperature changes influence food consumption in largemouth bass (*Huro floridana*). *Trans. Am. Fish. Soc.* **62**, 202–210.

McKONE D. (1971) Rate at which sockeye salmon alevins are evacuated from the stomach of mountain whitefish (*Prosopium williamsoni*). *J. Fish. Res. Bd Canada* **28**(1) 110–111.

MOLNÁR G. & TÖLG I. (1960) Rentgenologic investigation of the duration of gastric digestion in the pike perch (*Lucioperca lucioperca*). *Acta. Biol. Hung.* **11**, 103–108.

MOLNÁR G. & TÖLG I. (1962a) Experiments concerning gastric digestion of pike perch (*Lucioperca lucioperca* L.) in relation to water temperature. *Acta. Biol. Hung.* **13**, 231–239.

MOLNÁR G. & TÖLG I. (1962b) Relation between water temperature and gastric digestion of largemouth bass (*Micropterus salmoides* Lacépède). *J. Fish. Res. Bd Canada* **19**, 1005–1012.

MOLNÁR G., TAMMASEY E. & TÖLG I. (1967) The gastric digestion of living predatory fish. *The Biological Basis of Freshwater Fish Production*, (S.D. Gerking, ed.), pp. 137–149. Blackwell Scientific Publications, Oxford.

NICHOLLS J.V. (1931) The influence of temperature on digestion in *Fundulus heteroclitus*. *Contrib. Can. Biol. Fish.* N.S., **7**, 45–55.

NOBLE R.L. (1973) Evacuation rates of young yellow perch, *Perca flavescens* (Mitchell). *Trans. Am. Fish. Soc.* **102**(4), 759–763.

PANDIAN T.J. (1967a) Transformation of food in fish *Megalops cyprinoides*. I. Influence of quality of food. *Marine Biol.* **1**, 60–64.

PANDIAN T.J. (1967b) Intake, digestion, absorption and conversion of food in the fishes *Megalops cyprinoides* and *Ophiocephalus striatus*. *Mar. Biol.* **1**, 16–32.

PANOV D.A. & SOROKIN Y.I. (1962) Speed of digestion in bream larvae. *Byulleten Instituta Biologii Vodokhranilishch* **13**, 24–26.

PAPPAS, C.J., TIEMEIER O.W. & DEYOE C.W. (1973) Chromic sesquioxide as an indicator in digestion studies on channel catfish. *Prog. Fish-Cult.* **35**(2), 97–98.

PETERS D.S. & HOSS D.E. (1974) A radioisotopic method of measuring food evacuation time in fish. *Trans. Am. Fish. Soc.* **103**(3), 626–629.

PIERCE E.L. (1936) Rates of digestion in the yellowtail (*Ocyurus chrysurus*) and the white grunt (*Haemulon plumieri*). *Copeia* 1936(2), 123–124.

RICKER W.E. (1946) Production and utilization of fish populations. *Ecol. Monogr.* **16**, 373–391.

ROZIN P.N. & MAYER J. (1964) Some factors influencing short-term food intake of the goldfish. *Am. J. Physiol.* **206**(6), 1430–1436.

SAROKON J. (1975) Feeding frequency, evacuation, absorption, growth and energy balance in rainbow trout, *Salmo gaidneri*. Ph.D. Thesis, University of Colorado, Boulder.

SEABURG K.G. & MOYLE J.B. (1964) Feeding habits, digestion rates, and growth of some Minnesota warmwater fishes. *Trans. Am. Fish. Soc.* **93**, 269–285.

SHRABLE J.B., TIEMEIER O.W. & DEYOE C.W. (1969) Effects of temperature on rate of digestion by channel catfish. *Prog. Fish-Cult.* **31**, 131–138.

SMIT H. (1967) Influence of temperature on the rate of gastric juice secretion in the brown bullhead, *Ictalurus nebulosus*. *Comp. Biochem. Physiol.* **21**, 125–132.

STEIGENBERGER L.W. & LARKIN P.A. (1974) Feeding activity and rates of digestion of northern squawfish (*Ptychocheilus oregonensis*). *J. Fish. Res. Bd Canada* **31**(4), 411–420.

SWENSON W.A. & SMITH L.L. Jr. (1973) Gastric digestion, food consumption, feeding periodicity, and food conversion efficiency in walleye (*Stizostedion vitreum vitreum*). *J. Fish. Res. Bd Canada* **30**(9), 1327–1336.

SURBER E.W. (1930) A quantitative method of studying the food of small fishes. *Trans. Am. Fish. Soc.* **60**, 158–163.

TSUNIKOVA Y.P. (1969) Feeding and growth of young of the roach (*Rutilis rutilis heckeli* Nord) in the Kuban estuaries. *Prob. Ichthyol.* **9**, 555–563. (Translated by the Amer. Fish. Soc.]

TSUNIKOVA Y.P. (1970) Daily food consumption of young pike-perch and Azov roach (Taran) in the Kuban fish farms. *J. Ichthyol.* **10**, 658–662. [Translated by the Amer. Fish. Soc.]

TYLER A.V. (1970) Rates of gastric emptying in young cod. *J. Fish. Res. Bd Canada* **27**, 1177–1189.

WARREN C.E., WALES J.H., DAVIS G.E. & DOUDOROFF P. (1964) Trout production in an experimental stream enriched with sucrose. *J. Wildl. Manage.* **28**, 617–660.

WINDELL J.T. (1966) Rate of digestion in the bluegill sunfish. *Invest. Indiana Lakes Streams* **7**, 185–214.

WINDELL J.T. (1967). Rates of digestion in fishes. *The Biological Basis of Freshwater Fish Production* (S.D. Gerking, ed.), pp. 151–173. Blackwell Scientific Publications, Oxford.

WINDELL J.T. (1968). Food analysis and rate of digestion. *Methods for Assessment of Fish Production in Fresh Waters* (W.E. Ricker, ed.), pp. 197–203. Blackwell Scientific Publications, Oxford.

WINDELL J.T. & NORRIS D.O. (1969) Gastric digestion and evacuation in rainbow trout, *Salmo gairdneri*. *Prog. Fish-Cult.* **31**, 20–26.

WINDELL J.T., KITCHELL J.F., NORRIS D.O., NORRIS J.S. & FOLTZ J.W. (1976) Temperature and rate of gastric evacuation by rainbow trout, *Salmo qairdneri*. *Trans. Arm. Fish. Soc.* **105**, 712–717.

WINDELL J.T., NORRIS D.O., KITCHELL J.F. & NORRIS J.S. (1969) Digestive response of rainbow trout, *Salmo gairdneri*, to pellet diets. *J. Fish. Res. Bd Canada* **26**, 1801–1812.

WINDELL J.T. (1971) Food analysis and rate of digestion. *Methods for Assessment of Fish Production in Fresh Waters* (W.E. Ricker, ed.), pp. 215–226. Blackwell Scientific Publications, Oxford.

WINDELL J.T., HUBBARD J.D. & HORAK D.C. (1972) Rate of digestion in rainbow trout, *Salmo gairdneri*, fed three pelleted diets. *Prog. Fish-Cult.* **34**(3), 156–159.

Chapter 8: Partitioning of Energy into Metabolism and Growth

Paul W. Webb

8.1 Introduction

Production in fisheries is the product of the net increase in numbers of a population and the increase in mass of the fish making up that population. In natural populations, both parameters vary in accord with the complex of physical and biotic factors impinging on the fish. For managed populations in aquaculture systems, population parameters are less variable and emphasis is placed on the carrying capacity of culture ponds. In all cases, increase in mass is crucial to obtaining a marketable product. Increase in fish mass in turn depends on the energy made available to the fish and the way that energy is distributed and utilized within the body. This chapter concerns itself with such questions of energy usage. The traditional approach incorporating food, metabolism and growth in thermodynamically balanced equations is used to provide a framework within which the influence and effects of environmental and biotic parameters on energy usage can be discussed. The subject is large and space does not permit an exhaustive treatment. References are, therefore, selective and made primarily to reviews and to more recent papers. These may be consulted for earlier work.

8.2 Energetic principles

The utilization of food energy for metabolism and growth is described by the thermodynamically 'balanced energy equation' (Ivlev 1939, Winberg 1956). This states that all energy entering and leaving a system must be accounted for. Writing Q for rates of energy change;

$$p.Q_R = Q_M + Q_G \qquad (1)$$

where; Q_R = food consumed
Q_M = metabolism (catabolism)
Q_G = growth (anabolism)
p = proportion of food consumed that is assimilated.

Equation 1 is an inadequate model for the design of analytical experiments or for the generation of first principles production models. The equation is usually expanded on the basis of energy flow diagrams (e.g. Brett 1970a, Warren 1971) (Fig. 8.1). Then,

$$Q_R - (Q_F + Q_N) = Q_S + Q_L + Q_{SDA} + Q_G + Q_P \qquad (2)$$

where; Q_F = faecal loss
$\quad\quad Q_N$ = excretory (nitrogen) loss or non-faecal loss
$\quad\quad Q_S$ = standard metabolism
$\quad\quad Q_L$ = locomotor (activity) metabolic cost
$\quad\quad Q_{SDA}$ = apparent specific dynamic action
$\quad\quad Q_P$ = reproductive cost for gamete synthesis.

Relative energy distribution to growth and reproduction is important to understanding various life history strategies of fish, (Williams 1961), as well

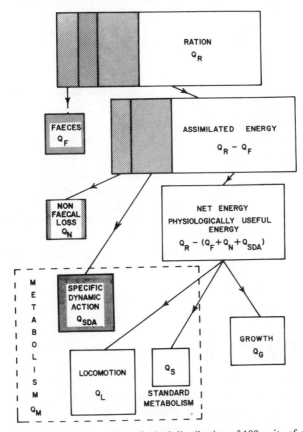

Fig. 8.1. Flow diagram illustrating a hypothetical distribution of 100 units of consumed energy by a fish. The size of each box illustrates likely proportions of energy loss or use by various components of the system under average conditions of food abundance and in the absence of stress.

as some growth effects (Iles 1974). The present chapter restricts itself to a discussion of somatic growth.

8.3 Food, metabolism and growth relations

Equations 1 and 2 provide the basis on which food, metabolism and growth relations can be evaluated. These interactions will be discussed in relation to size, the most important biotic factor of concern in production problems. The discussion provides the reference scheme within which effects of environmental variation can then be evaluated.

FOOD

The capacity for sustained metabolism and growth depends on food availability and on the proportion of ingested food that is assimilated. Therefore Q_R essentially dictates the sustainable limits for Q_M plus Q_G. Maximum food consumption $Q_{R\ max}$ is defined as appetite when food is unrestricted. In field situations, food is normally limiting and prescribed ration varying from zero to $Q_{R\ max}$ is, therefore, a key variable in experimental studies.

$Q_{R\ max}$ characteristically increases with fish mass (Pandian 1967a, Brett 1971a, Gerking 1971 and 1972, Beamish 1972, Niimi & Beamish 1974 and others). Allometric equations relating food consumption to mass have exponents less than 1 so that Q_R per unit fish mass decreases with mass (Fig. 8.2). Pandian (1967a) notes that exponents of allometric equations for various energetic parameters, Q_R, Q_S, and Q_G are frequently similar.

Not all food consumed is assimilated. Winberg (1956) considers that some 15% is lost as faeces. This value for unassimilated food is too high for artificial diets commonly used in fish culture and also for other high energy diets such as for piscivorous game fish. Numerous measurements of assimilation efficiencies (percent ration component assimilated) have shown that approximately 96 to 99% of dietary protein is assimilated (Gerking 1962 and 1971, Pandian 1967a, Birkett 1969, Beamish 1972, Niimi & Beamish 1974). Low melting point fats are 84 to 93% assimilated and high melting point fats 45 to 78% assimilated by brook trout, *Salvelinus fontinalis* (McCay & Tunison 1935). For a wide variety of fish, assimilation efficiencies of 80 to 90% are typical for fats in natural food items (Gerking 1972, Beamish 1972). Simple sugars, such as glucose and maltose, are up to 99% absorbed and complex carbohydrates, such as starch, are 38% assimilated in salmonids (Phillips 1969). Assimilation efficiencies are also high for energy with values of 89 to 96% for a wide variety of species (Pandian 1967, Beamish 1972, Niimi & Beamish 1974). Assimilation efficiencies may be lower for some natural diets. For example, Wissing (1974) found an average of 73% assimila-

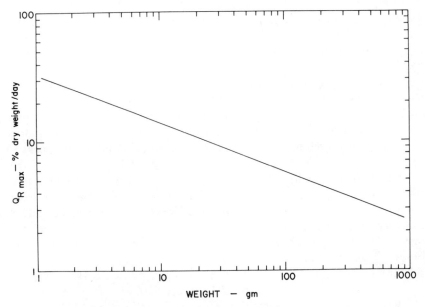

Fig. 8.2. Relations between maximum food consumption, $Q_{R\ max}$ (expressed as food dry weight consumed per unit fish body mass) and body mass for *Oncorhynchus nerka*. The figure is based on equations in Brett (1971b).

tion of food energy in white bass, *Morone chrysops*, fed crayfish with a high proportion of indigestible exoskeleton. Indiscriminate use of Winberg's (1956) value of 85% for assimilation efficiency is to be discouraged on the basis of present observations.

Assimilation efficiency varies little with ration size for most fish studied (Pandian 1967, Gerking 1971, Warren 1971, Beamish 1972, Niimi & Beamish 1974). This is of considerable importance in simplifying first principles production models, but there is some indication that the general relationship may not hold during certain seasons (Warren 1971) as discussed below.

Further energy losses, Q_N, from food consumed accrue from the metabolism of assimilated protein. Amino acids in excess of growth requirements are metabolized. Approximately 15% of the available protein energy is excreted by the gills and kidney as incompletely oxidized nitrogenous compounds, mainly ammonia (e.g. Forester & Goldstein 1969, Savitz 1969, 1971, Olson & Fromm 1971, Niimi & Beamish 1974). The magnitude of the nitrogenous loss depends on size and maturity, being lower for actively growing or reproducing fish (Winberg 1956).

METABOLISM

The food energy assimilated, minus nitrogenous losses after assimilation, is the energy available for metabolism and growth. Q_M must be satisfied first and will deplete stored energy when Q_R is very low. Metabolic energy expenditure is usually measured by indirect calorimetry (Kleiber 1975) using oxycalorific equivalents of oxygen consumption and assuming mixed substrate use. This method, developed for homeotherms, is applicable to fish (Brett 1973; see also Krueger *et al.* 1968) and is accurate to within $\pm 1 \cdot 5\%$ (Winberg 1956). Occasionally more precise oxycalorific equivalents are calculated using the respiratory quotient or the nitrogen modified respiratory quotient (Kutty 1968 and 1972). This level of precision does not appear to be necessary for most energetic studies relating to production.

Anaerobic metabolism is typically neglected, or can be estimated from oxygen debt repayment (Brett 1964). In long term studies a major anaerobic energy expenditure will not accrue. However, oxygen consumption data from short term experiments are commonly used in energetic modelling, and these may be in error through neglect of anaerobic contributions to total metabolism. There is an increasing literature implying some fish sustain a considerable oxygen debt (Blazka 1958, Connor *et al.* 1964, Kutty 1968, Pritchard *et al.* 1971, Johnston & Goldspink 1973, Greer-Walker & Pull 1973, Smit *et al.* 1971). Observations on three species of gobies (Congleton 1974) and on goldfish (*Carassius auratus*), Tilapia and salmonids (Kutty 1968, Smit *et al.* 1971) indicate that aerobic and anaerobic metabolic capacities may be inversely related as described for amphibia (Bennett & Licht 1973). This area is poorly researched for fish, but may be of considerable significance to energy budgeting.

Q_M varies within finite limits. The lower limit is Q_S which approximates the energy required to maintain a non-stressed fish in the post-absorptive state and at rest. The upper limit, $Q_{M\ max}$, is the active metabolic rate defined as the maximum sustained metabolic rate. In practice $Q_{M\ max}$ is defined in relation to a critical swimming speed determined in an increasing velocity test with speed increments at approximately 60-minute intervals (Brett 1964). The y-intercept of the regression equation relating Q_M to swimming speed gives Q_S.

The difference of $Q_{M\ max}$ minus Q_S is defined as the metabolic scope (Fry 1971). This is a measure of the energy that can be made available for all activities over and above basic maintenance, e.g., digestion, absorption, locomotion, regulation under stress, and growth.

There are well-known relations between size and Q_M. It is well established that Q_S is related to body mass M according to:

$$Q_S = \alpha M^\gamma \tag{3}$$

where α and γ are fitted constants for a given species. Over a wide range of mass for homeothermic and poikilothermic species, γ takes values of 0·75 to 0·8 (Hemmingsen 1960). This modal value applies to fish, but γ varies from 0·5 to 1·0 among species so that a value of 0·75 or 0·80 cannot be automatically assumed (Fry 1971). It is important to remember that equation (3) is an empirical description of relations between Q_S and M. Attempts to formulate theoretical explanations for the modal exponent and species variability remain problematic. Arguments of geometrical similarity are certainly inadequate (Kleiber 1975) although other criteria may prove to be appropriate (McMahon 1973).

$Q_{M\ max}$ also follows allometric relations with M. However, γ is often close to unity (Brett 1972, Brett & Glass 1973). The reasons for $Q_{M\ max}$ being almost independent of body mass are not clear. Jones (1971) considers the capacity for oxygen exchange limits $Q_{M\ max}$, but morphometric parameters do not scale in the appropriate manner (Hughes & Morgan 1973). Since $Q_{M\ max}$ is usually measured in increasing velocity tests, the relation with M may be confounded by a large anaerobic energy contribution, which may be independent of size.

Although metabolic scope defines the energetic limits for activity, not all this energy is available in feeding fish. A feeding fish must process its food and this requires energy for digestion, assimilation, transportation, biochemical treatment and incorporation. The sum of these energy requirements is apparent specific dynamic action, Q_{SDA} (Beamish 1974). Specific dynamic action alone, SDA, represents the biochemical costs of food treatment and is considered to be the major portion of apparent specific dynamic action (Beamish 1974, Kleiber 1975). SDA varies from approximately 5% of food intake for carbohydrates to 30% for protein, the high value for the latter being attributed to the cost of deaminating amino acids.

Several measurements of Q_{SDA} have now been made for fish (Warren & Davis 1967, Livingston 1968, Averett 1969, Warren 1971, Wissing & Hasler 1971, Muir & Niimi 1972, Beamish 1974, Niimi & Beamish 1974, Pierce & Wissing 1974, Wissing 1974). Q_{SDA} is measured from increases in oxygen consumption following a meal, which persists for several hours and up to two days. The currently most effective methods for determining Q_{SDA} exercise fish continuously at a fixed swimming speed to avoid possible confounding effects of spontaneous activity. The oxygen consumption of the fish is measured after acclimating the fish to the chamber and then during and following a meal until the prefeeding level is again obtained. The most comprehensive study to date is that by Beamish (1974) who measured Q_{SDA} of largemouth bass, *Micropterus salmoides*, swimming in a respirometer and fed freshly thawed emerald shiners. The duration of elevated metabolism following a meal increased with Q_R and M. Q_{SDA} also increased with Q_R and M (Fig. 8.3) but was independent of swimming speed. When expressed as a percentage of Q_R, Q_{SDA} was independent of Q_R, with a mean value of

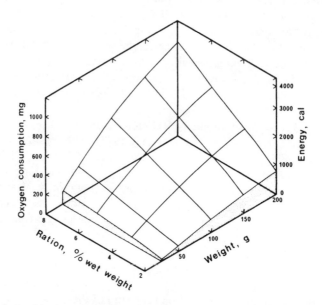

Fig. 8.3. Relationships between apparent specific dynamic action, Q_{SDA} (mg 0_2 consumed following a meal) ration, Q_R (expressed as percent of body mass) and body mass for *Micropterus salmoides*. From Beamish (1974).

$14.9 \pm 4.9\%$. Similar values have been obtained of 16 to 19% Q_R for aholehole, *Kulia sandvicensis*, (Muir & Niimi 1972) and 12.5% Q_R for bluegill sunfish, *Lepomis macrochirus* (Pierce & Wissing 1974).

GROWTH

Physiologically useful energy in excess of metabolic requirements is available for deposition as body tissue in growth or for synthesis of reproductive products. Warren & Davis (1967) consider Q_G is most appropriately measured as energy changes. Gerking (1955) favours measurement in terms of nitrogen changes since these are diagnostic of protein incorporation. Ideally, both measures should be used as the economics of fish production must take into account energy/protein ratios and advantage can be taken of protein sparing while maximizing production. Physiological questions relating to food technology are discussed elsewhere (see reviews in Halver 1972). Many growth experiments calculate Q_G from dry weight changes of samples of fish. This method is subject to some error in comparing groups of fish because of changes in proximate composition (e.g. Brett *et al.* 1969, Niimi 1972a, b).

The magnitude of the error for single species experiments can be obtained from data in Brett *et al.* (1969) and Brett (1971b) for sockeye salmon, *Oncorhynchus nerka*. The error in using dry weight instead of energy to measure growth was $\pm 4\%$ for fish fed at various Q_R and at various temperatures. For the same species fed various natural and artificial diets the error was $\pm 10\%$. Winberg (1956) estimated interspecific error would be of the order of $\pm 20\%$.

Irrespective of the method of measurement, growth of individuals, as well as populations, most commonly follows a logistic curve when mass is plotted against time. Such growth curves cover a large time period and obscure short term variations in mass that result from environmental variability. Warren (1971) emphasizes that a gross treatment of growth responses prevents fundamental interpretation and states that information is required for mass changes over short but frequent intervals throughout an animal's life. Such information can lead to detailed understanding of environmental and biotic effects on food consumption, metabolism and growth.

In order to analyze and understand growth phenomena, it is therefore convenient to consider short growth periods or stanzas for arbitrarily defined time periods. The theoretical basis for fish production questions is based on studies of fish growth over such stanzas. During a growth stanza, body mass usually increases with time according to the exponential function;

$$M = ae^{Q_G t} \tag{4}$$

where Q_G is the growth rate, often expressed as the specific growth rate per unit mass, and t is time.

In all studies of growth, Q_R is a key variable, and relations between Q_R and Q_G have been documented for several species (see for example Brett *et al.* 1969, Gerking 1971, Warren 1971, Niimi & Beamish 1974). An example of relations between Q_R and Q_G is provided (Fig. 8.4) for yearling sockeye salmon by Brett *et al.* (1969). From such relations, two ration levels of importance can be defined; maintenance ration $Q_{R\ maint}$ when growth is zero, and the maximum ration $Q_{R\ max}$ where growth is greatest. When $Q_R <$ $Q_{R\ maint}$, Q_G is negative. There is also an optimum Q_R when Q_R and Q_G are related curvilinearly (see Fig. 8.4). At the optimum ration, growth per unit ration is maximized. When Q_R and Q_G are linearly related there will be no optimum Q_R, as for bluegill sunfish (Gerking 1972) and largemouth bass (Niimi & Beamish 1974).

$Q_{R\ maint}$ and $Q_{R\ max}$ set limits where Q_G is zero and maximum respectively. $Q_{R\ maint}$ is the ration required to just maintain the feeding fish, and will be the sum of Q_S, Q_L, Q_{SDA}, Q_F and Q_N for that Q_R. Warren and Davis (1967) and Warren (1971) have defined the difference between $Q_{R\ max}$ and $Q_{R\ maint}$ as the scope for growth. Maximum scope for growth will be obtained in an optimum environment (minimizing Q_M and $Q_{R\ maint}$) with unrestricted

Chapter 8

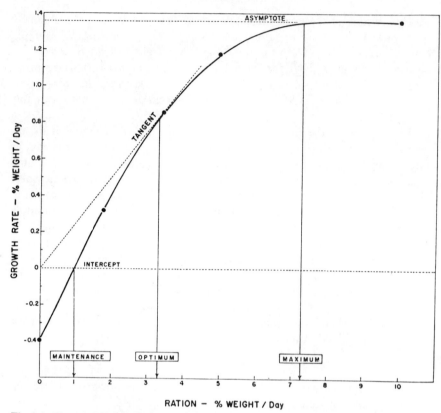

Fig. 8.4. The relationship between ration level, Q_R, and growth rate, Q_G, for yearling *Oncorhynchus nerka* at 10C. The maximum ration, $Q_{R\ max}$, is that giving greatest Q_G, the maintenance ration, Q_{maint}, that where Q_G is zero, and the optimum ration is that where growth per unit ration is maximised From Brett *et al.* (1969).

food. Since environmental manipulation seeks an optimum situation that will maximise scope for growth, the concept of scope for growth is of considerable theoretical and practical importance. In view of the continuing improvement in salmon rearing by environmental control (Brett 1974a, b) it seems likely that the maximum scope for growth, and hence Q_G has not been realized for fish. Warren and Davis (1967) caution that metabolic scope and scope for growth should not be confused. They are two useful operational tools but the various components are not physically or conceptually equivalent.

Growth and production studies are not concerned solely with growth or scope for growth, but also with the efficiency of food conversion into flesh (Kleiber 1975). Two major efficiencies are useful, gross (food) conversion efficiency, η_{gross}, and net (food) conversion efficiency, η_{net};

$$\eta_{gross} = Q_G/Q_R \qquad (5)$$

$$\eta_{net} = Q_G/(Q_R - Q_{R\ maint}) \qquad (6)$$

η_{gross} expresses the overall efficiency of converting food to flesh, the efficiency of greatest importance to production economics. η_{net} expresses the efficiency of converting food in excess of maintenance requirements to flesh, and interfaces with scope for growth. The value of η_{net} measurements has been questioned (Paloheimo & Dickie 1966a, b, Pandian 1967. However, it is apparent that no real understanding of growth phenomena in the face of environmental variability is possible without a more comprehensive analysis of both metabolism and growth energetics (Warren & Davis 1967, Brett *et al.* 1969, Warren 1971). Therefore, net conversion efficiency and scope for growth are the logical analytical levels for the evaluation of growth responses.

Both η_{gross} and η_{net} depend on Q_R. An example of relations between η_{gross} and Q_R has been elaborated for sockeye salmon at various temperatures (Fig. 8.5). At any temperature, η_{gross} increases with ration up to the optimum and decreases thereafter. This is the typical response when Q_R and

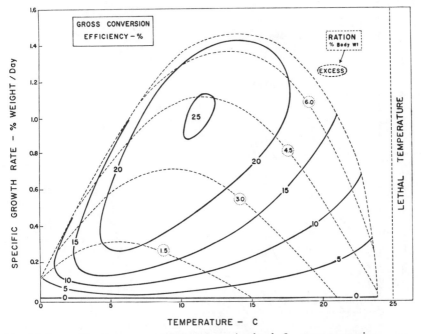

Fig. 8.5. Interactions between growth rate, Q_G, ration level, Q_R, gross conversion efficiency, η_{gross}, and temperature for yearling *Oncorhynchus nerka*. Dotted lines are isopleths for various Q_R levels and solid lines are isopleths for $\%\eta_{gross}$. From Brett *et al.* (1969).

Q_G are related curvilinearly as in Fig. 8.4 (e.g. Brett *et al.* 1969, Warren 1971, Andrews & Stickney 1972, Saksena *et al.* 1972, McCormick *et al.* 1972). When Q_R and Q_G are linearly related, η_{gross} increases continuously with ration (Gerking 1971, Warren 1971, Niimi & Beamish 1974).

Relationships between Q_R and η_{gross} have received much attention in the literature, particularly in relation to growth and production modelling (Paloheimo & Dickie 1966a, b, Kerr 1971a, b, c). Unfortunately, effort has been concentrated on one part of efficiency curves, referred to as '*k*-lines'. Paloheimo and Dickie (1966a, b) analyzed a variety of laboratory growth studies and found that the logarithm of η_{gross} decreased regularly with Q_R, this being the '*k*-line' where k is synonymous with η_{gross}. Warren (1971) has clearly pointed out that efficiency curves with η_{gross} as a function of Q_R must increase from zero and will only decrease at high ration levels (above the optimum Q_R), and then only for curved Q_R/Q_G relations (e.g. Fig. 8.4). The ecological significance of the '*k*-line' is questionable because food availability is likely to be sub-optimal under natural conditions. The discussion that has surrounded the '*k*-line' since Paloheimo and Dickie (Kerr 1971a, b) now appears to be of historical interest only.

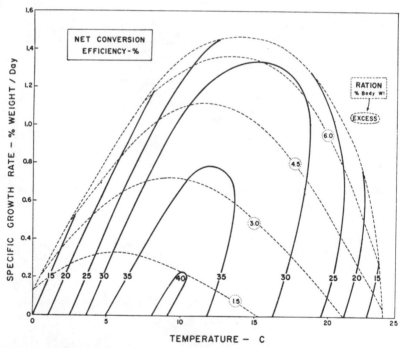

Fig. 8.6. Interactions between growth rate, Q_G, ration level, Q_R, net conversion efficiency, η_{net}, and temperature. Dotted lines show isopleths for various Q_R levels and solid lines are isopleths for various $\%\eta_{net}$. From Brett *et al.* (1969).

Relationships between η_{net} and Q_R as functions of temperature (Fig. 8.6) are also examined for sockeye salmon (Brett *et al.* 1969). Numerically, η_{net} always exceeds η_{gross} since Q_{Rmaint} is subtracted from Q_R in calculating the efficiency (Equation 6). At any temperature η_{net} increases with Q_R up to a maximum value at an optimum Q_R and thereafter decreases at higher rations. This again applies to curved relations between Q_R and Q_G. When these parameters are linearly related, η_{net} usually increases linearly with Q_R (Carline & Hall 1973).

η_{gross} and η_{net} only apply where Q_R exceeds Q_{Rmaint}. In field situations, food availability varies seasonally and is often less than Q_{Rmaint}. Then growth is discontinuous and fish metabolize stored material (Weatherley 1972). The efficiency of stored material usage, the partial maintenance efficiency, η_{part}, (Kleiber 1975) is therefore of interest, and;

$$\eta_{part} = Q_1/p.Q_R \tag{7}$$

where Q_1 is the weight loss prevented by the food available.

Although Q_R dictates energy availability for a given fish, the use of that energy varies with fish size. In larger fish proportionally more assimilated energy is used for metabolism and probably reproduction (Iles 1974) with consequent reduction in Q_G with increasing M (Brett & Shelbourn 1975). The importance of this is emphasized by Brett (1974a) in that greatest production can be obtained from younger fish. η_{gross} and η_{net} decrease with size for the same reasons as Q_G. These efficiences will approach zero for large fish. Food conversion efficiencies are highest in embryos when η_{gross} may be as high as 60 to 70% (Winberg 1956, Brett 1970b, and others). These well known growth phenomena in relation to size have recently been well illustrated in comprehensive experiments on largemouth bass by Niimi and Beamish (1974).

8.4 Environmental effects on food consumption, metabolism and growth

In the foregoing discussion, metabolism and growth responses were considered in relation to food consumption, the most important variable affecting overall energetics, and size. The energetic parameters are also profoundly affected by environmental conditions which may increase or decrease Q_R, Q_M and Q_G. Fry (1971) has pointed out that all environmental variations are ultimately expressed at the metabolic level, Q_M. According to equation (1), for a given Q_R, changes in Q_M will have reciprocal affects on Q_G. Therefore, equations (1) and (2) provide a framework within which the energetic effects of environmental variables can be quantitatively approached.

Fry (1971) has provided a classification of environmental factor effects on metabolism that may be used in conjunction with the energetic models.

The classification emphasizes animal responses to the environment, rather than responses to a specific component that is fortuitously measured with ease by an investigator. Fry identifies five response categories. The first of these, lethal factors, destroys the integrity of an organism and is not of concern here. The other four factors affect the amount of energy made available to an organism (controlling and limiting factors) and the spatial and temporal distribution of that energy within an animal (masking and directive factors). The effects of each of these on food consumption, metabolism and growth will be discussed in turn. The general form of interactions with growth energetics is summarized in Fig. 8.7.

It is important to note that the present extension of Fry's classification to discuss environmental effects on food consumption and growth is an operational usage. Environmental effects are ultimately expressed at the metabolic level, and the resultant classification of environmental factors is used as a descriptive reference framework to examine other responses. However, it is not intended to imply that the underlying mechanism of a response is the same. For example, temperature is a controlling factor affecting metabolic rate and these effects can be explained in terms of thermodynamic theory. Variation in appetite with temperature cannot be explained in this way. Food intake dictates net energy available for all activities, but the use of that energy for growth and reproduction is markedly affected by metabolism on which the environmental changes operate; anabolic use depends on the residual energy. The present extension of Fry's classification is, therefore, analogous to the extension of Fry's earlier concept of scope (for metabolism) to growth (Warren & Davis 1967). The response components are recognized as not being physically or conceptually equivalent.

Nevertheless, there is no doubt that Fry's classification of environmental factors is currently the most effective way of synthesizing physiological–ecology problems at the whole animal energetics level. As a result of research growing from Fry's stimulus, the approach will doubtless evolve and become modified.

CONTROLLING FACTORS

Environmental factors in this category are temperature and pressure; better known temperature effects are considered here. Controlling factors effect rates of chemical breakdown and hence set minimal energy requirements for repair and maintenance which approximate to Q_S. The same factors deter-

Fig. 8.7. Summary of effects of various environmental factors on energy distribution and scopes for metabolism and growth. Key; Q_{Rmax} the maximum food consumption; Q_F fecal loss; Q_N nitrogenous excretion loss; Q_{Mmax} maximum metabolic rate (active rate); Q_S standard metabolic rate; Q_{SDA} apparent specific dynamic action; Q_L costs of physical activity; Q_G growth. Further explanation is given in the text.

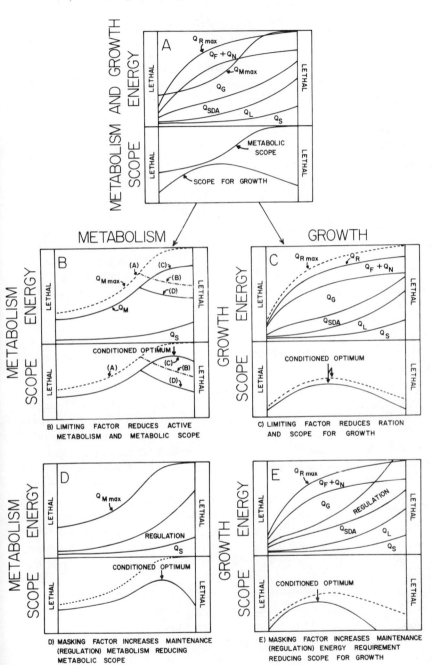

B) LIMITING FACTOR REDUCES ACTIVE
METABOLISM AND METABOLIC SCOPE

C) LIMITING FACTOR REDUCES RATION
AND SCOPE FOR GROWTH

D) MASKING FACTOR INCREASES MAINTENANCE
(REGULATION) METABOLISM REDUCING
METABOLIC SCOPE

E) MASKING FACTOR INCREASES MAINTENANCE
(REGULATION) ENERGY REQUIREMENT
REDUCING SCOPE FOR GROWTH

TEMPERATURE

mine reaction rate, setting a maximum Q_{Mmax} when there is no restraint on the supply or removal of metabolites. Controlling factors also affect food consumption (hence Q_R) and assimilation efficiencies.

Factors determining appetite for fish are not yet clear. Q_{Rmax} follows rates of stomach evacuation to some extent. These rates, and Q_{Rmax} increase with temperature (Windell 1967, Magnuson 1969, Brett & Higgs 1970, Brett 1971a, Beamish 1972, Elliott 1972, Niimi & Beamish 1974, Steigenberger & Larkin 1974, and others). However, appetite and stomach evacuation rates do not always increase in the same way with temperature. The rate of increase in Q_{Rmax} often decreases at higher temperatures to give a curved relation (Fig. 8.7a). The reasons for this type of relation are not certain but might reflect limiting interactions of oxygen availability (see below and Fry 1971, Warren 1971, Muir & Niimi 1972) or perhaps reduced metabolic scope (Tyler & Dunn 1976). Temperature also has some effect on dietary requirements. The protein requirement of chinook salmon, *Oncorhynchus tschawytscha*, increases from 40 to 55% Q_R for a temperature increase from 8 to 14C, (DeLong *et al.* 1958). Assimilation efficiency of rainbow trout, *Salmo gairdneri*, increases from 72 to 78% for a temperature increase from 5 to 20C (Brocksen & Bugge 1974). In both cases, however, the temperature effects were not statistically significant. It is probable, therefore, that temperature effects on assimilation efficiency can be neglected. However, the question of effects of controlling factor level on food intake and assimilation efficiency are poorly researched and require further investigation.

Effects of controlling factor level on metabolism have been extensively studied. Q_S increases logarithmically with temperature in the usual way for poikilotherms (Fig. 8.7a and Fig. 8.8). Q_{Mmax} similarly increases for some fish, for example the yellow bullhead, *Ameiurus nebulosus* (Fig. 8.8). In other fish, notably salmonids, Q_{Mmax} increases to a maximum at an optimum temperature and thereafter decreases (Fig. 8.7b curve B, and Fig. 8.8). The reason for the decrease in Q_{Mmax} at high temperatures is not certain. Experiments to determine if reduced dissolved oxygen concentrations at high temperatures are limiting have not proved decisive (Brett 1964, Jones 1971).

The Q_{10} for Q_S is variable. 'Krogh's standard curve' fits some standard metabolism data, but there are sufficient exceptions that it should be used with caution in modelling temperature effects for production predictions (Brett 1970b, Fry 1971).

Metabolic scope either increases continuously with temperature (Fig. 8.7a) or may be related to temperature by a peaked curve with an optimum at some temperature intermediate between upper and lower lethal levels (Fig. 8.7b, curve B). This latter shaped curve follows when Q_{Mmax} decreases at high temperatures. At the optimum temperature, fish may be considered to have the greatest energetic capacity for non-essential activities, above maintenance requirements.

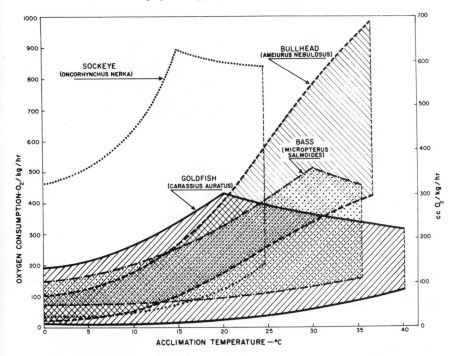

Fig. 8.8. Active and standard metabolic rates of four species of fish, shown as functions of temperature. From Brett (1972).

Growth responses to controlling factor levels follow from interactions with Q_R and Q_M. Two overall growth responses in relation to controlling factor level have been described. At Q_{Rmax}, Q_G may increase continuously with temperature as Niimi and Beamish (1974) found for largemouth bass. The more common response is for growth and scope for growth to increase to some optimum temperature and thereafter decrease (Fig. 8.7a). For example, optimum temperatures are 15C for salmonids (Brett *et al.* 1969, McCormick *et al.* 1972) and are independent of body size (Shelbourn *et al.* 1973). Horning and Pearson (1973) found a similar trend with an optimum temperature for growth in largemouth bass, contrasting with Niimi and Beamish (1974). This might reflect intraspecific differences among different stocks of fish.

The optimum temperature for growth depends on Q_R (Brett *et al.* 1969). In Fig. 8.5 it can be seen that the inflection points of the dotted lines for each Q_R shift to higher temperatures as Q_R increases. When Q_R was 1·5% body dry weight/day the optimum temperature for yearling sockeye salmon was 5C, increasing to 14 to 15C at a Q_R of 6% body dry weight/day.

Since scope for growth is usually related to temperature by a dome-shaped curve, it follows that η_{gross} will be similarly related to temperature, as exemplified by yearling sockeye salmon (Fig. 8.5) from Brett *et al.* (1969). The figure also illustrates interrelations between η_{gross}, Q_R and temperature. There is a distinct axis of increasing η_{gross} with temperature and Q_R. Maximum η_{gross} for this species was at 12C at a Q_R of 4·5% body dry weight/day. However, it should be noted that there is a plateau effect so that η_{gross} remains high for a range of temperatures from approximately 7·5 to 17·5C over a wide range of Q_R.

Somewhat different relations are found between η_{net}, Q_R and temperature (Fig. 8.6). For sockeye salmon η_{net} is greatest at an optimum temperature of approximately 13 to 17C at Q_{Rmax}. Maximum η_{net} occurs at very much lower Q_R and at lower temperatures. In addition, largest values of η_{net} occur over a narrower band of temperatures and rations than for η_{gross}. Thus the maximum η_{net} of 40% occurred at temperatures of 8 to 10C for Q_R up to 1·5% body dry weight/day.

When considering effects of controlling factors on rates of biological activities, it has frequently been pointed out that fish at extremes of major global temperature regimes might have metabolic rates too high or too low for efficient use of available resources if Q_M followed normal Q_{10} relations. Fish attain some measure of independence of controlling factors by acclimation over short time periods and by compensation over geological time spans (see Brett 1970b, Fry 1971). In the latter case, the phenomenon of 'cold adaptation' has received much attention (see Holeton 1974 for refs). Recently, Holeton (1974) has reviewed the two studies on which the concept of cold adaptation is based and has also provided new data. He was unable to support the concept, and attributes the discrepancy between his results and earlier studies to improved techniques. Holeton (1973), following Dunbar (1968), also points out that cold adaptation is disadvantageous when food is limited because a large proportion would be required for Q_S leaving little for growth and reproduction. Therefore, although there is some adaptation to thermal extremes, the effect appears to be less than previously considered, particularly for polar species.

LIMITING FACTORS

Full expression of metabolism and growth potential in response to controlling factor level is only possible in the absence of limiting factors. These are factors that restrict supply of metabolites (food, oxygen and water) or the removal of waste products (carbon dioxide, water and nitrogenous excreta).

Reference has already been made to the limiting role of food availability, which, of course, is itself regulated by the action of all environmental and biotic factors on forage populations (Weatherley 1972). However, it is also important to recognize food quality or diet as a potential limiting factor.

Certain essential amino acids and lipids cannot be synthesized and must be provided by the food (Ashley 1972). Consequently, an excess ration deficient in essential components will be limiting for growth. Vitamin and mineral supply can also be limiting. Dietary imbalance is not likely to be limiting under natural conditions, although this might become a problem as a result of long-term environmental changes that eliminate or reduce availability of some natural diet components (Ashley 1972). Advances in fish-feed technology are rapidly eliminating earlier dietary imbalance problems for fish culture (see Halver 1972).

Following Q_R, the most important limiting factors are respiratory gases, oxygen and carbon dioxide, and metabolic products. These interact with appetite to reduce Q_R below $Q_{R max}$ which in turn appears to affect Q_G. Thus, reduced oxygen levels decrease appetite for many species (see reviews by Warren & Davis 1967, Doudoroff & Shumway 1970; Warren 1971). Reduced appetite at high temperatures may result from decreased dissolved oxygen content, such that insufficient oxygen is available for Q_{SDA} in competition with other activities (Fry 1971, Warren 1971, Muir & Niimi 1972). This explanation is supported by observations by F. E. Hutchins (cited by Warren 1971) of reduced appetite in fish forced to swim continuously. It is considered that decreased appetite follows from reduced metabolic scope. However, appetite is also reduced by super-saturated oxygen levels which could increase metabolic scope (Doudoroff & Shumway 1970). Therefore, the limiting interactions of oxygen availability with oxygen demand, appetite and temperature are still unclear.

Elevated carbon dioxide concentrations reduce Q_G by depressing Q_R in sturgeon, *Acipenser* sp. In the gudgeon, *Gobio* sp., Q_R is unaffected by elevated carbon dioxide (see Doudoroff & Shumway 1970).

Environmental toxicants may also act as limiting factors on Q_R. The pesticide dieldrin reduces food intake in *Cottus perplexus* (R. Brocksen & T. Chadwick cited by Warren 1971). Fenitrothion causes regurgitation and hence reduces food consumption in salmonids (Symons 1973, Wildish & Lister 1973). In contrast, kraftmill effluent has little or no effect on palatability of food or on appetite (Tokar & Owens 1968, Warren 1971). The herbicide pentachlorophenol has a positive effect on Q_R in the cichlid *Cichlasoma bimaculatum*. It acts as a masking factor elevating metabolic rate and the fish compensate with increased appetite (Warren 1971). The ability of fish to increase food consumption when under stress and in the presence of excess food leads to unanswered questions of voluntary appetite control.

Other toxicants act indirectly to limit Q_R. Detergents damage taste and smell receptors in some fish (Bardach *et al.* 1965). Detergents, therefore, act as limiting factors reducing food consumption because of impaired ability of fish to successfully forage.

In contrast to the uncertain limiting effects of respiratory gases on Q_R, effects on metabolism are well-known. The literature is extensive and has been reviewed by Doudoroff and Shumway (1970), Dickson and Kramer (1971) and Fry (1971). Q_{Mmax} tends to decrease almost in proportion to reduced dissolved oxygen levels. In goldfish and carp (*Cyprinus carpio*), Q_{Mmax} becomes independent of oxygen concentrations above air saturation, while bullhead and salmonids show dependence at all levels (Basu 1959). Q^S is relatively independent of ambient oxygen levels but is elevated at very low concentrations (Basu 1959, Beamish 1964a, b, Kutty & Saunders 1973).

Elevated carbon dioxide and nitrogenous waste levels also limit Q_{Mmax}, while Q_S is again relatively insensitive (Basu 1959, Beamish 1964c). In addition, acclimation occurs within approximately 24 hours (Saunders 1962, Lloyd & Jordan 1964, Lloyd & White 1967, Mayer & Kramer 1973).

Other environmental factors act indirectly as limiting factors to oxygen supply. Suspended solids and solutes displace oxygen and reduce ambient levels. They may, in addition, damage gills and interfere with gas exchange. These substances behave in the same way as primary limiting factors by reducing Q_{Mmax} with Q_S being less sensitive (MacLeod & Smith 1966, Smith *et al.* 1967, Sprague 1971). Gas bubble disease is similarly an indirect limiting factor. Gas embolisms form in the blood and tissues of fish that have been exposed to atmospheric gases at pressures exceeding normal atmospheric pressure. Embolisms in the blood block arteries and veins resulting in local or systematic tissue hypoxia, or acidosis from the accumulation of toxic wastes (D'Aoust & Smith 1974, Newcomb 1974).

The net effect of any limiting factor on metabolism reduces Q_{Mmax} throughout the range of sublethal temperatures (Fig. 8.7b, curve C) and the interaction of temperature on oxygen solubility in water may further depress Q_{Mmax} at high temperatures in some fish (curve D). In either case, metabolic scope is reduced. When Q_{Mmax} increases continuously with temperature (curve A), there is an optimum temperature close to the upper lethal temperature. Limiting factors, however, shift this optimum to a lower temperature (curve C), this being a condition optimum (Fry 1971) resulting from interactions among environmental factors.

The effects of limiting factors on Q_{Mmax} are readily appreciated, but since Q_S is relatively insensitive except at very low levels, it is not clear how these factors interact to affect Q_G. Theory would predict that Q_R should be reduced by physical limiting factors, but as noted above empirical data are not completely convincing. Nevertheless, reductions on Q_G and η_{gross} have been described (see Doudoroff & Shumway 1970). Assuming that limiting factors affect Q_R, responses are illustrated in Fig. 8.7c. η_{gross} is not reduced with Q_R, except at very low oxygen levels, presumably as a result of increased Q_S in respiratory compensation (Swift 1964, Dorfman & Whitworth 1969, Doudoroff & Shumway 1970, Brungs 1971, Andrews *et al.* 1973). The response of η_{gross} to limiting factor level is generally consistent with the view that the factor reduces Q_R directly rather than acting through some intermediate function. Interaction of controlling and limiting factors results in a conditioned optimum temperature for growth. Present data however suggests this effect is small (Fig. 8.7c).

Almost all experiments with limiting factors have been restricted to steady state situations. In the field, many limiting factors will fluctuate. Studies of the effects of fluctuating oxygen levels on Q_G in largemouth bass have been made but the results are inconclusive. Fluctuating oxygen regimes have been shown to reduce Q_G in coho salmon, *Oncorhynchus kisutch*, compared to fish at an equivalent mean level (Whitworth 1968, Doudoroff & Shumway 1970).

MASKING FACTORS

Environmental changes necessitating changed metabolic regulatory costs are classified by Fry (1971) as masking and directive factors. The former consider largely environmental factors that perturb homeostatic systems. The metabolic costs of subsequent internal regulation are appropriately considered as an addition to Q_S over and above that required for an unstressed fish at rest.

Few experimental studies have been made of regulatory costs for fish, with the exception of responses to salinity. In general, Q_R is greatest and Q_S and Q_{maint} smallest under least stress, so that Q_G, η_{gross} and η_{net} are then maximal. For salinity changes, least stress usually occurs when the ambient water is isosmotic with the blood (Job 1967, Rao 1968, Farmer & Beamish 1969, Otto 1971, Brocksen & Cole 1972). Metabolism and growth responses to masking factors are illustrated in Fig. 8.7d and 8.7e. As with limiting factors, masking factors interact with temperature to not only affect energetics, but also to shift the animal's optimum temperature to a lower conditioned optimum. Present data suggests the shift in temperature optima may be large as a result of masking factor responses.

Many environmental factors not only act as limiting factors but also as masking factors that elevate Q_S. For example, particulate matter elevates metabolic rates of walleye, *Stizostedion vitreum* (Smith et al. 1967). Q_S was increased two-fold in sockeye salmon when exposed to bleached kraftmill effluent at a concentration of one-tenth the concentration that was lethal to 50% of fish after 96 hours of exposure in an acute lethal bioassay (P. W. Webb, unpublished data).

Since toxicants tend to reduce Q_R and increase Q_S, reduced Q_G and η_{gross} are predictable. Expected losses in growth and food conversion efficiencies have been described for fibre (Smith et al. 1967), kraftmill effluents (Tokar & Owens 1968, Webb & Brett 1972), the pesticides carbaryl

(Carlson 1971), endrin (Grant & Merle 1973), and pentachlorophenate (Krueger *et al.* 1968, Warren 1971, Webb & Brett 1973) and heavy metals (Mount & Stephan 1969, Pickering & Gast 1972, McKim & Benoit 1974 and others). Low pH from acid rain reduces Q_G in many species (R. J. Beamish 1974, R. J. Beamish *et al.* 1975). Low pH probably also acts as a limiting factor by deleteriously affecting blood/oxygen saturation kinetics.

Very low levels of toxicants can have a positive effect on Q_G (e.g. Webb & Brett 1972 and 1973, McLeay & Brown 1974). McLeay and Brown (1974) have discussed the phenomena of growth stimulation in relation to low level exposure of salmon to kraftmill effluents. They compared the observed growth stimulation with the chronic stress response of mammals, hypothesizing that a slightly altered hormone balance (pituitary–adrenal axis) improved food utilization and growth. The corollary is that fish do not normally use food resources as efficiently as possible so that improved efficiency is possible under subinhibitory concentrations of toxicants.

The effects of masking factors on energetic responses of fish is an area requiring substantial research. Environmental manipulation could potentially increase productivity (see Brett 1974a), at least for some species (Shaw *et al.* 1975). Hormonal changes associated with growth stimulation might provide further opportunities to increase food conversion efficiencies in fish culture. In addition, masking factor effects emphasize the importance of water quality for good production, not only for farming and culture practice, but also for coastal zone nursery areas and commercial fisheries. Research is needed in this area because coastal areas with good water flushing characteristics are attractive for both aquaculture and discharge of wastes. Optimum use of limited water resources requires better understanding of the impact of masking factors on fishes and animal communities.

DIRECTIVE FACTORS

In contrast to immediate regulatory responses imposed by masking factors, directive factors result in longer term energy redistributions. Responses involve *distance* reception of stimuli frequently characterized by spatial and/ or temporal gradients. One type of directive factor response involves interactions with other organisms (for example, behaviour with respect to food, territory and conspecifics) and chemical gradients (for example, oxygen and toxicants). In these cases, energy is typically redirected to cover metabolic costs of locomotor activity. A second type of directive factor response is to environmental periodicity which involves temporal gradients causing changes in metabolic strategy (for example, to low winter temperature regimes and for reproduction).

As with masking factor responses, little pertinent research has been done on directive factor effects. Appetite certainly varies seasonally. For example,

Warren and Davis (1967), Brocksen *et al.* (1968), Warren (1971), Kolehmainen (1974) and Small (1975) document numerous cases where Q_R was
reduced during fall and winter from the summer maximum, although food
was present in excess of $Q_{R\,max}$ at all times. Feeding is known to be inhibited
in fish with advanced parental care, particularly in mouth brooders such as
many cichlids and *Amia*. Q_S also varies seasonally, typically being reduced
in the winter. Seasonal maxima in Q_M occur from spring to fall and are
further elevated by reproductive costs, themselves regulated by directive
factors (Beamish 1964c, Dickson & Kramer 1971, Moore & Wohlschlag 1971,
Burns 1975). In addition to seasonal metabolic changes, diurnal fluctuations
are common, usually reflecting circadian rhythmicity in spontaneous activity
(Fry 1971). Q_G tends to vary in the same way as Q_M with seasonal changes,
rather than inversely as is the case for other environmental factors (Warren
& Davis 1967, Brocksen *et al.* 1968, Warren 1971, Otto 1971, Das 1972).
These effects are attributed to unspecified changes in metabolic strategy
(Iles 1974) and to normal seasonal variability in food availability and/or
unfavourable temperature regimes. The complex of effects of environmental
periodicity of growth and metabolism energetics is difficult to predict and
requires additional research.

Social interactions also effect metabolism and growth. Brett (1973) found
that social hierarchical interactions elevated Q_M of yearling sockeye salmon
swimming in groups. In contrast, schooling has been shown to reduce Q_M
in 21 species (Parker 1973). Although these experiments did not permit
adequate acclimation times to be able to fully explain the reduced metabolic
rate, schooling appears to exert a 'calming' effect and there may be further
hydrodynamic energy economies (Weihs 1973). When group behaviour
increases Q_M, Q_G is inversely affected. Overall Q_G and η_{gross} for a group of
fish were reduced as a result of hierarchical interaction in medaka (*Oryzias
latipes*) and zebra fish (*Bracydanio rerio*) (Magnuson 1962, Eaton & Farley
1974). In contrast to these experiments, Carline and Hall (1973) consider that
metabolic costs of agonistic behaviour in coho salmon has no effect on Q_G
or η_{gross}. It is well known that dominant fish tend to monopolize food and
show better Q_G, the phenomena being variously described as the size hierarchy effect (Brown 1957) or growth depensation (Ricker 1958). Kawanabe
(1969) has described changes in social behaviour of the ayu, *Plecoglossus
altivelis* in response to changing food availability that apparently results in
better utilization of the food resource by the population.

One net effect of the multitude of discrete spatially distributed environmental stimuli acting as directive factors is presumably reflected in a fish's
spontaneous activity. The ultimate effect is an increase in Q_M above Q_S in
the same way as for masking factors (Fig. 8.7d and e). The net costs of
spontaneous activity (sometimes described as routine metabolism) is of
considerable importance in growth and production modelling. These routine

metabolic costs may be as high as one-third to one-half of the metabolic scope (Beamish & Mookherjii 1964, Peterson & Anderson 1969, Brett 1973, Fry 1971, Rajagotal & Kramer 1974). Production models frequently follow Winberg (1956) in computing routine metabolism as twice Q_S. Soloman and Brafield (1972) and Ware (1975) have recently summarized criticisms of using such a multiple, and also suggests a factor of $3 \times Q_S$ may be more appropriate for young, actively growing fish.

BIOTIC FACTORS

A consideration of metabolism and growth energetics must include effects of biotic factors in addition to environmental factors. Many factors or responses, such as social interactions, training, oxygen debt, etc., have been classed as biotic factors (e.g. Brett 1970a). However, such factors may be viewed as components in a fish's environment to which the fish responds with the same metabolic consequence as a response to a physical component of the environment. If this approach is taken, the only purely biotic factors are species, size (and age), sex and disease. Size effects have already been discussed. Species variation is extensive, and will not be discussed here. Useful reviews are provided by Winberg (1956), Doudoroff and Shumway (1970), and Brett (1972). An example of species effects on Q_M is given in Fig. 8.8. It should be noted that the variability in responses to directive factors may be attributable to species variability.

Effects of sex on energy use are apparently small. Numerous metabolic studies have been made and do not record major effects of sex, other than those related to differences in body mass (see Burns 1975). Differences occasionally appear at maturity, as Claridge and Potter (1975) have recently shown for the lamprey, *Lampetra fluviatilis*.

Effects of disease are not well known and generalizations are not possible at this time. Parasitic gut nematodes reduced Q_G in rainbow trout, apparently by reducing assimilation efficiency and elevating Q_M (Hiscox & Brocksen 1973). In contrast, intestinal helminths had no effect on Q_G nor the assimilation of L-leucine and D-glucose in the same species (Ingham & Arme 1973).

CONCLUSIONS

An energetic approach to fish metabolism and growth that takes into account effects of environmental and biotic variability has application in two major areas relating to production. In the first case, research should provide the basis for first principles models that can lead to better management of a renewable resource. There are numerous models that generate production estimates (e.g. Paloheimo & Dickie 1966a, b, Kerr 1971a, b, c, Healey 1972, Feldmeth & Jenkins 1973, Hoss 1974, Morgan 1974, Ware 1975), although few couple these with fishery exploitation and optimization models (see Sharp & Francis 1976).

The second area of importance is application of energetics to intensive culture. Improved production of terrestrial animals has exploited this approach, particularly in terms of a controlled environment, including diet, and manipulation of hormonal profiles of animals. Essentially this involves utilizing normal regulation functions to minimize metabolic losses while maximizing food conversion efficiency. Proper understanding of fish responses to masking and directive factors is, therefore, an area that appears crucial and also requires further research if environmental manipulation and timing is to be economically used. It should be noted that timing questions are likely to be particularly important because the physical properties of water make environmental manipulation a potentially expensive method of increasing production. Therefore, increased production by manipulating regulation responses is expected to be dependent on optimum use of naturally occurring environmental variability.

At the present time, few species have been studied in sufficient detail to provide an adequate research base from which questions on appetite, metabolism and growth can be usefully approached. Major exceptions are sockeye salmon (see Brett *et al.* 1969, Brett 1973, 1974a, b, Brett & Glass 1973), and largemouth bass (see Niimi & Beamish 1974). As a result, there is still a large gap between much of the theory and its application to production questions. Only the research on salmon has effectively led into intensive culture practices (Brett 1974a, b). Comparative studies, focusing on the basic energetic information needed for production optimization or fishery exploitation are, therefore, imperative.

Reference

ANDREWS J.W., MURAL T. & GIBBONS G. (1973) The influence of dissolved oxygen on the growth of channel catfish. *Trans. Am. Fish. Soc.* **102**, 835–837.

ANDREWS J.W. & STICKNEY R.R. (1972) Interactions of feeding rates and environmental temperature on growth, food conversion efficiency and body composition of channel catfish. *Trans. Am. Fish. Soc.* **101**, 94–99.

ASHLEY L.M. (1972) Nutritional Pathology. *Fish Nutrition* (J.E. Halver, ed.), pp. 439–537. Academic Press, N.Y.

AVERETT, R.C. (1969) Influence of temperature on energy and material utilization by juvenile coho salmon. Ph.D. thesis, Oregon State University, Corvallis, Oregon.

BARDACH J.E., FUJIYA M. & HALL A. (1965) Detergents; Effects on the chemical senses of the fish *Ictalurus natalis*. *Science* **148**, 677.

BASU S.P. (1959) Active respiration of fish in relation to ambient oxygen and carbon dioxide. *J. Fish. Res. Bd. Canada* **16**, 175–212.

BEAMISH F.W.H. (1964a) Respiration of fishes with special emphasis on standard oxygen consumption. III. Influence of oxygen. *Can. J. Zool.* **42**, 355–366.

BEAMISH F.W.H. (1964b) IV. Influence of carbon dioxide and oxygen. *Can J. Zool.* **42.**, 847–856.

BEAMISH F.W.H. (1964c) Seasonal changes in the standard rate of oxygen consumption of fishes. *Can. J. Zool.* **42**, 189–194.

BEAMISH F.W.H. (1972) Ration size and digestion in largemouth bass, *Micropterus salmoides* Lacèpéde. *Can. J. Zool.* **50**, 153–164.

BEAMISH F.W.H. (1974) Apparent specific dynamic action of largemouth bass *Micropterus salmoides*. *J. Fish. Res. Bd. Canada* **31**, 1763–1769.

BEAMISH F.W.H. & MOOKHERJII P.S. (1964) Respiration of fishes with special emphasis on standard oxygen consumption. I. Influence of weight and temperature on respiration of goldfish, *Carassius auratus*. L. *Can. J. Zool.* **42**, 161–175.

BEAMISH R.J. (1974) Growth and survival of white suckers (*Catostomus commersoni*) in an acidified lake. *J. Fish. Res. Bd. Canada*

BEAMISH R.J., LOCKHART W.L., VAN LOON J.C. & HARVEY H.H. (1975) Long-term acidification of a lake and resulting effects on fish. *AMBIO* **4**, 98–102.

BENNETT A.F. & LICHT P. (1973) Relative contributions of anaerobic and aerobic energy production during activity in amphibia. *J. comp. Physiol.* **87**, 351–360.

BIRKETT L. (1969) The nitrogen balance in plaice, sole and perch. *J. Exp. Biol.* **50**, 375–386.

BLAZKA P. (1958) The anaerobic metabolism of fish. *Physiol. Zoöl.* **31**, 117–128.

BRETT J.R. (1964) The respiratory metabolism and swimming performance of young sockeye salmon. *J. Fish. Res. Bd. Canada* **21**, 1183–1226.

BRETT J.R. (1970a) Fish—the energy cost of living. *Marine Aquaculture*, (W.J. McNeil, ed.), pp. 37–52. Oregon State University Press, Corvallis, Oregon, U.S.A.

BRETT J.R. (1970b) 3. Temperature. 3.3 Animals. 3.32 Fishes. *Marine Ecology*, vol. 1. (O. Kinne, ed.), pp. 515–560. Wiley-Interscience, London.

BRETT J.R. (1971a) Satiation time, appetite and maximum food intake of sockeye salmon (*Oncorhynchus nerka*). *J. Fish. Res. Bd. Canada* **28**, 409–415.

BRETT J.R. (1971b) Growth responses of young sockeye salmon (*Oncorhynchus nerka*) to different diets and planes of nutrition. *J. Fish. Res. Bd. Canada* **28**, 1635–1643.

BRETT J.R. (1972) The metabolic demand for oxygen in fish, particularly salmonids, and a comparison with other vertebrates. *Resp. Physiol.* **14**, 151–170.

BRETT J.R. (1973) Energy expenditure of sockeye salmon, *Oncorhynchus nerka*, during sustained performance. *J. Fish. Res. Bd. Canada* **30**, 1799–1809.

BRETT J.R. (1974a) Marine fish aquaculture in Canada. p. 54–84 in *Aquaculture in Canada*, *Bull. Fish. Res. Bd. Canada* **188**, 84 p.

BRETT J.R. (1974b) Tank experiments on the culture of pan-size sockeye salmon (*Oncorhynchus nerka*) and pink salmon (*O. gorbuscha*) using environmental control. *Aquaculture* **4**, 341–352.

BRETT J.R. & GLASS N.R. (1973) Metabolic rates and critical swimming speeds of sockeye salmon (*Oncorhynchus nerka*) in relation to size and temperature. *J. Fish. Red. Bd. Canada* **30**, 379–387.

BRETT J.R. & HIGGS D.A. (1970) Effect of temperature on the rate of gastric digestion in fingerling sockeye salmon, *Oncorhynchus nerka*. *J. Fish. Res. Bd. Canada* **27**, 1767–1779.

BRETT J.R. & SHELBOURN J.E. (1975) Growth rate of young sockeye salmon, *Oncorhynchus nerka*, in relation to fish size and ration level. *J. Fish. Res. Bd. Canada* **32**, 2103–2110.

BRETT J.R., SHELBOURN J.E. & SHOOP C.T. (1969) Growth rate and body composition of fingerling sockeye salmon, *Oncorhynchus nerka*, in relation to temperature and ration size. *J. Fish. Res. Bd. Canada* **26**, 2363–2394.

BROCKSEN R.W. & BUGGE J.P. (1974) Preliminary investigations on the influence of temperature on food assimilation by rainbow trout, *Salmo gairdneri* Richardson. *J. Fish Biol.* **6**, 93–97.

BROCKSEN R.W. & COLE R.E. (1972) Physiological responses of three species of fish to various salinities. *J. Fish. Res. Bd. Canada* **29**, 399–405.

BROCKSEN R.W., DAVIS G.E. & WARREN C.E. (1968) Competition, food consumption, and production of sculpins and trout in laboratory stream communities. *J. Wildlife Manage.* **32**, 51–75.

BROWN M.E. (1957) Experimental studies on growth. *The Physiology of Fishes* Vol. 1. (M.E. Brown, ed.), pp. 361–400. Academic Press, N.Y.

BRUNGS W.A. (1971) Chronic effects of low dissolved oxygen concentrations on the fathead minnow (*Pimephales promelas*). *J. Fish. Res. Bd. Canada* **28**, 1119–1123.

BURNS J.R. (1975) Seasonal changes in the respiration of pumpkinseed, *Lepomis gibbosus*, correlated with temperature, day length, and stage of reproductive development. *Physiol. Zoöl.* **48**, 142–149.

CARLINE R.F. & HALL J.D. (1973) Evaluation of a method for estimating food consumption rates of fish. *J. Fish. Res. Bd. Canada* **30**, 623–629.

CARLSON A.R. (1971) Effects of long-term exposure to carbaryl (sevin) on survival, growth and reproduction of the fathead minnow (*Pimephales promelas*). *J. Fish. Res. Bd. Canada* **29**, 583–587.

CLARIDGE P.N. & POTTER I.C. (1975) Oxygen consumption, ventilatory frequency and heart rate of lampreys (*Lampetra fluviatilis*) during their spawning run. *J. Exp. Biol.* **63**, 193–206.

CONGLETON J.L. (1974) The respiratory response to asphyxia of *Typhlogobius californiensis* (Teleostei: Gobiidae) and some related gobies. *Biol. Bull.* **146**, 186–205.

CONNOR A.R., ELLENG C.H., BLACK E.C., COLLINS G.B., CAULEY J.R. & TREVOR-SMITH E. (1964) Changes in glycogen and lactate levels in migrating salmonid fish ascending experimental "endless" fishways. *J. Fish. Res. Bd. Canada* **21**, 255–290.

DAS N. (1972) Growth of larval herring (*Clupea harengus*) in the Bay of Fundy and Gulf of Maine area. *J. Fish. Res. Bd. Canada* **29**, 537–575.

D'AOUST B.G. & SMITH L.S. (1974) Bends in fish. *Comp. Biochem. Physiol.* **49A**, 311–321.

DELONG D.C., HALVER J.E. & MERTZ E.T. (1958) Nutrition of salmonids fishes. IV. Protein requirements of chinook salmon at two water temperatures. *J. Nutr.* **35**, 589–599.

DICKSON I.W. & KRAMER R.H. (1971) Factors influencing scope for activity and active and standard metabolism of rainbow trout (*Salmo gaidneri*) *J. Fish. Res. Bd. Canada* **28**, 587–596.

DORFMAN D. & WHITWORTH W.R. (1969) Effects of fluctuations of lead, temperature and dissolved oxygen on the growth of brook trout. *J. Fish. Res. Bd. Canada* **26**, 2493–2501,

DOUDOROFF P. & SHUMWAY D.L. (1970) Dissolved oxygen requirements of freshwater fishes. *FAO Fisheries Tech. Pap.* No. **86**, 291 p.

DUNBAR M.J. (1968) *Ecological Development in polar regions; a Study in Evolution.* Prentice Hall, Englewoods Cliffs, N.J.

EATON R.C. & FARLEY R.D. (1974) Growth and reduction of dispensation of zebrafish, *Brachydanio rerio*, reared in the laboratory. *Copeia* 1974, 204–209.

ELLIOTT J.M. (1972) Rates of gastric evacuation in brown trout, *Salmo trutta*. *Freshwat. Biol.* **2**, 1–18.

FARMER G.J. & BEAMISH W.H. (1969) Oxygen consumption of *Tilapia nilotica* in relation to swimming speed and salinity. *J. Fish. Res. Bd. Canada* **26**, 2807–2821.

FELDMETH C.R. & JENKINS T.M. (1973) An estimate of energy expenditure by rainbow trout (*Salmo gairdneri*) in a small mountain stream. *J. Fish. Res. Bd. Canada* **30**, 1755–1759.

FORESTER R.P. & GOLDSTEIN L. (1969) Formation of excretory products. *Fish Physiology*, Vol. 1 (W.S. Hoar & D.J. Randall, eds.), pp. 313–350. Academic Press, N.Y.

FRY F.E.J. (1971) The effect of environmental factors on the physiology of fish. *Fish Physiology*, Vol. 6, (W.S. Hoar & D.J. Randall, eds.), pp. 1–98. Academic Press, N.Y.

GERKING S.D. (1955) Influence of rate of feeding on body composition and protein metabolism of bluegill sunfish. *Physiol. Zoöl.* **28**, 267–282.

GERKING S.D. (1962) Production and food utilization in a population of bluegill sunfish. *Ecol. Monogr.* **32**, 31–78.

GERKING S.D. (1971) Influence of rate of feeding and body weight on protein metabolism of bluegill sunfish. *Physiol. Zoöl.* **44**, 9–19.

GERKING S.D. (1972) Revised food consumption estimate of a bluegill sunfish population in Wyland Lake, Indiana, U.S.A. *J. Fish Biol.* **4**, 301–308.

GRANT B.F. & MERLE P.M. (1973) Endrin toxicosis in rainbow trout (*Salmo gairdneri*). *F. Fish. Res. Bd. Canada* **30**, 31–40.

GREER-WALKER M. & PULL G. (1973) Skeletal muscle function and sustained swimming speeds in the coalfish *Gadus virens* L. *Comp. Biochem. Physiol.* **44A**, 494–501.

HALVER J.E. (1972) (Ed.). *Fish Nutrition.* Academic Press, New York, N.Y.

HEALEY M.C. (1972) Bioenergetics of the sand goby (*Gobius minutus*) population. *J. Fish. Res. Bd. Canada* **29**, 187–194.

HEMMINGSEN A.M. (1960) Energy metabolism as related to body size and respiratory surfaces, and its evolution. *Rep. Steno. Hosp. Copenhagen* **9** (part 2), 1–110.

HISCOX J.I. & BROCKSEN R.W. (1973) Effects of a parasitic gut nematode on consumption and growth in juvenile rainbow trout (*Salmo gairdneri*). *J. Fish. Res. Bd. Canada* **30**, 443–450.

HOLETON G.F. (1973) Respiration of Arctic char (*Salvelinus alpinus*) from a high arctic lake. *J. Fish. Res. Bd. Canada* **30**, 717–723.

HOLETON G.F. (1974) Metabolic cold adaptation of polar fish: Fact or artifact. *Physiol. Zoöl.* **47**, 137–152.

HORNING W.B. & PEARSON R.E. (1973) Growth temperature requirements and lower lethal temperature for juvenile smallmouth bass (*Micropterus dolomieui*). *J. Fish. Res. Bd. Canada* **30**, 1226–1230.

HOSS D.E. (1974) Energy requirements of a population of pinfish *Lagodon rhomboides* (Linnaeus). *Ecology* **55**, 848–855.

HUGHES G.M. & MORGAN M. (1973) The structure of fish gills in relation to their respiratory function. *Biol. Rev.* **48**, 419–475.

ILES T.D. (1974) The tatics and strategy of growth in fishes. *Sea Fisheries Research* (F.R. Harden Jones, ed.), pp. 331–345. John Wiley and Sons, New York, N.Y.

INGHAM L. & ARME C. (1973) Intestinal helminths in rainbow trout, *Salmo gardneri* (Richardson): Absence of effect on nutrient absorption and fish growth. *J. Fish Biol* **5**, 309–313.

IVLEV I. (1939) Energy balance in the carp. *Zool. Zh.* **18**, 303–318.

JOB S.V. (1967) The respiratory metabolism of *Tilapia mossambica* (Teleostei). I. The effect of size, temperature and salinity. *Marine Biol.* **2**, 121–126.

JOHNSTON L.A. & GOLDSPINK G. (1973) A study of the swimming performance of the Crucian carp, *Carassius carassius* (L.) in relation to the effects of exercise and recovery on biochemical changes in myotomal muscles and liver. *J. Fish Biol.* **5**, 249–260.

JONES D.J. (1971) The effect of hypoxia and anaemia on the swimming performance of rainbow trout (*Salmo gairdneri*). *J. Expt. Biol.* **55**, 541–551.

KAWANABE H. (1969) The significance of social structure in production of the "Ayu", *Plecoglossus altivelis. Symposium on salmon and trout in streams* (T.G. Northcote, ed.), pp. 243–251. H. R. Macmillan Lectures in Fisheries, Univ. British Columbia, Vancouver, B.C.

KERR S.R. (1971a) Analysis of laboratory experiments on growth efficiency of fishes. *J. Fish. Res. Bd. Canada* **28**, 801–808.

KERR S.R. (1971b) Prediction of growth efficiencies in nature. *J. Fish. Res. Bd. Canada* **28**, 809–814.

KERR S.R. (1971c) A simulation model of lake trout growth. *J. Fish. Res. Bd. Canada* **28**, 815–819.

KLEIBER M. (1975) *The Fire of Life: An Introduction to Animal Energetics* (Revised edition). R. E. Krieger Publ. Co., Huntingdon, N.Y.

KOLEHMAINEN S.E. (1974) Daily feeding rates of bluegill (*Lepomis macrochirus*) determined by a refined radioscope method. *J. Fish. Res. Bd. Canada* **31**, 67–74.

KRUEGER H.M., SADDLER J.B., CHAPMAN G.A., TINSLEY I.J. & LOWRY R.R. (1968) Bioenergetics, exercise, and fatty acids of fish. *Amer. Zool.* **8**, 119–129.

KUTTY M.N. (1968) Respiratory quotient in goldfish and rainbow trout. *J. Fish. Res Bd. Canada* **25**, 1687–1728.

KUTTY M.N. (1972) Respiratory quotient and ammonia excretion in *Tilipia mossambica*. *Marine Biol.* **16**, 126–133.

KUTTY M.N. & SAUNDERS R.L. (1973) Swimming performance of young Atlantic salmon (*Salmo salar*) as affected by reduced ambient oxygen concentration. *J. Fish. Res. Bd. Canada* **30**, 223–227.

LIVINGSTON R.J. (1968) A volumetric respirometer for long-term studies of small aquatic animals. *J. Mar. Biol. Ass. U. K.* **48**, 485–497.

LLOYD R.D. & JORDAN H.M. (1964) Some factors affecting the resistance of rainbow trout (*Salmo gairdneri* Richardson) to acid waters. *Air Water Pollution* **8**, 393–403.

LLOYD R. & WHITE W.R. (1967) Effects of high concentrations of carbon dioxide on the ionic composition of rainbow trout blood. *Nature* **216**, 1341–1342.

MCCAY C.M. & TUNISON A.V. (1935) Report of the experimental work at the Courtland Hatchery for the year 1934. *State of New York Conservation Department*, Albany, N.Y.

MCCORMICK J.H., HOKANSON K.E.F. & JONES B.R. (1972) Effects of temperature on growth and survival of young brook trout, *Salvelinus fontinalis*. *J. Fish. Res. Bd. Canada* **29**, 1107–1112.

MCKIM J.M. & BENOIT D.A. (1974) Duration of toxicity tests for establishing "no effect" concentrations for copper with brook trout (*Salvelinus fontinalis*) *J. Fish. Res. Bd. Canada* **31**, 449–452.

MCLEAY D.J. & BROWN D.A. (1974) Growth stimulation and biochemical changes in juvenile coho salmon (*Oncorhynchus nerka*) exposed to bleached kraft pulpmill effluent for 200 days. *J. Fish. Res. Bd. Canada* **31**, 1043–1049.

MCMAHON T.A. (1973) Size and shape in biology. *Science* **179**, 1201–1204.

MACLEOD J.C. & SMITH L.L. (1966) Effect of pulpwood fiber on oxygen consumption and swimming endurance of fathead minnow *Pimephales promelas*. *Trans. Am. Fish. Soc.* **95**, 71–84.

MAGNUSON J.J. (1962) An analysis of aggressive behaviour, growth and competition for food and space in medaka (*Oryzias latipedes* Pisces: Cyprinodontidae). *Can. J. Zool.* **40**, 313–363.

MAGNUSON J.J. (1969) Digestion and food consumption by skipjack tuna (*Katsuwonus pelamis*). *Trans. Am. Fish. Soc.* **98**, 379–392.

MAYER F.L. & KRAMER R.H. (1973) Effects of hatchery water reuse on rainbow trout metabolism. *Prog. Fish-Cult.* **35**, 9–10.

MOORE R.H. & WOHLSCHLAG D.E. (1971) Seasonal variations in the metabolism of the Atlantic midshipman, *Porichthys porosissimus* (Valenciennes). *J. Expt. Marine Biol. Ecol.* **7**, 163–172.

MORGAN R.I.G. (1974) The energy requirements of trout and perch populations in Loch Leven, Kinross, *Proc. R. Soc. Edinburgh* (B) **74**, 333–345.

MOUNT D.I. & STEPHEN C.E. (1969) Chronic toxicity of copper to the fathead minnow (*Pimephales promelas*) in soft water. *J. Fish. Res. Bd. Canada* **26**, 2449–2457.

MUIR B.S. & NIIMI A.J. (1972) Oxygen consumption of the euryhaline aholehole (*Kuhlia sandvicensis*) in relation to salinity, swimming and food consumption. *J. Fish. Res. Bd. Canada* **29**, 67–77.

NEWCOMB T.W. (1974) Changes in blood chemistry of juvenile steelhead trout, *Salmo gairdneri*, followed sublethal exposure to nitrogen supersaturation. *J. Fish. Res. Bd. Canada* **31**, 1953–1957.

NIIMI A.J. (1972a) Changes in the proximate body composition of largemouth bass (*Micropterus salmoides*) with starvation. *Can. J. Zool.* **50**, 815–819.

NIIMI A.J. (1972b) Total nitrogen, nonprotein nitrogen, and protein content in largemouth bass (*Micropterus salmoides*) with reference to quantitative protein estimates. *Can. J. Zool.* **50**, 1607–1610.

NIIMI A.J. & BEAMISH F.W.H. (1974) Bioenergetics and growth of largemouth bass (*Micropterus salmoides*) in relation to body weight and temperature. *Can. J. Zool.* **52**, 447–456.

OLSON K.R. & FROMM P.O. (1971) Excretion of urea by two teleosts exposed to different concentrations of ambient ammonia. *Comp. Biochem. Physiol.* **40A**, 999–1007.

OTTO R.G. (1971) Effects of salinity on the survival and growth of pre-smolt coho salmon (*Oncorhynchus kisutch*). *J. Fish. Res. Bd. Canada* **28**, 343–349.

PALOHEIMO J.E. & DICKIE L.M. (1966a). Food and growth of fishes. II. Effects of food and temperature on the relation between metabolism and body weight. *J. Fish. Res. Bd. Canada* **23**, 869–908.

PALOHEIMO J.E. & DICKIE L.M. (1966b) III. Relations among food, body size and growth efficiency. *J. Fish. Res. Bd. Canada* **23**, 1209–1248.

PANDIAN T.J. (1967) Intake, digestion, absorption and conversion of food in the fishes *Megalops cyprinoides* and *Ophiocephalus striatus*. *Marine Biol.* **1**, 16–32.

PARKER F.R. (1973) Reduced metabolic rates in fishes as a result of induced schooling. *Trans. Am. Fish. Soc.* **102**, 125–131.

PETERSON R.H. & ANDERSON J.M. (1969) Influence of temperature change on spontaneous locomotor activity and oxygen consumption of Atlantic salmon, *Salmo salar*, acclimated to two temperatures. *J. Fish. Res. Bd. Canada* **26**, 93–109.

PHILLIPS A.M. (1969) Nutrition, digestion and energy utilization. *Fish Physiology* Vol. 1 (W.S. Hoar & D.J. Randall, eds.), pp. 391–432. Academic Press, New York, N.Y.

PICKERING Q.H. & GAST M.H. (1972) Acute toxicity of cadmium to the fathead minnow (*Pimephales promelas*). *J. Fish. Res. Bd. Canada* **29**, 1099–1106.

PIERCE R.J. & WISSING T.E. (1974) Energy cost of food utilization in the bluegill (*Lepomis macrochirus*). *Trans. Am. Fish. Soc.* **103**, 38–45.

PRITCHARD A.W., HUNTER J.R. & LASKER R. (1971) The relation between exercise and biochemical changes in red and white muscle and liver in the jack mackerel, *Trachurus symmetricus*. *U.S. Fish and Wildl Serv. Fish. Bull.* **69**, 379–386.

RAJAGOPAL R.K. & KRAMER R.H. (1974) Respiratory metabolism of Utah chub, *Gila atraria* (Girard) and specled dace, *Rhinichthys osculus* (Girard). *J. Fish Biol.* **6**, 215–222.

RAO G.M.M. (1968) Oxygen consumption of rainbow trout (*Salmo gairdneri*) in relation to activity and salinity. *Can. J. Zool.* **46**, 781–786.

RICKER W.E. (1958) Handbook of computations for biological statistics of fish populations. *Fish. Res. Bd. Canada Bull.* **119**.

SAKSENA V.P., STEINMETZ C. & HOUDE E.D. (1972) Effects of temperature on growth and survival of laboratory-reared larval sardine, *Harengus pensacolae*. *Trans. Am. Fish. Soc.* **101**, 691–695.

SAUNDERS R.L. (1962) The irrigation of gills in fishes. II. Efficiency of oxygen uptake in relation to respiratory flow activity and concentrations of oxygen and carbon dioxide. *Can. J. Zool.* **40**, 817–862.

SAVITZ J. (1969) Effects of temperature and body weight on endogenous nitrogen excretion in the bluegill sunfish (*Lepomis macrochirus*). *J. Fish. Res. Bd. Canada* **26**, 1813–1821.

SAVITZ J. (1971) Effects of starvation on body protein utilization of bluegill sunfish (*Lepomis macrochirus*) with a calculation of caloric requirements. *Trans. Am. Fish. Soc.* **100**, 18–21.

SHARP G.D. & FRANCIS R.C. (1976) An energetics model for the exploited yellowfin tuna population in the eastern Pacific ocean. *U.S. Nat. Mar. Fish. Serv. NOAA, Fish. Bull.* **74**, 36–51.

SHAW H.M., SAUNDERS R.L. & HALL H.C. (1975) Environmental salinity: its failure to influence growth of Atlantic salmon (*Salmo salar*) parr. *J. Fish. Res. Bd. Canada* **32**, 1821–1824.

SHELBOURN J.E., BRETT J.R. & SHIRAHATA S. (1973) Effect of temperature and feeding regime on the specific growth rate of sockeye salmon fry (*Oncorhynchus nerka*), with a consideration of size effect. *J. Fish Res. Bd. Canada* **30**, 1191–1194.

SMALL J.W. (1975) Energy dynamics of benthic fishes in a small Kentucky stream. *Ecology* **56**, 827–840.

SMIT H., AMELINK-KOUTSTALL J.M., VIJVERBERG J. & VON VAUPEL-KLEIN J.C. (1971) Oxygen consumption and efficiency of swimming goldfish. *Comp. Biochem. Physiol.* **39A**, 1–28.

SMITH L.L., KRAMER R.H. & OSEID D.M. (1967) Long-term effects of conifer-groundwood paper fibre on walleyes. *Trans. Am. Fish. Soc.* **95**, 60–70.

SOLOMAN D.J. & BRAFIELD A.E. (1972) The energetics of feeding metabolism and growth of perch (*Perca fluviatilis*). *J. Anim. Ecol.* **41**, 699–718.

SPRAGUE J.B. (1971) Measurement of pollutant toxicity to fish. III. Sublethal effects and "safe" concentrations. *Water Research* **5**, 245–266.

STEIGENBERGER L.W. & LARKIN P.A. (1974) Feeding activity and rates of digestion of northern squawfish (*Ptychocheilus oregonensis*). *J. Fish. Res. Bd. Canada* **31**, 411–420.

SWIFT D.R. (1964) The effect of temperature and of oxygen on the growth rate of the Windemere char (*Salvelinus alpinus willughbii*). *Comp. Biochem. Physiol.* **12**, 179–183.

SYMONS P.E.K. (1973) Behaviour of young Atlantic salmon (*Salmo salar*) exposed to or force-fed fenitrothion, and organo-phosphate insecticide. *J. Fish. Res. Bd. Canada* **30**, 651–655.

TOKAR E.M. & OWENS E.L. (1968) The effects of unbleached kraft mill effluents on salmon. I. Growth, food consumption and swimming ability of juvenile chinook salmon. *Nat. Counc. Air Stream Improv. Tech. Bull.* **217**, 1–46.

TYLER A.V. & DUNN R.S. (1976) Ration, growth, and measures of somatic and organ condition in relation to meal frequency in winter flounder, *Pseudopleuronectes americanus*, with hypotheses regarding population homeostasis. *J. Fish. Res. Bd. Canada* **33**, 63–75.

WARE D.M. (1975) Growth, metabolism and optimum swimming speed of a pelagic fish. *J. Fish. Res. Bd. Canada* **32**, 33–41.

WARREN C.E. (1971) *Biology and Water Pollution Control*. W. B. Saunders, Philadelphia.

WARREN C.E. & DAVIS G.E. (1967) Laboratory studies on the feeding bioenergetics, and growth of fish. *The Biological Basis of Freshwater Fish Production* (S.D. Gerking, ed.), pp. 175–214. Blackwell Scientific Publications.

WEATHERLEY A.H. (1972) *Growth and Ecology of Fish Populations*. Academic Press, New York, N.Y.

WEBB P.W. & BRETT J.R. (1972) The effects of sublethal concentrations of whole bleached kraftmill effluent on the growth and good conversion efficiency of underyearling sockeye salmon (*Oncorhynchus nerka*). *J. Fish. Res. Bd. Canada* **29**, 1555–1563.

WEBB P.W. & BRETT J.R. (1973) Effects of sublethal concentrations of sodium pentachlorophenate on growth rate, food conversion efficiency and swimming performance in underyearling sockeye salmon (*Oncorhynchus nerka*). *J. Fish. Res. Bd. Canada* **30**, 499–507.

WEIHS D. (1973) Hydromechanics of fish schooling. *Nature* (Lond.) **241**, 290–291.

WHITWORTH W.R. (1968) Effects of diurnal fluctuations of dissolved oxygen on the growth of brook trout. *J. Fish. Res. Bd. Canada* **25**, 579–584.

WILDISH D.J. & LISTER N.A. (1973) Biological effects of fenitrothion in the diet of brook trout. *Bull. Environ. Contam. Toxicol.* **10**, 333–339.

WILLIAMS G.C. (1961) *Adaptation and Natural Selection; a Critique of Some Current Evolutionary Thought.* Princeton University Press, Princeton, N.J.

WINBERG G.G. (1956) Rate of metabolism and food requirements of fishes. Belorussion State Univ. Minsk. *Fish Res. Bd. Canada Trans. Ser.* No. 194. 1960.

WINDELL J.T. (1967) Rates of digestion in fishes. *The Biological Basis of Freshwater Production* (S.D. Gerking, ed.), pp. 151–173. Blackwell Scientific Publications, Oxford.

WISSING T.E. (1974) Energy transformations by young-of-the-year white bass *Morone chrysops* (Rafinesque) in Lake Mendota, Wisconsin. *Trans. Am. Fish. Soc.* **103**, 32–37.

WISSING T.E. & HASLER A.D. (1971) Effects of swimming activity and food intake on the respiration of young-of-the-year white bass *Morone chrysops*. *Trans. Am. Fish. Soc.* **100**, 537–543.

Chapter 9: The Role of Predaceous* Fish in Ecosystems

O. A. Popova

9.1. Introduction

The study of freshwater ecological systems is inextricably linked with the development of the theory of biological productivity and with investigations of the basic principles governing the reproduction and distribution of organic matter. In the final result, an ecosystem is studied with a view to rational management of its biological resources for obtaining products useful to man. One approach to ecosystem analysis is the study of the distribution of energy in various segments of the food chain. In such an analysis the relation between the predator and its prey is an important element.

In every aquatic ecosystem, the predator has complicated food relationships with other fish species, both 'peaceful' and predaceous, constituting one component of the 'triotroph', i.e., a triple food relationship in which every animal population has both adaptations promoting successful feeding upon its own food organisms, and adaptations protecting it successfully from its enemies which in their turn possess the same system of adaptations (Gaevskaya 1955, Manteifel 1961). Food relationships are thus inseparable from defensive ones.

Although in this paper all the relevant problems are discussed with reference to the true predaceous fish, i.e. ichthyophages, the main principles governing the predator–prey relationships among this group of fish are also applicable to other groups, such as zooplankton feeders, as well.

9.2. Structural features of predaceous fish

Predaceous fish have a number of distinctive structural features in their mouth apparatus and digestive tract (Andriyashev 1944a, 1944b, 1945, 1948, Svetovidov 1932 and 1953, Reshetnikov 1961). Freshwater predators usually swallow their prey whole†. They have a big mouth which, however, widely

---

* Dr. Popova uses the word 'predaceous' as equivalent to 'piscivorous'. This is made clear when 'true predaceous' fishes are referred to as 'ichthyophages'.
† Only a few predaceous fishes (marine and freshwater) tear their prey into pieces biting out the piece by sharp saw-like teeth during school hunting (piranha, barracuda, sharks).

varies in shape and position. The head is about 30% of body length, and the horizontal width of mouth opening is about 40–58% of the head length (16–24% in 'peaceful' fishes). The great extension of the jaws enables the predator to increase sharply the volume of the mouth cavity in order to snatch the prey and to suck in considerable water at the same time. These actions enable the prey to be drawn rapidly into the gill- and mouth-cavity.

Different types of prey capture depend upon, firstly, the prey's size, and secondly, the mouth's shape and the method of hunting. The predator either catches large prey by attacking it from the head end or by grasping it first from the side and then recatching it from the head end. This type of hunting is characteristic among predators that ambush their prey, such as pike, *Esox lucius*, and sheatfish, *Silurus glanis*. Pelagial predators that pursue their prey, such as the zander, *Stizostedion lucioperca*, and asp, *Aspius aspius*, catch and swallow the prey tail first (Pihu & Pihu 1971). A few groups of predators capture small fish without a preferred body orientation.

Some predators (Percidae, Esocidae, Siluridae, Salmonidae) hold the prey in their mouth with sharp spear-like teeth on the jaws and spine-like teeth which cover, like a brush, the jaws, vomer, palatine bones, and pharyngeal tracts. The teeth are continuously shed the year round, one after another; a new tooth usually grows behind the old one which falls out. The prey is manipulated into position for swallowing by means of spiny gill rakers. The movement cycle of the gullet apparatus follows a common pattern. That is, a large prey is seized and gradually pushed to the gullet where it is held during a pause in contraction; then it is moved farther into the gullet by a considerable longitudinal shift of the upper pharyngeal tract. These predators do not subject the food to any preliminary processing in the mouth cavity.

Other predators from the family Cyprinidae have no teeth on their jaws but only on the lower pharyngeal bones. They have a double row of thin, elongated, curved teeth with hooks and indentations on the tips. Predators of this type work over their food first in the gullet where it is broken up to some extent by the pharyngeal teeth before it enters the intestine.

In the first type of fish, the food enters a voluminous stomach, while in the second type it enters the intestine directly for there is no well-defined stomach (Fig. 9.1). The total length of the digestive tract is shorter than in other fish: an average 76–106% of body length in predators, and 110–420% in benthos-feeding fish.

As a rule, predaceous fish spawn early in the spring and their fingerlings have fast growth rates so that they exceed their prey in size and begin feeding on the larvae and fry of other species upon reaching a length of 4–6 cm. The transition from plankton feeding to feeding on fish is associated with a number of morphological changes. The teeth make their appearance and

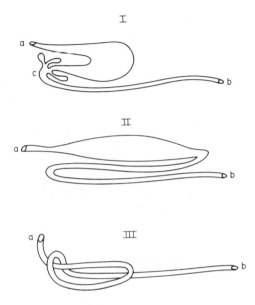

Fig. 9.1. Schematic diagram of the digestive tract in predaceous fish with well developed stomach (I—perch, II—pike) and in those without stomach (III—asp).
a, pharynx; b, anal orifice; c, pyloric caeca.

increase in number, the stomach is formed, the intestine becomes longer, the number of pyloric caeca increases, and the gill rakers change in shape (Shamardina 1957, Vasnetzov *et al.* 1957, Kostomarova 1962).

9.3. Availability of prey

The composition of food depends on the predator's morphological features, its mode of life and its behaviour. Such ambushing predators as the sheatfish and pike, having a big mouth and voluminous stomach, can feed on relatively large fish inhabiting the vegetated shoreline. The pelagic predators (zander and especially asp) feed on smaller fish, predominantly the young. Thus, selection of food items is associated with their availability.

The availability of prey to predators is determined by a set of factors, the principal ones being the presence and relative abundance of particular kinds of food. The great abundance of a given prey, however, does not always mean that it is utilized greatly as food. Cases are known where a prey is present in maximum numbers but is not used by the predator since another, more preferable, object is present at the same time. When the latter disappears, the predator shifts to other food items even if their abundance drops. Apart from abundance, the availability of food also depends on the distribution of the prey species in the water, the coincidence of the eco-areas of the predator and its prey, and on the behaviour, activity and size both of the prey and the predator.

The prey fish have a number of adaptations for protecting themselves from predators: strong outer covering, an armament of body spines and barbs, protective colouration, toxicity, rapidity of movement, and formation of schools. These adaptations do not afford complete protection but do increase the chance of survival. For example, the presence of a spine in the dorsal fin of the spiny bitterling, *Acanthorhodeus asmussi* decreases its consumption by predators as compared with the common bitterling, *Rhodeus sericeus*, which lacks this spine (Table 9.1). Spines or barbs cannot, however,

Table 9.1. Ratio of spiny bitterlings (*Acanthorhodeus asmussi*) to common bitterlings (*Rhodeus sericeus*) in the rations of predators and in catches with a small-mesh seine, expressed as percentage of the total numbers of these two species (after Lishev 1950b)

Fish species	Spine	In rations of predators	In seine catches
		%	%
Spiny bitterling	yes	19	70
Bitterling	no	81	30

completely protect the prey against predators; they only promote an increase in the overall size of the prey and thus make it less available to smaller predators. Among many other similar adaptations, the Pacific Cottidae increases the headwidth twofold by spreading the gill covers, (Paraketzov 1958). Naturally, with increase in the size of the predator its percentage consumption of 'armed' fish increases (Frost 1954, Fortunatova 1959).

On the other hand, the presence of an armament is often a less reliable protective factor than is mobility, protective colouration, or availability of shelter (Ivlev 1955). Thus the three-spined stickleback, *Gasterosteus aculeatus*, which has large and sharp spines, are consumed by predators in greater numbers than the nine-spined stickelback, *Pygosteus pungitius*, which has small and weak spines but exhibits better mobility and better protective colouration (Hoogland *et al.* 1956–1957).

Another important protective measure is the formation of a school. It is much more difficult for a predator to catch a fish in a school than as an isolated individual. At the same time, a school of predators hunts more efficiently than a single predator (Radakov 1961, 1965, 1972, Girsa 1962).

9.4. Methods of investigating the feeding of predaceous fish

The foregoing discussion shows that the feeding of predaceous fish involves practically all aspects of their relationships with other fish species. Hence, great importance is attached to the study of feeding and to the development of methods for its study. Procedures for studying the feeding of predaceous fish have been developed along the following main lines: (1) precise determination of the species and size composition of the organisms found in the food bolus; (2) study of the physiology of digestion in predaceous fish, in particular of digestion rates; and (3) determination of the quantity of food consumed by the predators. Since these procedures are described in greater detail elsewhere (Fortunatova 1951, 1955, 1964, Fortunatova & Popova 1973, Borutsky 1974), we shall confine ourselves to a brief consideration of each of these three lines of work.

SPECIES AND SIZE DISTRIBUTION

To determine the species and to reconstruct the initial weight of the eaten organisms, use is made of various bone fragments, which are better preserved in predator stomachs than are the soft parts. These include otoliths, vertebrae, and lower pharyngeal or mandibular bones. Special determinative tables, graphs, and nomograms of ratios between the length and weight of a fish and the length of its bone fragment have been prepared (Lishev 1950a, Kovalev 1958, Vasarheleyi 1958, Horoszewicz 1960, Skalkin 1961 and 1965, Schmidt 1968, Pihu & Pihu 1970, Fortunatova & Popova 1973, Borutsky 1974). Fig. 9.2 shows that the size and weight of a swallowed prey fish can be reconstructed from the length of a pharyngeal bone.

DIGESTION RATE

Digestion rates in predators are measured experimentally in artificial ponds and aquaria using natural or force feeding and then sacrificing the fish at certain intervals (Karpevich & Bokova 1936 and 1937, Arnol'di & Fortunatova 1937, Fortunatova 1940 and 1950, Karpevich 1941, Reshetnikov *et al.* 1972, 1974, 1975, Fortunatova & Popova 1973). All these studies have established that the digestion rate and feeding intervals depend on water temperature (Table 9.2, Fig. 9.3). By now, such studies have been carried out both on fish of temperate latitudes and on tropical and artic fish (Karpevich 1941, Ivlev 1955, Winberg 1956, Hunt 1960, Manteifel *et al.* 1965, Popova 1965 and 1967, Novikova 1966, Windell 1967, Edwards 1971, Edwards *et al.* 1971, Pérès *et al.* 1973, Swenson & Smith 1973, Smith & Paulson 1974, Reshetnikov *et al.* 1975, Popova *et al.* 1976). Similar data on digestion rates of predatory fishes were obtained by an X-ray method (Molnar *et al.* 1967).

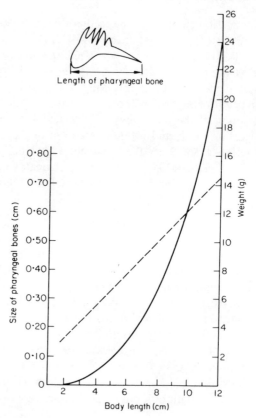

Fig. 9.2. Relationship between the length of lower pharyngeal bone and the size and weight of bleak *Alburnus alburnus* (from Popova 1967).

Table 9.2. Comparison of food digestion rates by scorpion fish (*Scorpaena porcus*) and Cuban predators from the genus *Lutjanus* feeding on fish at different water temperatures (time is shown in hours) (After Fortunatova 1950, Reshetnikov *et al.* 1975)

Digestion stage	Amount of digested food, %	*Scorpaena porcus*				*Lutjanus* spp.		
		7–13C	14–20C	20–23C	22–25C	21–22C	25–26C	28–29C
I	0–6	0–24	0–18	0–18	0–15	1–8	1–5	1–3
II	7–15					2–11	2–9	2–10
III	16–50	24–72	18–48	18–48	15–24	7–32	7–24	4–12
IV	51–80	72–96	48–68	48–68	24–40	12–32	11–31	8–22
V	81–100	94–144	68–96	68–96	40–48	19–43	22–48	20–33

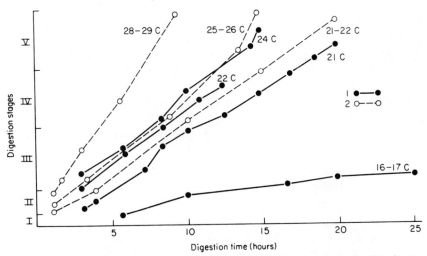

Fig. 9.3. Digestion rate at various temperatures: 1, of bleak by river perch; 2, of *Jenkinsia* by *Lutjanus* sp. (from Reshetnikov *et al.* 1975).

Adult zander, sheatfish and pike change their average digestion rates with temperature in the following way:

Months	Water temperature °C	Digestion rate, in days
July, August	over 25	1
June, September	18–25	2
May, October	8–18	3
April, October	4–8	4
November–March	2–4	7
November–March	0, 1–2, 0	9

It has been found that fish with well developed stomach motor activity digest food more rapidly than those with weak activity, because a larger surface of the food bolus is exposed to the gastric juice during motility. Digestion rate has also been shown by many investigators to depend on the size of the prey, the density of its scales and skin covering, as well as on the species, size and physiological condition of the predator itself. Several small fishes are digested several times more rapidly than one large fish covered with dense, scaly armour and having a rigid skeleton. Because of this, the digestion rate in the asp, which feeds exclusively on fingerlings, is much faster than in other predators. Other information on digestive rates can be found in Windell (this book).

QUANTITY OF FOOD CONSUMED

By using experimental data on digestion rates as well as data from direct observations in nature of seasonal changes in the qualitative and quantitative

Chapter 9

composition of predator food throughout the year, it has been possible to determine the daily (24-hour), monthly and annual rations of predators. Eventually this should permit us to establish the effect of predaceous fish on the ichthyofauna as a whole or on its individual species.

On the basis of studies of digestion rates at various temperatures, a scale of stages of digestion has been developed (Table 9.3) by which it is

Table 9.3. Process of destruction of food (stages of digestion) depending on the rate of digestion. Roman numerals refer to stages of digestion, rated from I to V.

Digestion rate in days, V	Number of days from the first-day ingestion				
	1	2	3	4	5
1 day	I-V				
2 days	I	II-V			
3 days	I	II	III-V		
4 days	I	II	III	IV-V	
5 days	I	II	III	IV	V

easy to estimate the time that the prey was captured (this scale is partly shown in Table 2). Knowing the digestion rate in days at a given temperature and the magnitude of 24-hour portions of food (index of consumption), the daily ration is calculated from the formula:

$$DR = \frac{\sum\limits_{i=1}^{i=V} (S_i/n_i)}{V} \times \frac{n}{N}$$

where: DR = daily ration in % body weight;
N = the total number of predaceous fishes in the sample;
n = the number of feeding fish in the sample;
V = digestion rate in days;
n_i = the number of fish that fed in each of V-days;
S_i = the amount of food eaten by all fish in each of V-days (in index of consumption, i.e. percent of the reconstructed weight of food to the weight of the predator).

A numerical example follows:

In May at 8–18C of water temperature, the digestion rate (V) equals 3 days. Among 30 fishes in the sample (N) there are 20 feeding (n). The food, discovered in the stomachs, has different rates of digestion (see line 3 in Table 9.3). 10 fishes (n_1) have the food in the I stage of digestion ($S_1 = 13.50$). 15 fishes (n_2) have the food in the stage II ($S_2 = 19.50$), eaten the day before.

7 fishes (n_3) have the food in the stage III–V) ($S_3 = 15\cdot35$), eaten the two days before. Then according the formula we have

$$DR = \frac{\dfrac{13\cdot50}{10} + \dfrac{19\cdot50}{15} + \dfrac{15\cdot35}{7}}{3} \cdot \frac{20}{30} = 1\cdot07\%$$

If the digestion rate equals 1 day, then $i = V = 1$, $n_1 = n$, and $DR = S/N$. Note once more than S_i is expressed as the amount of food *in index of consumption*, i.e. *percent* of the reconstructed weight of food to the weight of the predator. Therefore, DR is also expressed as percent of predator body weight.

The intensity of feeding of predaceous fish and the size of their daily rations change with the seasonal changes in ecological conditions. Thus, in the Volga delta, the daily ration of a mature pike amounts to 3·89% of body weight in the spring, 0 to 0·84% in the summer, 3·04% in the autumn, and 0·16 to 0·88% in the winter. The average monthly food consumption by a predator also changes accordingly during the year. Every predaceous species has its own characteristic annual feeding rhythm (Fig. 9.4), and its own annual ration (Table 9.4). It is of interest to note that daily rations are

Table 9.4. Annual rations in percent body weight of mature (age 3+ and older) predaceous fish in the Volga delta (Fortunatova & Popova 1973) and in the Rybinsk Reservoir (Ivanova 1968)

Species	Size of predator, cm	Volga delta	Rybinsk reservoir
Pike, *Esox lucius*	40–90	341–344	270–340
Zander, *Stizostedion lucioperca*	35–70	199–220	170–230
Perch, *Perca fluviatilis*	20–40	175–220	180–280
Sheatfish, *Silurus glanis*	65–110	171–300	–
Asp, *Aspius aspius*	30–60	174	–
Burbot, *Lota lota*	35–70	–	280–320

highest in young fish during the transition to predation: 9–40% an average of 21·9% of body weight in new broods of 4–5 cm sheatfish; 9·5% in fry 5–7 cm long; 7·7% in individuals 7–9 cm long; while in adult fish it hardly reaches 2 to 3% and only rarely exceeds 5%.

There is very little data in the literature on the daily rations of tropical fish. According to our estimates, the daily rations of Cuban predaceous fish from the family Serranidae were 2·41 to 5·70% of the body weight in the summer at water temperature 28–29C (Reshetnikov & Popova 1975, Reshetnikov *et al.* 1975). From figures given in a paper by Graeber (1974)

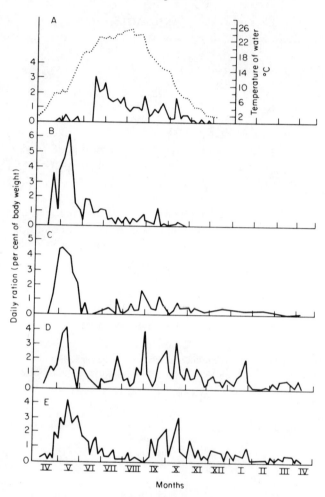

Fig. 9.4. Changes in daily rations of predaceous fish during the year (by five-day periods). A, asp; B, sheatfish; C, zander; D, pike; E, perch.

showing initial data on the feeding of young sharks, *Negaprion brevirostris*, reared artificially in ponds in the area of the Bahama Islands and fed on fish, we have estimated that their daily ration was 0·8–1·5% in August at a water temperature of 30·2C and 1·4–1,7% in September–October at 28·8C. When feeding on Polychaeta, 5 species of Arabian Sea plaice consumed 25% of their body weight per week at 28C, i.e. their daily ration was about 3·5% of body weight (Edwards *et al.* 1971).

The annual rations of the Black Sea ruffe (*Gymnocephalus cernua*) correspond to those species in Table 9.4, i.e. 386–410% for sexually mature individuals. The annual ration changes with the age of the fish as does the daily ration. In Rybinsk Reservoir the annual ration of pike fingerlings up to 10 cm in length is 532%, of a one-year old 10–20 cm length 350%; of 2–3 years old and 20–40 cm long 206%; and of a mature fish over 40 cm long 287% of body weight (Ivanova 1968).

With increasing age the feeding rhythm also changes (Fig. 9.5). Thus, in

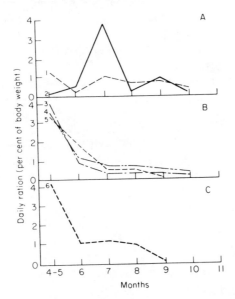

Fig. 9.5. Daily rations of sheatfish of different size groups. Length of fish (cm): 1=under 55; 2=55–64; 3=65–74; 4=75–84; 5=85–94; 6=over 95. A=immature fish; B=mature fish of intermediate size; C=mature, older fish.

one- and two-year old sheatfish the main feeding occurs in the summer months, when the growth rate is at its maximum and when much food is available in the form of young cyprinoid fishes (Fig. 9.5A). Adult mature individuals (Fig. 9.5B) mainly feed in the spring months, during a period of maximum energy expenditure for the maturation of sex products and for the spawning itself, when much food is available in the form of spawning vobla (*Rutilus rutilus caspicus*). Only three-year old individuals (2+) feed uniformly and intensively the year round (Fig. 9.5C), which is manifested by continuing intensive growth and attaining sexual maturity.

Careful longitudinal studies of seasonal variations in the feeding of predaceous fish in the Volga delta have provided an approach to evaluating the effects of predaceous fish on the structure and abundance of prey populations, i.e., on individual elements of the ecosystem and on the system as a whole.

9.5. Effect of predators on populations of other species

PREDATOR SIZE AND PREY SIZE

Most predators selectively remove individuals from a prey population, choosing fish of a particular size, age and sex according to their availability. With increasing predator size, the range of prey sizes also, of course, increases, since the capacity of the mouth apparatus becomes greater. However, it would be wrong to believe that it is an invariable rule that the larger the predator the greater the size of the prey it eats. When the variety of available prey is sufficient, predators feed on those organisms which meet their requirements most fully. The latter are determined by the size of predator mouth, the length and volume of its stomach, and the energy expended on searching, seizing and digesting the prey, as well as by its caloric value.

In most bodies of water of the middle part of Europe inhabited by percid-cyprinid and percid-smelt-cyprinid communities, predaceous fish predominantly feed on readily available and abundant small non-migratory and semi-migratory fish 8–15 cm in length and their young 2–8 cm long (Fig. 9.6). It is only in the lower reaches of northern rivers with whitefish-salmon communities, where no large accumulations of small-sized fish are present, that large freshwater predators have to consume mainly adult individuals of valuable semi-migratory species 25–30 cm in length.

Some ratios of predator size to prey size are shown in Fig. 9.6. The curves of the absolute average size of prey for perch usually occupy an intermediary position between the curves of minimum and maximum prey sizes, while for other predators (pike, sheatfish, asp) they lie closer to the curve of minimum prey size. In view of this, the relative prey size (percent of predator body length) falls from 57–52% at the time of transition to predation to 20–15% for all predaceous fish at the time of maturity. Subsequently, the relative prey size again rises to 30% for percid fish and stabilizes at 10% for other fish. This fact is connected with the transition to new food; young of the predator feed mostly on cyprinid fingerlings, but mature predators eat adult cyprinids, percids and other fish. This transition is very sharp for small-size predators. Similar prey fish sizes have been reported for North American garpike, *Lepisosteus* sp., (Crumpton 1971) and Nile perch, *Lates niloticus*, (Hopson 1972). The above relationships hold not only for freshwater predators but also for marine ones. In particular, we have found similar relationships among cod (*Gadus morhua*) in the Atlantic and among predaceous fish on the Cuban shelf (Popova 1962, 1968, Reshetnikov *et al.* 1972, 1974, 1975, Reshetnikov & Popova 1975).

THE EFFECT OF PREDACEOUS FISH ON PREY POPULATIONS

The majority of predaceous fish feed on the immature part of the prey stock i.e., on that part of the prey population not used by a fishery. However,

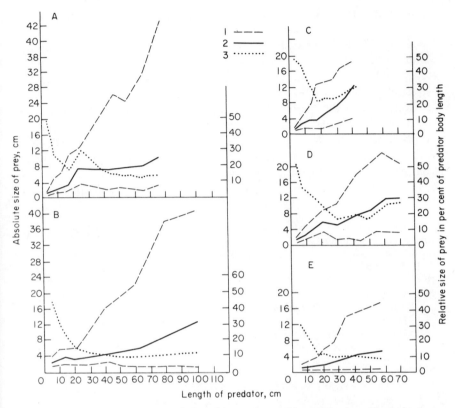

Fig. 9.6. Relationship between sizes of predator and prey. The graphs show (1) maximum and minimum absolute size of prey; (2) average absolute size of prey; and (3) relative size of prey (in percent of predator body length). Species represented are: A, pike; B, sheatfish; C, perch; D, zander; E, asp.

when prey is very abundant, predators may consume mature individuals, choosing from a population those fish whose growth rate and condition factor are reduced and which are weak, mainly males. They may feed predominantly during the period of spawning when prey availability is highest. In this respect the effect of predators differs from that of a fishery, which removes from a population mainly fish that are mature and in good condition (Fortunatova 1957, 1961, Nikol'skii 1974a, 1974b).

Quantitative data concerning what fraction of a population is eaten by predators are scarce because, first, there is very little information in the literature on the size and composition of the annual rations of predators and, second, because it is not always possible to determine the abundance of both the predator and prey populations at the same time. However, some authors have made attempts of such determination. Thus Crossman (1959) indicates that rainbow trout (*Salmo gairdneri*) in Paul Lake consume from 0·15 to 5% of the population of redside shiner (*Richardsonius balteatus*). Pliszka (1953) has estimated that the new brood of pike in Lake Garsh eats about 20–25% of the total new brood of roach. However, in all cases the isolated effect of only one predator on one prey was considered, and the estimates are based on limited evidence.

Consider now the combined effect of all predaceous fish on the population structure of individual species as exemplified by the Volga delta. All these estimates are dealt with in greater detail in our book (Fortunatova & Popova 1973). First let us consider the effect of predaceous fish on the *vobla* population. Vobla constitutes the greatest part of the annual ration of nearly all predators (Table 9.5). Knowledge of the share of vobla in the annual ration

Table 9.5. Composition of food used by predaceous fish in the lower zone of the Volga delta (in percent of annual ration)

Prey species	Sheatfish	Pike	Zander	Asp
Carp, *Cyprinus carpio*	11·1	16·0	1·2	3·5
Vobla, *Rutilus rutilus caspicus*	46·5	28·9	61·7	23·0
Bream, *Abramis brama*	1·6	3·6	2·3	2·4
Redfin, *Scardinius erythrophthalmus*	13·0	3·7	1·6	0·2
White bream, *Blicca björkna*	10·0	8·0	18·9	19·1
Bleak, *Alburnus alburnus*	1·1	3·7	8·1	19·4
Blue bream, *Albramis ballerus*	0·3	0·2	1·0	1·2
Chekhon, *Pelecus cultratus*	–	–	–	13·6
Sheatfish, *Silurus glanis*	–	4·3	0·2	0·6
Frogs, *Rana*	10·3	26·1	1·7	9·3

and of predator abundance may be used to calculate the removal of mature vobla by predators (Table 9.6). Between 1945 and 1954, approximately 60·8 thousand tons of vobla were removed through fishing and 51·78 thousand tons by all predators, the greatest quantity of vobla (70%) being eaten by zander (35·74 thousand tons). At that time, the abundance of predators was estimated at 32·1 million, and they ate 1441·7 million voblas annually while only 500 million voblas were taken by the fishery; in other words, predaceous fish took three times as many voblas as the fishery (Fig. 9.7). The fishery, however, took fishes with an average length of 18 cm and 2–3 years old, while predators ate individuals 12 cm long and aged 1 year. We have found that predaceous fish eat both adult individuals during spawning and the new brood in the summer during downstream migrations from spawning grounds, as well as in the autumn, during pre-hibernation concentrations. The young of vobla is mainly used by adult asp and immature predators of other species (Popova 1965 and 1971). The total consumption of vobla by

Table 9.6. Removal of mature vobla by fishery and predaceous fish from the Volga delta (Fortunatova & Popova 1973)

Index	Fishery	Sheat-fish	Pike	Zander	Asp	Perch	All predators
Catch, thousand tons	60,8	7,9	6,4	28,5	1,3	5,95	50,0·5
Number of individuals, millions		2·8	4·0	14·2	1·3	9·8	32·1
Annual ration of own weights, %		200	340	220	170	170	–
Total amount of fish eaten, thousand tons		15·80	21·76	62·70	2·21	10·11	112·58
Proportion of vobla in annual ration (%)		50·0	27·0	57·0	2·3	22·0	–
Total number of voblas removed thousand tons	60·80	7·90	5·87	35·74	0·05	2·22	51·78
Millions of fish	500	159·8	235·3	965·6	2·3	78·7	1441·7

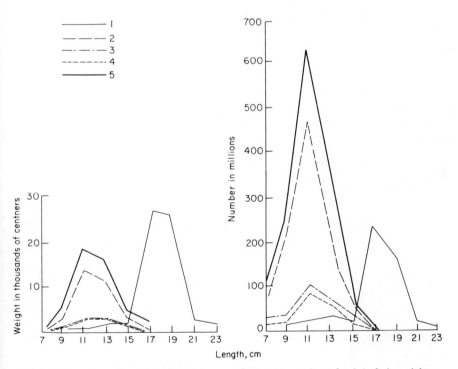

Fig. 9.7. Size composition of vobla in fishery catches and in predator food. Left, in weight; right, in number; 1, in fishery catches; 2, in pike food; 3, in sheatfish food; 4, in zander food; 5, in food of all predaceous fish.

predaceous fish from bodies of water in the Volga delta is shown in Table 9.7. If, according to Monastyrsky (1940), the return of vobla by the fishery is 0·6% of the downstream-migrating fry, then the average yield of a vobla

Table 9.7. Annual removal of vobla from the Volga delta

| Index | Yield of a vobla generation | Vobla removed by predators | | | Taken by fishery |
		Aged 0+	Aged 1+ and 2+	Total	
Number of individuals, millions	83	8·6	1·5	10·1	0·5
%	100	10·6	1·6	12·2	0·4
Biomass, thousand tons		45·4	517·8	563·2	608·0

generation in the 1950's was estimated at 83 billion individuals. Of this number, predaceous fish ate 8·6 billion voblas of the new brood (10·6%) and 1·5 billion aged 1+ and 2+, i.e., a total of 12·2% of the whole generation. For comparison, we may note that the fishery takes 0·5 billion voblas, i.e. 0·4%. In terms of the entire vobla population, predaceous fish annually eliminate 22% of the biomass, or 24% of the vobla population to the end of the feeding period. In terms of biomass, the effect of predaceous fish on the vobla population is similar to that of the fishery (Fig. 9.7).

Similar estimates were made of the numbers and biomass of other fish species annually consumed by predaceous fish in the Volga delta (Table 9.8).

Table 9.8. Number of food organisms annually eaten by a single immature (Im) and mature (M) predator in the Volga delta

| Predator | | Commercial species | | Non-commercial species | | |
		Vobla	Young of cyprinid fishes	Small coarse fish	Young of predaceous fish	Frogs
Sheatfish	Im	—	438	275	25	50
	M	40	46	195	39	36
Zander	Im	15	39	300	5	–
	M	48	112	218	55	6
Pike	Im	8	484	275	25	10
	M	15	97	181	52	85
Asp	Im	–	1184	525	75	–
	M	3	475	730	52	10
Perch	Im	–	212	125	5	–
	M	4	85	52	11	1

Of other valuable commercial fish, apart from vobla, predators have been found to consume annually 14·7 billion wild carp and 1·8 billion bream of fingerling size. At the same time they eliminate about 30 billion small coarse fish, some 2·5 billion young of predaceous fishes, and about 1 billion frogs, which, in terms of biomass, is equal to 50·4 thousand tons.

9.6 Abundance of predators and their prey

The problem of the effect of a predator on the structure and abundance of a prey population is closely associated with that of the numerical relationship between these two components of the system. It is difficult to take into account all non-biotic and biotic factors involved in considering the relationship between the abundance of a predator and that of its prey. Because of this, many of the attempts to express the predator–prey relationships by a universal formula have been doomed to failure. Some of the more interesting attempts are described below.

A mathematical interpretation of the predator–prey relationships was first given by Volterra (1928) who, by allowing for a constancy of environmental conditions and by changing only the voracity of a predator and the reproductive capacities of the two species involved, has formulated the following laws:

1 the periodic cycle law according to which the fluctuations in the abundances of both species are periodic;

2 the law of preservation of means: given the constant growth coefficients for species abundance, the degree of prey protection, and the efficiency of predator hunting, the mean sizes of the populations of the both species are constant and do not depend on environmental conditions;

3 the law of disturbance of means: if the populations of both species are reduced in proportion to their abundance, then the mean number of individuals of the prey species begins to increase and that of the predator species to decrease.

To verify these theoretical propositions of Volterra, simple experimental models of Gause (1934a, b) have been used and further developed by Williamson (1972). In a general biological form, similar propositions concerning fluctuations in animal population sizes have been formulated by Elton (1933). Elton's hypothesis that a very low population density of a species may be biologically inadequate and result in depression or even disappearance of that species, has been confirmed in practice by ichthyologists (Tyurin 1973, 1974). Tyurin (1974) has proposed the existence of a relatively critical threshold of abundance above which any fish species passes from a state of inhibition to a state in which its abundance sharply increases. Below that threshold, there occurs a depression in population abundance. For fish found in bodies of water in the northwest part of the USSR (zander, pike, roach, ruffe, perch) the threshold is estimated at 7% of the total catch in a given body of water. A shift in the abundance of a particular predator species below that limit as a result of over-fishing leads to a drastic change in the composition of the fish fauna; predator abundance declines, while the abundance of small inferior prey species increases. These latter species compete with and supplant commercially important species.

Ricker (1952) suggests three types of possible quantitative relationships between a predator and its prey:

1 The bulk of the prey population avoids the effect of the predator because of the great abundance of the prey and the short duration of its contact with the predator. This is exemplified by certain predators that feed on spawning accumulations of herring (*Clupea harengus*) and others on the downstream migrating young of salmon.

2 All-year-round contact with the prey, where the magnitude of prey removal by the predator is independent of the duration of contact but is determined by the abundance of both species. Example: predation by squawfish (*Ptychocheilus oregonensis*) on the young of sockeye salmon (*Oncorhynchus nerka*) in Cultus Lake, British Columbia.

3 The quantity of prey eaten by the predator is not dependent on prey abundance but is determined by the availability of shelter, i.e., depends on the degree of prey availability. Example: fish-eating birds preying on young salmon in rivers.

Many aspects of the predator–prey relationship have been developed under experimental conditions by Ivlev (1955). On the basis of experiments, mathematical formulae have been worked out to describe (1) the intensity of predator feeding as a function of prey concentration and distribution and (2) the electivity of predator feeding as a function of predator satiation and population density and the presence of protective devices in the prey. Ivlev was the first to introduce the notion of a quantitative index of predation as the ratio of the optimum prey size to predator size. For predaceous fish such as pike and perch this index is $18 \cdot 10^{-2}$ to $32 \cdot 10^{-2}$, and for 'peaceful' fish such as carp, bream and bleak, it is $13 \cdot 10^{-4}$ to $38 \cdot 10^{-5}$. His equation relating the maximum ration (R) to the actual ration (r) and the density of the prey population (p)

$$r = R(1 - e^{-\varepsilon p}) \quad \text{or}$$
$$r = R(1 - 10^{-kp}),$$

where ε and k are proportionality coefficients, has been used in many mathematical models of the 'predator–prey' type.

It is usually assumed that the relationship between a predator (A) and its prey (B) proceeds along the feedback line of the type $A \underset{\pm}{\overset{-}{\rightleftarrows}} B$, i.e., the prey improves conditions for the predator while the latter depresses the prey.

When the reaction of the predator is delayed, such a relationship leads to undamped fluctuations in the abundances of the predator and its prey through increases in predator abundance. Actually, however, the predator–prey relationship is much more complex; Nikol'skii (1974a) believes that it has the form: $A \quad B$. By eating the prey and reducing its abundance (operation minus) the predator at the same time creates more favourable conditions for, and increases the growth rate and reproduction of, the remaining part of the prey by thinning out its population (operation plus). In its turn, the large abundance of the prey leads to an increase in predator number (operation plus) but small abundance of the prey slows down the reproduction rate of the predator and reduces its abundance (operation minus). Moreover, in nature the prey always depends on the availability of food (C) and usually acts as the middle member in the tritoroph system:

$$A \quad B \quad C.$$

The predator also experiences a triple relationship of this kind since it is under pressure of the second-order predators.

For land animals, there is usually a close relationship between the abundances of the predator and its prey (Dajoz 1972, Williamson 1972). In fish, however, such a synchronization is difficult to establish because in most cases predators do not eat only one kind of food. When the prey species becomes less abundant, the predator shifts to other available foods and, sometimes, to cannibalism of their own young, so that predaceous fish do not perish on a mass scale in a body of water. The fact that smelt (*Osmerus eperlanus*), many species of salmon, cod (*Gadus morhua*), navaga (*Eleginus navaga*), river perch (*Perca fluviatilis*), pike and other species feed on their own young in years when the latter are abundant may be regarded as a means of regulating abundance when food-relationships become more severe because of possible over-crowding (Nikol'skii 1953, Popova 1965, 1971, 1975).

The regulation of predaceous fish in a body of water is a continuous process occuring at all stages of ontogeny. The young of predators is subject to considerable predation by fish-eating birds, reptiles and amphibia (Skokova 1955, 1966a, 1966b, Borodulina 1958, Markuze 1964a, 1964b, 1966, Nikol'skii 1974b). But in practice large predaceous fish, such as pike, are preyed upon only during the first year of life, while small predaceous fish, such as perch, are preyed upon for 3 years. The abundance of a mature predator stock is influenced, apart from non-biotic factors, only by man through fishing. This effect, however, proves to be so great that sometimes it results in a drastic reduction of the stock and disrupts the normal relationships between the populations of predators and their prey species.

The most striking demonstration of the relationship between the abundances of a predator and its prey can be provided by means of mathematical models using an electronic computer. The simplest model has been described by Menshutkin (1964, 1971) for a perch population in Lake Tiulenye where only one fish species is present. The perch is represented by two forms in the lake: small plankton-feeding perch and large predaceous perch. An increase in abundance of the young perch favours the feeding of the large predaceous perch whose abundance increases owing to increases of younger age groups with good growth rates. However, when a certain level of abundance is reached by the predator, the population of young perch is reduced. Subsequently, the abundance and biomass of both small and large perch is then reduced in the lake, and this means a reduction in egg number and consequently less young.

The reduced pressure by the predators and the thinning out of the population favour the growth of a small, littoral form of plankton-feeding perch, some of which turn to cannibalism of their own young and thus behave like the large predaceous perch.* When the abundance of large predators increases at some future time, it marks the beginning of a new fall in the abundance of small perch, and so on. Thus the predator–prey system of a single perch population involves cycles of fluctuations in abundance. Such auto-fluctuations were obtained first with a 17-year periodicity on a computer simulated model and then similar fluctuations were found in nature. Experimental poisoning of small forest lakes with the subsequent stocking of a perch monoculture have confirmed the validy of computer-based conclusions. In some cases whole generations of perch were absent from such lakes, having been eaten by the large predaceous element of the population (Burmakin 1961, Burmakin & Zhakov 1961, Zhakov 1968).

The block diagram of a perch population model is given in Fig. 9.8. The system's state is determined by the numbers of fish in 9 age-groups $(N_1 \ldots N_9)$. The modelling of cannibalism, which is a particular case of predator–prey relationship, was based on the concepts of Shorygin (1952) and Ivlev (1955, 1961). The numbers of small perch $(N_1–N_4)$ depend on the food supply (P_{pb}), while those of predator perch depend on plankton-feeding perch. From the 6th year of age, perch become the object of fishing (Y) with a certain intensity (F). The maximum ration (Rm) of a predator is distributed in proportion to the probability of catching its prey, and the overall effect of all predators on the prey of each age as well as other quantities are determined using appropriate formulae (Menshutkin 1971). Attaining maturity at 5 years of age, all individuals have a reproductive potential (E) and

* Experimental works by Shentyakova (1959) and Illyina (1970)have demonstrated that plankton-feeding, benthos-feeding and predaceous perch can develop from a single egg mass of a small, littoral perch.

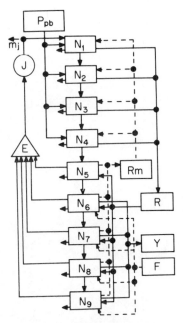

Fig. 9.8. Block-schema of a model of perch population in Lake Tiulenye (from Menshutkin 1971). Explanation of symbols is given in the text.

deposit a certain amount of eggs (J) which, at a certain level of mortality (m_j), produces recruitment in the form of the new generation (N_1). The numerical ratio of immature to mature (predaceous) perch is shown in the form of a phase diagram (Fig. 9.9).

Even in such a simple case when only one fish species is present in a body of water, four trophic levels can be observed: phytoplankton, zooplankton and benthos, small plankton-feeding perch and large predaceous perch.

An example of a more complex model is the model of a pelagic fish community in Dalnee Lake (Krogius *et al.* 1969, Menshutkin 1971). This lake contains stickleback (*Gasterosteus aculeatus*) and the young of sockeye salmon (*Oncorhynchus nerka*) which feed on plankton. There is also another predator, a char of the genus *Salvelinus*, which eats both sockeye salmon young and stickleback. The state of this complex ecosystem is determined by 54 variables while the number of parameters reaches 400. Given the wide range of changes in the system's parameters, no limiting stable point, but only auto-fluctuations, have been found in the model. A similar effect has been described for sockeye salmon earlier (Ricker 1954).

The fluctuations in the ichthyomass of the three fish species in Dalnee

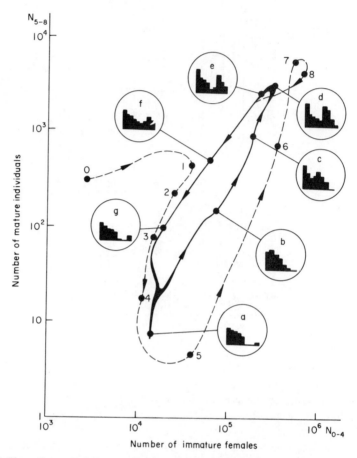

Fig. 9.9. Phase diagram of perch population dynamics. Unbroken line: established cycle; broken line: transitional process; *a* to *g*: histograms of the age composition of the population.

Lake are indicated in Fig. 9.10. Stickleback and sockeye salmon fluctuations are more or less synchronous, i.e., the competitive relationships are secondary in importance, the primary role being played by predator–prey relationships between char, on the one hand, and stickleback and sockeye salmon young, on the other.

The maximum of stickleback biomass (WSSK) either coincides or precedes by 1–2 years that of sockeye salmon biomass (WSSN). This is due to the fact that the stickleback, being a short-cycle fish, is able to respond more rapidly by more intensive reproduction to a decrease in predator pressure

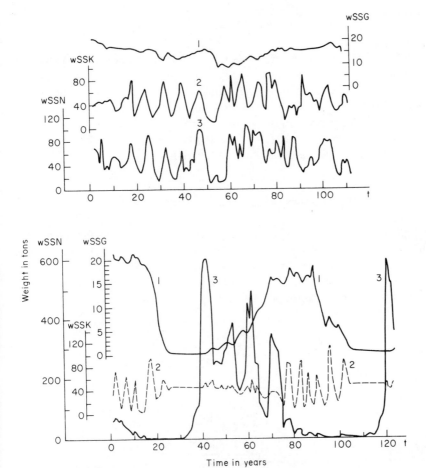

Fig. 9.10. Dynamics of the ichtyomass of populations of char (1), stickleback (2), and the young of sockeye salmon (3) in the absence of fishing (above) and during periods of increased rations and food supply (below). WSSN = Sockeye salmon (*Oncorhynchus nerka*) WSSK = stickleback (*Gasterosteus aculeatus*); WSSG = char (*Salvelinus alpinus*) (from Menshutkin 1971).

than is the sockeye salmon. The maxima of char biomass (WSSG) are associated with those of sockeye salmon biomass (Menshutkin 1971). When the system is upset by disturbing the constant food supply, the char and sockeye salmon biomasses show dramatic changes.

Therefore, predaceous fish can be considered to be a stabilizing factor in water bodies and help to achieve a balance of the abundance of different species in the ecosystem. The predators eat mostly the more abundant fish;

therefore, a sharp decrease in the predator population from overfishing results in an imbalance in the system and an increase in the abundance of small forage fish.

The use of mathematical models enables more detailed analyses to be made of fluctuations in predator and prey abundance. They also give an idea of the ecosystem's behaviour under extreme conditions (Menshutkin 1964, 1971, Larkin 1966, Andreev 1968, Winberg & Menshutkin 1974). The models allow us to visualize some of the principal problems of the dynamics of fish stocks, to systematize our accumulated knowledge, and to analyse the fundamental mechanisms of ecosystem dynamics.

9.7 The role of predaceous fish in a body of water

Using the predaceous fish of the Volga delta as an example, let us consider now a range of questions relating to food relationships. On the whole, predaceous fish have a wide feeding spectrum (up to 30 prey species), although the basic ration includes only few prey species (Table 9.5). This is demonstrated more clearly in the scheme of food relationships (Fig. 9.11).

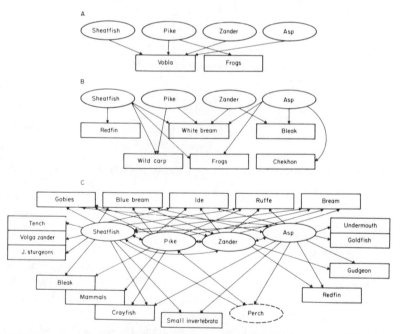

Fig. 9.11. Diagram of food relationships of predaceous fish in the lower zone of the Volga delta. Prey species: A, prey species making up over 20% of annual ration; B, prey species making up 5 to 20%; C, prey species making up under 5%.

The primary food components (over 20%) in the ration of predaceous fish are vobla in the lower zone and vobla, redfin and gobies in the upper delta. A large proportion of pike and sheatfish rations constitute non-fish foods (mainly frogs). In the lower zone, secondary food organisms, which make up 5 to 20% of the annual ration, are young of wild carp and fresh-water fish such as redfin (*Scardinius erythrophthalmus*), white bream (*Blicca björkna*), bleak, chekhon (*Pelecus cultratus*), and frogs. Occasional food objects (under 5%) include all other freshwater fish species, the young of predators and non-commercial fish.

At first sight it may seem that severe competition exists among predaceous fish in the Volga delta; however, there are many ways of decreasing this competition (Furtunatova & Popova 1973). On one hand, the similarity of the food of the various predators is due to the abundance of many vobla, white bream, redfin and frogs. Between 1940 and 1960 the abundance of vobla outnumbered the common abundance of all predators by 15·5 times. On the other hand, there are considerable differences in feeding habits among predator species, some size specific. Morpho-ecological characters of preda-tors diversify the habitat selection (eco-areas), type of hunting, prey selection, and daily and annual rhythms of feeding (Figs. 9.4, 5, 6, 7, 11, 12). All of these differences decrease the competition for food, but at the same time, increase the complexity of the ecosystem.

The role of predaceous fish differs in different types of water. In small waters of the river and lake type, and in the upper reaches of rivers where there is a small food supply and relatively stable ecological conditions, such predators as pike, sheatfish and zander are usually low in abundance and their role is that of 'improvers' of the biological environment. At the same time they are themselves the objects of a fishery. In larger waters with a large food supply, characterized by greater diversity and variability of ecological conditions (large lakes, lower reaches and deltas of rivers, es-tuaries), the abundance and growth rate of predators is high, they rapidly attain maturity, and are more fecund. In bodies of water of this type preda-tors frequently eat large numbers of commercial fish as well as small coarse fish, for the abundance of the former is high and they are readily available (Popova 1965). In this case too, predators play an advantageous role since they consume mainly small, weak individuals, thereby increasing the food supply and feeding condition for the rest of the fish.

The economic importance of predators may change in a body of water when ecological conditions undergo drastic changes. In recent years, because of regulation of the flow of many rivers in the USSR, large-scale investiga-tions have been carried out to study changes in the composition of the fish fauna of newly constructed reservoirs.

Owing to the flooding of large areas in the first year after a reservoir is filled, conditions for spawning and feeding of phytophilus species are usually

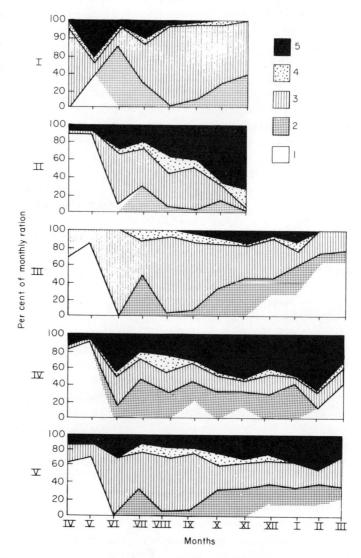

Fig. 9.12. Relationship of important food-organisms in the lower zone of the Volga delta in the course of a year (in % of month ration). 1, mature commercial fish; 2, the young of commercial fish; 3, non-commercial fish; 4, the young of predaceous fish; 5, non-fish objects. I, asp; II, sheatfish; III, zander; IV, pike; V, perch.

very favourable (roach, perch, white bream, and also pike). As a result, the abundance of coarse fish and fish of low economic value, as well as that of the predators, increases sharply. The growth rate of predators also increases, the gain in length of pike often attaining 40 cm during the first year. With the establishment of the permanent water regime for the reservoir, conditions for the spawning of phytophilus species deteriorate and the abundance and growth rate of predators sharply decrease. The main food organisms in most reservoirs are roach, ruffe, young of perch and smelt. As in large lakes, when commercially valuable species are very abundant in reservoirs they are often consumed by the pike and zander (Vashchenko 1958, Domanevsky 1964, Dyuzhikov 1959, Zadulskaya 1960). Even under these conditions predators play a favourable role. By thinning out prey populations, they contribute to an increase in their growth rate and to the improvement of the commercial qualities of the surviving part of the stock. Hence, with respect to pike and zander, which are important objects of the fishery and appear as biological meliorators, the generally accepted view of ichthyologists is that their abundance must be maintained or even increased in newly formed reservoirs. They are almost the sole means available to control the abundance of coarse fish and fish of low economic value which attain high population densities during the first years after filling the reservoir and which, in the absence of interference by man, may supplant other fish species of greater economic importance.

The economic importance of a facultative predator such as perch is in most cases quite different from that of pike, zander, and sheatfish. Unlike these latter, the perch has a much slower growth rate, and in shallow bodies of water with a poor food supply it is usually itself a coarse fish, unused by the fishery and often eaten by larger predators.

In larger waters such as lakes, especially reservoirs, and in the delta areas of rivers, perch usually form ecological groups that differ in their growth rate, maturation time, and fecundity. Because of that the perch can utilize fully the food supply and inhabit all biotops of the waterbody. In such bodies of water the rapidly-growing group of perch is usually small and plays the same role as pike, i.e., they improve the biological situation. At the same time large perch are the object of a fishery and sport angling. Slow growing perch are a coarse fish and a food for a large predator in these situations. Hence, in a majority of waters management is required to control the abundance of slow-growing perch by increasing the abundance of larger predator species.

In order to determine the role that particular predaceous fish play in an ecosystem, it is necessary to know: (1) the annual ration of the predator and the importance of individual fish species in it; (2) the economic value of the predator and of its prey species; (3) the food coefficient of the predator.

The main index of the economic value of a predaceous fish in a body of water may be the food coefficient, i.e. the ratio of the weight of food consumed to the weight gain of the predator. This index has a number of disadvantages and is of little value for theoretical calculations when there are sharp differences between the chemical composition of the food and the

Chapter 9

fish body (Ivlev 1955, Winberg 1956).* The food coefficient varies considerably from season to season, especially when there are sharp seasonal changes in foods or in weight gain. The coefficient also depends on the mobility of the fish. Among sedentary benthic fish, food coefficients are, as a rule, lower than those of pelagic fish. Nikol'skii (1974c) believes that the predaceous fish of southern latitudes should have lower food coefficients that those of northern latitudes. In our studies, the ratio of the annual ration to the annual weight gain appears to adequately reflect food consumption when the differences in the energy equivalents of the prey and predator species are small and when the weight gain is judged from the annual gain in body mass.

It may be noted that the food coefficient changes as the fish grows. For example, the annual ration (percent of body weight) and the food coefficient of zander of Volga delta change with age in the following way:

Age	1	2	3	4	5	6
Annual ration	264	264	224	224	157	157
Food coefficient	2·6	3·1	4·1	5·1	4·7	6·3

The average values of food coefficients for predaceous fish are shown in Table 9.9. The food coefficients of predaceous fish are in the range from 5 to 10, i.e., are much lower than for all other fish species: zooplankton-eating fish have food coefficients around 10; crayfish-eaters, 9 to 14; mollusc-eaters, 30; and plant-eaters, 40–50.

In the last few years, wide use has been made of the energy principle in studying food relationships and productivity of ecological systems (Winberg 1962, 1964, Odum 1971, Dajoz 1972, Royce 1972, Williamson 1972, Zaika

Table 9.9. Food coefficients of mature (3+ and more) predaceous fish (from Fortunatova & Popova 1973)

Fish species	Food coefficient
Rainbow trout, *Salmo gairdneri*	8·0
Pike, *Esox lucius*	8·8
Perch, *Perca fluviatilis*	5·5
Zander, *Stizostedion lucioperca*	5·1
Scorpion fish, *Scorpaena porcus*	6·4
Asp, *Aspius aspius*	6·3
Sheatfish, *Silurus glanis*	6·2
Sea burbot, *Gaidropsarus mediterraneus*	6·0
Burbot, *Lota lota*	10·0
Cod, *Gadus morhua*	7·5

* For that reason, it is often replaced by the 'trophic' coefficient which is the ratio of the energy of consumed food to that of weight gain.

1972, Winberg & Menshutkin 1974). Predators usually occupy the terminal fourth or fifth link in the food chains of an ecosystem. They make up about 30% of the fish production and they eat about 30% of all fish production. In the Rybinsk Reservoir, the annual increase in fish biomass is estimated at 215×10^3 tons or 470 kg ha^{-1}; of this quantity, 300 kg ha^{-1} or 63% perish from such causes as hunger, diseases, birds, asphyxiation etc.; the predators eat 34% but only 2·5% are taken by man. The catches make up $5·35 \times 10^3$ tons or 11 kg ha^{-1}; expressed in another way, the fishery takes 0·04% of the primary production of the reservoir which amounts to 0·8 cal m^{-2} (Gordeev *et al.* 1974).

In Lake Baikal, 1·6 kg ha^{-1} are caught, which is 3·7% of all the fish produced, or 0·08% of the primary production (Moskalenko 1971). In a number of small lakes of the USSR, the fishery is 30–50 kg ha^{-1} or 0·12–0·41% of phyto-plankton production (Winberg 1962). Similar figures of annual catches have been reported by Odum (1971) for inland North American waters:

	kg ha^{-1}	kcal m$^{2-1}$
Great Lakes of America	1·1– 7·8	0·2– 1·6
Small lakes	2 –178	0·4–36·0
Ponds in which predators are reared	44 –166	9 – 34
Fertilized ponds	220 –550	45 –112

The rearing of plant-eating fish in southern countries of the Far East, of course, yield a much higher harvest (1125–1750 kg ha^{-1}). However, shortening the food chains and obtaining maximum yields does not always result in high-quality products. The methods used by fisheries depend on goals and living standards. Many countries prefer to rear for sport those predaceous fish (pike, perch) which are at the end of a long food chain.

As mentioned above, the pronounced changes in one link of food chain leads to quantitative transformations in the rest of the links. Similar transformations occur during sharp changes in the abundance of predators due to overfishing or to their introduction into a new water. In another example, when the sockeye salmon numbers fall, the quantity of phosphate (P_2O_5) entering Dalnee Lake from fish corpses, decreases from 6·8 mg m^{-3} during 1937–1947 (or 26% of the annual phosphate influx) to 1·0 mg m^{-3} (or 5%) in 1957–1969. The disturbance of the supply of phosphate nutrients of marine origin brought about profound changes in all links of the ecosystem of Dalnee Lake. According to preliminary calculations (Krokhin 1975) the changes that occurred over the period 1937–1969 were the following:

1 annual primary production has decreased by 20% on average, from 3010 to 2435 kcal m^{-2};
2 total annual production of zooplankton decreased by 30% from 400 to 280 kcal m^{-2};

3   total annual production of plankton-eating fish decreased by 45% from 67 to 37 kcal m^{-2};

4   the annual production of predaceous fish decreased nearly 60% from 7 to 3 kcal m^{-2}.

Therefore, the role of predaceous fish in a body of water is multipurpose. The predator influences the quantitative and qualitative composition of prey species. Predaceous fish are a necessary element of the ichtyofauna of any body of water, because they enhance the stability of the ecosystem by regulating the abundance of different prey species as well as themselves. At the same time, they can be regarded as biological meliorators by eliminating sick, weak and abnormal individuals, i.e., that part of the population which die first under unfavourable conditions. In this manner, they promote an increase of the rate of genetic adaptation of prey species to the influence of stress factors. The latter role is most important in modern pollution conditions.

Study of the biology of predaceous fish and their importance in ecosystems provides a basis for the introduction of fish faunas into newly created or already existing reservoirs. This work involves active interference by man in individual links of trophic chains. Some predaceous fish may be replaced with others which are more rational and have lower food coefficients, and which inhibit undesirable or coarse prey species.

References

ANDREEV V.L. (1968) A simple model of open controlled system of predator–prey. *Uch. Zap. Dalnevostochnogo Univ.* **15**, (2) 145–153.

ANDRIYASHEV A.P. (1944a) On the methods of studying morphology and function of the pharyngeal teeth of teleosts. *Zool. Zh.* **23** (6), 316–329.

ANDRIYASHEV A.P. (1944b) Procedure of food procuring in *Scorpaena porcus* L. *Zhurn. obshch. Biol.* **5** (1), 56–59.

ANDRIYASHEV A.P. (1945) On the function of the pharyngeal apparatus in some predaceous fish. *Priroda* **2**, 68–70.

ANDRIYASHEV A.P. (1948) The role of the pharyngeal apparatus in grey mullet feeding. *Sbornik pam. akad. A.A.Zernova, Akad. Nauk SSSR* 108–112.

ARNOL'DI L.V. & FORTUNATOVA K.R. (1937) On the experimental studying of fish feeding in the Black Sea. *Dokl. Akad. Nauk SSSR* **15** (3), 505–508.

BORODULINA T.L. (1958) On the significance of gull (birds) in the spawn-growing fish-farm. *Vopr. Ikhtiol.* **11**, 205–209.

BORUTSKY E.V. (ed.) (1974) *Methodical Manual for Study of Food and Feeding Relationships Among Fish Under Environmental Conditions.* Nauka Press, Moscow. 254 p.

BURMAKIN E.V. (1961) Absolute number and biomass of perch in small lakes. *Trudy Soveshch. po dinamike chislennosty ryb Izdat. Akad. Nauk SSSR* 235–237.

BURMAKIN E.V. & ZHAKOV L.A. (1961) Experimental determination of fish productivity of the perch lake. *Nauchno-teknhich. bull. GosNIORKh* **13–14**, 25–27.

CROSSMAN E.J. (1959) A predator-prey interaction in freshwater fish. *J. Fish. Res. Bd. Canada* **16** (3), 269–281.

CRUMPTON J. (1971) Food habits of longnose gar (*Lepisosteus osseus*) and Florida gar (*Lepisosteus platyrhinchus*) collected from five central Florida lakes. *Proc. 24th Annu. Conf. South-east. Assoc. Game & Fish Comiss.*, Atlanta, Ga., 1970. S. **1**, 419–424.

DAJOZ R. (1972) *Précis d'Ecologie*. Deuxième edit. Paris. (Cit. in Russian edit. Moscow, "Progress", 1975.

DOMANEVSKY L.N. (1964) Some features of the interspecific relationships between pike and the dominant fish species in Tsimlyansk reservoir. *Zool. Zh.* **43** (1), 71–79.

DYUZHIKOV A.T. (1959) On the accumulation of pike below the Volga hydroelectric station. *Ryb. Khoz.* **3**, 15–16.

EDWARDS D.J. (1971) Effect of temperature on rate of passage of food through the alimentary canal of the plaice, *Pleuronectes platessa* L. *J. Fish Biol.* **3** (4), 433–439.

EDWARDS R.R.C., BLAXTER J.H.S., GOPALAN U.K., MATHEW C.V. & FINLAYSON D.M. (1971) Feeding metabolism and growth of tropical flat fish. *J. expm. Mar. Biol. Ecol.* **6**, 279–300.

ELTON C. (1933) *The Ecology of Animals*. Methuen, London. (Cit. in Russian edit. M.-L. Biomedgiz, 1934).

FORTUNATOVA K.R. (1940) Feeding of *Scorpaena porcus* L. (On methods for quantitative study of the dynamics of feeding of marine predaecous fish). *Dokl. Akad. Nauk SSSR* **29** (3), 244–248.

FORTUNATOVA K.R. (1950) Biology of feeding of *Scorpaena porcus* L. *Trudy Sevastopol'. biol. Sta.* **7**, 193–235.

FORTUNATOVA K.R. (1951) Methods of studying feeding of predaceous fishes. I. *Zool. Zh.* **30** (6), 562–571.

FORTUNATOVA K.R. (1955) Methods of studying feeding of predaceous fishés. II. *Trudy Soveshch. po metodam izuchen. korm. bazi i pitania ryb. Akad. Nauk SSSR* 62–84.

FORTUNATOVA K.R. (1957) Some data on effect of predators upon the size composition of fish–prey populations. *Zool. Zh.* **36** (4), 575–586.

FORTUNATOVA K.R. (1959) Availability of sticklebacks as food for predatory fish of the Volga delta. *Zool. Zh.* **38** (11), 1689–1701.

FORTUNATOVA K.R. (1961) Effect of predaceous fish upon the structure of commercial fish populations. *Trudy Soveshch. po dinamike chislennosti ryb. Akad. Nauk SSSR* 108–116.

FORTUNATOVA K.R. (1964) On food indices in fish. *Vopr. Ikhtiol.* **4** (1), 188–190.

FORTUNATOVA K.R. & POPOVA O.A. (1973) *Feeding and Food Relationships of Predaceous Fish in Volga Delta*. Nauka Press, Moscow.

FROST W.E. (1954) The food of the pike *Esox lucius* L. in Windermere. *J. Anim. Ecol.* **23**, 339–360.

GAEVSKAYA N.S. (1955) Aspects of the food reserve and fish feeding with reference to the main problems of the biological basis of fisheries. *Trudy Soveshch. po metodam izuch. korm. bazi i pitania ryb. Akad. Nauk SSSR* 6–21.

GAUSE G.V. (1934a) *The Struggle for Existence*. Williams and Wilkins, Baltimore.

GAUSE G.V. (1934b) Experimental analysis of Vito Volterra's mathematical theory of the struggle for existence. *Science* **79**, 16–17.

GIRSA I.I. (1962) On the decrease in availability of small fish for predators in connection with the formation of defensive conditioned reflexes. *Vopr. Ikhtiol.* **2** (4), 747–749.

GORDEEV N.A., PODDUBNY A.G. & ILJINA L.K. (1974) Experience of estimating potential fish productivity of a reservoir. *Vopr. Ikhtiol.* **14** (1), 20–25.

GRAEBER R.C. (1974) Food intake patterns in captive juvenile lemon sharks, *Negaprion brevirostris*. *Copeia* **2**, 554–556.

HOOGLAND R., MORRIS Q. & TINBERGEN N. (1956–1957) The spines of sticklebacks (*Gasterosteus* and *Pygosteus*) as means of defence against predators (*Perca* and *Esox*). *Behaviour* **10** (3–4), 205–236.

Hopson A.J. (1972) A study of the Nile perch (*Lates niloticus* L.), Pisces: Centropomidae in Lake Chad. *Foreign & Commonwealth Office Overseas Dev. Admin. Overseas Res. Publ.* **19**, 1–93.

Horoszewisz L. (1960) Wartosc kosjardlowych dolnych (ossa pharyngea inferiora) jako kryteriiw gatunkowego oznaczania ryb carpiowatych (*Cyprinidae*). *Roczniki nauk rolnicznych* **75** (2), 237–258.

Hunt B.P. (1960) Digestion rate and food consumption of Florida gar, warmouth and largemouth bass. *Trans. Amer. Fish. Soc.* **89** (2), 206–211.

Iljina L.K. (1970) On the qualitative difference of the young fish and the irregular growth of scales in the fingerlings of perch. *Perca fluviatilis* L. *Vopr. Ikhtiol.* **10** (1), 170–175.

Ivanova M.N. (1968) Food rations and food coefficients of predaceous fishes of Rybinsk reservoir. *Trudy Inst. Biol. Vnutr. Vod* (Borok) **17** (20), 180–198.

Ivlev V.S. (1955) *Experimental Ecology of the Feeding of Fishes*. Pishchepromizdat, Moscow, 252 pp. (Translat. by Douglas Scott (1961), New Haven, Yale Univ.).

Ivlev V.S. (1961) Principles of mathematical representation of the dynamics of commercial fish populations. *Trudy Soveshch. po dinamike chislennosti ryb. Akad. Nauk SSSR*, 330–336.

Karpevich A.F. (1941) The rate of digestion in some Black Sea fishes. *Zool. Zh.* **20** (2), 252–257.

Karpevich A.F. & Bokova E.N. (1936) The rate of digestion in marine fishes. Part I. *Zool. Zh.* **15** (1), 143–168.

Karpevich A.F. & Bokova E.N. (1937) The rate of digestion in marine fishes. Part II. *Zool. Zh.* **16** (1), 28–44.

Kostomarova A.A. (1962) Effect of starvation on the development of fry of teleost fishes. *Trudy Inst. Morf. Zhivot.* **40**, 4–77.

Kovalev K.G. (1958) Determination of body weight and body length of certain species of fishes in the Volga delta from the suprapharyngeal and submaxillary bones. *Trudy Astrakh. zapovednika* **4**, 237–267.

Krogius F.V., Krokhin E.M. & Menshutkin V.V. (1969) *The Community of Pelagic Fishes of Lake Dalnee*. Nauka Press, Leningrad.

Krokhin E.M. (1975) Transport of nutrients by salmon migrating from the sea into lakes. *Ecol. Stud.* **10**, Berlin e.a., 153–156.

Larkin P.A. (1966) Exploitation in a type of predator–prey relationship. *J. Fish. Res. Bd. Canada* **23** (3), 349–356.

Lishev M.N. (1950a) On the method of studying food composition of predaceous fish. *Izv. TINRO*, **32**, 121–128.

Lishev M.N. (1950b) Feeding and food relationships among predaceous fishes of the Amur basin. *Trudy Amur. ikhtiol. expedit.*, 1945–1949, **1**, *Izd. MOIP*, 19–146.

Manteifel B.P. (1961) Vertical migrations of marine organisms. II. On the adaptive importance of vertical migrations of fish-planktophages. *Trudy Inst. Morf. Zhivot.* **39**, 5–46.

Manteifel B.P., Girsa I.I., Lesheva T.S. & Pavlov D.S. (1965) Diurnal rhythms of feeding and locomotor activity of some freshwater predaceous fishes. *Sbornik "Pitanie khishchnikh ryb i ikh vzaimootnosheniya s kormovymi organismami" Akad. Nauk SSSR*, Moscow, 3–81.

Markuse V.K. (1964a) The role of the water ringed snake in the rearing hatcheries in the Volga delta. *Vopr. ikhtiol.* **4** (4), 736–745.

Markuse V.K. (1964b) Lake frog (*Rana ridibunda* Pallas) and its role in the rearing hatcheries in the Volga delta. *Zool. Zh.* **43** (10), 1511–1516.

Markuse V.K. (1966) *Fish-eating Birds in the Rearing Hatcheries in the Volga delta and Their Significance*. Sbornik "Ryboyadnye pticy i ikh znachenie v rybnom khozyaistive" Nauka Press, Moscow, 71–92.

MENSHUTKIN V.V. (1964) A study of fish population dynamics on the basis of a fish population conceived as a cybernetic system. *Vopr. Ikhtiol.* **4** (1), 23–33.

MENSHUTKIN V.V. (1971) *Mathematical Modelling of Populations and Aquatic Faunal Communities.* Nauka Publ. Leningrad.

MOLNÅR G., TAMÅSSY E. & TÖLG I. (1967) The gastric digestion of living predatory fish. *The Biological Basis of Freshwater Fish Production* pp. 135–149 (S.D. Gerking, ed.), Blackwell Scientific Publications, Oxford.

MONASTYRSKY G.N. (1940) Methods of evaluating the roach stocks of the North Caspian. *Trudy VNIRO* **11**, 115–170.

MOSKALENKO B.K. (1971) Biological productivity of Baikal Lake. *Gidrobiolog. Zh.* **7** (5), 5–14.

NIKOL'SKII G.V. (1950) On the biological basis of the rate of exploitation and the ways to regulate the population of a fish stock. *Zool. Zh.* **29** (1), 16–26.

NIKOL'SKII G.V. (1953) On some peculiarities of feeding relations in freshwater fish. *Sbornik "Ocherki po obschim voprosam ikhtiologii" Akad. Nauk SSSR. M.-L.* 261–281.

NIKOL'SKII G.V. (1974a) *Theory of the Dynamics of Fish Stocks.* 2nd Edit. Moscow, (English Version 1969. Oliver & Boyd, Edinburgh).

NIKOLSKII, G.V. (1974b) *The Ecology of Fishes.* Moscow, (English Version, London & New York, 1963, Academic Press.

NIKOL'SKII G.V. (1974c) On the cause of greater predator pressure on the non-predatory fish populations in low latitude. *Zh. obshch. Biol.* **35** (3), 346–352.

NOVIKOVA N.S. (1966) On the relation of the diurnal course of feeding of Barents Sea cod, *Gadus morhua morhua* L., to tidal phenomena. *Vopr. Ikhtiol.* **6** (1), 91–97.

ODUM E.P. (1971) *Fundamentals of Ecology.* 3rd Edit. W.B. Saunders, Philadelphia-London-Toronto (Cit. in Russian, Izd. MIR, Moscow, 1975).

PARAKETZOV I.A. (1958) On the protective role of body spines and sharp fin rays in fishes. *Lh. obstich. Biol.* **19**, 449–456.

PÉRÈS Y., BOGE Y., COLIN D. & RIGAL A. (1973) See p. 218, 219. Effects de la temperature sur les processus digestifs des poissons, Activiles enzymatiques et absorption intestinale. *Rev. trav. Inst. perches mar.* **37**, 223–232.

PIHU E.H. & PIHU E.R. (1970) The reconstruction of the size of fishes swallowed by predators from the vertebral fragments. *Vopr. Ikhtiol.* **10** (5), 929–932.

PIHU E.H. & PIHU E.R. (1971) On the swallowing of a prey by predators. *Izv. ENSV Tead. Akad. toimetised. Biologia* **20** (2), 127–132.

PLISZKA F. (1953) Spostrzeżenia nad wplywem warunków rozody ryb jeziorowych na liczeboność populacji ich stadiow modocianych. *Polskie arch. Hydrobiol.* **1** (14). 165–188.

POPOVA O.A. (1962) Some data on feeding of cod in the Newfoundland area of the North-West Atlantic. *Sbornik "Sov. rybokhoz. issledovaniya v severo-zapadnoy chasti Atlantich okeana".* Moscow, VNIRO-PINRO 235–253.

POPOVA O.A. (1965) Ecology of pike and perch in the Volga delta. *Sbornik "Pitanie khish-chnikh ryb i ikh vzaimootnosheniya s kormovymi organismami".* Akad. Nauk SSSR 91–172.

POPOVA O.A. (1967) The 'predator–prey' relationship among fish. *The Biological Basis of Freshwater Fish Production.* (S.D. Gerking, ed.), pp. 359–376. Blackwell Scientific Publications, Oxford & Edinburgh.

POPOVA O.A. (1968) Morphological characters of different populations of atlantic cod in connection with feeding. *Sbornik "Materialy po ekologii treski Severnoy Atlantiki"* Nauka Press, Moscow 26–68.

POPOVA O.A. (1971) Biological peculiarities of pike and perch in waterbodies with different hydrological regimes. *Sbornik "Zakonomernosti rosta i sozrevaniya ryb".* Nauka Press, Moscow, 102–152.

POPOVA O.A. (1975) Some peculiarities of feeding of predatory fish in Pskov-Chudsky Lake during a sharp decline in the abundance of a smelt. *Sbornik "Osnovy bioproduktivnosti vnutrennikh vodoemov Pribaltiki", Vilnyus* 90–93.

POPOVA O.A., RESHETNIKOV Yu.S. & TRJAPITZINA L.N. (1976) Peculiarities of biological fish cycles in tropical, temperate and artic waters. *Tez. dokl. "Ecologicheskaya fisiologiya ryb"*. Naukowa dumka Press, Kiev. 41–42.

RADAKOV D.V. (1961) Aspects of the defensive behaviour of shoals of some pelagic fish. *Trudy Inst. Morf. Zhivotn.* **39**, 47–71.

RADAKOV D.V. (1965) On the role of the school among predaceous fishes in catching their prey. *Sbornik "Pitanie khishchnikh ryb i ikh vzaimootnosheniya s kormovymi organismami". Akad. Nauk SSSR* 173–178.

RADAKOV D.V. (1972) *Shoals of Fish as Ecological Fact.* Nauka Press, Moscow.

RESHETNIKOV Yu.S. (1961) On the connection between the number of gill rakers and the character of feeding in chars of the genus *Salvelinus. Zool. Zh.* **40** (10), 1574–1576.

RESHETNIKOV Yu.S., CLARO R. & SILVA A. (1972) The rate of digestion in some tropical predaceous fish. *Vopr. Ikhtiol.* **12** (5), 893–900.

RESHETNIKOV Yu.S., CLARO R. & SILVA A. (1974) Ritmo alimentario y velocidad de digestion de algunos peces depredadores tropicales. *Acad. Sci. Cuba, Ser. Oceanologica*, La Habana, **21**, 3–13.

RESHETNIKOV Yu.S. & POPOVA O.A. (1975) The daily rations and rate of food digestion in tropical fish. *Sbornik "Biologiya shelfa", Vladyvostok* 144–145.

RESHETNIKOV Yu.S., SILVA A., CLARO R. & POPOVA O.A. (1975) Rates of food digestion in tropical fishes. *Zool. Zh.* **54** (10), 1506–1514.

RICKER W.E. (1952) Numerical relations between abundance of predators and survival of prey. *Canad. Fish Culturist* **13**, 5–9.

RICKER W.E. (1954) Stock and recruitment. *J. Fish. Res. Bd. Canada* **11** (5), 559–692.

ROYCE W.F. (1972) *Introduction to the Fishery Sciences.* Academic Press, New York & London. (Cit. in Russian, "Pishchevaya promyshchlennost", Moscow, 1975, 272 pp.).

SCHMIDT W. (1968) Vergleichend morphologische Studie über die Otolithen mariner Knochenfische. *Arch. Fishereiwiss.* **19** (1), 1–95.

SHAMARDINA I.P. (1957) Developmental stages in pike. *Trudy Inst. Morf. Zhivot.* **16**, 237–298.

SHENTYAKOVA L.F. (1959) Some peculiarities of perch growth. *Trudy Inst. Biol. Vodokhranilisch (Borok)* **1** (4), 298–308.

SHORYGIN A.A. (1952) *Feeding and Food Relationships of Caspian Fish.* "Pishchepromizdat", Moscow.

SKALKIN V.A. (1961) Otoliths of far-eastern *Gadidae. Vopr. Ikhtiol.* **1** (2), 286–289.

SKALKIN V.A. (1965) On methods of working up material on food and feeding of fishes. *Vopr. Ikhtiol.* **5** (4), 735–737.

SKOKOVA N.N. (1955) Feeding of common cormorant in the Volga delta. *Vopr. Ikhtiol.* **5**, 170–185.

SKOKOVA N.N. (1966a) Feeding of heron, great white heron and a little egret in the Volga delta in connection with their fishery significance. *Sbornik "Ryboyadnye ptichy i ikh znachenie y rvbnom khozyaistve"*, Nauka Press, Moscow, 93–124.

SKOKOVA N.N. (1966b) Effect of a common cormorant and heron-birds upon the fauna of the Volga delta and their fishery significance. *Ibidem*, 55–70.

SMITH R.L. & PAULSON A.C. (1974) Food transit times and gut pH in two pacific parrotfishes. *Copeia* **3**, 796–799.

SVETOVIDOV A.N. (1932) On the relation between the character of food and the number of pyloric caeca in herrings. *Dokl. Akad. Nauk SSSR* **8**, 202–204.

SVETOVIDOV A.N. (1953) On the relation between the number of pyloric caeca and the character of feeding among fishes. *Sbornik "Ocherki po obshchim voprosam ikhtiologii"*. *Akad. Nauk SSSR, M.L.* 282–289.

SWENSON W.A. & SMITH L.L.J. (1973) Gastric digestion, food consumption, feeding periodicity, and food conversion efficiency in walleye (*Stizostedion vitreum vitreum*). *J. Fish. Res. Bd. Canada* **30** (9), 1327–1336.

TYURIN P.V. (1973) Theoretical background of the rational management of fishery. *Izv. GosNIORKh* **86**, 7–25.

TYURIN P.V. (1974) Biological principles of fish stock reconstruction in Pskovsky-Chudskoy Lake. *Izv.GosNIORh* **83**, 153–187.

VASARHELEYI I. (1958) Beiträge zur Bestimmung der Karpfenartigen mit Hilfe der Schlund-knochen. *Arch. Fischereiwiss.* **9** (3), 187–199.

VASHCHENKO D.M. (1958) The carp as food of pike during the first year after the filling of Kakhovka reservoir. *Zool. Zh.* **37** (11), 1745–1748.

VASNETSOV V.V., EREMEEVA E.F., LANGE N.O., DMITRIEVA E.N. & BRAGINSKAYA R.Ya. (1957) Developmental stages in commercial anadromous fishes of the Volga and Don rivers. *Trudy Inst. Morf. Zhivot.* **16**, 8–76.

VOLTERRA V. (1928) Variations and fluctuations of the number of individuals in animal species living together. *J. Conseil.* **3** (1), 3–51.

WILLIAMSON M. (1972) *The Analysis of Biological Populations*. Edward Arnold, London (Cit. in Russian, edit. "Mir", Moscow, 1975.).

WINBERG G.G. (1956) *Rate of Metabolism and Food Requirements of Fishes*. Minsk. Belo-russk. Gos. Univ. (*Fish. Res. Bd. Canada, Trans. Ser.*

WINBERG G.G. (1962) Energetic principle for studying trophic relations and productivity of ecological systems. *Zool. Zh.* **41** (11), 1618–1630.

WINBERG G.G. (1964) Quantitative means for studying consumption and assimilation of food by aquatic animals. *Zh. obshch. Biol.* **25** (4). 254–266.

WINBERG G.G. & MENSHUTKIN V.V. (1974) Significance of mathematical modeling for the development of a scientific basis for the rational utilization of freshwater biological resources. *Sbornik "Problemy dolgosrochnogo planirovaniya biolog issledovanii"*. *Zoologiya*, **1**, "Nauka", Leningrad 25–44.

WINDELL J.T. (1967) Rates of digestion in fishes. *The Biological Basis of Freshwater Fish Production* (S.D. Gerking, ed.), pp. 151–173. Blackwell Scientific Publications, Oxford.

ZADULSKAYA E.S. (1960) Food and food interrelations among predaceous fishes of North Rybinsk reservoir. *Trudy Darvinsk gosud. zapovednika* **6**, 345–406.

ZAIKA V.E. (1972) *Specific Production of Aquatic Intervebrates*. Naukova Dumka, Publ., Kiev.

ZHAKOV L.A. (1968) On regulation of number and age structure of lake fish populations. *Sbornik "Gidrobiolog. i ikhtiolog. issledovaniya vnutrennikh vodoemov Privaltiki"*. *Vilnyus* 137–143.

Chapter 10: Estimating the Food Consumption of Fish in Nature

K. H. Mann

10.1 Introduction

If a fish population is to be effectively managed for optimum production, it is necessary to estimate its food consumption. Only then is it possible to calculate the efficiency with which food is being converted to fish flesh and to judge whether the density and age structure of the population is in proper relation to the food resources available. It is also essential to know food consumption for the calculation of energy flow or nutrient cycling through a fish population in connection with trophic models. This chapter is concerned with the various attempts that have been made to estimate the food consumption of fish populations in nature.

There are two basic approaches. The first is to take field data on stomach contents, and convert these to estimates of daily food consumption. The second involves laboratory experiments on feeding or metabolism and subsequent extrapolation to field conditions. Several recent papers have compared the results of two approaches (Elliott 1975a, Morgan 1973). In attempting to judge the success of various methods, it is important to remember that since the object is to determine food consumption of whole populations in nature, it is necessary to have data applicable to a good range of size classes, temperatures and seasons. Methods which yield data for less than the full range are of limited usefulness.

10.2 Methods involving mainly analysis of gut contents

Swenson and Smith (1973) determined the food consumption of walleye (*Stizostedion vitreum vitreum*), using analyses of stomach contents. They began by determining in the laboratory rates of digestion at various fish sizes, temperatures, meal sizes and food particle sizes. This was done by pumping stomachs at various intervals after feeding with fathead minnows. They used two temperatures (14·5C and 20C) 3 sizes of food (0·8 g, 1·1–1·9 g, and 3·1–5·0 g) and a good range of fish sizes and meal sizes. After calculating regression equations for the relationships between variables, they extrapolated them to cover the whole range encountered in nature.

Field studies consisted of collecting fish samples by day and by night, and analysing stomach contents. They determined the length of each partially digested food item, and from this calculated its initial weight. Comparison of partially digested weight with initial weight gave a figure for percentage digestion. The previous experiments on rates of digestion were then used to determine the time required to reach the observed state of digestion, and hence the time of feeding.

Information on the initial weight of food consumed and the time it was consumed enabled Swenson and Smith to reconstruct a record of the diurnal pattern of feeding. They found that in July, August and September the greatest part of the food was consumed at night and early in the morning. During June, when less food was obtained, feeding activity was spread fairly evenly over the 24 hours. Food consumption calculated from June to September in 1969 and again in 1970, for the 1966 year class of walleye, increased from a mean of 40·9 g per fish in June to 247·8 g in August, then fell to 219·2 g in September. During that period, mean food conversion efficiency fell from 30·1 % in June to 21·2 % in August.

A similar technique, but with fewer refinements was used by Elliott (1973) to calculate the food uptake of brown trout (*Salmo trutta*) and rainbow trout (*S. gairdneri*) in a mountain stream. Stomach samples were taken every 6 h for 24 h. When the mean water temperature was 4·7C and 7·3C the stomachs were full only at the collections made between 1·30 and 2·30 a.m. It was inferred that the fish could digest only 1 meal per day, and that the contents of a full stomach represented a daily ration. However, when the temperature was 10·8C the stomachs were full between 1·30 and 2.30 and also between 13.30 and 14.30, so it was inferred that a daily ration consisted of the contents of a stomach filled twice over. The explanation offered was that the higher temperature permitted a higher rate of digestion, and hence allowed a second meal to be taken. The method permitted calculation of the daily ration for a range of body sizes of the two species, at three specific temperatures which occurred in different parts of the same mountain stream in July/August. Elliott (1975a and b) explored further the maximum food consumption of *Salmo trutta* in captivity at a range of temperatures. He was able to show that food consumption in one meal increased with rising temperature up to 13·6C, remained approximately constant to 18C, and then decreased sharply. However, the rate of digestion also increased, permitting up to 3 meals a day to be taken at higher temperatures, so that maximum daily food consumption increased at all temperatures studied. This information, while valuable as a guide to feeding levels in aquaculture, is of limited value for estimating food consumption in nature.

Backiel (1971) calculated the food consumption of six species of predatory fish in the Vistula River. One of the methods he used was based on the amount of food in the stomach, its state of digestion, and estimates of the rate of food passage through the gut. The study was broad in its scope and the variance on the observations was rather high. Nevertheless, data were obtained for many size classes of fish, in four seasons, and useful approximations of the food consumption were obtained.

A second method of utilizing analyses of stomach contents is that employed by Staples (1975). Working with the upland bully, *Philypnodon breviceps*, in New Zealand, he took samples every 4 h for 48 h. One half of each sample was killed immediately while the remaining fish were left for 4 h without food in a tank. At the end of a 4 h period, the difference in gut contents between the fish taken from the lake and the fish held without food was taken as the food consumption in nature during that 4 h period. Daily rations, calculated from such data, were determined once in each season; spring, summer, autumn and winter. Food conversion efficiencies ranged from 35% for age 2+ fish in summer, to negative values for several age groups in winter.

10.3 Methods involving mainly extrapolation from laboratory measurements

An obvious approach to estimating food consumption in nature is to simulate natural conditions in the laboratory and measure food consumption directly. Provided care is taken to simulate the whole range of environmental variables, including temperature, day length, food availability, water movements, etc., and provided some effort is made to match the growth rates of the fish in the laboratory to growth rates observed in nature, this method has much to recommend it. However, in practice the direct methods prove to be difficult and tedious, and people have seldom succeeded in simulating all the required ranges of environmental variables.

The alternative is to calculate food requirements indirectly from energy or materials budgets. These methods involve more assumptions than the direct methods, but they permit greater flexibility for incorporating the effects of environmental variables, and are particularly suitable for numerical modelling.

DIRECT MEASUREMENTS OF FOOD CONSUMPTION

The food requirements of fish in captivity have been measured many times for purposes of extrapolating to nature. Surber (1935) fed fish with *Gammarus*, Brown (1946) fed brown trout (*Salmo trutta*) with a mixture of minced meat and liver, and Johnson (1960) fed pike (*Esox lucius*) on live minnows. In each case, two parameters were determined: (1) the food required to keep the body weight constant, which is the maintenance ration and (2) the efficiency of food utilization for growth, i.e. the ratio between food consumed above the maintenance ration and the corresponding gain in weight. The first person to attempt to extrapolate such data to the field situation was Allen (1951) who measured the vital statistics of a population of brown trout in the Horokiwi Stream, New Zealand, and adapted Pentelow's data to give an estimate of the yearly food requirements of the fish. A similar technique was used by Horton (1961) for a population of brown trout in a stream in England, and by Johnson (1960) who extrapolated his own results to obtain an estimate of the food requirements of a population of pike in Windermere, in the English Lake District.

The weakness of the method is that it takes weeks to obtain one figure for the maintenance requirements of a fish, so that the number of determinations made on a particular species is limited. Yet the maintenance requirements per unit body weight vary according to the size of the fish, temperature, activity and season. No one has succeeded in exploring these variables to the full. The efficiency of conversion is also highly variable. If the amount of food is limited, then any variation in the maintenance requirements will lead to variation in the amount of food available for growth. Moreover, a young, fast-growing fish usually has a higher efficiency of conversion than an ageing one, and when food is plentiful a fish will expend more energy on digestion and assimilation than at times of food scarcity, so that its efficiency of conversion will change.

Yet the extrapolations from laboratory feeding experiments to natural populations have been made on the basis of a constant efficiency of food utilization, and many have also assumed a constant maintenance ration per unit body weight. Gerking (1962) showed that Allen's (1951) approximations resulted in an overestimate of the food requirements of the trout population by a factor of two to three.

The tendency, more recently, has been to avoid the concept of maintenance ration, and express growth efficiency, or food conversion efficiency as a simple ratio of increase in fish biomass to biomass consumed. This ratio may be expressed in wet weight, dry weight, or some component of these, such as carbon, nitrogen or calories. The ratio may be positive, or negative in the case of fish that lose weight. A value of zero corresponds to the old idea of maintenance. The studies of Staples (1975) and Warren *et al.* (1964) are examples of the use of a ratio based on calories. Swenson and Smith (1973) and Prinslow and Valiela (1974) used a biomass ratio. Even with this simplification in the approach, the amount of work needed to make a predictive model of food requirements for a range of size classes, temperatures, seasons and levels of activity is formidable indeed.

INDIRECT ESTIMATES OF FOOD CONSUMPTION

With the aim of trying to formulate generalizations about feeding, metabolism and growth, numerous workers have turned to indirect methods of estimating food consumption. The technique is to construct a budget, usually in terms of nitrogen or energy, by means of which food requirements can be predicted under a range of conditions.

Nitrogen budgets

Gerking (1962) produced cogent arguments for concentrating attention on the nitrogen budgets of fish. The fat content of organisms (and hence their calorific value) is extremely variable, while their protein content is relatively stable. Protein synthesis is the most characteristic feature of growth and protein digestion is so efficient in many fish that ingestion and absorption can virtually be equated. Gerking determined for bluegill sunfish (*Lepomis macrochirus*) the protein required to replace that broken down 'during the normal course of living', i.e. the maintenance requirement, and the efficiency of protein utilization for growth. Much of the experimental work involved feeding experiments similar to those mentioned in the previous section, except that food and growth were recorded in terms of their nitrogen content. An important advantage of the nitrogen method over direct estimates of food uptake is that the maintenance requirement can be measured relatively quickly by determining the rate of nitrogen excretion when the animal is in a resting state on a nitrogen-free diet. Gerking made a thorough study of the relationship between food consumed and food retained at various levels of feeding, but he was able to cover only a limited range of temperature, body size and season. He extrapolated his results to give an estimate of the food turnover of the bluegill population of Wyland Lake, Indiana, during the summer growing season, May to September. Subsequently he revised those estimates (Gerking 1972), on the basis of new data on the length of the growing season (Gerking 1966), new information on the influence of rate of feeding and body weight on protein metabolism, and a correction for the effect of activity on metabolic rate (Winberg 1956 see p. 256). By chance, the various corrections tended to compensate for one another, and the revised estimate was only about 6% higher than the first.

This method is really intermediate between direct and indirect methods of estimating food requirements. The estimates of the nitrogen requirements for routine maintenance are indirect, being calculated from rates of excretion. On the other hand, the calculations of food requirements above the maintenance level are based on feeding experiments much like those described in the previous section.

While Gerking is the prime exponent of the use of this method for estimating food consumption, there are several other studies of nitrogen conversion efficiency in fish. Birkett (1969) calculated the gross efficiency of nitrogen conversion, and the maintenance level of absorption and excretion for perch (*Perca fluviatilis*), plaice (*Pleuronectes platessa*) and sole (*Solea vulgaris*), at temperatures around 17C. He compared his results with those of Ivlev (1939) on carp, Menzel (1960) on red hind, and with Gerking's (1952, 1955) work on sunfish. All the data could be fitted to linear regressions of growth on amount of nitrogen absorbed but there was no indication of how the variables of these regressions might be related to temperature, activity or quality of the diet.

The work of Pandian (1967a, b, c; 1970) is an interesting link between this section and the next, for he constructed and compared the energy and nitrogen budgets of several species of fish during aquarium feeding experiments. He studied the effect of varying body size, temperature, quality of food and quantity of food, and documented differences between species. The work gave important insights into fish physiology but had too few replications to permit extrapolation to a range of natural conditions.

Energy budgets

Ivlev (1939) pioneered the idea of energy budgets for fish, but the method did not come into prominence until Winberg (1956) published his exhaustive review of the subject. The basic assumption is that the energy content of the food equals the sum of the energy contents of (1) the materials lost in egestion and excretion, (2) the materials retained in growth and (3) the material broken down in metabolism. Egestion, excretion and growth may be measured directly, and metabolic rate may be determined indirectly by respirometry. The calorific equivalent of 1 ml of oxygen consumed varies according to the material being respired, from about 5·0 cal for carbohydrates to about 4·5 cal for proteins. If we assume that a diet contains a mixture of fats, carbohydrates and proteins, it is reasonable to assume that 1 ml oxygen is equivalent to about 4·8 cal. The basic principles of ecological energetics are now well established. Winberg's main contribution was to reduce the mass of published data on respiration, egestion and excretion in fish to a form in which they could be applied to whole populations. This involved making a number of simplifying approximations. Taking as the basic equation for resting metabolism in relation to weight as

$$Q = a\,w^b$$

where Q is rate of metabolism, w is body weight and a and b are constants for the species, Winberg showed that b normally lies between 0·71 and 0·81, but a is much more variable, the value for salmonids being about 2·5 times that for cyprinodonts. In other words, when log respiration is plotted against log weight, the slope of the line is always close to 0·8, but the intercept, defined by a, is variable from one species to another. A high value of a implies a high general level of metabolism. He proposed that the effect of temperature on metabolic rate could be predicted from the curve of Krogh (1916). Hence, by determining a value for a, it should be possible to predict with a fair degree of accuracy the rate of resting metabolism of a fish at any combination of body size and temperature. If the *a* values calculated by Winberg are accepted at their face value, there is no need to make any measurements at all for the species which have been investigated.

Winberg also considered published data on the proportion of the food that is assimilated and the proportion egested. The number of determinations that has been made is much smaller than for metabolic rate, but even in 1956 they covered 10 species of fish and 18 species of food organisms. He found that figures for assimilation efficiency varied between 76 and 96·6%, but considered that much of the variation was due to experimental error, and that freshwater fish feeding on benthic animals normally assimilated about 85% of the calories in the diet, egesting 15%. He likewise argued that the calorific value of the urine normally lies in the range 3–7% of the energy of the diet, and that egestion and excretion together can be taken to account for 20% of the intake calories. The basic equation can then be simplified to read:

Energy of growth + energy of metabolism = 0·80 energy of the diet

or:

Energy of the diet = 1·25 (energy of growth + energy of metabolism)

Since growth of fish is routinely measured in connection with fisheries management, a good assessment of food requirements, by Winberg's method, requires only an assessment of metabolic rate. Winberg's (1956) review of the literature used only laboratory measurements of the metabolic rate of fish at rest (Winberg I), but he proposed that when extrapolating to field conditions one should double the resting level of metabolism (Winberg II) to take account of activity in nature. In more recent literature, Winberg I would be referred to as routine metabolism, i.e. the metabolism of fish showing spontaneous activity but no locomotion (see p. 188). Mann (1965) argued that the metabolic rate of animals in nature ranges from about 1·5 to about 2·5 times the resting level, that Winberg's factor of 2 is about right and that there is a high probability that the true level will lie within the range ± 25% of the calculated value. At this time, about 20 years after Winberg made his proposal, it is useful to review the evidence for and against his method.

A good starting point is the study by Solomon and Brafield (1974) of the energy relationships of perch, *Perca fluviatilis*, in which all categories of energy flow were continuously measured in growing fish. They concluded that 'the majority of the assumptions often made in energy studies are valid'. They found that assimilation efficiencies ranged from 83·5% to 87%, thus confirming Winberg's approximation of 85%. They found that although small amounts of urea were produced, it was reasonable to calculate the energy content of the urine on the assumption that only ammonia was produced, and this normally represented less than 10% of the energy consumed. They found that the rate of respiration of fish fed a maintenance ration was almost exactly that predicted by Winberg's equation for 'resting metabolism' (Winberg I). Fish on a restricted ration above maintenance had a metabolic rate intermediate between that predicted by Winberg I, and by double that figure (Winberg II). In this respect, their results confirm the early work by Pentelow (1939) on brown trout (*Salmo trutta*) (Fig. 10.1).

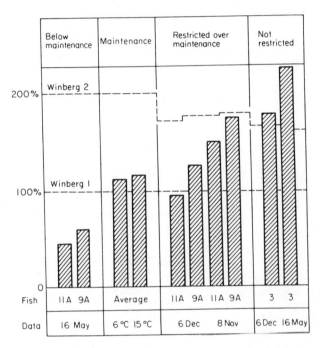

Fig. 10.1. Food consumption by trout (*Salmo trutta*) in the experiments of Pentelow (1939) compared with Winberg's predictions based on resting metabolism (Winberg 1) and twice the rate of resting metabolism (Winberg 2). From Mann (1965), reproduced by permission of the British Ecological Society.

There are several other kinds of evidence to suggest that the range of Winberg I to Winberg II spans the metabolic rate of fish from the maintenance level of feeding to the condition of unlimited food. Elliott's (1973) regressions of daily ration in trout, based on stomach contents, corresponded very closely with the prediction of Winberg I at 4·7C, with Winberg II at 7·3C, and with twice Winberg II at 10·8C (Fig. 10.2). Elliott suggested that only at the higher temperature did the fish have energy to spare for growth. Small (1975) compared the daily diet of the fantail darter *Etheostoma flabellare*, estimated by extrapolation from laboratory feeding experiments, with the

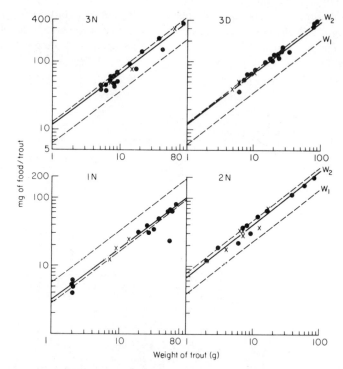

Fig. 10.2. Relationship between dry weight of food in the stomachs and the live weight of *Salmo gairdneri* (●) and *S. trutta* (X) from a Pyrenean mountain stream. *3N*, 10·8C at night; *3D*, 10·8C by day; *1N* 4·7C at night; *2N* 7·3C at night. Broken lines: W_1 and W_2 correspond with values given by Winberg I and Winberg II formulations. From Elliott (1973).

predictions of Winberg I and Winberg II, and concluded that the feeding rate in nature was close to Winberg I. He also showed that in nature these fish are feeding at a level only slightly above maintenance.

In retrospect, it appears that Winberg's basic equations for 'resting metabolism' of fish are an excellent representation of the metabolic rate of fish which are feeding at a maintenance level and that fish feeding at a higher level have a higher metabolic rate, associated with searching for, capturing, digesting and assimilating the food. Winberg (1956) proposed to allow for these factors by doubling the 'resting' level of metabolism, and justified this increase by reference to increased activity in nature. Warren and Davis (1967) drew attention to the evidence for the importance of Specific Dynamic Action, SDA, in determining the total metabolic rate of a fish. By this, they meant the increase in oxygen consumption associated with digestion, assimilation and storage of materials consumed. They suggested that it is unlikely that swimming activity in search of food would be a major influence in determining metabolic rate, by comparison with SDA. Hence 'there remains doubt as to how reliable estimates of the food consumption of fish in nature may be when based on growth and on routine metabolic rate multiplied by any constant'.

We shall return below to the debate on the relative importance of SDA and foraging activity, but it is worth noting that there are two independent pieces of work which support the idea that the metabolic rate of fish in nature is about twice the routine or resting level. Ivlev (1961) studied the feeding of bleak (*Alburnus alburnus*) on zooplankton in a hatchery pond. He first established experimentally a relationship between food density and feeding rate. Then, by observing the amount of time spent feeding and the density of the food populations, he was able to calculate the energy of food consumed. He also established the rate of respiration of feeding and of resting fish, and balanced the energy equation. The 24 h budget yielded an average rate of respiration close to twice the resting metabolism. Ivlev attributed the increase over resting rate to the activity involved in pursuing food. It is, however, possible that it also involved the SDA of the food.

Edwards (1968) correlated the rate of respiration of young plaice, *Pleuronectes platessa*, with the rate of elimination of Zn^{65} from the body; he then used the rate of loss of Zn^{65} as an index of metabolic rate of the fish in a natural environment. He showed in the laboratory that the rates of elimination of the isotope were reasonably constant in the period 7–17 days after administration. He then placed the fish in a large netting cage in the sea and determined the rate of respiration from the rate of loss of the isotope. It came out at 2·03 times the resting level (resting level being defined as the respiration rate of fish fed 24 h before measurement and at rest during the period of measurement).

A more radical discussion of these questions is contained in papers by Kerr (1971a, b and c). He analyzed the total metabolism T_T of fishes by identifying its components: T_F the cost of foraging activity; T_S the cost of standard metabolism; T_C the energy cost specific to utilization of food; and T_R the energy expended on spontaneous activity. Thus:

$$T_T = T_F + T_S + T_C + T_R$$

He pointed out that in laboratory feeding experiments T_F is often trivial. However, it can be described for any situation by

$$T_F = Z(R/g)$$

where R is the ration, Z is the average energy cost of acquiring a single diet item, and g is the weight of a single diet item. Some trends consequent on this formulation are that Z will increase as food items become scarce and widely dispersed, or capable of behaving in a fugitive manner. As food items become larger, i.e. g increases, the energy cost of acquiring a specific level of ration will decrease.

T_S is standard metabolism, and is measured by relating metabolic rate to rate of swimming, then extrapolating back to zero locomotion. Routine metabolism, measured while keeping unfed fish in a resting condition in a closed container, is usually higher than standard metabolism. For example, Muir and Niimi (1972) obtained figures for routine metabolism of aholehole, *Kuhlia sandvicensis* which were 22–68% higher than standard. Morgan (1973) obtained figures for trout in which routine metabolism was 51–132% higher, and for perch in which it was 23–141% higher, depending on the temperature.

T_C which Kerr called the energy cost specific to the utilization of food, is the *SDA* of other authors. Kerr (1971a) calculated T_C for brown trout from a variety of published data, and concluded that it was well described by the equation $T_C = 0.2882R$. That is to say, the energy cost is a constant 28·8% of the food intake, independent of the body size of the fish. (This generalization does not necessarily apply to all species. See for instance, Warren and Davis (1967), and Webb, this volume).

The energy expended on spontaneous activity T_R has not been systematically explored. In a starved fish under experimental conditions any difference between routine and standard metabolism may be attributed to spontaneous activity. However, in a fish which has been fed before the experiment, a proportion of the routine metabolism may be attributable to *SDA*. Kerr (1971a) has assembled evidence to indicate that some fish are appreciably more active when fed than when starved and suggested that this is an explanation of the decrease in growth efficiency with increasing rations which has been so frequently observed (Paloheimo & Dickie 1966). However, in modelling growth in nature, Kerr (1971b, c) thought it advisable to drop the term for spontaneous activity, since it was likely to be small compared with *SDA* and foraging activity. The equation for metabolism thus appears as:

$$T_T = T_F + T_C + T_S$$

The cost of foraging was evaluated by Kerr in terms of the distance travelled and the energetic cost of swimming that distance. When food particles were sufficiently close that the fish could perceive them without random search, he calculated the mean distance travelled between particles Q, from

$$Q = 0.554 \left(\frac{g}{D}\right)^{1/3}$$

where g is the mean size of a food particle and D the food density in grams per cubic metre. When particles were more widely dispersed, so that random search was necessary, he used

$$Q = \frac{g}{\pi r^2 D}$$

where r is the distance at which prey can be perceived. Then for a given ration R, the total distance travelled is RQ/g. With an appropriate evaluation of the energy cost of travelling that distance, he was able to define the cost of foraging T_F in terms of food density and particle size. He displayed his results (Fig. 10.3) in terms of growth efficiency, K. It is clear that, in this

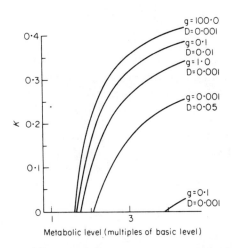

Fig. 10.3. Growth efficiency (K) (=weight gain per unit of food consumed) as a function of metabolic level for a 1000 g trout, for different combinations of prey size (g, in grams) and prey density (D, in grams/m³). From Kerr (1971b).

model, food density and particle size have a profound effect upon growth, through their effect on energy cost of foraging. The parameters used are those appropriate to brook trout, *Salvelinus fontinalis*, and the model predicts that over a range of food density and particle size the metabolic rate for zero growth (maintenance) will be 1·5–2·1 times the basal, or standard level. In Fig. 10.4 are shown the curves for growth efficiency per unit ration, for lake trout, *Salvelinus namaycush*. The maxima on these curves predict the optimum level of food utilization for the species. Over a range of food densities and particle sizes, optimum growth efficiency is achieved at 2 to 4 times standard metabolism.

Two conclusions may be drawn from Kerr's models. The first is that while for trout the SDA is significant amounting to more than 25% of the energy of the ration, the metabolic cost of foraging for food is also significant, and makes the difference between high and low growth efficiencies, according to the food density and particle size. The second conculsion is that for lake

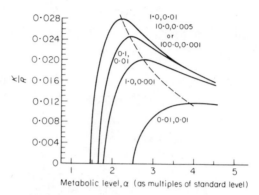

Fig. 10.4. Response to changes in metabolic level of the ratio growth efficiency/ration (K/R) for a 1000 g trout. The first number in each pair indicates mean particle size in grams, the second is prey density in grams/m². From Kerr (1971c).

trout the strategy for optimizing growth calls for a metabolic expenditure about 75% higher than the maintenance level. Since the maintenance level of metabolism has been shown many times to be approximated by the resting or routine metabolism of unfed fish in a respirometer (Winberg I), a doubling of this rate (Winberg II) is a useful approximation of the metabolism of a fish which optimizes its growth rate in nature. Kerr's (1971a, b, c) papers indicate the large number of parameters which must be evaluated to give a more precise prediction.

More recently, Ware (1975) re-analyzed the data in Ivlev's (1961) study of young bleak. On the basis of probable chemical composition of the food (taken from published analyses of copepods) and literature values for assimilation efficiency and *SDA* factor of each component, he concluded that the food conversion efficiency into physiologically useful energy used by Ivlev was too high, and that it should be changed from 0·8 to 0·7. He recalculated the food intake of the fish using the model of Holling (1966):

$$I = \frac{avc}{1 + avch}$$

where *I* is food intake, *a* is the area successfully searched, *v* is the swimming speed, *c* is the food concentration and *h* is the time required to capture and consume one calorie of food.

Using this formulation, he showed that bleak can operate with an energy surplus at swimming speeds between 25 and 210 m h^{-1}. He then showed that at the observed swimming speed (107 m h^{-1}) growth was optimized, whereas growth efficiency, and growth efficiency per unit ration were not optimized.

Ware (1975) next used the assumption that bleak attempt to maximize their growth rate by foraging at the appropriate speed for each food concentration. A simulation model incorporating all of the assumptions mentioned, indicated that, at the observed food density, the observed growth rate of 5 cal day^{-1} requires a metabolic level of 2·67 times standard. When the food supply is better than that observed, the fish could maintain the same growth rate on 2·0–2·5 times standard metabolism.

10.4 The energy method applied to fish in the River Thames

An energy budget has been compiled for populations of fish in the River Thames at Reading, England. The results have been published in detail elsewhere (Mann 1965) and will be given here only in outline to illustrate the extent and value of the information yielded. Winberg produced a general equation for the resting metabolic rate of freshwater cypriniform fishes other than goldfish, carp, and tench which was $Q = 0.336W^{0.80}$ at 20C. This was checked against the metabolic rate of the four most abundant species in the river in a respirometer with flowing river water at the actual river temperature. The measurements were repeated in spring, summer and winter. There was good agreement with Winberg's equation for bleak (*Alburnus alburnus*), perch (*Perca fluviatilis*) and gudgeon (*Gobio gobio*) but the metabolic rate of the roach (*Rutilus rutilus*) was significantly lower.

The vital statistics of the fish population had been determined by Williams (1963, 1965). For each age group of each species a calculation was then made of the energy of metabolism at the temperature appropriate to each month of the year. The energy of growth was calculated on the assumption that all the observed annual growth increment was made during a summer growing season of 6 months. Both growth and mortality were assumed to occur at a constant relative rate during the summer and were allowed for by a process of integration carried out by computer.

A stock of fish totaling kcal m^{-2} was found to require 705 kcal m^{-2} of food and to have an annual production of 42·6 kcal m^{-2}. Stated in terms of the equivalent biomass, this is a stock of 659 kg ha^{-1} consuming something of the order of 10–15 times its own weight in a year and giving an annual production of about 65% of the mean standing crop. This is a very high population density, one of the highest on record for a river and seldom exceeded in any natural habitat in a temperate region. As one might expect, it is also a slow-growing population so that most of the available food energy is channelled into maintaining a large biomass and comparatively little is available for production. Assuming that the food available is not to be increased, it would be a comparatively straightforward matter to work out a management programme for this stock to give optimum yield. The data for the individual age groups show that ecological efficiency—the steady state ratio of production to food ingested—varies from about 13% in young fish to about 2% in ageing fish. A weighted mean for the roach population is 5·5% and for bleak 6·6% (Table 10.1).

Table 10.1. Ecological efficiency of each year class of roach and bleak in the River Thames From Mann (1965). Ecological efficiency = production/food ingested.

Species	Age-group									
	I	II	III	IV	V	VI	VII	VIII	IX	Overall
Roach	13·8	8·5	7·2	5·3	2·7	3·0	3·3	2·7	3·1	5·5
Bleak	12·9	7·7	6·1	2·9	2·4	1·6	–	–	–	6·6
All species	–	–	–	–	–	–	–	–	–	6·0

If the old, inefficient fish were eliminated from the stocks and the density of the younger age groups were reduced, a substantial improvement in production would be obtained. The improvement might be quite striking, for when Warren *et al.* (1964) doubled the food supply to a population of cutthroat trout, (*Salmo clarki*), by enriching a stream with sucrose, there was a seven-fold increase in production.

10.5 The energy method applied to a marine benthic fish population

MacKinnon (1972, 1973a, b) applied the energy flow approach to the study of an unfished population of American plaice, (*Hippoglossoides platessoides*) on the east coast of Canada. In assessing total energy uptake, he used the balanced energy equation, with the usual components for growth, metabolism, egestion and excretion. The metabolic component consisted of standard metabolism, energy expended on activity and energy of *SDA*. The assumptions made in calculating these values were: (i) standard metabolism can be approximated by low routine metabolism determined in the laboratory (ii) activity metabolism for foraging and swimming is a fraction of standard metabolism at starvation and increases linearly with total metabolic level, and (iii) food is processed with a constant conversion efficiency which is similar for standard and activity metabolism and for somatic and gonad energy storage requirements. An unusual and interesting aspect of the study was measurement and subsequent modelling of summer storage of energy and its utilization for winter metabolism. Equations representing the assumptions listed above are:

$$I = \frac{1}{E_A} (M + G_B + G_G)$$

$$M = M_S + M_A + Q_P$$
$$M_A = a_1 M_S + a_2 (M - M_S)$$
$$Q_p = \{(1 - E_p)/E_p\} (M_S + M_A + G_B + G_G)$$

where: I is energy intake as food; E_A is assimilation efficiency; M is total metabolism, M_S is standard metabolism, M_A is activity metabolism, a_1 and a_2 are constants, G_B and G_G are the amounts of energy storage in body and gonad components respectively, E_P is the ratio of an energy requirement to the energy of food that must be processed to meet that requirement, and Q_p is the amount of heat released during food processing.

Using this energy model, MacKinnon (1973b) fed into it (i) field observations on population dynamics and environmental temperatures, (ii) laboratory analyses of tissues which illustrated summer storage of energy and its use for winter metabolism and gonad maturation and (iii) laboratory measurements of low routine metabolism (Fry 1971) at a variety of temperatures and body sizes. The output of the model enabled him to plot on an annual basis (Fig. 10.5) the variation in food intake, metabolism, body and

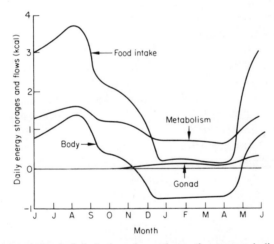

Fig. 10.5. Seasonal variation in daily body and gonad growth rate, metabolism and food intake of a mature age 11 American plaice, (*Hippoglossoides platessoides*) female. From MacKinnon (1973b).

gonad growth for each age class of fish. Utilization of stored body energy for winter metabolism is clearly shown, and was verified in nature. The final figure for annual food consumption by the population, 12·3 kcal·m⁻² is about half the estimated yearly production (25 kcal·m⁻²) of the invertebrate food populations. This is a reasonable figure, considering that the fish are the major resident population, but are sharing the food resource with other migratory groundfish species such as cod and haddock. The food conversion efficiency of the population as a whole was calculated to be 16·7%.

The advantage of such a formulation is that it enables a total picture to be obtained of the feeding and growth strategy of all age classes at all times of year. While it may lack precision, on account of the assumptions made, it gains enormously in generality and power as a tool for management. Using this model, MacKinnon was able to make predictions about a variety of harvesting strategies that could be applied to the population, and was able to compare the natural growth strategies of populations of the same species studied in a variety of situations on both east and west sides of the North Atlantic.

10.6 Comparison of methods used and suggestions for further work

In the previous review of this subject (Mann 1967) it was stated that all the viable methods of estimating food consumption by fish in nature were based on extrapolations from laboratory data. However, the careful and detailed analyses of stomach contents exemplified by the work of Swenson and Smith (1973) and Staples (1975) give reasons for modifying this statement. As empirical methods for determining food consumption in specific conditions, the methods used by these authors appear to be adequate, if laborious. The resulting estimates embody fewer assumptions than indirect estimates involving extrapolation from laboratory conditions. However, they lack generality, for there have emerged from them no rules to indicate how food consumption might be expected to change with changing conditions. In fact, Swenson and Smith (1973), after a great deal of work, arrived at estimates for only one year class for four months of the year.

Similar remarks apply to attempts to extrapolate from laboratory feeding experiments. Any relationship between food and growth requires many days to establish one point for one temperature, size of fish and level of feeding. On the other hand, models involving estimates of metabolism can be related in short-term experiments, to variations in temperature, activity, body size and feeding. It is significant that several authors (Elliott 1973, 1975a, b, Solomon & Brafield 1974, Small 1975) who used extrapolation from laboratory feeding experiments turned to Winberg's formulations when attempting to generalize their results.

Winberg's formulation has stood up reasonably well to critical review almost 20 years after it was proposed. Evidence obtained by other methods suggest that fish in nature do indeed take up food at a rate which is at some level not too far above or below the figure given by Winberg II. The exact level depends on the amount of energy expended by the fish in foraging for food, which is in turn determined by the abundance and particle size of that food. What is now needed is a method of refining the estimate for a particular situation in nature. Those who extrapolate from laboratory studies on food and growth would argue that the growth rate of fish in nature is an indication of their level of food availability and hence of their food consumption. However, the models of Kerr (1971b, c) and Ware (1975) indicate that foraging activity is likely to be a significant item in determining efficiency of utilization of food for growth. Hence, unless the foraging activities are faithfully reproduced in the laboratory, extrapolation to field conditions from static feeding experiments should be made only with appropriate qualifications.

10.7 Food requirements in relation to food available

From the various estimates of the food requirements of natural populations the same conclusion has emerged: that the food stocks are very intensively grazed. Allen (1951) estimated that the trout of the Horokiwi stream required an annual diet of benthos which was 40–150 times the amount present at any one time. We have seen that this figure is certainly too high by a factor of 2–3, but even so the gap between annual requirements and amount present is remarkable. Horton (1961) estimated the annual food requirement of the trout as 8·7–25·9 times the mean biomass of the bottom fauna. Both these workers assumed a constant efficiency of utilization of food for growth, of the order of 20%, so that the weight increase of the fish was taken to be about 20% of the weight of the food consumed above the maintenance ration. This is a reasonable figure for a fish in middle life but most of the production of these two trout populations was by fish in the lower age groups where the efficiency of conversion is likely to be nearer 40%. Hence both authors have given an inflated estimate of the food requirements of the populations.

Gerking (1962), on the other hand, may have erred a little on the low side when he estimated that the bluegills of Wyland Lake consumed about 6 times the July standing crop of invertebrates in a summer growing season of 5 months. He avoided the errors inherent in Allen's and Horton's estimates but he extrapolated from aquarium tanks to field conditions with no correction for activity, and we have seen (p. 254) that his fish seem to have been near a resting rate of metabolism. This omission was corrected in Gerking (1972).

The gap between the annual food requirements of a fish population and the amount present at any one time is bridged in a number of ways. Most important is the short life histories of most invertebrates compared with fish, so that there is often a high ratio of production to standing crop. Table 10.2 shows the results of intensive studies of the productivity of various kinds of benthic organisms and indicates that annual production may be as much as 9 times the mean standing crop. A study of the benthos of the Thames, (see

Table 10.2. Factors by which production exceeds standing crop in benthos

Author	Animals	Factor
Lundbeck, 1926	midges	3–4
	tubificids	3
	Chaoborus	4
	molluscs	0·3
	others	2
Borutzsky, 1939	all benthos	1·7
Miller, 1941	littoral midges	8–9
Anderson & Hooper, 1956	midges	4

Edmondson & Winberg 1971, p. 164) suggests that animals taking two years to complete their life histories may have a ratio of production to standing crop of about 2 to 1, annuals may have a ratio of 5 to 1 and those completing more than one generation per year may produce annually about 10 times the average standing crop.

In certain circumstances these figures may be exceeded by a large margin. Hayne and Ball (1956) made a detailed study of the benthos of two artificial ponds for a summer season. A fish population was in one pond for half the summer and in the other pond for the remainder. As expected, the biomass of benthos was much lower when the fish were present. The remarkable feature was that when the fish were removed from the first pond, the benthos increased its biomass at a rate up to 18% per day; in a growing season of 150 days it was potentially capable of producing 27 times its own weight of material. In fact the rapid rate of production was not maintained for long because the various populations expanded in a logistic manner until resources became limiting, and they settled down at a high density and a low rate of production. This may well be a model of the situation in nature. Where the benthos is lightly grazed, as in deep parts of a lake below the thermocline, the ratio of production to standing crop may be about 2 to 1 but in the littoral zone of lakes, and in rivers and streams where the fish population is at the maximum level permitted by food resources, the food may be turned over at a rate which leads to annual production being 10 or more times the average standing crop.

Zaika (1970, 1972) analyzed the ratio of production to biomass for a variety of aquatic invertebrates, most of which comprise the food of fish. He expressed his results on a daily basis, i.e. daily production of a population, divided by the biomass present, but he stated that his data were average daily rates measured over a year. It is, therefore, permissible to extrapolate them to annual values, for comparison with the ratios quoted above. He found that aquatic molluscs had a daily rate ranging from 0·0003 to 0·03, that is to say 0·1 to 10·9 on an annual basis. The value of the P/B ratio is closely correlated with length of life (Fig. 10.6) and Zaika's figures are in fair agreement with those quoted earlier. For example, a P/B ratio of 2 for animals living 2 years, or a ratio of 5 for annuals, would fit well with the trend shown in the figure.

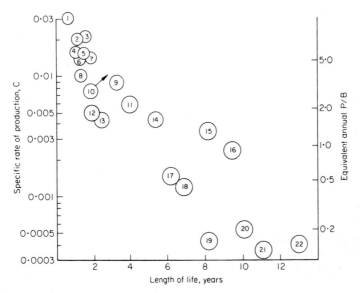

Fig. 10.6. Relation between specific daily production, C, and length of life, for 22 species of aquatic molluscs. (From Zaika 1972). For comparison, annual P/B ratios, calculated as $C \times 365$ are indicated on right side. $1 = Lacuna\ pallidula$ (da Costa); $2 = Spisula\ elliptica$ (Brown); $3 = Margarita\ helicina$ (Fabricius); $4 = Anisus\ vortex$ (L); $5 = Gyraulus\ albus$ (Muller); $6 = Valvata\ pulchella$ (Studer); $7 = Adacna\ vitrea$ (Eichwald); $8 = Rissoa\ splendida$ (Eichwald); $9 = Mytilaster\ lineatus$ (Gmelin); $10 = Margarita\ helicina$; $11 = Abra\ ovata$ (Sowerby); $12 = Bithynia\ tentaculata$ (L); $13 = Sphaerium\ corneum$ (L); $14 = Cardium\ edule$ (L); $15 = Acmaea\ digitalis$ (Rathke); $16 = Mytilus\ galloprovincialis$ (Lamarck); $17 = Dreissena\ polymorpha$ (Pallas); $18 = Acmaea\ testudinalis$ (Muller); $19 = Modiolus\ demissus$ (Dillwyn); $20 = Anodonta\ anatina$ (L); $21 = Unio\ tumidus$ (Philipsson); $22 = Unio\ pictorum$ (L).

Support for the same general trend was given in a paper by Waters (1969). He constructed a theoretical model and showed that most organisms would be expected to have a P/B ratio of 3 to 6 over the life span of a cohort. If a cohort lives for 1 year, the annual P/B ratio would be 3 to 6, but if there are two generations in a year the annual P/B ratio would be 6 to 12, and so on.

10.8 Availability of food and selection by fish

The debt which fish biology owes to Ivlev is deepened by his pioneer work on *The Experimental Ecology of the Feeding of Fishes* (Ivlev 1961). Starting from the premise that a knowledge of the total biomass of food available to a population is not the same as a knowledge of its nutritive value since some organisms may be unacceptable or unavailable, he has made a detailed quantitative study of the relationship of fish to their food under experimental conditions. He has shown that when a fish is offered a mixture of food organisms it exhibits a marked preference for certain species. When the preferred species are not available in the required amounts, the total intake by the fish usually declines, although adequate food of other kinds may be available. The preference may also extend to certain size groups of organisms and to organisms that move at particular speeds.

Fish have marked effects on each other when they feed in groups. Among fish of the same species, each tends to consume less food than when feeding alone, even when preferred food is not scarce. It is possible that being in a group conveys some sense of security which results in a lower metabolic rate. A subsidiary effect, observed particularly in active predators, is that the proximity of two or three fish of the same kind stimulates each to increase its rate of feeding, probably in a spirit of competition.

In mixed species groups Ivlev found that the food intake of one species often increased while that of another declined, and these effects were of a greater magnitude than the intraspecific effects. Moreover, in mixed groups some species tended to change their feeding habits so as to cause minimum overlap with competing species, and this was associated with a reduced food consumption. Clearly, it is wrong to generalize about the feeding habits of a species without reference to the food organisms available and the species of fish which might be competing for them. A case in point is the diet of roach in Britain. Various observers have differed in their accounts of the extent to which roach are herbivorous. In the Thames, roach show a marked improvement in growth rate when they become large enough to consume fully grown gastropod molluscs, which are not exploited by any other species (Williams 1963). It seems likely that roach take large quantities of plant food only when they are in strong competition with other species for benthic invertebrates. The adjustments made by fish when in mixed communities may lead to reduced rates of growth by individuals, but they also lead to effective use of resources by the community as a whole.

References

ALLEN K.R. (1951) The Horokiwi Stream: a study of a trout population. *Fish. Bull. N. Z.* **10**, 1–238.

ANDERSON R.O. & HOOPER F.F. (1956) Seasonal abundance and production of littoral bottom fauna in a southern Michigan lake. *Trans. Am. Micros. Soc.* **75**, 259–270.

BACKIEL T. (1971) Production and food consumption of predatory fish in the Vistula River. *J. Fish Biol.* **3**, 369–405.

BIRKETT L. (1969) The nitrogen balance in plaice, sole and perch. *J. Exp. Biol.* **50**, 375–386.

BORUTZSKY E.V. (1939) Dynamics of the total benthic biomass in the profundal of Lake Beloie (In Russian). *Trudy limnol. sta. Kosine*, **22**, 196–218. (Translation supplied by Mich. Dept. Conservation.)

BROWN M.E. (1946) The growth of brown trout (*Salmo trutta* L.) *J. exp. Biol.* **22**, 118–155.

EDMONDSON W.T. & WINBERG G.G. (1971) *A Manual on Methods for the Assessment of Secondary Productivity in Fresh Waters*. International Biological Programme. Blackwell Scientific Publications, Oxford.

EDWARDS R.R.C. (1968) Estimation of the respiratory rate of young plaice (*Pleuronectes platessa* L.) under natural conditions, using zinc-65. *Nature, Lond.* **216**, 1335–1337.

ELLIOTT J.M. (1973) The food of brown and rainbow trout (*Salmo trutta* and *S. gairdneri*) in relation to the abundance of drifting invertebrates in a mountain stream. *Oecologia* (Berl.) **12**, 329–347.

ELLIOTT J.M. (1975a) Weight of food and time required to satiate brown trout, *Salmo trutta* L. *Freshwat. Biol.* **5**, 51–64.

ELLIOTT J.M. (1975b) Number of meals in a day, maximum weight of food consumed in a day and maximum rate of feeding for brown trout, *Salmo trutta* L. *Freshwat. Biol.* **5**, 287–303.

FRY F.E.J. (1971) The effect of environmental factors on the physiology of fish. *Fish Physiology*, **6**, (W.S. Hoar & D.J. Randell, eds.). pp. 1–98 Academic Press, New York.

GERKING S.D. (1952) The protein metabolism of sunfishes of different ages. *Physiol. Zoöl.* **25**, 358–372.

GERKING S.D. (1955) Influence of rate of feeding on body composition and protein metabolism of bluegill sunfish. *Physiol. Zoöl.* **28**, 267–282.

GERKING S.D. (1962) Production and food utilization in a population of bluegill sunfish. *Ecol. Monogr* **32**, 31–78.

GERKING S.D. (1966) Annual growth cycle, growth potential, and growth compensation in the bluegill sunfish in northern Indiana lakes. *J. Fish. Res. Bd. Canada* **23**, 1923–1956.

GERKING S.D. (1972) Revised food consumption estimate of a bluegill sunfish population in Wyland Lake, Indiana, U.S.A. *J. Fish Biol.* **4**, 301–309.

HAYNE D.W. & BALL R.C. (1956) Benthic productivity as influenced by fish predation. *Limnol. Oceanogr.* **1**, 162–175.

HOLLING C.S. (1966) The functional response of invertebrate predators to prey density. *Mem. Entomol. Soc. Can.* **48**, 1–86.

HORTON P.A. (1961) The bionomics of brown trout in a Dartmoor stream. *J. Anim. Ecol.* **30**, 311–338.

IVLEV V.S. (1939) Energy balance in the carp (in Russian). *Zool. Zh.* **18**, 303–318.

IVLEV V.S. (1961) *Experimental Ecology of the Feeding of Fishes*. (Translated from the Russian by Douglas Scott.) New Haven, Yale Univ.

JOHNSON L. (1960) *Studies in the Behaviour and Nutrition of Pike (Esox lucius L.)* Ph.D. Thesis, University of Leeds.

KERR S.R. (1971a) Analysis of laboratory experiments on growth efficiency of fishes. *J. Fish. Res. Bd. Canada* **28**, 801–909.

KERR S.R. (1971b) Prediction of fish growth efficiency in nature. *J. Fish. Res. Bd. Canada* **28**, 809–814.

KERR S.R. (1971c) A simulation model of lake trout growth. *J. Fish. Res. Bd. Canada* **28**, 815–819.

KROGH A. (1916) *Respiratory Exchange of Animals and Man*. Longmans Green & Co., London.

LUNDBECK J. (1926) Die Bodentierwelt norddeutscher Seen. *Arch. Hydrobiol. Suppl.* **7**, 1–473.

MacKinnon J.C. (1972) Summer storage of energy and its use for winter metabolism and gonad maturation in American plaice. *J. Fish. Res. Bd. Canada* **29**, 1749–1759.

MacKinnon J.C. (1973a) Metabolism and its relationship with growth of American plaice, *Hippoglossoides platessoides* Fabr. *J. Exp. Mar. Biol. Ecol.* **11**, 297–310.

MacKinnon J.C. (1973b) Analysis of energy flow in an unexploited marine flatfish population. *J. Fish. Res. Bd. Canada* **30**, 1717–1728.

Mann K.H. (1965) Energy transformations by a population of fish in the River Thames. *J. Anim. Ecol.* **34**, 253–275.

Mann K.H. (1967) The cropping of the food supply. *The Biological Basis of Freshwater Fish Production*, p. 243–257, (S.D. Gerking, ed.). Blackwell Scientific Publications, Oxford.

Menzel D.W. (1960) Utilization of food by a Bermuda reef fish, *Epinephelus guttatus. J. Conseil* **25**, 216–222.

Miller R.B. (1941) A contribution to the ecology of the chironomidae of Costello Lake, Algonquin Park, Ontario. Univ. Toronto Studies, biol. Ser. No. 49 *Publs. Ont. Fish. Res. Lab.* **60**, 7–63.

Morgan R.I.G. (1973) The energy requirements of trout and perch populations in Loch Leven, Kinross. Proc. Roy. Soc. Edinburgh (B), **74, 22**, 333–345.

Muir B.S. & Niimi A.J. (1972) Oxygen consumption of the euryhaline fish aholehole (*Kuhlia sanvicensis*) with reference to salinity, swimming and food consumption. *J. Fish. Res. Bd. Canada* **29**, 67–77.

Paloheimo J.E. & Dickie L.M. (1966) Food and growth of fishes III. Relations among food, body size and growth efficiency. *J. Fish. Res. Bd. Canada* **23**, 1209–1248.

Pandian T.J. (1967a) Intake, digestion, absorption and conversion of food in the fishes *Megalops cyprinoides* and *Ophiocephalus striatus. Mar. Biol.* **1**, 16–32.

Pandian T.J. (1967b) Transformations of food in the fish *Megalops cyprinoides*. I. Influence of quality of food. *Mar. Biol.* **1**, 60–64.

Pandian T.J. (1967c) Transformations of food in the fish *Megalops cyprinoides* II. Influence of quantity of food. *Mar. Biol.* **1**, 107–109.

Pandian T.J. (1970) Intake and conversion of food in the fish *Limanda limanda* exposed to different temperatures. *Mar. Biol.* **5**, 1–17.

Pentelow F.T.K. (1939) The relation between growth and food consumption in the brown trout (*Salmo trutta* L.) *J. Exp. Biol.* **16**, 446–473.

Prinslow T.E. & Valiela I. (1974) The effect of detritus and ration size on the growth of *Fundulus heteroclitus* (L.) *J. exp. mar. Biol. Ecol.* **16**, 1–10.

Small J.W. Jr. (1975) Energy dynamics of benthic fishes in a small Kentucky stream. *Ecology* **56**, 827–840.

Solomon D.J. & Brafield A.E. (1974) The energetics of feeding, metabolism and growth of perch (*Perca fluviatilis* L.). *J. Anim. Ecol.* **41**, 699–718.

Staples D.J. (1975) Production biology of the upland bully *Philypnodon breviceps* Stokell in a small New Zealand lake. III. Production, food consumption and efficiency of food utilization. *J. Fish Biol.* **7**, 47–69.

Surber E.W. (1935) Trout feeding experiments with natural food (*Gammarus fasciatus*). *Trans. Am. Fish. Soc.* **65**, 300–304.

Swenson W.A. & Smith L.L. Jr. (1973) Gastric digestion, food consumption, feeding periodicity, and food conversion efficiency in walleye (*Stizostedion vitreum vitreum*). *J. Fish. Res. Bd. Canada* **30**, 1327–1336.

Ware D.M. (1975) Growth, metabolism and optimal swimming speed of a pelagic fish. *J. Fish. Res. Bd. Canada* **32**, 33–41.

Warren C.E., Wales J.H., Davis G.E. & Doudoroff P. (1964) Trout production in an experimental stream enriched with sucrose. *J. Wildl. Manage.* **28**, 617–660.

Warren C.E. & Davis G.E. (1967) Laboratory studies on the feeding, bioenergetics and growth of fish. *The Biological Basis of Freshwater Fish Production*, (S.D. Gerking, ed.), pp. 175–214. Blackwell Scientific Publications, Oxford.

WATERS T.F. (1969) The turnover ratio in production ecology of freshwater invertebrates. *Am. Nat.* **103**, 173–185.

WILLIAMS W.P. (1963) *A Study of Freshwater Fish in the Thames*. Ph.D. Thesis, University of Reading.

WILLIAMS W.P. (1965) The population density of four species of freshwater fish, roach (*Rutilus rutilus* (L.)), bleak (*Alburnus alburnus* (L.)), dace (*Leuciscus leuciscus* (L.)), and perch (*Perca fluviatilis* (L.)) in the river Thames at Reading. *J. Anim. Ecol.* **34**, 173–185.

WINBERG G.G. (1956) *Rate of Metabolism and Food Requirements of Fish*. Fish Res. Bd. Can. Transl. Ser. 194 (from: Intensivnost obmena i ouscgevte oetre vristu rtv, Baycgbte Trudy Belorusskovo Gosudarstvennovo Universiteta imeni. V. I. Lenina, Minsk.)

ZAIKA V.E. (1970) Relationship between the productivity of marine molluscs and their life span. *Oceanology* **10**, 547–552.

ZAIKA V.E. (1972) *Specific Production of Aquatic Invertebrates*. (In Russian). Naukova Dumka Publishers, Kiev. 147 p.

COMPETITION AND SOCIAL BEHAVIOUR INFLUENCING PRODUCTIONS

This series of chapters takes a fresh look at several types of behaviour and gains some new insights into their relations to production. Two reasons motivate a broad coverage of competition and social behaviour. First, many biologists believe that growth is to some degree an expression of intra- and interspecific interactions and that a clear elucidation of these interactions would partially explain variations of growth that are observed in nature. Second, there is a strong feeling that the successful culture of certain favoured species will depend on an intimate knowledge of their behaviour in an artificial situation. Both of these viewpoints are strengthened by these chapters. Nevertheless, there are still some types of behaviour that would appear to be adapted for more efficient feeding but for which no substantiating data exist. The image of schooling, for example, as a more efficient foraging device than these same individuals might achieve separately still remains in the realm of theory as it applies to fishes.

Fish growth stands as one of the best examples of a density-dependent process in the animal kingdom. This conclusion has come about from nearly 100 years of growth calculations. It is largely a deductive conclusion, backed up by masses of data collected from natural populations rather than by experimental work. Probably the best examples of the latter come from an increased growth response when a population is thinned and from growth achieved in artificial ponds stocked at different densities. Fishery biologists now rely on growth as a sensitive measure of the general condition of the habitat to support a population. It is disconcerting to confess, however, that the mechanism by which density-dependent growth manifests itself is not always clear. Under certain conditions it seems to be direct response to the amount of food available and under other conditions it seems to be a more complex interplay between behaviour, such as hierarchical or territorial, and food. Our views on this question have not changed substantially in the last ten years, and the first chapter in this series draws on still more recent data to reaffirm the reality of density-dependent growth.

The tremendous thrust of research on size-biased predation of fish on zooplankton has demonstrated the pervasive impact of predation on aquatic communities. Tracing the change in size and species composition of zooplankton in the once fishless lakes of Sweden, stocked for the first time, shows how the biota is influenced by the higher trophic levels. The theoretical ecologist will be delighted to have a clear confirmation of the oft-repeated statement that a change at one trophic level reverberates throughout the ecosystem. Evidence is also advanced that two or more newly introduced fish species practicing size-biased feeding on a common food organism can lead ultimately to their segregation into separate niches.

Our traditional view of migration to a feeding ground as an adaptation to augment production is given added weight from a broadly based review of migration in tropical, temperate and arctic species. In addition, the review results in a new conception and definition of migration because several other motivations for migration besides feeding receive the attention they deserve for the first time. Apparently our previous views have been overly influenced by the dramatic migrations of the salmon, eel, and shad, and the relative scarcity of knowledge about fish life histories in many parts of the world. The evolutionary thread running through the discussion lends added significance to the conclusions.

We are forced to admit that the study of social behaviour in fishes is lagging behind that of the higher vertebrates. Bird behaviourists, for example, have highly developed ideas about how behaviour, evolution and ecology meld together. These same ideas as applied to fish behaviour should pay off handsomely, since fishes exhibit the basic patterns of vertebrate behaviour, such as hierarchies, dominance-subordination, territoriality and schooling. This field should be fertile ground for research from both evolutionary and practical standpoints. Our chapter on social behaviour shows that this has only begun.

Chapter 11: Some Density Relationships for Fish Populations Parameters*

Tadeusz Backiel

and

E. David Le Cren

11.1 Introduction

The very definition of the biological production of fish populations implies consideration of two basic parameters: growth of individuals and their number. They both vary in every known population; thus, production should also vary but there are certain limits to these variations. While extreme limits of growth are rather species specific, though primarily dependent on food supply, the obvious upper limit for numbers is space which can harbour not more than a certain quota. The space available for a population must have the properties which allow the population to live there. If these properties change, we may say that the available, or the best available, space changes, too. When the density, i.e. number or biomass per unit area, varies, the area, or space, for each individual varies accordingly. But how do the other characteristics of populations closely related to the production process behave if population density changes within the extreme limits? This is the question to which we would like to draw the reader's attention here.

Our aim is to illustrate several interrelationships among feeding, growth, mortality, production and density in, first of all, freshwater fish. We will also try to show some comparisons and a few possible interpretations. We do not pretend to have made an exhaustive review of literature for all relevant information on fish production; a selection of those data most readily available and known to us have been made. We are especially conscious that there remains hidden in the published and unpublished literature many data which with a little more analysis and processing could yield valuable information of fish production processes.

* This is a revised version of our paper published in the book *The Biological Basis of Freshwater Fish Production* (1967) (Shelby D. Gerking, ed.), Blackwell Scientific Publications, Oxford). The outline and the main ideas remain the same although a few new facts are referred to here and some conclusions are modified. This version was accomplished during the service of the first author at the Institute of Biology and Geology, University of Tromsø, Norway. We are indebted to this most northern university in the World for the creative atmosphere and most helpful attitude.

11.2 Food production and feeding

The relationship between fish as predators and other organisms is analogous to the relationship between fishermen and fish. As the population density of the fish and, hence, their food consumption increases, the amount of food consumed per unit feeding effort (catch per unit effort) will tend to decrease.

Ivlev (1955, 1961), by introducing the concepts of hunger and of maximum ration* was able to mathematically describe the results of his experiments in which the density of food varied, viz.:

$$\text{Food ration} = \text{Maximum food ration} \ (1 - e^{-kp})$$

where p is food concentration or density and k is a coefficient. These concepts, modified to allow for varying fish density and for the dynamics of food organisms, were incorporated into the models of the dynamics of plaice (*Pleuronectes platessa*) and haddock (*Melanogrammus aeglefinus*) populations by Beverton and Holt (1957) and were found to fit empirical data. However, in natural conditions the relationship between fish density and food organisms is more complicated.

In lakes, experiments with the removal and then re-introduction of fish populations while populations of invertebrate fauna were being observed, have been carried out by Hayne and Ball (1956) and by Macan (1966). Macan shows a selective effect of predation on different species of prey animals and also changes in the distribution of some species in relation to the vegetation cover which protects them from predation by fish. He concludes that some species, important as fish food, must have increased their production after the introduction of fish, because although many were evidently eaten by the fish the same numbers were present as adults as there were before fish were introduced.

The role of hiding places for prey organisms was confirmed in experiments by Ware (1972) who observed the number of successful attacks of rainbow trout, *Salmo gairdneri*, on prey organisms. He also confirmed the significance of the concept of hunger and emphasized the role of size of prey and learning by the predator to hunt and capture the prey.

Observations on the numbers of food organisms present in ponds stocked with different densities of carp (*Cyprinus carpio*) have been reported by Grygierek (1962, 1965), Wojcik-Migala (1965) and Grygierek and Wolny (1962). Ponds with dense fish stocks (15,000 to 30,000 fish/ha) had 2 to 4 times the abundance of planktonic crustacea than was present in unstocked ponds. There was also a difference in the proportions of the species present; ponds

---

* Maximum ration is analogous to maximum catch by a fishing gear and hunger can be compared to the 'free space' in a gear: the degree of gear saturation follows the same mathematical formulation as that of food ration (Kennedy 1951). These concepts, however, have not been incorporated in the most common models of the dynamics of fish populations.

without fish contained the larger cladocera, whereas the presence of fish encouraged smaller cladocera and copepods. The fecundity of *Daphnia longispina* was higher in ponds with dense fish stocks. Similar, but rather less marked, relationships were found between the densities of benthos, particularly chironomid larvae, and fish density. The influence of fish predation on the population dynamics of zooplankton and benthos has also been studied by Hrbáček and colleagues (e.g. Hrbáček *et al.* 1961, Straškraba 1965, Lellak 1965) and it is clear from these papers as well as the experimental work of Slobodkin (1962) that many organisms preyed on by fish can respond to the increased mortality by increased production, and further that the more rapid turnover of essential nutrients resulting from egestion and excretion by fish can stimulate primary productivity.

There are also important direct relationships between fish density and the feeding behaviour of the fish themselves. This has been effectively explored experimentally by Ivlev and his colleagues (Ivlev 1955, 1961). In studies on the food ration eaten by carp and other species and the interaction due to the presence of other individuals (over and above direct competition due to the total ration being shared among a greater number of mouths). Ivlev showed that while there was normally a reduction in the ration, the presence of other fish could sometimes stimulate feeding. Nikol'skii (1955) states that plankton-eating fish feed more intensively when in schools than when dispersed; in fish feeding on benthos a reversed situation was observed.

The role of schooling of prey was mathematically explored by Brock and Riffenburgh (1960). They conclude that this phenomenon may reduce predation on the schooling organisms. Radakov (1973) in his analysis of schooling of fish emphasized its defensive role in some species but also, with predators, the significance of schools in organizing the search for food. Thus, behaviour of both prey and predator, usually species specific, determines the relationship between the 'number' of food items and the success of feeding. For example, Ivlev (1955, 1961) showed that the food ration of fish increased when the distribution of food organisms, in this case sedentary, was changed from uniform to clumped.

LeCren (1965) postulates that some of the reduction in growth rate he found with increase in the population density of young trout may be due to the increased time spent in defence of territories detracting from feeding activity. A number of examples concerning the behavioural aspects of feeding and their interpretation are also given by Braum and by Nilsson in this book.

Other papers in this volume have discussed the digestion and assimilation of food, and here we will only point out that the density of feeding fish can affect the rate of feeding and the efficiency of utilization of that food. Certainly it is an unwarranted assumption to consider that there is a certain rate of production of food organisms that can be merely divided by the number of fish in order to ascertain the ration eaten by each fish; the situation is more complex due to interaction among the organisms involved.

11.3 The effects of population density on growth

The determination of age and growth in fish from scales has been popular for the last fifty years so there is a considerable body of information available on the growth rates and the effects on growth rate of various factors including population density, though the latter has rarely been measured with anything like the same accuracy as the growth rate.

First the experimental work that has shown conditioning effects on growth connected with population density should be mentioned. Allee *et al.* (1940) have demonstrated a 'growth promoting substance' that can be derived from the skin of goldfish (*Carassius auratus*) and Rose and Rose (1965) have shown growth depression in tadpoles living in conditioned water. Pfuderer *et al.* (1974) were able to partially purify the 'crowding factor', a hormone inhibiting growth and reproduction and depressing the heart rate, secreted by goldfish and by carp in small bodies of water.

After demonstrating a negative correlation between abundance and growth in the sockeye salmon (*Oncorhynchus nerka*) in lakes with different productivities, Johnson (1965) concludes that a space-limiting or conditioning factor could be involved.

However, Meske (1973) grew common carp up to a density of 1 kg fish per 3 l in aquaria without any adverse effects. He provided a high flushing rate: the total volume of water changed every 8 minutes.

Parker (1973) measured metabolism in 14 species of fish and found that 13 of the species showed significantly lower oxygen consumption in groups than in isolation. This was explained partly by hydrodynamic advantages of schools and partly by a 'calming effect' of groups in non-schooling species.

The experimental work of Brown (1946) showed the importance of hierarchies in the growth rate of brown trout (*Salmo trutta*) in aquaria. In streams these hierarchies might not occur, however, (Kalleberg 1958). One of us (Backiel 1954) stocked 7 ponds, each with two groups of carp fry. The groups, distinctly different in size, were selected from the same original stock. After 3 or 4 months, the difference in size almost disappeared in spite of the various densities and the presence or absence of older carp; apparently there were no hierarchy effects. LeCren in one small experiment found that the presence of small numbers of yearling trout depressed the growth of fry of the same species, but further experiments are needed to confirm this finding.

Whatever mechanisms may be involved in the effects of varying density on growth the relationship has been well known to fish culturists for centuries, and we present here just one example from Walter's (1934) experiments (Fig. 11.1). Beverton and Holt (1957) say that 'the variation of growth with density in fish populations is perhaps the best established of the density-

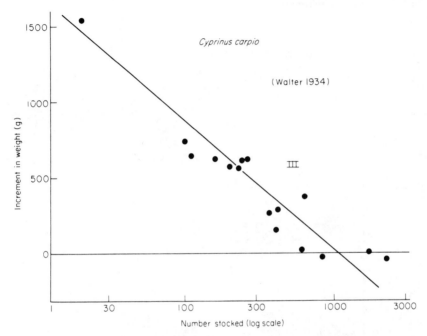

Fig. 11.1. Effects of population density (number stocked per hectare) on individual weight increments in carp (*Cyprinus carpio*) during its third year of life in ponds. (After Walter 1934.)

dependent effects which we consider in this paper . . . ' They refer to a review of this subject made by Hile (1936). Their analysis of data on plaice and haddock in the North Sea confirmed the existence of an inverse growth-density relationship.

The growth of haddock could quite often be reasonably described by a direct inverse relationship between maximum length (L_∞ in the von Bertalanffy growth equation) and the density of the stock in numbers (N). For plaice a better relationship was found between L_∞ and the biomass of the stock (B). Very roughly Beverton and Holt's data show that a ten-fold change in population numbers for haddock can result in a three-fold change in growth rate. For plaice a two-fold change in density causes a 50% change in growth rate. These are very rough approximations, and are about the maximum found in the North Sea. Growth rate in this case is described by L_∞, but this may be roughly proportional to the average annual increment of growth in length by individuals. The perch (*Perca fluviatilis*) in Windermere showed a change of the same order as that of the haddock; a reduction to 3 % of the original density in numbers was accompanied by four-fold increase in annual weight increment, resulting in a biomass reduction to about 10 % of the original (LeCren 1958).

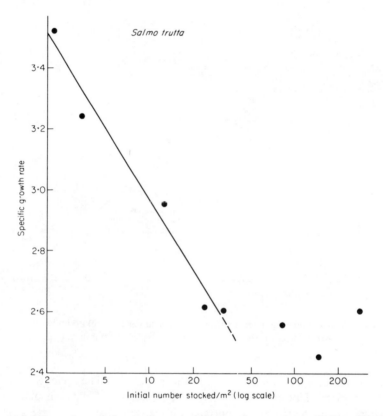

Fig. 11.2. Effects of population density on specific growth rates of trout (*Salmo trutta*) in a screened natural beck. (After LeCren 1965.)

Walter's (1934) data relating average individual weight increment to number stocked exhibit a direct and inversely proportional relationship between the increment and the logarithm of population density (Fig. 11.1). The data of LeCren (1962, 1965 and unpublished) for young brown trout frequently yielded a directly proportional relationship between specific growth rate and the logarithm of population density (Fig. 11.2). One typical relationship obtained for the specific growth rate (G) between April and September was:

$$G = 4 \cdot 5 - 0 \cdot 8 \log N$$

where N was the initial number of trout per square metre. This relationship held only for population densities up to about $50/m^2$; above this density there was no further decrease in growth. High density-dependent mortality probably reduced populations to below $50/m^2$ before any growth occurred. In similar later experiments under rather different conditions of environmental productivity the relationship was not so consistent nor quite the same, but there was always a depression of growth with increasing density.

The form of the relationship between growth rate and population density implies that variations of density (N) at a low level brings about much greater growth changes than the variations when the abundance is high. One may expect that fish populations in many natural environments reach high levels of density near the carrying capacity. That is, perhaps, why a quantitative relationship between the two parameters is not often clear cut; the lack of precision in the estimates of abundance contributes to the difficulty.

Changes in growth rate do not always accompany changes in abundance in short-term studies or in studies pursued in insufficient detail for slight effects to be noticed. LeCren (1949) first reported that a great reduction in the population of perch in Windermere had led to no significant increase in growth rate, but a more prolonged and detailed study (LeCren 1958) showed that the earlier work had overlooked the early stages of what was eventually to be a dramatic increase. This study, too, revealed the importance of taking into account the influence of other complicating factors such as the year-to-year variations caused by temperature fluctuations.

Zawisza (1961) found considerable 'growth stability' in studies of 8 species in lakes in northern Poland in spite of intensive exploitation leading to considerable changes in population density. A major increase in growth did occur, however, after serious population reduction from 'winter kill' in two shallow eutrophic lakes. The subsequent catches from these lakes showed that the population quickly reverted to its original density and growth rate. Faster growth is often observed in newly formed reservoirs (e.g. Sebentsov 1953, Driagin *et al.* 1954). A decrease in population density with the expansion of the area of water is not the only factor involved; often a temporary high rate of general productivity results from the decomposition of newly submerged terrestrial vegetation as well. However, after several years of the reservoir's existence the growth rate of fish falls down to a level observed before damming (Ostroumov 1956).

In observations on the growth of young perch in Windermere (LeCren 1958) very little correlation between brood strength and growth was found. If anything, the more abundant broods seemed to have grown faster. That kind of relationship was also found by Lishev and Rimsh (1961) in Baltic salmon (*Salmo salar*). The more abundant year classes grew faster in their first year of river life, but temperature and hydrological regime was favourable during those years. Van Oosten and Hile (1949) observed faster growth of some year classes of whitefish (*Coregonus clupeaformis*) in Lake Erie during years of greater abundance. Nümann (1958) found a similar coincidence in *Coregonus wartmannii* in the Lake of Constance. These parallel changes of growth and abundance may have been due to changes of abiotic factors and of food supply which were influencing both growth and survival in the same direction. Certain fast-growing year classes of pike (*Esox lucius*) in the first year of life exhibited high survival in Lake Windermere (Kipling & Frost 1970) when temperature conditions were favourable.

In some situations, such as the dystrophic perch tarns studied by Alm (1946), where no increase in growth was obtained after population reduction, growth might have been limited by factors other than competition for food, or possibly growth was so slow that a small, but significant increase would go unnoticed. A further complication could arise in such situations where recruitment was severely depressed by cannibalism; any attempt to reduce the population of older fish might then produce an immediate higher survival of young resulting in crowding and stunting similar to that originally present.

Three other observations deserve mention. First, Pentelow (1944) has suggested that the well known slower growth of brown trout in 'acid' waters may be due at least in part to the denser populations correlated with better spawning conditions in such waters. What follows is that, perhaps, certain physico-chemical conditions decrease both growth rate and mortality thus creating seemingly density-dependent growth. Secondly, the effect of changes in the population density of one species may be damped in more complex associations of fish, as far as that species is concerned, by reactions with other species (Zawisza 1961, Parker 1962). Finally, Watt (1959) has implied, as a result of his studies on the smallmouth bass (*Micropterus dolomieui*) in South Bay, Ontario, that the importance of factors other than population density is likely to be greater towards the boundaries of the normal ecological range of the species.

It can be inferred from the few examples quoted above that in less complex communities, as in pond and laboratory experiments, major changes in population density are nearly always accompanied by inverse changes in growth rate. Where it does not occur, some other density independent factors may be involved, and should be sought. We have indicated a number of these. Because the relationship is by no means simple, it follows that when searching for management practices to improve fish growth it is unwise to base these practices on growth studies alone; it is necessary to have a wider understanding of the role of the species considered in the ecosystem.

As a corollary to this, we may say that studies of the environmental influences on fish growth are of little relevance without data on population density. If methods of population estimation had been developed with the same enthusiasm as methods of reading fish scales fifty years ago, our understanding of the factors influencing freshwater fish production would be a good deal further ahead now.

11.4 The effects of population density upon mortality

It is now widely accepted among animal ecologists that some form of regulation occurs in populations, at least at extremes of density, and that this usually involves some kind of density-dependent mortality (e.g. Solomon 1949, Lack 1954, Krebs 1972). Population regulation in fish has been discussed by many authors (e.g. Ricker 1954, Beverton & Holt 1957, Larkin 1963, Cushing 1971 and 1974), who have introduced the concepts of 'reproduction curves' or 'recruitment curves'. In these curves the numbers of resultant recruits or adult fish are plotted against the numbers of parents or eggs from which they were derived. Variations in such curves can take the form of the numbers of fish at one age plotted against the number at an earlier age. Any departure of these plots from the straight line—convex or concave—indicates density-dependent mortality.

The authors quoted above dealt mainly with marine and migratory fishes and so did Johnson (1965) who gave a series of recruitment curves for various parts of the life history of sockeye salmon, the combination of which yielded a curve with a high dome. A similar curve appears to best fit the data on the Vistula 'vimba' (*Vimba vimba*), a migratory cyprinid (Fig. 11.3).

Beverton and Holt (1957) did not find any evidence showing the dependence of natural mortality on density in adult fish populations. Contrary to that statement, we can quote evidence on a brook trout, *Salvelinus fontinalis*, population (Saunders & Smith 1962) the number of which depended on the number of hiding places available. Hence, mortality depended upon the size of the population present. A similar situation exists where trout (usually artificially reared) are added to an existing wild population; there is often an almost complete mortality among stocked fish (Miller 1958). The mortality is presumably caused by the aggressive behaviour of the native stock in defending well established territories.

When species are introduced into waters where they had not previously occurred, they may find a favorable milieu and exhibit low mortality at low initial density. Bernatowicz (1953) introduced *Coregonus albula* alevins into Lake Isag and obtained a minimum survival rate of 14·2% over the first three years. Similar high survivals were reported by Backiel (1965) with pike-perch (*Stizostedion* sp.) in reservoirs and by Burmakin (1963) with bream (*Abramis brama*) in many Siberian waters. In these cases, to which also contradicting examples can be added, the young fish rather than adults responded to the low density.

The evidence for density-dependent mortality is more clearly demonstrated

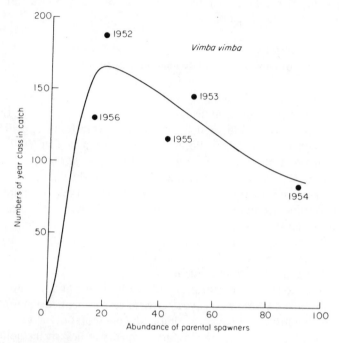

Fig. 11.3. Numbers of spawners of *Vimba vimba* from Vistula River versus resultant abundance of year classes; reproduction curve drawn free-hand. (Data of Backiel 1966 and unpublished data.)

in young fish rather than adults (LeCren 1965). Data on young bream plotted against potential numbers of eggs laid (Fig. 11.4). show that very low and very high 'reproduction potential' did not produce the largest numbers of the young, thus indicating some dependence of mortality among eggs, larvae and fry on their densities. The abundance of a year class in the catchable stock is, however, determined at early stages of that year class, since the year class strength is positively correlated with the numbers of young of the same year class caught per unit effort (Fig. 11.5 and Monastyrsky 1952).

Experimental populations of young trout showed a recruitment curve rising rapidly to a horizontal asymptote at a maximum density of about 9 fry/m² (LeCren 1965). It is evident that this is due to aggressive territorial behaviour at the stage when the fish are first feeding. Excess fry, which do not manage to obtain territories, die of starvation, and the mortality is strongly density-dependent above quite low population densities (Fig. 11.6). In these experiments the coefficient of mortality (Z) appeared to be directly proportional to the logarithm of initial population density. In LeCren's experiment the relationship is approximately:

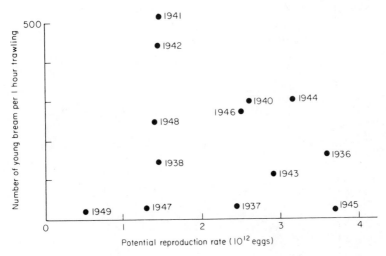

Fig. 11.4. Relative abundance of young bream (*Abramis* sp.) in Volga estuary versus potential number of eggs deposited in the same year. (Data from Zemskaia 1961.)

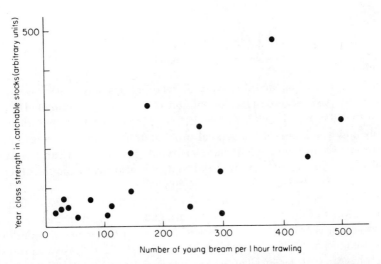

Fig. 11.5. Year class strength in catchable stock of the bream (*Abramis* sp.) of Volga estuary versus relative abundance of young of the same year class. (Data from Dementeva 1952, and Zemskaia 1961.)

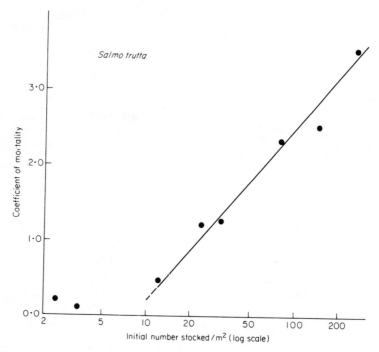

Fig. 11.6. Effects of population density on instantaneous mortality rate of trout (*Salmo trutta*) in a screened beck. (After LeCren 1965.)

$$Z = (2 \log n) - 1 \cdot 5$$

where n = the number of newly hatched fry/m².

Similar territorial and behavioural relationships have been found by Chapman (1965) for young coho salmon (*Oncorhynchus kisutch*). He also found that the size of the territories depended to some extent upon the food supply and has discussed the general problem of food and space as limiting factors for young salmonids (Chapman 1966). The part that can be played by physical objects that prevent visual contact is also clear from the observations of Stuart (1953) and Kalleberg (1958).

Experimental data on the survival of non-salmonid freshwater fish fry are rare. Most of the experiments with carp and other pond species start with fish at least a year old. However, Wolny (1962), starting with one-month-old carp, obtained no clear relationship between mortality and density. Zuromska (1967) has studied the survival of the larvae of the roach (*Rutilus rutilus*) in natural lakes, experimental ponds and aquaria. The mortality of the larvae was high and largely due to predation by invertebrate animals, but there

was no evidence of any density-dependent relationships even though observations had been made over several years and a wide range of egg
densities.

Although the reproduction curves and some experiments indicate a
direct positive relationship between mortality and density, there are cases
when an inverse relationship occurs. Johnson (1965) found this in his studies
on sockeye salmon and Forney (1971) quotes a few other cases of this type
in the discussion of his own data on yellow perch (*Perca flavescens*). The
juveniles of the perch showed a greater mortality when they were less abundant; mortality coefficients and indices of abundance were negatively correlated ($r = -0.67$). Such a depensatory mortality occurs where a relatively
constant number of predators feed on schools of young fish. Similar relationships can exist between fishing mortality and density of a catchable stock:
e.g. in the migrating vimba (Backiel 1966). The fraction taken by the fishery
was greatest at the lowest abundance of the stock and vice versa.

If such a depensatory mortality occurs at some stages of the life history
then a compensatory mortality must occur at another stage; otherwise the
population would become extinct (Forney 1971).

It is clear from the few quoted examples that any density-dependent
effect is either complex or obscured by the effects of other factors. This is
also the case with the perch in Windermere (LeCren 1955) where no
definite shape for a reproduction curve can yet be determined. There is a
suggestion, however, that the average survival rate of young perch (from
egg to two years of age) may have been less after the experimental reduction
in adult population. Perhaps, if this really is so, it can be explained by the
increased growth rate allowing more adults to reach a size at which they can
prey on their own young, or that this cannibalism lasts for a longer period
of juvenile life. The year class strength of pike in the same lake correlated
well with the number of degree-days during the first year of life and no
density-dependence was discovered (Kipling & Frost 1970). As with the
growth effects the effect of density on mortality can be damped by interactions
with other species.

An example of inter-specific density effects was provided by a small
experiment with mixed populations of young trout and salmon. The mortality
coefficient of the salmon appeared to be proportional to the total population
density of salmon plus trout, whereas the mortality coefficient for the trout
was not influenced by the numbers of salmon present (LeCren, unpublished
data).

Cushing (1974) states that the numbers of fish in the sea are probably
regulated during larval stages by a combination of density-dependent
mortality and density-dependent growth. This may also be true in fresh
waters, but with the exception of experiments on young trout the direct
evidence is confusing.

11.5 The effects of population density upon production

The true production of natural fish populations has been measured on only a few occasions, and data on the effects of population density upon production are even rarer.

With varying density and with constant growth and mortality rates, production should vary in proportion to the density. Allen (1951) divided the Horokiwi River into 6 sections for the purpose of studying brown trout production and observed different population densities, growth rates and production estimates for each section. The production per unit area in each of the 6 sections was roughly proportional to the biomass present per unit area in the sections.

Hunt's (1966) study of the production of brook trout covered in detail 3 year classes in 2 sections of Lawrence Creek, Wisconsin. A plot of the total production of each year class in each section against the estimate of the population density of the year classes on emergence from the gravel shows a remarkably constant proportional relationship. The production of trout in Lawrence Creek was thus directly proportional to the population density.

In these two cases production was not dependent on density in quite the same sense discussed in the previous sections, because production is measured as a balance between growth and mortality.

Ricker and Foerster (1948) estimated the production of young sockeye salmon in Cultus Lake for 9 year classes before and 3 year classes after predator control. An analysis of their data (LeCren 1962) shows that the production was roughly proportional to the 0·8 power of the biomass of eggs giving rise to each year class. Here there was an indication of density dependence, and it probably operated through an inverse effect on growth rate and thus mortality as well (Johnson 1965).

LeCren (1962, 1965 and unpublished data) has studied experimental populations of young trout in a small soft-water stream in the English Lake District. The effect of increasing the population density was to increase the mortality rate and decrease the growth rate (as explained above), and the result was a rather constant biomass and production per unit area, regardless of population density above a certain minimum (Fig. 11.7). It is apparent that the general form of this relationship was determined early in life, within 40 days of the time when the fry started to feed. It is also clear that space was the factor of overriding importance; under the conditions of the experiment, young trout could produce 12·6 gm^{-2} by September and no more (Fig. 11.8). Below a density of about 8 fish m^{-2} this potential could not be reached. Rough estimates of the food consumed by these populations, based on Winberg's method (Winberg 1960) gave remarkably constant results for all but one of the 6 denser populations. Other experiments similar to the one described here gave similar results.

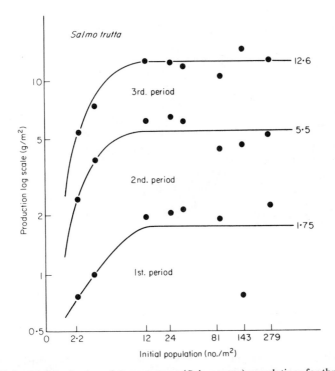

Fig. 11.7. Sums of production of O group trout (*Salmo trutta*) populations for three 40-day periods between April and September versus initial numbers of the fish (on log-scales). (Data of LeCren.)

This rather constant production per unit area is almost certainly due to the aggressive territorial behaviour of the brown trout fry soon after emergence from the gravel. Many of the fry that died in the first 3 weeks did so before making any growth or eating any significant quantity of food. They died of starvation, but in a fully natural situation many would probably have emigrated downstream to find territories or, if unsuccessful, to be eaten by predators in a moribund condition. Territorial behaviour thus acts as a population controlling mechanism before any trophic demands have been made on the environment.

These observations and experiments with brown trout indicate that Allen (1951) and possibly other authors, have overestimated juvenile production; most of the high mortality observed in the first 3 or 4 months occurs in the first 2 or 3 weeks, and little or no production should be attributed to these fish before they die.

These data for brown trout conflict with those obtained by Hunt (1966)

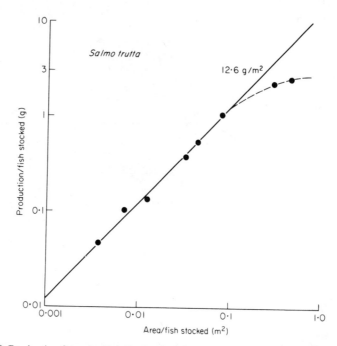

Fig. 11.8. Production from April to September of O group trout per fish initially stocked versus the area of stream available per fish initially stocked (on log-scales). (Data of LeCren.)

for brook trout where there was no apparent 'flattening out' of production at high population levels. Perhaps the brook trout does not possess the same rigid territorial behaviour, or perhaps other factors hold the population density in Lawrence Creek below a level at which density effects begin to operate in this rather productive water. Chapman (1966) has observed territorial population limitation among young coho salmon, and he obtained an increase in the density of population with an increase in food supply.

Another set of data on the relationships of production to population density is provided by the literature on pond culture. Among older carp, at least, the rate of mortality in ponds is so low that the yield of carp obtained when the ponds are drained is at least proportional to and often virtually identical with production in the ecological sense in which we are using this term. The data published by Walter (1934) can be used as an illustration of the kind of results obtained in several such experiments (Fig. 11.9). Production (yield) rises to a maximum at an optimum stocking density. This optimum density is considerably greater for yearling carp (and the resultant yield

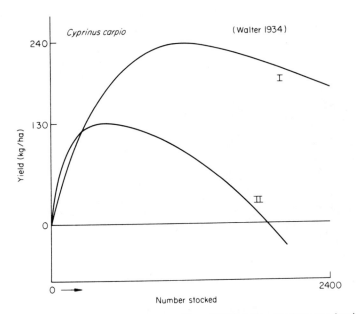

Fig. 11.9. Walter's (1934) schematic representation of the relationship between density of carp (*Cyprinus carpio*) in its second (I) and third (II) year of life in ponds and the annual yield (biomass recovered minus biomass stocked) per hectare.

greater, too) than it is for two-year-olds. Experiments with carp fry carried out by Wolny (1962) have shown a similar curve.

In these pond experiments with carp all, or nearly all, of the changes in production are due to decreases in growth rate with increasing population density (Fig. 11.1). Above the optimum the increase in numbers present cannot compensate for the slower growth of the crowded fish and a decrease in total production results. It has generally been assumed that the decrease in growth is the result of less efficient utilization of the food eaten, since the amount available to each fish becomes little more than that needed for maintenance alone. In some of Walter's experiments with very crowded carp, the fish actually lost weight.

When plotting biomass increments (which we may identify with production here) against the initial biomass of carp, the same initial biomasses of age-group 0 (K_0), age-group I (K_1) and age-group II (K_2) carp resulted in quite different production values; the production of younger and smaller fish was much greater than that of older (Fig. 11.10). According to Winberg's (1960) formula for the maintenance requirement, the biomass of K_1 (consisting of 25 g individuals) required 1·6 times more food for maintenance than an equal biomass of K_2 (270 g). It would seem, therefore, that less food

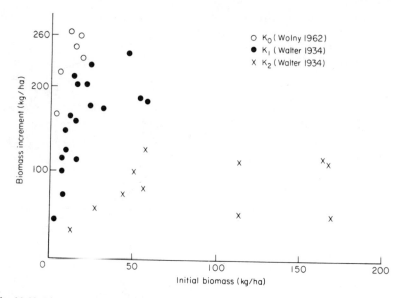

Fig. 11.10. Biomass increments versus biomass stocked in Walter's (1934) and Wolny's (1962) pond experiments with carp.

was left for K_1 growth, but the data show just an opposite situation. One can infer that the effect of numbers of grazing fish was much greater than their biomass as far as production is concerned. An explanation of this trend may be sought in the response of food organisms to grazing which has been shown by Grygierek (1962, 1965), Wojcik-Migala (1965) and others mentioned in the section on feeding.

Variations of production of a fish population in lakes are exemplified by the data on pike in Windermere (Kipling & Frost 1970). Production of adults and that in the second year of life studied over a period of over 20 years, was directly dependent on the year class strength. Changes in the growth rate had only a minor effect on production and, as mentioned in the previous sections, there was no dependence of growth and mortality on density.

We can infer from the few quoted data that whenever growth or mortality or both are density dependent, then production is not directly proportional to numbers or biomass. With young salmonids a rather constant production, over a wide range of densities above a certain minimum, resulted from interaction of antagonistic changes of growth and mortality rates. The domed curves in carp culture were due to density dependent growth only. It is possible to obtain a similar curve when mortality is depensatory and growth rate constant though we do not know of any such situation. If neither growth nor mortality vary with population density, production will mainly depend on the biomass according to the simple formula:

$$P = (\text{Coefficient of growth}) \times (\text{Mean biomass})$$

11.6 Discussion

The rather conflicting results we have outlined concerning the effects of population density on growth, mortality and especially production make it rather premature to attempt to draw any general conclusions from this survey. Nevertheless, it may be worthwhile to try to see how the results observed might have arisen and at least to suggest some framework of hypotheses on which our plans for future work can be hung.

In stream-living salmonids an aggressive territorial behaviour in very early life leads to mortality or emigration before much growth takes place and, as a result, the population density is often regulated below the level at which density effects on subsequent growth or mortality will be very great. Observations and experiments which include this early regulatory phase in the life history will show density-dependent effects whereas experiments starting later in life may not reveal these. With cyprinids and schooling species we do not yet know if and when population regulation through density-dependent mortality occurs, though the effects of density on growth may be obvious enough. Certainly in many temperate waters it would appear to be natural for many species to exist in dense populations with growth rates well below the potential for the species. The equilibrium situation is that of a dense slow-growing population with a low production rate relative to the biomass.

We must also beware of drawing general conclusions from our rather biased experience with the oversimplified situations present in temperate waters with a limited fish fauna. In other waters there is often a much more complex community of fishes the interactions between which may tend to cushion the effects in the population density of only one of them.

Perhaps, however, we can allow ourselves the general conclusion that there are two main density effects operating on fish populations. The first, that operates through mortality, will be strongest in early life, particularly the larval and immediately post-larval stages. This density-dependent mortality may operate through intraspecific competition for a limited resource such as food or space and may even have become 'conventional' as is probably the case with young salmonids. It may be an indirect effect of competition for food operating by lengthening the time taken for the fish to grow through a stage vulnerable to predation. In other situations there may be predation by other organisms or by adult fish of the same species and then depensatory mortality can occur. There will, however, be a tendency for density-dependent mortality to become less important with increasing age and size. Cushing (1974) employed this assumption in his model of population dynamics.

The second effect, the one that operates through growth, will tend to become of increasing importance as the fish gets older. There will be density-dependent growth effects early in life, some of which will affect survival as well, but it is among adult fish that the influence of population density on growth is most apparent if it is not obscured by other factors which can often be the case. These influences on adult growth rate may also affect fecundity, but only rarely will the scale of the effect due to differences in fecundity be as large as those due to density-dependent growth.

A simple, but perhaps sometimes useful approach would be to divide the life span of a fish into two phases: (1) early stages when mortality responds strongly to the changes of population density while growth changes may be neglected and (2) later stages when mortality changes are of minor importance but growth varies with density. Thus, we may consider production of phase 1 (P_g1) as proportional to the steady growth rate and to the varying mortality, and production of phase 2 (P_g2) as:

$$(P_g2) \propto N_2(a - b \log N_2)$$

where N_2 is the number of fish at the beginning of phase 2, and a and b are coefficients. This empirical equation appears to fit very well to Walter's (1934) data for carp (Fig. 11.11). It should be borne in mind, however, that

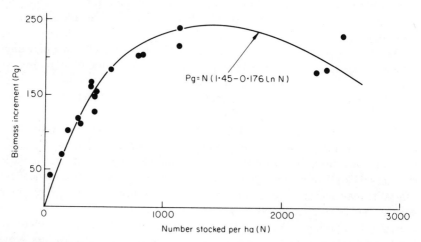

Fig. 11.11. Walter's (1934) data for density-dependent production of carp in second year of rearing and their approximation by the empirical equation:

$$P_g = N(1\cdot4 - 0\cdot176 \ln N)$$

other factors play an important role, e.g., temperature which greatly influences growth (LeCren 1958).

Since a large part of the total production of a fish population is produced by the younger age groups, it is likely that the greatest influence on the amount of production will be the density of such young fish and those factors that limit the survival of eggs, larvae and young. Variations of growth whether dependent on density or on other factors are never as great as those of numbers; thus their effect on production is probably seldom significant in natural populations.

It is clear that our knowledge of population densities and mortality rates lags behind that of growth rate. Many more quantitative data on the young stages and the whole 'recruitment' process are needed. This is the period of maximum rate of production and maximum effect of mortality factors. Information which would provide not only estimates of natural mortality rates but also allow their partitioning between various factors would be extremely valuable.

Another great hope for progress lies in the development of population experiments. A review of the history of fish ecology soon highlights the importance of the experimental approach; either deliberate experiments or the careful long-term observation of the effects on population and other changes produced unwittingly by such factors as fishermen and the weather. The scope for using experimental fish ponds and model streams for such experiments is immense.

Finally, it is to be hoped that simultaneous studies on all the species present can be attempted. There are manifest advantages in concentrating effort on one species at a time, but we believe that interspecific reactions can no longer be neglected in studies of fish populations and production.

References

ALLEE W.C., FINKEL A.J. & HOSKINS W.H. (1940) The growth of goldfish in homotypically conditioned water: A population study in mass physiology. *J. Exp. Zool.* **84**, 417–443.

ALLEN K.R. (1951) The Horokiwi stream. A study of a trout population. *Bull. mar. Dep. N.Z. Fish.* **10**, 1–231.

ALM G. (1946) Reasons for the occurrence of stunted fish populations with special regard to the perch. *Meddn. St. Unders.-o. FörsAnst. Sötvatt. Fisk.* **25**, 1–146.

BACKIEL T. (1954) Changes in the degree of length differentiation in the carp fingerling population (Polish, Eng. & Rus. summ.) *Roczn. Naukroln.* **67B**(4), 481–494.

BACKIEL T. (1965) River fisheries. (T. Backiel, ed.), *Freshwater Fisheries of Poland.* Cracow, Pol. Ac. Sci. Limnol. Committee.

BACKIEL T. (1966) On the dynamics of an intensively exploited fish population. *Verh. int. Verein. theor. angew.* **16**, 1237–1244.

BERNATOWICZ S. (1953) Preliminary analysis of the small whitefish population (*Coregonus albula* L.) introduced into Lake Isag in Mazurian district. (Polish, Eng. & Rus. summ.) *Roczn. Nauk. roln.* **67B**(1), 53–60.

BEVERTON R.J. & HOLT S.J. (1957) On the dynamics of exploited fish populations. *Fishery Invest., Lond.* (II) **19**, 1–533.

BROCK E.V. & RIFFENBURGH R.H. (1960) Fish schooling; a possible factor in reducing predation. *J. Cons. perm. int. Explor. Mer* **25**, 307–317.

BROWN M.E. (1946) The growth of brown trout (*Salmo trutta* L.) I. Factors influencing the growth of trout fry. *J. exp. Biol.* **22**, 118–129.

BURMAKIN E.W. (1963) Akklimatizatsia presnovodnych ryb w SSSR. (Rus.) *Izv. gosud. nauctno-issled. Inst. Ozern. Rechn. Ryb. Khoz.* **53**, 5–316.

CHAPMAN D.W. (1965) Net production of juvenile coho salmon in three Oregon streams. *Trans. Am. Fish. Soc.* **94**, 40–52.

CHAPMAN D.W. (1966) Food and space as regulators of salmonid populations in streams. *Amer. Nat.* **100**(973), 345–358.

CUSHING D.H. (1971) The dependence of recruitment on parent stock in different groups of fishes. *J. Cons. perm. int. Explor. Mer.* **33**, 340–362.

CUSHING D.H. (1974) The possibly density-dependence of larval mortality and adult mortality in fishes. *The Early Life History of Fish* (J.H.S. Blaxter, ed.), pp. 103–111. Springer Verlag, Berlin.

DEMENTEVA T.F. (1952) Metodika sostavleniia prognozov ulovov lescha severnogo Kaspia. (Rus.) *Trudy vses. nauchno-issled. Inst. morsk. ryb. Khoz. Okeanogr.* **21**, 163–182.

DRIAGIN P.A., GALKIN G.G. & SOROKIN S.M. (1954) Usloviia rozmnozheniia i rost ryb v Tsimlianskom vodokhranilishche v pervyi god ego sushchestovovaniia. *Izv. gosud. nuchno-issled. Inst. Ozern. Rechn. Ryb. Khoz.* **34**, 134–155.

FORNEY J.L. (1971) Development of dominant year classes in a yellow perch population. *Trans. Am. Fish. Soc.* **100**(4), 739–749.

GRYGIEREK E. (1962) The influence of increasing carp fry population on crustacean plankton. (Polish, Eng. summ.) *Roczn. Nauk. roln.* **81B**(2), 189–210.

GRYGIEREK E. (1965) The effect of fish on crustacean plankton. (Polish, Eng. summ.) *Roczn. Nauk. roln.* **86B**(2), 147–168.

GRYGIEREK E. & WOLNY P. (1962) The influence of carp fry on the quality and abundance of snails in small ponds. (Polish, Eng. summ.) *Roczn. Naukrolnroln.* **81B**(2), 211–230.

HAYNE D.W. & BALL R.C. (1956) Benthic productivity as influenced by fish predation. *Limnol. Oceanogr.* **1**, 162–175.

HILE R. (1936) Age and growth of cisco, *Leucichthys artedi* (Le Sueur), in the lakes of the North-eastern highlands, Wisconsin. *Bull. Bur. Fish. Wash.* **48**, 211–317.

HRBÁČEK J., DVORAKOVA M., KOŘINEK V. & PROCHÁZKÓVA L. (1961) Demonstration of the effect of the fish stock on the species composition of zooplankton and the intensity of metabolism of the whole plankton association. *Verh. int. Verein. theor. angew. Limnol.* **14**, 192–195.

HUNT R.L. (1966) Production and angler harvest of wild brook trout in Lawrence Creek, Wisconsin. *Tech. Bull. Wis. Conserv. Dep.* **35**, 1–52.

IVLEV V.S. (1955) *Eksperimentalnaya ekologiya pitaniya ryb.* Moskva, Pishchepromizdat.

IVLEV V.S. (1961) *Experimental Ecology of Feeding of Fishes.* New Haven, Yale Univ. Press, viii + 302 pp. (Eng. transl. of Ivlev 1955).

JOHNSON W.E. (1965) On mechanisms of self-regulation of population abundance in *Oncorhynchus nerka. Mitt. int. Verein. theor. angew. Limnol.* **13**, 66–87.

KALLEBERG H. (1958) Observations in a stream tank of territoriality and competition in juvenile salmon and trout (*Salmon salar* L. and *Salmo trutta* L.). *Rep. Inst. Freshwat. Res. Drottningholm* **39**, 55–98.

KENNEDY W.A. (1951) The relationship of fishing effort by gill nets to the interval between lifts. *J. Fish. Res. Bd. Canada* **8**, 268–274.

KIPLING C. & FROST W.E. (1970) A study of the mortality, population numbers, year class strength, production and food consumption of pike (*Esox lucius* L.) in Windermere from 1944 to 1962. *J. Anim. Ecol.* **39**, 115–157.

KREBS C.J. (1972) *Ecology*. Harper & Row Publishers, New York.

LACK D. (1954) *The Natural Regulation of Animal Numbers*. Oxford, Clarendon.

LARKIN P.A. (1963) Interspecific competition and exploitation. *J. Fish. Res. Bd. Canada* **20**, 647–678.

LECREN E.D. (1949) The effect of reducing the population on the growth rate of *Perca fluviatilis*. *Verh. int. Verein. theor. angew. Limnol.* **10**, 258.

LECREN E.D. (1955) Year to year variation in the year-class strength of *Perca fluviatilis*. *Verh. int. Verein. theor. angew. Limnol.* **12**, 187–192.

LECREN E.D. (1958) Observations on the growth of perch (*Perca fluviatilia* L.) over twenty-two years with special reference to the effects of temperature and changes in population density. *J. Anim. Ecol.* **27**, 287–334.

LECREN E.D. (1962) The efficiency of reproduction and recruitment in freshwater fish, *The Exploitation of Natural Animal Populations* (E.D. LeCren and M.W. Holdgate, eds.), pp. 283–296. Blackwell Scientific Publications, Oxford.

LECREN E.D. (1965) Some factors regulating the size of populations of freshwater fish. *Mitt. Verein. theor. angew. Limnol.* **13**, 88–105.

LELLAK J. (1965) The food supply as a factor regulating the population dynamics of bottom animals. *Mitt. int. Verein. theor. angew. Limnol.* **13**, 128–138.

LISHEV M.N. & RIMSH E.I. (1961) Nekotorye zakonomernosti dinamiki chislennosti Baltiiskogo lososia. *Trudy Latv. nauchno-issled. Inst. ryb. Khoz.* **3**, 5–103.

MACAN T.T. (1966) The influence of predation on the fauna of a mooeland fishpond. *Arch. Hydrobiol.* **61**, 432–452.

MESKE C. (1973) *Aquakultur von Warmwasser Nutzfischen*. E. Ulmer, Stuttgart, 163 p.

MILLER R.B. (1958) The role of competition in the mortality of hatchery trout. *J. Fish. Res. Bd. Canada* **15**, 27–45.

MONASTYRSKY G.N. (1952) Dinamika chislennosti promyslovych ryb, *Trudy, vses. nauchno-issled. inst. morsk. ryb. Khoz. Okeanogr.* **21**, 3–162.

NIKOL'SKII G.V. (1955) O biologicheskom znachenii stai u ryb. *Trudy Sovesheh. ikhtiol. Kom.* **5**, 104–107.

NÜMANN W. (1958) Vorläufiger Bericht über stark veränderte Wachstum der Blaufelchen im Bodensee und Versuch einer Fangprognose für das Jahr 1958. *Allg. Fischztg.* **83**(5), 88–91.

OSTROUMOV A.A. (1956) O vozrastnom sostave stada i roste leshcha rybinskogo vodo-khranilishcha. *Trudy Inst. Biol. Vodokhran.* **2**, 166–183.

PARKER F.R. Jr. (1973) Reduced metabolic rates in fishes as a result of induced schooling. *Trans. Am. Fish. Soc.* **102**, 125–131.

PARKER R.R. (1962) A concept of the dynamics of pink salmon populations. *Symposium on Pink Salmon*. Macmillan Lect. Br. Columb. Univ. 203–211.

PENTELOW F.T.K. (1944) Nature of acid in soft water in relation to the growth of brown trout. *Nature, Lond.* **153**, 464.

PFUDERER P., WILLIAMS P. & FRANCIS A.A. (1974) Partial purification of the crowding factor from *Carassius auratus* and *Cyprinus carpio*. *J. Exp. Zool.* **187**(3), 375–382.

RADAKOV D.V. (1973) *Schooling in the Ecology of Fish*. J. Wiley & Sons, New York, Toronto.

RICKER W.E. (1954) Stock and recruitment. *J. Fish. Res. Bd. Canada* **11**, 559–623.

RICKER W.E. & FOERSTER R.E. (1948) Computation of fish production. *Bull. Bingham oceanogr. Coll.* **11**, 173–211.

ROSE S.M. & ROSE F.C. (1965) The control of growth and reproduction in freshwater organisms by specific products. *Mit. int Verein. theor. angew. Limnol.* **13**, 21–34.

SAUNDERS J.W. & SMITH M.W. (1962) Physical alteration of stream habitat to improve trout production. *Trans. Amer. Fish. Soc.* **91**, 185–188.

SEBENTSOV B.M. (1953) Razvitie rybnogo naseleniia v pervyi god sushchestvovaniia vodokhranilishch Wolgo-Donskogo Kanala. *Trudy vses nauchno-issled. Inst. prud. ryb. Khoz.* **6**, 193–230.

SLOBODKIN L.B. (1962) Predation and efficiency in laboratory populations, *The Exploitation of Natural Animal Populations* (E.D. LeCren and M.W. Holdgate, eds.), pp. 223–241. Blackwell Scientific Publications, Oxford.

SOLOMON M.E. (1949) The natural control of animal populations. *J. Anim. Ecol.* **18**, 1–35.

STRAŠKRABA M. (1965) The effect of fish on the number of invertebrates in ponds and streams. *Mitt. Int. Verein. theor. angew. Limnol.* **13**, 106–127.

STUART T.A. (1953) Spawning migration, reproduction and young stages of loch trout (*Salmo trutta* L.). *Freshwat. Salm. Fish Res.* **5**, 1–39.

VAN OOSTEN J. & HILE R. (1949) Age and growth of the lake whitefish *Coregonus clupeaformis* (Mitchill) in Lake Erie. *Trans. Am. Fish. Soc.* **77**, 178–249.

WALTER E. (1934) Grundlagen der allgemeinen fischeilichen Produktionslehre. *Handb. Binnenfisch. Mitteleur.* **4**(5), 480–662.

WATT K.E.F. (1959) Studies on population productivity. II. Factors governing productivity in a population of smallmouth bass. *Ecol. Monogr.* **29**, 367–392.

WARE D.M. (1972) Predation by rainbow trout *Salmo gairdneri*: the influence of hunger, prey density and prey size. *J. Fish. Res. Bd. Canada* **29**, 1193–1201.

WINBERG G.G. (1960) Rate of metabolism and food requirements of fish. *Fish Res. Bd. Canada*, Transl. Ser. 194. 253 pp.

WOJCIK-MIGALA I. (1965) The effect of young carp on the dynamics of bottom fauna. (Polish, Eng. & Rus. summ.). *Roczn. Nauk. roln.* **86B**(2), 195–214.

WOLNY P. (1962) The influence of increasing the density of stocked fish population on the growth and survival of carp fry. (Polish, Eng. & Rus. summ.) *Roczn. Nauk. roln.* **81B**(2), 171–188.

ZAWISZA J. (1961) The growth of fishes in lakes of Wegorzewo district. (Polish, Eng. & Rus. summ.) *Roczn. Nauk. roln.* **77B**(2), 731–748.

ZEMSKAIA K.A. (1961) O vliianii nagula i chislennosti proizvoditelei na velichinu potomstva kaspiiskogo leshcha. *Trudy Soveshch. ikhtiol. Kom.* **13**, 307–313.

ZUROMSKA H. (1967) Mortality estimation of roach (*Rutilus rutlus* L.) eggs and larvae on lacustrine spawning grounds. (Polish, Eng. & Rus. summ.) *Roczn. Nauk. roln. H*, **90**(3), 538–556.

Chapter 12: The role of Size-biased predation in Competition and Interactive Segregation in Fish

Nils-Arvid Nilsson

12.1 Introduction

The profound effect that fish predation has on the composition, size, and many other aspects of population ecology of the prey species has only been appreciated for the last 20 years. The body of new literature that has grown on the subject provides a sound basis for a theoretical explanation of the size and species composition of zooplankton communities but also the species composition of fish communities which use the same food resource. This chapter will discuss both these aspects of predation and competition.

Theoretically the predator fish search for a maximum reward which is expressed as the 'optimal prey selection' (Krebs 1973). Such parameters as density of prey (cf Ricker 1954), patchiness (cf Ivlev 1961), general behaviour (for instance escape instincts, 'availability', etc.) physical and chemical 'optima', colour, structure and size of prey affects the predator when searching for the most 'rewarding' prey. The phenomenon called 'hunting by searching image' (Tinbergen 1960) helps the predator to persist in seeking a certain prey as long as the prey gives a satisfactory profit in terms of energy. When a prey is no longer profitable from this viewpoint, the predator may switch to a more rewarding prey. It seems obvious that all of these factors interact when the predator searches for a dependable source of food. Density has long since been the main factor studied in fishes (cf Backiel & LeCren, Chapter 11), but during the last decade size-biased predation has become more and more prominent in the research of both zooplankton specialists and fishery biologists.

12.2 Morphology of the predator

It may seem a truism that the morphology of the predator is a major factor in size-biased predation and yet it is not always taken into account. Experimental studies, initiated by Ivlev's (1961) 'electivity index', have revealed innate preferences for, or avoidance of, certain food items. The size and structure of the mouth plays an important part in food selection. For example, Northcote (1954) in connection with the study for the interrelationships of two cottid species, pointed out the significance of minor differences in mouth size as to the availability of prey (also see Popova, Chapter 9).

303

It has long been accepted that the number of gillrakers of whitefish can be correlated with the main size of the prey. Species with long and numerous gillrakers with little space between them tend to feed on small prey. This tendency, however, is not without exceptions. Svärdson (1950) found that small-sized whitefish (*Coregonus* spp.) with dense gillraker sets transplanted into fishless lakes increased their growth rate drastically, presumably because they changed their diet from crustacean plankton to more substantial food. Nilsson (1958) studied three sympatric species of whitefish in northern Sweden and found that in a lake where only two of the species lived together, the one with few gillrakers (mean = 20) lived on bottom animals and grew rapidly, the one with numerous gillrakers (mean = 45) fed on crustacean plankton and grew slowly. In another lake a third species with an intermediate number of gillrakers (mean = 35) occupied the 'plankton niche' and the one with numerous gillrakers (mean = 45) lived on surface food (winged insects) and grew rapidly. Moreover, Kliewer (1970), also studying whitefish in Manitoba, found a negative correlation between gillraker number and gill-raker length, but a positive relation between food size and gillraker number and gillraker space. In the genus *Alosa* the difference is clear-cut. According to Brooks and Dodson (1965) the two species *A. pseudoharengus* (with closely spaced gillrakers), and *A. mediocris* (with widely spread gillrakers) distinctly differ in their food habits, the former being a plankton-feeder, the latter being piscivorous (Fig. 12.1).

Turning to tropical species, such as the African cichlids, the significance of the structure of the oral tract becomes still more obvious. Fryer and Iles (1972) demonstrated that cichlids 'fill' most niches of the great lakes of East Africa, being phytoplankton feeders, deposit feeders, Aufwuchs eaters (with many strange specializations), periphyton collectors (species feeding on algae growing on plants), leaf choppers (species tearing pieces from higher plants), mollusc feeders, feeders on insects and other benthic arthropods, zooplankton feeders, scale eaters (feeding on scales of other fish), fin biters, piscivores, egg, embryo and larval fish eaters, and finally the bizarre eye-biter (*Haplochromis compressipes*), feeding on the eyes of other fishes. All the species have mouths and other organs specialized for their feeding habits.

To sum up, it is quite obvious that the morphology of the fish cannot be ignored when discussing size-biased feeding, since in some cases large prey, in other cases small prey or at least prey with a very specific size may be selected.

12.3 Some aspects of behaviour

Together with the morphology of the fish, general behaviour patterns naturally play an important role in size-biased predation. It is not possible to review the vast literature on the predatory behaviour of fish within the

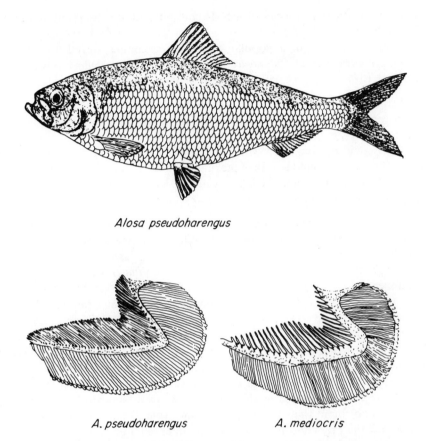

Alosa pseudoharengus

A. pseudoharengus   *A. mediocris*

Fig. 12.1. Above, mature specimen of *Alosa pseudoharengus*. Below comparison between gillrakers of the planktophage *A. pseudoharengus* and the piscivore *A. mediocris*. (From Brooks 1968.)

frame of this book, but some of the intricacies may be appreciated by referring to Chapter 9. Here I will select certain ideas that are particularly relevant to our topic.

Krebs (1973) discussed four hypotheses concerning mechanisms of importance in the behaviour of predators: (1) hunting by searching image, (2) hunting by expectation, (3) area restricted search, and (4) 'niche' hunting. The theory of 'hunting by searching image' has been adopted especially by sport fishermen ever since the time of Izaac Walton. Sportsmen depend on the fish to respond to a lure which is a reasonable imitation of its natural food. If the theories of 'hunting by searching image' and 'optimal prey selection' are accepted, different types of stimuli obviously should apply to prey selection for different species of fish. I shall illustrate this by pointing out a few examples.

Fabricius (1953), in a popular paper, made comparisons between the feeding of brown trout (*Salmo trutta*), Arctic char (*Salvelinus alpinus*) and whitefish (*Coregonus* sp.) with birds in the following way:

1   Brown trout could be compared with flycatchers (*Muscicapidae*), keeping a territory or home range, and, from this site making short excursions to catch suitable prey,
2   Char could be compared with nighthawks (*Caprimulgidae*) resting for a long time on the ground and then picking up food in a rapid succession, and,
3   Whitefish were compared with swallows, who continuously fly around searching for prey without resting.

This, of course, is a very simplified model but could serve as an example of how even taxonomically and ecologically related fish behave differently when searching for prey.

The feeding strategy of 'trout' (*Salmo* and *Salvelinus*) is characterized by a number of steps (Fig. 12.2), each often lasting just a few tenths of a second. The 'search' stage of the trout is synonymous with what Craig (1918) and Tinbergen (1951) called 'appetitive behaviour', that is a variable, exploratory, striving behaviour, which leads to the 'consummatory action', in this case starting with the recognition ('encounter') of the prey and ending up with ingestion. The prime physiological motivation should be hunger, as described by, for instance Ivlev (1961) and Ware (1971a).

The role of experience in the predation of trout was studied by Ware (1971b), who found that the reactive distance (the distance from which the fish attack the prey) and feeding time (the time required to capture a prey) of rainbow trout (*Salmo gairdneri*) varied with the time that the fish spent in contact with the prey (time of experience). The relation between time of experience and reactive distance could be described by an equation fitted to a sigmoid curve, which was inverse to the curve obtained for feeding time. This supports the theory of the searching image. Ivlev (1961) obtained similar results when training carp to two different types of prey (chironomid larvae and molluscs).

The choice that a fish makes in actually selecting a prey item is influenced very much by the same factors as mentioned earlier, such as experience of the predator, size, colour and other morphological as well as behavioural characteristics of the prey. Nilsson (1965) offered brown trout (*Salmo trutta*) and lake trout (*Salvelinus namaycush*) two types of food (red liver cubes and white fish cubes), and found that both species preferred the liver cubes. By changing the density (frequency) of the 'prey' in favour of fish cubes, the trout learned to choose fish cubes and still chose them even during a short period when liver cubes were more frequent. Subordinate fish even learned

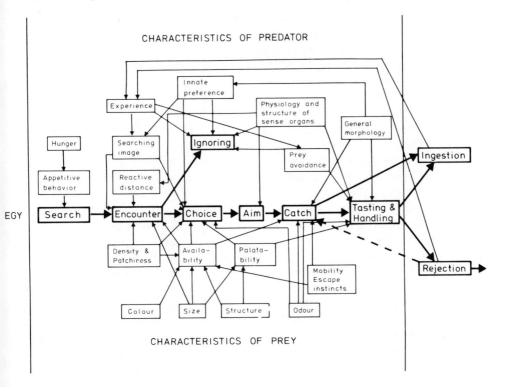

Fig. 12.2. Feeding strategy of trout, with suggested interacting parameters between predator and prey.

to pick up the 'prey' from the bottom of the aquarium, an act which was never performed by the dominant fish.

As most fish species have a very restricted field of binocular vision, they have to direct the whole body towards the prey before catching it. In this respect the structure of the eye (distribution of light and colour receptors in retina etc.) certainly is important in the case of predators like trout. Ahlbert (1969, 1976) found specific distributions of the cones in the retina, which could be correlated to specific feeding habits. For example, the best visual acuity of the 'predatory' perch (*Perca fluviatilis*) is found temporally in the horizontal plane of the retina, but in the cisco (*Coregonus albula*) in the ventro-temporal part (Ahlbert 1969). Also young brown trout, which catch their prey by making excursions from the bottom towards the surface of the water have a similar distribution of the cones in the retina (Ahlbert 1976).

The fixation time ('aim') can be very short, for example when a trout in running water makes a sudden dash from its territory towards a prey floating on the surface of the water. After the fixation phase the prey is seized ('catch'), which can lead to immediate ingestion or rejection, but often is followed by a 'tasting and handling' phase. During this phase, structure, odour and size of the prey is tested, and piscivorous fish, for instance, manipulate the prey so as to pass the oesophagus 'head first'. The tasting can also lead to rejection, if the structure, odour or size of the prey is not suitable. Sometimes, however, the rejected prey is recaptured and swallowed (cf Hoogland *et al.* 1957).

The feeding strategy of planktivorous fish should basically follow a similar pattern although the schooling instincts of this type of feeder, linked with learning from other members of the school, probably plays an important part. The mouths of genuine plankton feeders open upward, the eyes are placed above the middle of the head, and the highest density of their colour receptors is placed around the ventro-temporal part of the retina (Ahlbert 1969). Brooks (1968) observed a group of alewives (*Alosa pseudoharengus*) kept in a tank and provided with plankton netted in a nearby lake. He found that the fish usually moved slightly upwards to feed. For the first few minutes after the plankton was introduced into the tank 'the school broke up, each fish making small darting movements after its prey, but after about 10 minutes the school had completely reformed' (*op. cit.* p. 278).

Species dwelling most of their time in darkness, or feeding on prey hidden in the mud, etc., have to rely primarily on sense organs other than the eyes. The importance of the olfactory organs, which are extremely sensitive in fish (cf Höglund *et al.* 1975) has probably been underestimated in this context, as well as the ability of hearing. On the whole the size of the prey would not play such an important part for these species, as, for instance, planktivorous or piscivorous species.

12.4 Evidence of size-biased predation

The significance of the size of prey in predator–prey relationships in fish has been stressed for a long time. Of early works, those by Ivlev (1961), Lindström (1955), Hrbáček (1958), Hrbáček & Hrbáčkova-Esslová (1960), Berg and Grimaldi (1966), Hutchinson (1967, with references) and Brooks (1968) may be mentioned.

Many purely experimental methods of studying size selection have been tried (cf Ivlev 1961, Brooks 1968, Ware 1971). More dominant are those works which comprise parallel studies of the invertebrate fauna and food habits of fish in ponds or lakes (e.g. Lindström 1955, Hrbáček 1958, Hrbáček *et al.* 1961, Brooks & Dodson 1965, Galbraith 1967, Anderson 1971, 1975, Wells 1970, Hutchinson 1971, Stenson 1972, 1973, Berg & Grimaldi 1966, Grimaldi 1969, Giussani 1974). Other studies compare fenced areas which exclude fish with areas subject to the grazing of fish as well as before-and-after studies of poisoned waters (Lellak 1966, Kajak 1964, Straskraba 1965, Galbraith 1967, Berglund 1968, Tuunainen 1970, Stenson 1972). Comparisons between prey fauna and food habits of fish before and after the introduction of new fish species have also been done (Brooks & Dodson 1965, Galbraith 1967, Johannes & Larkin 1961, Wells 1970, Nilsson 1972). Finally attempts at finding correlations between fish fauna and zooplankton composition by surveying regionally many lakes in that respect have begun to appear (Anderson 1971, Nilsson & Pejler 1973, Northcote & Clarotto 1975, Ekström 1975).

LABORATORY FEEDING EXPERIMENTS

Ivlev (1961) studied experimentally the selective feeding of five species of fish with regard to the size of their prey. He used a group of 'predatory' species: pike (*Esox lucius*) and perch (*Perca fluviatilis*), to feed on roach (*Ratilus rutilus*), a group of benthophage species: carp (*Cyprinus carpio*) feeding on chironomid larvae, bream (*Abramis brama*) feeding on amphipods, and the planktivorous bleak (*Alburnus alburnus*) feeding on *Cladocera*. A range of sizes of prey was offered. The 'curves' of body-size of prey against frequency of capture indicated that all five species of fish had preferences for optimum size of the prey, but only the curves for the benthophage species were symmetrical. The curves for 'predatory' species were asymmetrical to the right (as large prey as possible), and the curve for bleak fed on *Cladocera* to the left. The latter phenomenon was explained by Ivlev as a selection of 'material basically suitable for the structural features of their filtration apparatus, but as a rule more dispersed in particle size than the maximum possible for the particular animal' (*op. cit.* p. 85).

Since most studies on size-biased feeding have been done on planktivorous species, the more recent work by Brooks (1968) should shed more light on this problem. Brooks kept a group of alewives in a tank and fed them with zooplankton collected in vertical tows from the nearby Bashan Lake. Immediately after the catch the live plankton was carefully mixed with the water in the tank, and then sampled at 15-minute intervals, each experiment lasting one hour. The most abundant species of planktonic crustacea were *Daphnia catawba*, *Epischura nordenskiöldi* and *Diaptomus minutus*. *Daphnia* disappeared after 15 minutes, while *Epischura* survived slightly longer (30 minutes). *Diaptomus* survived longest, and Brooks thus chose that species for a quantitative study. After counting and measuring the individuals remaining after each sampling, the survival sequence of five age-size stages of the species could be followed: *nauplii* and *metanauplii*, *copepodite I*, *copepodte II*, *copepodite III*, and adults.

Figure 12.3 shows the result of this investigation. In that graph the mortality due to predation by alewives of each of the stages is plotted against size. The adults were the first to be grazed down (at the end of 12 minutes 55% had been consumed). They were followed by each larval stage in order of size. It appeared that the curve thus obtained fitted to a hyperbole which meant that time (T) of a given level of mortality caused by predation would be inversely proportional to size measured as body length (L).

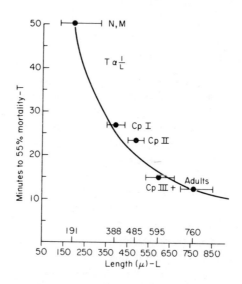

Fig. 12.3. Relationship between body size of *Diaptomus minutus* and predation by *Alosa pseudoharengus*. (From Brooks 1968.)

FIELD STUDIES

Brooks' experimental work was subsequent to a field investigation of the zooplankton composition of the small Crystal Lake, Wisconsin, before and after the introduction of another species of alewife (*Alosa aestivalis*) (Brooks & Dodson 1965). The introduction of the planktivorous fish species resulted in an alteration of the composition of the zooplankton. Small species, like *Bosmina longirostris*, *Tropocyclops* and *Ceriodaphnia*, replaced the larger species *Diaptomus minutus*, *Daphnia catawba*, *Mesocyclops edax*, *Epischura nordenskiöldi* and *Leptodora kindti* (Fig. 12.4). Several investigations in different parts of the world have detected similar phenomena.

For instance, very early, the team working at the Istituto Italiano di Idrobiologia, Pallanza (Berg & Grimaldi 1965, 1966, Grimaldi 1969, 1972, Giussani 1974, De Bernardi & Giussani 1975) studied the effects of the

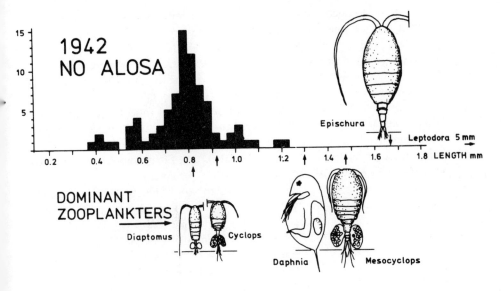

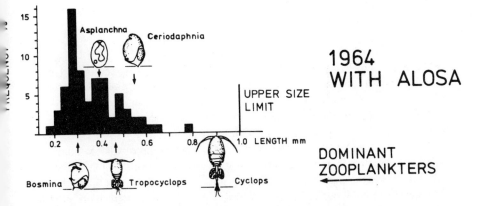

Fig. 12.4. The composition of zooplankton of Crystal Lake before and after the introduction of *Alosa aestivalis*. (From Brooks 1968.)

introduction of 'bondella', a *Coregonus* species (*C. macrophtalmus* according to Wagler 1941) from Lac du Neuchatel to Lago Maggiore in 1949. Earlier other *Coregonus* species had been introduced around the turn of the century, resulting in one population locally called 'lavarello' (a hybrid between *C. wartmanni* and *C. schinzii*, Berg & Grimaldi 1965). The introduction of 'bondella' resulted in many interesting interactions with other planktivorous fishes—not only with the 'lavarello' but also with bleak (*Alburnus*) and the landlocked Mediterranean shad (*Alosa ficta*). The main work as concerns the present subject was focused on the selection of food by the 'bondella'. In short the investigators found marked food preferences, the big *Bythotrephes longimanus* and *Leptodora kindti* and *Daphnia hyalina* larger than 1·500 μ being obviously selected, whereas copepods, e.g. the abundant diaptomoids, were rejected, as well as *Diaphanosoma brachyurum*.

Galbraith's work (1967) fitted well with the above mentioned observations. He studied two Michigan lakes, one 'combination lake' (with rainbow trout and warmwater game fish) and another reclaimed with toxaphene and used solely for rainbow trout. He found that both yellow perch (*Perca flavescens*) and rainbow trout (*Salmo gairdneri*) were very size selective, usually picking out *Daphnia* over 1·3 mm but ignoring more numerous smaller zooplankters. In the reclaimed lake the introduction of rainbow trout, smelt (*Osmerus mordax*) and fathead minnows (*Pimephales promelas*) had a drastic effect on the composition of the zooplankton:

1   *Daphnia pulex* disappeared completely and was replaced by two smaller species within 4 years,
2   The average size of the daphnids decreased from 1·4 mm to 0·8 mm, and
3   The percentage and volume of the daphnids was reduced although there was no reduction in numbers.

Very similar results were obtained by Hutchinson (1971), Stenson (1972, 1973), Karlsson and Nilsson (1968), Nilsson (1972), and Anderson (1975).

Hutchinson studied the effects of the introduction of alewife in a small lake in northern New York, and also compared the zooplankton composition of 10 other lakes in the Adirondack area. He found that the introduction of alewives resulted in a changed zooplankton community within three years in Black Pond, favouring smaller species. Comparing lakes with none or few planktivorous fish with those containing large populations of planktivorous fish, he found a significant difference in the size of both copepods and cladocerans, small species being predominant in the lakes with large populations of planktivorous fish.

Stenson compared in a similar way, 8 small forest lakes in southwest Sweden, four of which were treated with rotenone. Of the reclaimed lakes one sustained the treatment and retained the original fish community, the others were restocked with salmonids (*Salmo trutta*, *S. gairdneri* and *Salvelinus fontinalis*). It appeared that in the lakes with low predation on zooplankton (the 'trout lakes') large species, *Bythotrephes longimanus* and *Daphnia longispina* were present, but in the lakes with the original fish fauna these species were not present, *D. longispina* being replaced by the smaller *D. cristata*. Consequently a clear-cut difference in the size distribution of the zooplankters could be demonstrated.

Karlsson and Nilsson (1968) and Nilsson (1972) studied a large (60 km²) subarctic lake, Lake Pieskejaure in northern Sweden which had been practically devoid of fish, until Arctic char (*Salvelinus alpinus*) was introduced in 1961 and 1962. The study revealed that the anostracan fairy shrimp, *Polyartemia forcipata*, which had only been known to occur in small bodies of water—often frozen to the bottom in winter—apparently was a common member of the zooplankton community of the lake. The char fed preferentially on these slow-moving creatures and, as far as could be judged from the continuous sampling, eliminated them (Fig. 12.5). The char first displayed a spectacular growth, but then tended to get stunted along with a change in density and diet, turning from *Polyartemia* to smaller crustaceans. The older fish ate snails, chironomids and sculpin. The lakes hitherto mentioned are rather small bodies of water, and it is therefore important to point out that even in such a large lake as Lake Michigan similar changes in zooplankton composition were exhibited after the introduction of an alewife planktivore (*Alosa pseudoharengus*). Wells (1970) found that the zooplankton community of Lake Michigan underwent striking changes between 1954 when the alewife invasion and population explosion began, and 1966 when there was a decline in alewife abundance. The species that declined sharply were the large cladocerans *Leptodora kindti*, *Daphnia galeata* and *D. retrocurva*, and the largest copepods *Limnocalanus macrurus*, *Epischura lacustris*, *Diaptomus sicilis* and *Mesocyclops edax*. On the contrary medium-sized or small species such as *Daphnia longiremis*, *Holopedium gibberum*, *Polyphemus pediculus*, *Bosmina* spp. *Ceriodaphnia* and small *Cyclops* and *Diaptomus* species increased in numbers. When the alewife population declined in 1966 the zooplankton composition on the whole shifted back towards its condition in 1954.

12.5 Visible size

So far a common pattern has been revealed as to the selectivity by size in predator–prey relationships, at least between fish and zooplankton. A frequent conclusion is that large species of crustacean zooplankton are grazed down before the smaller ones, and that *Cladocera* are mainly more vulnerable than *Copepoda*. Apparently many factors interact to make a prey vulnerable to a predator. Size, however, is just one of the factors, and what is called 'size' by a human being very probably means something else to the predator fish.

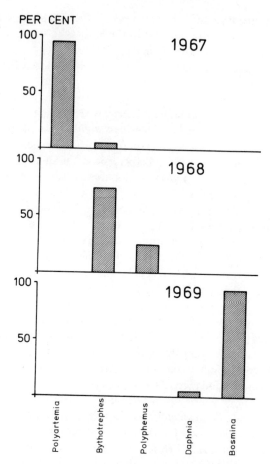

Fig. 12.5. The consumption of zooplankters by Arctic char (*Salvelinus alpinus*) after the introduction into a fishless lake (Lake Pieskejaure, Lapland).

The term 'cyclomophosis', first used by Lauterborn (1904) refers to a seasonal change in morphology, the phenomenon being displayed by planktonic animals. This polymorphism is exhibited by successive generations by increasing helmets, spines and other protruding parts of the body during certain parts of the season. For a thorough discussion of this phenomenon the reader is referred to Hutchinson (1967).

Brooks (1965, 1968), Jacobs (1965), Zaret (1972) and Nilsson and Pejler (1973) have discussed the significance of cyclomorphosis in relation to the 'visible size' of cladocerans. As Brooks (1968) expressed it: '. . . . the significance of this seasonal cycle of body shape (cyclomorphosis) lies in its relation to the closely correlated changes in body size at the onset of maturity. The midsummer individuals of *Daphnia* bearing the tallest helmets begin to produce eggs when the size of the visible part of the body (including eye, eggs and brood pouch) is at its annual minimum.' And: '. . . many plank-tivores rapidly lose interest as its (*Daphnia*) size decrease toward 1 mm; any mechanism that would depress the size at maturity to about 1 mm or slightly below would be of considerable adaptive value.' Many species are completely colourless and transparent except for their eyes (*Bythotrephes, Leptodora*) which often constitute the only means of detecting them macroscopically (Zaret 1972, Nilsson & Pejler 1973). Northcote (personal communication) has pointed out that ephippial *Daphnia* are selected by rainbow trout in a New Zealand lake, and this observation is certainly confirmed by data from Scandinavian lakes (Nilsson 1960, p. 195).

Very convincing experiments carried out by Zaret (1972) and Zaret and Kerfoot (1975) showed that the pigmented parts, that is the eyes, of clado-cerans (*Bosmina* and *Ceriodaphnia*) rather than the size of their bodies influence the food selection of fish. For instance, Zaret (1972) demonstrated that the food selection of a planktivore fish *Melaniris chagresi* in Gatun Lake, Panama, was positively correlated with the eye-pigmented area of *Ceriodaphnia cornuta*, on which it fed. On the basis of their findings Zaret and Kerfoot (1975) introduced the term 'visibility-selective predation'.

12.6 Movement of prey

Movement is another complicating factor. With regard to copepods, they apparently have escape instincts that save them from over-grazing (e.g. Lindström 1955). In short this behaviour means that the prey tends to escape by jumping out of the binocular field of the predator.

In this context it should be mentioned that the escape behaviour of some species of zooplankters also complicates comparisons between the 'zoo-plankton community' as measured by plankton nets or other samplers, and the food composition of their predators. To take just one example, Nilsson (1974), when studying the food habits of the off-shore fish community of Lake Vänern (South Sweden), found that the 'rare' species *Bythotrephes cederstroemi* was very important as food for, cisco (*Coregonus albula*), whitefish (*Coregonus* spp.), roach (*Leuciscus*) and smelt (*Osmerus eperlanus*). This small zooplankter, however, avoided nearly all attempts to catch it with plankton nets. Wesenberg-Lund (1937) compared the behaviour and morphology of *Bythotrephes* with that of dragonflies, and it is logical to suspect that their good visual abilities (large eye) and escape behaviour allow them to avoid 'blind predators' like plankton nets.

O'Brien and Vinyard (1974) concluded from similar premises that the use of Ivlev's electivity index in field research can be influenced by sampling bias, and should, therefore, necessitate very accurate sampling data. If one wishes to study the importance of a certain prey to fish, the old method of stomach analysis obviously is still necessary.

12.7 Interspecific relations within the plankton community

The predator–prey interactions within the plankton community itself should not be underestimated. De Bernardi and Giussani (1975), for instance, showed that the predatory cladocerans *Leptodora* and *Bythotrephes* in Lago Maggiore are responsible for a 'summer collapse' of the *Daphnia hyalina* population by feeding on young individuals of that species, while the whitefish (*Coregonus*) control the population growth of the adults. These interrelationships get still more complicated by the fact that the whitefish feed both on the aforementioned predatory zooplankters and the *Daphnia* as well as other filter feeding zooplankters.

Northcote and Clarotto (1975), who studied six lakes in British Columbia —three of them fish-less, and three with cutthroat trout (*Salmo clarki*) and Dolly Varden (*Salvelinus malma*)—found that chaoborids play an important part in the ecosystem by feeding on *Bosmina* and *Cyclops*. They concluded: 'Size-selective predation and its consequences seem by no means restricted to relatively large, active vertebrate planktivores preying on large forms but extend downwards in a size continuum including aquatic insects such as the chaoborid larvae implicated herein.'

The study of the relationships between vertebrate predators, invertebrate predators and herbivores seems to be an especially fruitful area for future research.

12.8 Interspecific food competition between fish populations

Park (1948) and Brian (1956) distinguished between two components in interspecific competition: interference and exploitation. Interference means direct harm to one or both competing species, for instance by aggressive behaviour such as fighting for territories, threat display etc. We will not discuss this aspect of interspecific competition here. Exploitation, on the other hand, is of direct interest to us, because it means an interaction may 'develop whenever one species is more efficient in a habitat than another, perhaps because it can find and use vacant resources more easily and quickly than the other' (Brian *op. cit.*).

An example of the exploitation type of competition in fish was described by Crossman (1959) and Johannes and Larkin (1961). In Paul Lake, British Columbia, redside shiner (*Richardsonius balteatus*) was introduced to serve as food for rainbow trout. This lead to a complicated situation involving,

(1) competition for food between the introduced species and the rainbow trout, (2) predation by redside shiner on trout fry, and (3) predation by trout on adult shiners.

Of special interest in this study of competition is the way in which, by selective feeding, the shiners could graze down the *Gammarus* population. Johannes and Larkin (op. cit.) found that the shiners were quicker to notice the prey and quicker to start feeding on it than were the trout. They fed by catching the *Gammarus* deep in the *Chara* beds where the trout either could not reach them or did not notice them. After about 25 years of common exploitation of the *Gammarus* population, the trout had almost stopped feeding on them (Fig. 12.6), and, as the authors remarked: 'an observer

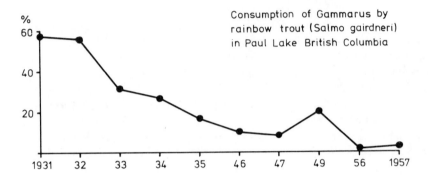

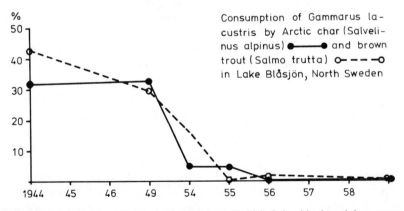

Fig. 12.6. Consumption of *Gammarus* in Paul Lake, British Columbia, by rainbow trout after the introduction of redside shiner (*Richardsonius balteatus* (above), and in Lake Blåsjön, Sweden, by Arctic char (*Salvelinus alpinus*) and brown trout (*Salmo trutta*) after impoundment of the lake. (After Johannes & Larkin 1961, Nilsson 1961.)

would hardly suspect that amphipods had been the most important item of competition'.

A similar case was reported by Nilsson (1961). In an impounded lake (Blåsjön), North Sweden, the *Gammarus* population lost the shelter provided by the shoreline vegetation because of the water-level fluctuation. The final extinction of the *Gammarus* population was caused by brown trout and Artic char which competed for that food item and probably selected for it as long as it was still available (Fig. 12.6). After the *Gammarus* disappeared, the two species segregated, the trout mainly feeding on winged insects and remnants of the bottom fauna, the char mainly on zooplankton (cf Nilsson 1967).

When a new species of fish is introduced into a lake several alternative events might happen:

1 *rejection*, which means that the introduced species cannot compete successfully with the original fish community for food, cover, spawning areas, etc.
2 *displacement* of one or more of the original species, which means that the new species is more successful in the environment than one of its competitors and 'fills the niche' of that species, or,
3 the new species could be *segregated* into an empty niche, allowing it to become a new member of the fish community. This process has been called *interactive segregation* by Brian (1956), working with ants (Myrmica), and was earlier also observed by Svärdson (1949) with respect to birds, and later used by Nilsson (1967) in interpreting the interactions between fish species.

Although exploitative competition by means of size-biased predation has hitherto not been fully proven to be a mechanism leading to rejection, displacement or interactive segregation of fish species, much evidence points toward it. From the Scandinavian literature the example of reinforced segregation of brown trout and char through increased exploitation of *Gammarus* has already been mentioned. The fact that the introduction of whitefish (*Coregonus*) often has led to the elimination of Arctic char in Scandinavian lakes was interpreted as a result of exploitative competition (Svärdson 1961). Lötmarker (1964) and Nilsson and Pejler (1973) supported that theory by showing that there is a difference in the zooplankton composition between lakes containing trout and char, and lakes containing whitefish alone. By examining the zooplankton composition of 64 lakes in northern Sweden, Nilsson and Pejler demonstrated that there exists 'a rough but logical correlation between colour, behaviour and size of the zooplankters and the presence of zooplankton-feeding fish' (*op. cit.* p. 74). By dividing the investigated lakes into 6 categories—fish-less, brown trout only, char only, char-trout, char-trout-whitefish, and whitefish—it became obvious that large-sized crustaceans such as *Daphnia longispina* and *Heterocope saliens* dominate in fishless lakes and in lakes with brown trout alone. In lakes containing char or whitefish the big *D. longispina* is replaced by the smaller, more transparent *D. galeata*, and in lakes with whitefish alone replacement is made by the very small species *D. cristata* and *Ceriodaphnia quadrangula*. Likewise the large-sized copepod *Heterocope saliens* is replaced by the smaller *H. appendiculata* in lakes where whitefish is the dominating species but accompanied by, for instance, brown trout, pike, roach and other species. (Fig. 12.7). Arguing from theory, the displacement of char by whitefish should occur in the early stages of the interaction when the species involved are all planktivorous. Under these conditions the whitefish graze down a plankton crop which is essential for the survival of the char, but survive themselves on smaller plankton species that they could exploit more efficiently than the char. Some whitefish species seem to be more successful competitors with the char than others, but it still remains to study in detail what mechanisms are involved in the displacement process. For instance both the morphology of the oral tract and general behaviour of the fish should play important parts.

Svärdson (1976) reviewed several similar situations from Scandinavian literature. For instance, he quoted Ekström's (1975) study of 35 oligotrophic lakes in southern Sweden, all of them containing Arctic char, but only 9 containing cisco (*Coregonus albula*). The examination of the zooplankton communities of the two groups of lakes gave a species composition symbolized by Fig. 12.8. It is obvious that the relatively large species *Daphnia longispina* and *Heterocope appendiculata* are abundant in lakes with no cisco, whereas the small species *Daphnia cristata* dominates in lakes with cisco. Svärdson concluded from this and other evidence that the European cisco, by being a very efficient plankton feeder, is a dominant species suppressing for example whitefish (*Coregonus* spp.) and smelt (*Osmerus eperlanus*).

A very strong competitive dominance due to the grazing of zooplankton (mainly *Daphnia cristata* and *Bosmina*) by roach was suggested as a means of suppressing the perch populations in southern Sweden (Stenson 1976, Svärdson 1976). One piece of evidence pointed out by Svärdson is the fact that the roach is extremely sensitive to acidification. In many cases the acidification of Western Swedish lakes has led to an increase in the abundance of the perch, which is not due to the perch being favoured by acidification per se, but to the fact that the roach populations decline.

12.9 Final comments

The evidences of the reciprocal effects of grazing in freshwater communities on both predator and prey populations are obvious and important, not only from a theoretical point of view, but even more with regard to management. Special stress has been put on the effects of grazing on zooplankton and the consequences in the relations between competing fish species. Svärdson (1976) suggested that the specialization to pelagic life contributes to dominance among lake-living fish, and 'the dominance of some fish species seems correlated to the capacity to catch small-sized zooplankton, which has far-reaching consequences for the plankton fauna and hence primary production'.

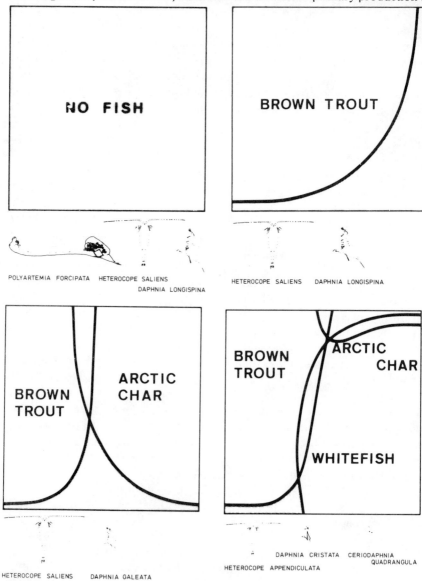

Fig. 12.7. Model of the 'dimensions' of the niches of brown trout, Arctic char and whitefish in allopatry and sympatry, and the dominant species of zooplankton. (After Nilsson & Pejler 1973.)

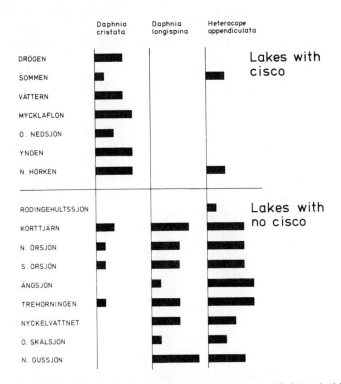

Fig. 12.8. Zooplankton composition in lakes with cisco (*Coregonus albula*) and with no cisco in 16 lakes in south Sweden. (After Ekström 1975.)

Grazing on zooplankton has a decided effect on primary production. An experiment recently performed by Stenson (1976), for instance, showed that a lake treated with rotenone exhibited a drop of alkalinity in years subsequent to the treatment. Moreover, the transparency was doubled and the primary production decreased from 15–20 mg C m^{-3} per hour to only 1–3 mg C m^{-3} within four years. These drastic effects were interpreted as the result of the absence of fish predation on the zooplankton. This is in good agreement with the findings of Hrbáček *et al.* (1961). Obviously, we will have to reconsider the role of the fish populations on the production of the lakes as a whole. For example, eutrophy or oligotrophy theoretically could depend on the presence or absence of certain fish species rather than physico-chemical parameters in some cases.

The science of the role of grazing in limnetic ecosystems is rather young, say twenty years, and many areas remain to be explored. Of special interest is how competition for food and interactive segregation between species are influenced by grazing. The possibility of 'stable niches', that is species segregation that has evolved through natural selection, is also an important area for future research. On the whole our policy concerning introductions of new species, eradicating undesirable species and the understanding of the interactions between species, should be basically founded on such research.

References

AHLBERT I.-B. (1969) The organization of the cone cells in the retinae of four teleosts with different feeding habits (*Perca fluviatilis* L., *Lucioperca lucioperca* L., *Acerina cernua* L. and *Coregonus albula* L.). *Ark. Zool. Ser. 2*, **22**(11), 445–481.

AHLBERT I.-B. (1976) Organization of the cone cells in the retinae of salmon (*Salmo salar*) and trout (*Salmo trutta trutta*) in relation to their feeding habits. *Acta Zool., Stockh.* **57**, 13–35.

ANDERSON R.S. (1971) Crustacean plankton of 146 alpine and subalpine lakes and ponds in western Canada. *J. Fish. Res. Bd. Canada.* **28**(3), 311–321.

ANDERSON R.S. (1975) An assessment of sport-fish production potential in two small alpine waters in Alberta, Canada. *Symp. Biol. Hung.* **15**, 205–214.

BERG A. & GRIMALDI E. (1965) Biologia delle due forme di Coregone (*Coregonus* sp.) del Lago Maggiore. *Mem. Ist. Ital. Idrobiol.* **18**, 25–196.

BERG A. & GRIMALDI E. (1966) Ecological relationships between planktophagic fish species in the Lago Maggiore. *Verh. int. Ver. Limnol.* **16**, 1065–1073.

BERGLUND T. (1968) The influence of predation by brown trout on *Asellus* in a pond. *Rep. Inst. Freshwat. Res. Drottningholm* **48**, 77–101.

BRIAN M.V. (1956) Segregation of species of the ant genus *Myrmica*. *J. Anim. Ecol.* **25**, 319–337.

BROOKS J.L. (1965) Predation and relative helmet size in cyclomorphic Daphnia. *Proc. Nat. Acad. Sci.* **53**, 119–126.

BROOKS J.L. (1968) The effects of prey size selection by lake planktivores. *Syst. Zool.* **17**(3), 273–291.

BROOKS J.L. & DODSON S.I. (1965) Predation, body size and composition of plankton. *Science* **150**, 28–35.

CRAIG W. (1918) Appetites and aversions as constituents of instincts. *Biol. Bull.* **34**, 91–107.

CROSSMAN E.J. (1959) A predator–prey interaction in freshwater fish. *J. Fish. Res. Bd. Canada* **16**(3), 269–281.

DE BERNARDI R. & GUISSANI G. (1975) Populations dynamics of three cladocerans of Lago Maggiore related to the predation pressure by a planktophagous fish. *Verh. int. Ver. Limnol.* **19**, 2906–2912.

EKSTRÖM Ch. (1975) Djurplankton. Rödingsjöar söder om Dalälven. *Inform. Inst. Freshw. Res.* (7). (Mimeographed in Swedish.)

GALBRAITH M.G. Jr. (1967) Size-selective predation on *Daphnia* by rainbow trout and yellow perch. *Trans. Amer. Fish. Soc.* **96**(1), 1–10.

GIUSSANI G. (1974) Planctofagia selettiva del coregone 'Bondella' (*Coregonus* sp.) del Lago Maggiore. *Mem. Ist. Ital. Idrobiol.* **31**, 181–203.

GRIMALDI E. (1969) Planctofagia sellettiva in specie ittiche lacustri: quesiti ed implicazioni. *Boll. Zool.* **36**.

GRIMALDI E. (1972) Lago Maggiore: Effects of exploitation and introductions on the salmonid community. *J. Fish. Res. Bd. Canada* **29**(6), 777–785.

FABRICIUS E. (1953) Öringen, vår vanligaste laxfisk. p. 31–55. *Lax och Öring. Sportfiske i svenska vatten.* (B. Haglund, ed.). P.A. Norstedt & Söners Förlag, Stockholm. (In Swedish.)

FRYER G. & ILES T.D. (1972) *The Cichlid Fishes of the Great Lakes of Africa, Their Biology and Evolution.* Oliver & Boyd, Edinburgh.

HOOGLAND R., MORRIS B. & TINBERGEN N. (1957) The spines of sticklebacks (*Gasterosteus* and *Pygosteus*) as means of defence against predators (*Perca* and *Esox*). *Behaviour* **10**, 205–236.

HRBÁČEK J. (1958) Typologie und Produktivität der teichartigen Gewässer. *Verh. int. Ver. Limnol.* **13**, 394–399.

HRBÁČEK J. & HRBÁČKOVA-ESSLOVÁ M. (1960) Fish stock as a protective agent in the occurrence of slow-developing dwarf species and strains of the genus *Daphnia. Int. Rev. Ges. Hydrobiol.* **45**(3), 355–358.

HRBÁČEK J., DVOŘAKOVA M., KOŘÍNEK V. & PROCHÁZKÓVA L. (1961) Demonstration of the effect of the fish stock on the species composition of zooplankton and the intensity of metabolism of the whole plankton association. *Verh. int. Ver. Limnol.* **14**, 192–195.

HUTCHINSON B.P. (1971) The effect of fish predation on the zooplankton of ten Adirondack lakes, with particular reference to the alewife, *Alosa pseudoharengus. Trans. Amer. Fish. Soc.* **100**(2), 325–335.

HUTCHINSON G.E. (1967) *A Treatise on Limnology. Vol. II. Introduction to Lake Biology and the Limnoplankton.* John Wiley & Sons, New York-London-Sydney.

HÖGLUND L.B. BOHMAN A. & NILSSON N.-A. (1975) Possible odour responses of juvenile Artic char (*Salvelinus alpinus* (L.)) to three other species of subartic fish. *Rep. Inst. Freshwat. Res. Drottningholm* **54**, 21–35.

IVLEV V.S. (1961) *Experimental Ecology of the Feeding of Fishes.* Yale Univ. Press, New Haven.

JACOBS J. (1965) Significance of morphology and physiology of *Daphnia* for its survival in predator–prey experiments. *Naturwissenschaften* **52**, 141–142.

JOHANNES R.E. & LARKIN P.A. (1961) Competition for food between redside shiners (*Richardsonius balteatus*) and rainbow trout (*Salmo gairdneri*) in two British Columbia lakes. *J. Fish. Res. Bd. Canada* **18**(2), 203–220.

KAJAK Z. (1964) The effect of experimentally induced variations in the abundance of *Tendipes plumosus*, L. larvae on intraspecific and interspecific relations. *Ekol. Polsk.* (A) **11**(15), 355–367.

KARLSSON R. & NILSSON N.-A. (1968) Rödingen och öringen i Pieskejaure. Nedbetning av näringsfaunan i en tidigare fisktom sjö. *Inform. Inst. Fresh. Res. Drottningholm* (14), 19 p. (Mimeographed in Swedish.)

KLIEWER E.V. (1970) Gillraker variation and diet in lake whitefish *Coregonus clupeaformis* in northern Manitoba. *Biology of Coregonid Fishes* (C.C. Lindsey and C.S. Woods, eds.), pp. 147–165. Univ. Manitoba Press, Winnipeg, Canada.

KREBS J.R. (1973) Behavioural aspects of predation. *Perspectives in Ethology* (P.P.G. Bateson and P.H. Klopfer, eds.), pp. 73–111. Plenum Press, New York-London.

LAUTERBORN R. (1904) Der formenkreis von *Anuraea cochlearis.* Ein Beitrag zur Kenntnis der Variabilität bei Rotatorens. II Teil. Die cyklishe oder temporale Variation von *Auraea cochlearis. Verh. naturh.-med. Ver. Heidelb.* n.f. **7**, 529–621.

LELLAK J. (1966) Zur Frage der experimentellen Untersuchung des Einflusses der Fresstätigkeit des Fischbestandes auf die Bodenfauna der Teiche. *Verh. int. Ver. Limnol.* **16**, 1383–1391.

LINDSTRÖM T. (1955) On the relation fish size–food size. *Rep. Inst. Freshw. Res. Drottningholm* **36**, 133–147.

LÖTMARKER T. (1964) Studies on planktonic crustacea in thirteen lakes in northern Sweden *Rep. Inst. Freshw. Res. Drottningholm* **45**, 113–189.

NILSSON N.-A. (1958) On the food competition between two species of *Coregonus* in a North-Swedish lake. *Rep. Inst. Freshw. Res. Drottningholm* **39**, 146–161.

NILSSON N.-A. (1960) Seasonal fluctuations in the food segregation of trout, char and whitefish in 14 North-Swedish lakes. *Rep. Inst. Freshw. Res. Drottningholm* **41**, 185–205.

NILSSON N.-A. (1961) The effect of water-level fluctuations on the feeding habits of trout and char in the Lakes Blåsjön and Jormsjön, North Sweden. *Rep. Inst. Freshw. Res. Drottningholm* **42**, 238–261.

NILSSON N.-A. (1965) Food segregation between salmonoid species in North Sweden. *Rep. Inst. Freshw. Res. Drottningholm* **46**, 58–78.

NILSSON N.-A. (1967) Interactive segregation between fish species. *The Biological Basis of Freshwater Fish Production* (S.D. Gerking, ed.), pp. 295–313. Blackwell Scientific Publications Oxford.

NILSSON N.-A. (1972) Effects of introductions of salmonids into barren lakes. *J. Fish. Res. Bd. Canada* **29**(6), 693–697.

NILSSON N.-A. (1974) Fiskens näringsval i öppna Vänern. (Food relationships of the fish community in the offshore region of Lake Vänern, Sweden.) *Inform. Freshw. Res. Drottningholm* (17). 51 p. (Mimeographed in Swedish with English summary.)

NILSSON N.-A. & PEJLER B. (1973) On the relation between fish fauna and zooplankton composition in north Swedish lakes. *Rep. Inst. Freshw. Res. Drottningholm* **53**, 51–56.

NORTHCOTE T.G. (1954) Observations on the comparative ecology of two species of fish, *Cottus asper* and *Cottus rhotheus*, in British Columbia. *Copeia* (**1**), 25–28.

NORTHCOTE T.G. & CLAROTTO R. (1975) Limnetic macrozooplankton and fish predation in some coastal British Columbia lakes. *Verh. int. Ver. Limnol.* **19**, 2378–2393.

O'BRIEN W.J. & VINYARD G.L. (1974) Comment on the use of Ivlev's electivity index with planktivorous fish. *J. Fish. Res. Bd. Canada* **31**(8), 1427–1429.

PARK T. (1948) Experimental studies on interspecies competition. I. Competition between populations of the flour beetles, *Tribolium confusum* Duval and *Tribolium castaneum* Herbst. *Ecol. Monogr.* **18**, 265–307.

RICKER W.E. (1954) Stock and recruitment. *J. Fish. Res. Bd. Canada* **2**(5), 559–623.

STENSON J.A.E. (1972) Fish predation effects on the species composition of the zooplankton community in eight small forest lakes. *Rep. Inst. Freshw. Res. Drottningholm* **52**, 132–148.

STENSON J.A.E. (1973) On predation and *Holopedium gibberum* (Zaddach) distribution. *Limnol. A Oceanogr.* **18**(6), 1005–1010.

STENSON J.A.E. (1976) Predator influence on zooplankton and benthos in some small forest lakes (Bohuslän, southwest Sweden). Dissertation. Dep. Zool., Univ. Gothenburg. 18 p. (Mimeographed.)

STRASKRABA M. (1965) The effect of fish on the number of invertebrates in ponds and streams. *Mitt. int. Ver. Limnol.* **13**, 106–127.

SVÄRDSON G. (1949) Competition and habitat selection in birds. *Oikos* **1**, 157–174.

SVÄRDSON G. (1950) The Coregonid problem. II. Morphology of two Coregonid species in different environments. *Rep. Inst. Freshw. Res. Drottningholm* **31**, 151–162.

SVÄRDSON G. (1961) Young sibling fish in northwestern Europe. *Vertebrate Speciation* (W.F. Blair, ed.), pp. 498–513. Univ. Texas Press.

SVÄRDSON G. (1976) Interspecific population dominance in fish communities of Scandinavian lakes. *Rep. Inst. Freshw. Res. Drottningholm* **55**, 144–171.

TINBERGEN L. (1960) The natural control of insects in pine woods. I. Factors influencing the intensity of predation by songbirds. *Arch. Néerl. Zool.* **13**(3), 265–343.

TINBERGEN N. (1951) *The Study of Instinct.* Clarendon Press, Oxford, England.

TUUNAINEN P. (1970) Relations between the benthic fauna and two species of trout in some small Finnish lakes treated with rotenon. *Ann. Zool. Fenn.* **7**(1), 67–120.

WAGLER E. (1941) Die Coregonen. p. 371–501. *Handbuch der Binnenfischerei Mitteleuropas* **3**.

WARE D.M. (1971a) Predation by rainbow trout (*Salmo gairdneri*). The effect of experience. *J. Fish. Res. Bd. Canada* **28**(12) 1847–1852.

WARE D.M. (1971b) Predation by rainbow trout (*Salmo gairdneri*): the influence of hunger, prey density, and prey size. *J. Fish. Res. Bd Canada* **29**(8), 1193–1201.

WELLS L.R. (1970) Effects of alewife predation on zooplankton populations in Lake Michigan. *Limnol. & Oceanogr.* **15**(4), 556–565.

WESENBERG-LUND C. (1937) *Ferskvandsfaunaen Biologisk Belyst. Invertebrata Andet Bind.* p. 415–837. Gyldendalske Boghandel-Nordisk Forlag, København. (In Danish.)

ZARET Th.M. (1972) Predators, invisible prey, and the nature of polymorphism in the *Cladocera* (Class *Crustacea*). *Limnol. & Oceanogr.* **17**(2), 171–184.

ZARET Th.M. & KERFOOT W.Ch. (1975) Fish predation on *Bosmina longirostris*: body-size selection versus visibility selection. *Ecology* **56**, 232–237.

Chapter 13: Migratory Strategies and Production in Freshwater Fishes

T. G. Northcote

13.1 Introduction

Migration, particularly in freshwater fishes, often has been regarded as an adaptive phenomenon for increasing growth, survival and abundance, which may combine to increase production. Such an argument was presented by Nikol'skii (1963) and its relevance to production of fishes important to man first pointed out by Gerbilsky (1958) and later more explicitly by Harden Jones (1968) in his statement 'If migration is an adaptation towards abundance, it would explain why the important commercial species are migratory: they are of commercial interest because they are abundant and abundant because they are migratory'. Yet, like many other seemingly obvious generalities in ecology, there is by no means a mass of well documented and conclusive evidence in its support.

In addition to general books on migration in the last decade there have been several dealing solely with fish migration such as Hasler (1966) and Harden Jones (1968) as well as many specific publications on the subject. Although these and other studies have furthered our understanding of factors initiating, orienting and controlling migration (see reviews by Banks 1969, Northcote 1969a, Brunel 1973, Arnold 1974, Shulman 1974) there have been few attempts to look widely for general migratory patterns or to examine in depth the adaptive significance of the phenomenon in fish. Exceptions include brief sections in the books by Nikol'skii (1963) and Harden Jones (1968) or some studies such as those by Hall (1972) and Schaffer and Elson (1975). Considerably more has been done in this respect for other migratory animals notably insects (Southwood 1962, Johnson 1969, Dingle 1972) and birds (Cox 1968, Lack 1968, Morse 1971, Stephan 1973, Katz 1974).

Before progressing further it should be noted that migration as used herein is given a more restrictive meaning than some might prefer (Harden Jones 1968, Dingle 1972) or than I have implied previously (Northcote 1967). Migration will refer to movements resulting in an alternation between two or more separate habitats (i.e., a movement away from one habitat followed eventually by a return again) occurring with regular periodicity (usually seasonal or annual, but certainly within the lifespan of an individual) and involving a large fraction of the population. Movement at some stage in this cycle is *directed* rather than a random wandering or a passive drift, although these may form part or one leg of a migration.

Largely for lack of space little emphasis will be given to 'micro-migrations' which occur over short periods of time (e.g. diel) or small distances, even though in some cases these may be as significant to production as are the more extensive and spectacular 'macro-migrations'. The latter consist essentially of the cycle of movements to and from the feeding habitat. Indeed Heape (1931) suggested that the primary cause of migration was almost invariably associated with food supply. Production, of course, may be affected by limitations or events in either the reproductive habitat or the feeding habitat as well as by those occurring en route between them. In a sense the reproductive habitat might be considered the 'original' one and migration as a strategy to 'escape' from its food restrictions, often thereby increasing growth, fecundity, survival and thus production. Furthermore, the best reproductive habitat rarely coincides with the best feeding habitat, particularly for fishes occupying environments subject to marked seasonal or spatial changes in productivity. Migration may simply be a more formalized or exaggerated pattern of behaviour evolved from the small scale, less regular movements which most animals undertake to optimize utilization of basic resources during feeding or reproductive phases of their life history.

Previously my discussion of migration in relation to production focused largely on salmonid fishes in north temperate freshwaters (Northcote 1967). Although there were, and are, good reasons for this, I now intend to broaden coverage of fish groups and world regions, and thereby hope to provide more insight into the general patterns and ecological significance of migration.

13.2 Patterns of migration

GENERAL FEATURES

Many reviews of fish migration, especially those by Meek (1916), Heape (1931) and Russell (1937) give major consideration to the spawning migration of adult stages and much less to the primary feeding migrations of juvenile or older stages which, in fact, produce the situation necessitating a return movement of mature adults. Although the importance of reproductive migrations cannot be denied, especially as they establish the founding population for the next year class, once the initial reduction in numbers occurs prior to or shortly after larval emergence, it is usually in the feeding and wintering stages that the major changes occur in biomass which so markedly affect production. Thus, in this section of the review emphasis will be given to the range and diversity in patterns of movement particularly in feeding and wintering migration shown by temperate, arctic and tropical freshwater fishes. Many of the examples will be used later in discussing strategies of migration and some readers may prefer to move directly to that section, referring back where necessary for more detail.

The patterns of movement between the spawning habitat and the feeding or 'winter' habitats typical of many freshwater fish migrations are cyclic (Fig. 13.1). In some cases the adult spawning habitat and the early juvenile

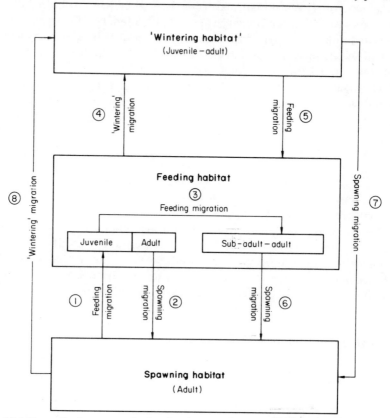

Fig. 13.1 Generalized patterns of migration between the three basic habitats utilized by many migratory freshwater fishes.

feeding habitat coincide, and in others the feeding and 'wintering' habitats are the same. A simple migratory cycle would involve movement of young from the adult spawning habitat to a single feeding habitat, from which at maturity the adults would move back to the spawning habitat, following the 1-2 sequence in Fig. 13.1. A somewhat more complex cycle might involve two feeding habitats, juvenile and sub-adult, resulting in a 1-3-6 sequence. If a wintering habitat was utilized the pattern would include steps 1-4-5-6. In other cases fish may reach maturity in their wintering habitat and move directly to the spawning habitat (step 7) or those surviving after spawning may move to a wintering habitat (step 8). Whatever the details of the migratory sequence, at some stage non-mature fish leave the area where they originated as young, move away to more suitable feeding grounds and later return to a habitat similar to that utilized by their parents for reproduction. It is this cyclic alteration between contrasting habitats which classify such movements as migrations. 'Wintering' has been commonly used (Nikol'skii 1963, Harden Jones 1968) to designate migrations in arctic and temperate fishes to habitats occupied during unfavourable conditions brought on by winter—the so-called 'climatic migrations' of Heape (1931). The term is, of course, inappropriate in reference to such migrations by tropical fishes but will be retained herein to denote movements away from unfavourable conditions, however induced, which form part of an overall migratory cycle.

FEEDING MIGRATIONS

Temperate regions

Patterns of feeding migrations evident in temperate freshwater fishes range from those in which the young leave the reproductive habitat of their parents long before feeding starts to those in which the young remain there for several years before migrating to a second feeding habitat.

Offspring of some species leave the area where they hatch and move passively, or largely so, to feeding habitats very early in their life history, even at egg or larval stages. Such is the case for several species of smelt, certainly the rainbow smelt of North America (*Osmerus mordax*) and its Eurasian counterpart as well as the eulachon (*Thaleichthys pacificus*) in several large rivers of the eastern Pacific coast (Scott & Crossman 1973). A similar pattern probably holds for river spawning populations of goldeye (*Hiodon alosoides*)—a species with semi-bouyant eggs and larvae (Scott & Crossman 1973), for larval catostomids emerging from inlet stream spawning sites (Geen *et al.* 1966) and for several species in rivers of eastern Asia (Nikol'skii 1963). Virtually no feeding occurs prior to or during migration in such forms.

Other fishes first migrate to feeding areas as fry after a brief residence period of perhaps only a few hours or days following emergence from their spawning habitat. Examples of this second type of feeding migration would include most pink salmon (*Oncorhynchus gorbuscha*), many eastern Pacific chum salmon (*O. keta*), a few sockeye salmon (*O. nerka*) populations, and many inlet stream spawning populations of lake resident rainbow trout, *Salmo gairdneri* (Northcote 1962, 1969a, b, Alexander & MacCrimmon 1974) and cutthroat trout, *S. clarki* (Northcote 1969a, Raleigh & Chapman 1971). At Loon Lake, British Columbia, young rainbow trout fry from the inlet spawning stream may reach feeding areas along the lake margins before all vestiges of their yolk sacs are gone. Young walleye (*Stizostedion vitreum*) in the Great Lakes region (Ferguson & Derksen 1971) and pike-perch (*Stizostedion lucioperca*) in the Dnieper (Belyi 1972) also migrate to feeding areas shortly after hatching as do the landlocked form of *Galaxias maculatus* studied by Pollard (1971, 1972). Similar to those in the first category, these species feed very little before or during migration, relying largely on energy reserves remaining from the egg. Some forms however, notably chum salmon, may spend several days or weeks in estuarine areas before moving to offshore marine feeding areas (Mason 1974a, Dunford 1975).

In a third type the young seem to disperse gradually from the spawning area of the adults without a massive or well-timed migration, eventually reaching their final feeding habitat after several months. The grayling (*Thymallus thymallus*) in the Sundsvall region of Sweden described by Peterson (1968) apparently would fit into this category (although in some streams they may move down to lakes soon after hatching; Mills 1971), as would several species of sturgeon in the Ural River (Bekeshev & Peseridi 1966) where the young collect temporarily in pools feeding on the rich benthic growth before moving on seaward. The feeding migrations of sculpins (*Cottus asper*) in coastal British Columbia rivers noted previously (Northcote 1967) also would be included here (Mason 1974b, Ringstad 1974). Migrants in this category may forage within the adult spawning habitat, and also along the way before they reach the main feeding areas.

Young of several other species of temperate fishes, largely salmonoids but also anadromous alosids (W. C. Leggett, pers. comm.), remain for several weeks, months or sometimes even years near the parental spawning habitat before migrating to their second feeding area, forming a fourth type of feeding migration. Again we find rainbow and steelhead trout which exhibit this pattern (Northcote 1969a, Bjornn 1971, Stauffer 1972, Alexander & MacCrimmon 1974) as well as Kamchatka trout, *Salmo mykiss* (Maksimov 1972), cutthroat trout (Averett & MacPhee 1971, Armstrong 1971), brown trout, *Salmo trutta* (Mills 1971, Thorpe, 1974, Williams & Harcup 1974), Atlantic salmon, *Salmo salar* (Symons 1970, Mills 1971), Dolly Varden, *Salvelinus malma* (Armstrong 1970, 1974), chinook salmon, *Oncorhynchus tshawytscha* (Bjornn 1971, Becker 1973, Reimers 1973), coho salmon, *O. kisutch* (Chapman 1965, Hartman 1965), Danube salmon, *Hucho hucho* (Mills 1971), American shad, *Alosa sapidissima* (Leggett & Whitney 1972, Marcy 1976), alewives, *Alosa pseudoharengus* (Richkus 1975), mountain whitefish, *Prosopium williamsoni* (Pettit & Wallace 1975), ide, *Idus idus* (Cala 1970) and even some populations of barbel, *Barbus barbus* (Bogatu & Stancioiu 1968). With many of these species, survival is apparently greatly enhanced by attaining a certain minimum size before migrating to the secondary feeding habitat. This is most clearly illustrated by coho salmon. Although many of these young salmon migrate to sea during their first summer, virtually all of the adults returning show one or more years of freshwater growth on their scales.

In a fifth group of temperate fishes there are two or more distinct feeding migrations to separate feeding habitats. This pattern is shown by most sockeye salmon populations whose young migrate shortly after emergence to a lake feeding habitat and remain there for several months to years before again migrating as smolts to the final marine feeding habitat (Hartman *et al.* 1967, Foerster 1968). Some chinook salmon populations also undergo a similar two-stage feeding migration (Reimers 1973, Northcote 1974), the first taking the recently emerged young from the spawning grounds to a freshwater feeding area in mainstem river tributaries, marshes or lakes and the second from there to the ocean several months or more later as smolts. Up to five species of galaxids might be included in this group even though the initial feeding migration from freshwater or estuarine spawning habitats to the ocean apparently is passive and the young only feed there for about six months before actively migrating as juveniles back to their main freshwater feeding habitat (McDowall 1968a, b, 1970, 1971, Campos 1973).

In the above groupings only the initial migration from the spawning habitat to the feeding habitat or between different feeding habitats has been considered. Within any particular feeding habitat there may be complex horizontal movements or vertical 'micro-migrations' which are essential to procurement of the food itself (Northcote 1967, Proddubny *et al.* 1968, Royce *et al.* 1968). Diel vertical feeding movements of sockeye salmon fry in rearing lakes are good examples of the latter (Narver 1970, McDonald 1973). Furthermore, in those species which undergo wintering migrations there may be post-winter movements back to feeding habitats. These are well illustrated by immature Dolly Varden in southeastern Alaska which make a return feeding migration to the sea from lakes or large river wintering habitats (Armstrong 1970, 1974). Also adults which survive after spawning usually return to feeding habitats in rivers, in lakes or in the ocean. Several species of trout, char and whitefish show this form of feeding migration in temperate waters.

For any one species several of the above types of feeding migration may be found in different populations or even within one (Reimers 1973). Usually only the most common pattern has been referred to herein.

Arctic regions

Few strikingly different patterns of feeding migrations are evident in arctic freshwater fishes in addition to those noted previously for temperate forms. However, those arctic species which first establish a feeding habitat in freshwater usually remain there much longer than do temperate forms before migrating to their second feeding habitat, such as an estuary or the ocean.

well as for migratory arctic char, *Salvelinus alpinus* (Gullestad 1973, Moore 1975a, b, Moore & Moore 1974), Atlantic salmon (Power 1973) and omul, *Coregonus autumnalis* (Dushin 1966), but apparently not for Siberian cisco This holds for several northern stocks of sockeye salmon (Foerster 1968) as *Coregonus albula* (Ustyugov 1972). The latter are represented by two forms which reach the Yenisei estuary from upriver spawning areas as yearlings or older stages. From then until they spawn several years later they occupy different feeding areas off the mouth of the river. One form, the large Kara cisco, is more euryhaline and feeds on mysids over a wide area away from the river mouth, whereas the smaller Turukhan form feeds on copepods and cladocerans close to the river mouth.

Arctic grayling (*Thymallus arcticus*) inhabiting river systems tributary to the Beaufort Sea in Alaska exhibit complex feeding migrations (Craig & Poulin 1975) which differ in several ways from the temperate types outlined previously. The fry, after emergence, remain in the spawning tributaries where they feed until early autumn and then all migrate downstream. Next year the juvenile and subadult grayling (1 to 7 years old) leave their over-wintering habitat in streams after ice break-up in early June and make extensive migrations, in some cases downstream and then upstream to reach feeding areas in rivers, lakes or even the same tributary from which they originated as fry. Grayling of the Pechora River in northern Russia also make a late spring feeding migration to upper reaches and tributaries where they stay until autumn (Zakharchenko 1973).

The seaward feeding migration of subadult fish from freshwater wintering habitats, noted previously in some north temperate Dolly Varden, also seems to be very common in arctic char (Mathisen & Berg 1968, Moore, 1975a, b, G. Power pers. comm.). Even though many of these fish may spend only a month or less feeding in the sea, this time is critical to their growth as they increase little in length during the long winter period and may decrease appreciably in weight (Mathisen & Berg 1968).

Most species inhabiting waters in arctic regions either have well defined anadromous stocks which make feeding migrations to the sea or representatives which at times feed well out into river estuaries and occasionally along the coast. At least 30 of over 40 species in arctic regions of north-western Canada and Alaska (McPhail & Lindsey 1970) make such movements, certainly a much higher proportion than among temperate freshwater fish of North America.

Tropical regions

Although tropical freshwaters do not have the marked summer and winter seasonality characteristic of temperate and arctic regions, the alternation between dry and wet seasons produce effects on fish which may in some ways be similar (Lowe-McConnell 1964). Hence it is not surprising to find that most patterns of feeding migration found in temperate fishes also have tropical counterparts. In several species which spawn in tributaries to Lake Victoria either the eggs or fry are washed back downstream to the lake (Van Someren 1963) so that there is little or no feeding in stream habitats. Welcomme (1969) notes that such behaviour only occasionally occurred in the Kafunta system at Bugungu on the lake. Nevertheless feeding migrations described for Amazon affluents by Fittkau (1970, 1973) seem to be similar to several of the patterns recognised for temperate fishes.

The early seaward migration of larval *Hilsa ilisha* down the Indus and Ganges river systems (Sujansingani 1957, Pillay 1958) as well as several rivers of southern India (Ganapati 1973) fits that shown by north temperate smelt and eulachon. Young shad (*Ethmalosa sp.*) make similar feeding migrations from West African lagoons and estuaries as do young pellonuline clupeids in some rivers there (J. D. Reynolds pers. comm.). These are some of the few examples of seaward feeding migrations in tropical regions.

Just as in temperate waters, young of some tropical fishes remain to feed at or near the parental spawning habitat for some time before migrating to their main feeding area. This is exemplified by three species of *Barbus* described by Welcomme (1969) in a Lake Victoria tributary. A similar migration may be shown by two species of *Barilius* in tributaries of Lake Malawi (Jackson *et al.* 1963).

There are several examples of two-stage feeding migrations in South American rivers and probably these occur in other tropical regions. Bayley (1973) notes that the semi-pelagic eggs of the migratory characins (*Prochilodus*) are swept downstream to river backwaters where the young remain to feed on algae and entomostracans before moving on downstream a further 500 km to their major feeding grounds in the River Grande. Apparently characins and other fishes in the Parana River system perform a similar migration (Bonetto *et al.* 1969, Godoy 1972).

As might be expected, considering the great diversity of freshwater fishes in tropical regions (Fryer & Iles 1972, Lowe-McConnell 1975), there are special forms of feeding migrations shown by fishes there. The most common of these is associated with the seasonal inundation of river floodplains (Welcomme 1974, 1975), whereby the young hatching out in areas along the flooded rivers, as well as adults, migrate laterally into rich feeding habitats (Roberts 1972, Petr 1974, 1975). Early stages of *Hemihaplochromis multicolor* are carried by mouth-brooding females from stream spawning areas to swamp feeding areas, where the young grow and later migrate downstream to river feeding areas (Welcomme 1969). Fryer and Iles (1972) describe for *Sarotherodon variabilis* another variant of such a migration over considerable distances within Lake Victoria. Somewhat similar feeding migrations are shown by *Sarotherodon mossambicus* in the South African Lake Sibaya

(Bruton & Boltt 1975). Some Lake Malawi fishes, for example the Usipa (*Engraulicypris sardella*), have pelagic larvae, which first feed well out in the lake but later migrate inshore in large schools to feed along sandy beaches (Fryer & Iles 1972).

WINTERING MIGRATIONS

Temperate regions

The subadult and adult stages of many anadromous fishes in temperate waters may leave feeding areas in the sea during autumn and migrate to over-wintering habitats in rivers or lakes. Nikol'skii (1963) records such behaviour in roach (*Rutilus rutilus*), bream (*Abramis* sp.), pike-perch and other fishes of the North Caspian and Aral seas. In autumn the ide which have been feeding in the sea return to the Swedish river Kävlingeån, the smallest ones first, and by late November all age groups are found in the lower reaches where they remain until next spring (Cala 1970). Maksimov (1972) notes that Kamchatka trout move in autumn and winter from the Sea of Okhotsk to pools of the Utkholok River. Armstrong (1971) suggests that sea-run cut-throat trout in southeastern Alaska make annual wintering migrations to a lake and has conclusive evidence of similar migrations for anadromous Dolly Varden in that area (Armstrong 1970, 1974).

Although Nikol'skii (1963) notes that wintering migrations from lakes to lower reaches of rivers are common in freshwater fish, giving grass carp (*Ctenopharyngodon idellus*) as an example, there seem to be few well docu-mented cases.

Wintering migrations also occur in lake resident fishes which leave littoral feeding areas in autumn and move to offshore, deeper habitats. Examples include cutthroat trout in coastal British Columbia lakes (Andrusak & Northcote 1971) and brown trout in Loch Leven (Thorpe 1974). Adult walleye apparently migrate to overwintering areas in several of the Canadian Great Lakes (Ferguson & Derksen 1971).

Some fish which spend their entire lives in rivers nevertheless may make extensive wintering migrations within them. Bream in the upper Don River do so (Fedorov *et al*. 1966). After spawning, the population of mountain whitefish in the North Fork of the Clearwater River, Idaho make a down-stream migration (up to 88 km) to overwinter in deep pools of the lower river (Pettit & Wallace 1975) where winter conditions are less severe and presum-ably survival is better.

Arctic regions

Surprisingly, there seem to be few examples of wintering migrations in arctic freshwater fish probably because of the paucity of detailed studies in arctic regions. Wintering migrations are illustrated by the anadromous arctic char of northern Labrador (Andrews & Lear 1956), Baffin Island (Grainger 1953, Moore & Moore 1974, Moore 1975a, b) and northern Norway (Mathisen & Berg 1968). Each year in late summer the subadult and adult char leave their feeding habitat in the sea and migrate back to rivers or lakes where they remain, often with very minimal feeding possibilities, up to eight months (September–May) in the Canadian arctic and nearly 11 months in northern Norway. Two forms of Siberian cisco have different overwintering migrations at the mouth of the Yenisei River but neither apparently move into the river proper during winter (Ustyugov 1972).

Spring-fed tributaries to the Canning River in Alaska form overwintering habitat critical to both anadromous and stream resident populations of arctic char in Alaska (McCart & Craig 1973, Craig & Poulin 1975). In Scandinavian reservoirs arctic char migrate long distances during mid to late winter to concentrate in regions where there is current (Aass 1970).

River populations of arctic grayling leave summer feeding habitats in small tributary streams of watersheds in northern Alaska and Canada, and migrate in autumn to widely separated and very localized overwintering areas in the watershed, in some cases over 80 km from their summer feeding habitat (Craig & Poulin 1975). The whole population may be concentrated in small spring-fed overwintering sites for 8 months of the year. Grayling of the Upper Pechora River in northern Russia make wintering migrations as far as 37 km downstream to deep pools (Zakharchenko 1973), and once there continue to make diel migrations of up to 2 km back and forth from rapids from feeding.

Tropical regions

In tropical freshwaters conditions which develop during the dry seasons may induce migrations which in some respects are the counterpart of the wintering migrations of temperate and arctic regions. When marshes, backwaters and tributaries which form rich feeding habitats dry up, young and adults alike are forced to move back into the mainstem river channels or lakes (Roberts 1972) where food resources are probably much more limited. Studies by Welcomme (1969, 1974, 1975) document this sequence in African fishes utilizing floodplain feeding habitats, and there are similar examples to be found in other tropical regions (Lowe-McConnell 1975). Unlike the overwintering period in many temperate and most arctic habitats, in tropical freshwaters there still may be considerable growth during the 'winter' although feeding usually is reduced. In Lake Sibaya, South Africa, *Sarotherodon mossambicus* adults move to offshore deeper waters in 'winter' and little feeding occurs there (Bruton & Allanson 1974, Bruton & Boltt 1975).

SPAWNING MIGRATIONS

Temperate regions

Most migratory freshwater fishes in temperate regions, after accumulating adequate energy reserves in their final feeding habitat for both the production of gametes and their transport, move to another region more suitable for reproduction. A wide range of migratory patterns is evident, depending on when such migrations begin, when spawning actually occurs, and what type of spawning habitat is utilized. Nevertheless, in many cases the essential feature seems to be a more or less obligatory return from a habitat rich in food but unsuitable for spawning to an area which can provide the requirements for reproduction, even though possibilities for feeding by adults and perhaps even young stages may be far from optimal. Thus, we find many examples in temperate fishes of spawning migrations from lake or marine feeding habitats to stream or river spawning habitats. The trout, salmon and char exhibit perhaps the greatest diversity in this respect but such behaviour also occurs in lampreys, sturgeon, thymallids and coregonids as well as in some clupeids, gasterosteids and even cyprinids such as ide (Cala 1970). While anadromous forms, which are to be found in all these groups, face additional problems of osmoregulation during the spawning migration (as their young also do during their feeding migration) as well as longer migratory routes, the general patterns of migration in anadromous and non-anadromous forms are not strikingly different. Many of the features described previously for spawning migrations from lakes to streams (Northcote 1967) apply with minor changes to forms, often of the same species, migrating from the sea to rivers. Commonly two or more seasonal races of an anadromous form enter spawning habitats while in different physiological conditions. One of these races may overwinter before spawning, but the other spawns almost at once (Berg 1934, Nikol'skii 1963, Smith 1969, Shilov *et al.* 1970). Such differentiation does not seem to be common in strictly freshwater fishes although seasonal races of different size do occur in river spawning runs of ayu (*Plecoglossus altivelis*) in Lake Biwa, Japan (Azuma 1973). The vimba (*Vimba vimba*) which spawns on stony substrate must migrate long distances up the Danube River or its tributaries to find a suitable spawning habitat (Moroz 1970) illustrating how spatial separation between rich feeding habitats and suitable spawning habitats may bring about extensive migrations in some freshwater fishes.

The well known catadromous migration of eels (Harden Jones 1968, Tesch 1973, 1975) wherein the adults leave freshwater feeding areas in rivers or lakes and move to an oceanic spawning habitat, is found to a lesser degree in a few other marine fishes such as mullet (*Mugil* spp.). Although galaxids are not catadromous (McDowall 1968a) one species does leave its stream feeding habitat and makes a spawning migration at least to estuarine areas (McDowall

1968b, Benzie 1968, Campos 1973), showing a pattern perhaps intermediate between the common one for temperate freshwater fishes and that of eels.

Arctic regions

Most types of spawning migrations found in temperate fishes are exemplified in arctic species as well. Anadromous Arctic char in marine waters along coastal northern Alaska, Canada and Eurasia migrate in late summer and autumn to freshwater spawning areas usually not far inland (McPhail & Lindsey 1970). The omul in the Kara sea may migrate over 1,000 km up the Yenisei River to spawn but in other rivers moves only a few kilometers (Podlesnyy 1968). The Kara form of the Siberian cisco migrates over 400 km up the Yenisei whereas the main spawning ground of the Turukhan form is some 1,550 km from the river mouth. Arctic fishes also undergo spawning migrations entirely within streams, similar to their temperate counterparts. Craig and Poulin (1975) describe long distance spawning migrations of arctic grayling in several river systems of northern Alaska and Canada. The general features of spawning migrations of arctic and temperate fishes are much the same, but spawning usually first occurs at an older age in the arctic, and at more infrequent intervals in species which spawn several times. Only a fraction of adults within a particular year class may spawn in any one year (Kennedy 1953, Johnson 1972).

Tropical regions

Reproductive migrations of fish from marine feeding areas to freshwater spawning habitats do not seem to be nearly so common in tropical waters as in temperate and arctic regions. One of the few tropical examples is that of the Indian *Hilsa ilisha* which migrates in large numbers up rivers during the monsoon period of flooding (Jones 1957, Islam & Talbot 1968, Rajyalakshmi 1973). J. D. Reynolds (pers. comm.) suggests that the spawning migration of shad and sea catfish into some West African estuaries might be other examples. The clupeid, *Pellonula vorax*, also make such migrations (Reynolds 1974).

Most spawning migrations of tropical freshwater fishes are potadromous movements within rivers and their floodplains or from lakes into rivers. These seem to be timed so that adults and young may exploit the rich resources which result from flooding. Many African examples are provided by Welcomme (1975) some involving distances up to 400 km. For South America Bayley (1973), Bonetto *et al.* (1969) and Godoy (1972) describe spawning migrations well over 600 km for some species of characins and catfishes. Several cyprinid fishes migrate to upper reaches of rivers in southern India for breeding purposes (Ganapati 1973), and others such as the carps time spawning to the monsoons when flooding occurs (Sinha *et al.* 1974).

Jackson *et al.* (1963), Welcomme (1969), Fryer and Iles (1972), Reynolds

(1971, 1974) and Lowe-McConnell (1975) describe spawning migrations from African lakes to rivers, and Santos (1973) records several species which move at spawning time from varzea lakes into the Amazon River at the beginning of the flood period.

Some tropical freshwater fish also may make extensive and regular spawning migrations entirely within lakes as noted by Fryer and Iles (1972), Scott (1974), Bruton and Allanson (1974) and Bruton and Boltt (1975). Not all tropical fishes time spawning movements with the period of floods. Reynolds (1974) discusses a few cases where spawning occurs during low waters of the dry season.

13.3 Strategies of migration

OPTIMIZE FEEDING

Many of the patterns of migration previously outlined appear to be strategies for young stages to optimize feeding by moving to habitats more productive than those where they hatched. In several types of feeding migrations a permanent switch is made from the parental spawning habitat to one for feeding which is usually much more productive the year round and where the juveniles and subadults remain until maturity. The migration may occur at early stages, or after a period of feeding at or near the spawning habitat. If indeed the prime strategy in these examples is feeding optimization to increase abundance via changes in growth, fecundity or survival, one might expect selection towards reducing the period of the life history spent in poor feeding habitats. In some populations of pink salmon which spawn in the lower reaches of streams or even at the mouths of rivers entering the sea the freshwater stage is reduced almost solely to egg incubation. Also sockeye salmon and kokanee which are typically two-stage feeding migrants may spawn in lakes thus eliminating one of the migratory phases, i.e. from the spawning habitat to the first feeding habitat. Nevertheless, selection for early migration to rich marine feeding habitats may be balanced by the lower saltwater mortality that larger young, having stayed longer in freshwater, would experience once they migrate to sea. Drucker (1972) suggests such a compensatory mechanism in coho salmon smolts of the Karluk River system in Alaska which have a longer freshwater residence than most stocks and are larger as seaward migrants.

In many feeding migrations the arrival of young at their feeding habitats seems to be timed so that they optimize foraging possibilities there. For example, sockeye salmon fry reach their lake feeding habitat in early spring at the beginning of the growing season when water temperatures and plankton are increasing and other conditions for rapid growth and survival are best. The smolts arrive at marine feeding areas at a similar time a year or more later. Obviously there is much geographical and other variation in the time at which conditions usually become optimal for feeding in particular habitats. In north temperate species occupying a wide latitudinal range, one would expect optimal feeding conditions earliest in the southern portion of their range. The seaward migration of sockeye smolts from Cultus Lake (about 49°N) is over half completed by late April on most years whereas that from Tazlina Lake (about 62°N) in Alaska does not reach this stage until late June (Hartman *et al.* 1967).

Highly productive feeding habitats are often found in rich coastal marine waters or estuaries, and these have been exploited by many temperate and arctic anadromous species. Numerous cases were noted of feeding migrations from stream spawning to lake feeding habitats as well as from up-river spawning to lower reach feeding habitats, as in vimba of the Danube (Moroz 1970). Although the evidence is not entirely convincing that all of these represent movements to richer feeding areas, many probably are. In northern Pacific waters juvenile salmon after seaward migration capitalize initially on the rich food resources present in estuarine and near-shore regions but most eventually move long distances offshore (Royce *et al.* 1968). This probably prevents rapid exhaustion of food supplies which might occur if the enormous numbers of young involved did not disperse, leaving the estuaries and coastal waters available as temporary nursery areas for juveniles subsequently migrating downstream from the rivers. The particular species, stocks and age classes of juvenile and subadult salmon do not seem to form discrete schools in the high-seas but instead overlap greatly in their ranges. Nevertheless, they do occupy distinctive, if large, regions of the ocean and apparently move continuously but not passively in a large, looped pathway which terminates in a well timed, predictable pattern near the mouth of the home spawning river (Royce *et al.* 1968). The net result is that the various age (size) classes, stocks and species are exploiting different parts of the oceanic feeding habitat at different times. This migration may well optimize production for an aggregate of populations and species, as well as for individual populations (P. A. Larkin, pers. comm.).

Some freshwater fishes, notably the arctic grayling, appear to have optimized feeding possibilities within cold arctic rivers by utilizing the rich food resources which develop during the short summer in small, warm tributaries to the mainstem rivers. Adults migrate long distances to spawn in certain tributaries, where the young emerge shortly after and feed until autumn in the warm waters (up to 17C) with abundant food (Craig & Poulin 1975). Although the adults do not stay long in these tributaries after spawning, yearling and older juveniles which also migrate there shortly after the adult spawners may remain to feed for much of the summer. We see here, and in the Pechora River of northern Russia (Zakharchenko 1973), an arctic variant of the temperate example described by Hall (1972) where upper, more productive, reaches of some streams serve as the important feeding habitat for young or even adult stages of migrating fishes. Interestingly enough in marine fishes there are also species which for various reasons spawn in rather unproductive offshore regions but whose young migrate onshore to rich coastal and estuarine feeding habitats (Hall 1975).

The strategy of another group of feeding migrations is to rapidly exploit a temporarily rich food source. General aspects of this type of migration in birds have been noted by Schoener (1971) and discussed more explicitly by Katz (1974). In fishes the massive short-term migrations of many tropical species onto river flood-plains to take advantage of a sudden increase in the food base brought about by flooding and washing out of terrestrial sources or the elaboration of new production with the upsurge in nutrient supply provide good examples (Santos 1973, Welcomme 1975). A similar strategy appears to have been adopted by fish utilizing temporarily flooded littoral zones of tropical lakes (Bruton & Allanson 1974, Bruton & Boltt 1975), a region of major importance to these ecosystems (Howard-Williams & Lenton 1975). Comparable migrations occur in some estuaries (Hoss 1974) and other examples for temperate freshwater fishes also can be found. Several species apparently follow or move onto spawning grounds used by salmon or trout and prey on eggs during the spawning season. Mountain whitefish schools move downstream to feed intensively on eggs of the large Kootenay Lake rainbow trout spawning in a restricted area at the outlet of Trout Lake in British Columbia. Rainbow trout predation on young sockeye salmon may be concentrated during brief periods when the fry move en masse into rearing lakes (Ward & Larkin 1964).

Environments which characteristically are highly productive, even if temporarily so, also seem to be ones which can be stressful (physiologically, physically or otherwise). Hall (1972) noted this in New Hope Creek where upper, more productive regions were also those subject to extreme diel variations in oxygen and seasonal changes in water depth compared to its less productive but more stable lower reaches. Migration here could optimize both fish production and ecosystem energy expenditures 'by sending armies of young to the highly productive regions to get a quick start in life, followed by dispersal of those that survive to more stable regions' (Hall 1972). Estuary environments may share many of these features (C.A.S. Hall, pers. comm.). Indeed the greater stresses imposed in highly productive environments combined with the often temporary nature of this production may explain why they are not permanently colonized by migratory species.

There seems to be good evidence that migratory strategies for optimizing feeding enhance growth rates, fecundity or survival. Clear demonstration that these increases similarly affect production is usually lacking. Migration from poor to rich feeding habitats is often reflected in rapid growth patterns on scales or other structures of individual fish, permitting age distinction even in many tropical species. Nikol'skii (1963) gives two examples of marine fishes which have migratory and non-migratory forms wherein the former, with larger and presumably richer feeding areas, are considerably more abundant. There are many more examples for freshwater fishes from various parts of the world. Anadromus arctic char are much larger and more fecund than their landlocked counterparts from the same region (Sprules 1952) or even if compared with non-migratory forms from temperate lakes (Everhart & Waters 1965). When anadromous artic char leave the rich marine feeding areas on their apparently obligatory wintering migrations to freshwater their growth rate is severly reduced (Mathisen & Berg 1968, Moore 1975b). Anadromous Dolly Varden in southeastern Alaska are much larger and very much more fecund than are the resident forms (Blackett 1973). In brown trout Bagenal (1969) notes that forms from mountain streams (presumably non-migratory) have far fewer eggs than in lowland waters, and a similar observation could be made for many headwater vs. below-falls trout populations in British Columbia streams (Northcote 1969a, Northcote *et al.* 1970). Anadromous species of lamprey are much larger than those restricted entirely to freshwater (Hardisty & Potter, 1971). Growth rate and fecundity of several anadromous or semi-anadromous cyprinids are much greater than in resident forms (Shikhshabekov 1969, Kunin 1974). Iles (1971) notes that the migratory cichlids (*Haplochromis*) in Lake Nyasa are also those which are most abundant and draws a causal relationship between these characteristics.

In migratory freshwater fishes one might expect that the females should migrate to a greater degree and spend longer in the rich feeding habitats in order to maximize egg production. Nikol'skii (1963) observes that in migratory fishes, females are often larger than males, and notes that in several species the males in the population, or at least many of them, do not migrate. Maksimov (1972) has found that in the Utkholok River a proportion of the anadromous Kamchatka trout, represented only by males, do not migrate to the estuary or the sea but mature within the river. In the upper Yukon there are small, mature but non-anadromous male chinook salmon, but it is not known if these spawn with the large anadromous females, some of which have migrated upstream over 2,000 km from the ocean (McPhail & Lindsey 1970). In at least one Swedish river the small, non-migratory male Atlantic salmon are common and spawn successfully with the large migratory females (Österdahl 1969). In many migratory salmonids, the males usually spend less time than females in feeding habitats before returning to spawn, and in several species precocious males constitute a significant fraction of the population, as for example in sockeye salmon (Foerster 1968) and rainbow trout (Hartman *et al.* 1962). Most males of Masu salmon (*Oncorhynchus masou*) are much less migratory than females and mature in freshwater.

It was noted earlier that there seemed to be few examples of anadromous migrations by freshwater fish in tropical regions. This may be a result of the oceanic waters in such areas usually being much less productive than in temperate or polar areas. Whereas the seas on temperate shelf areas are characterized by average annual primary production levels of 50–300 gC m^{-2} and boreal areas are even higher, vast areas of the tropical and subtropical oceans produce only 1–50 gC m^{-2} annually (Hempel 1973). Also tropical oceanic waters may not have such seasonally predictable periods of blooms in primary production as do temperate regions, which permit selection for regular migratory patterns (C.A.S. Hall, pers. comm.). One important anadromous fish in tropical regions, the clupeid *Hilsa ilisha*, may not move far offshore but instead feed in the rich estuarine areas as, for example, at the mouth of the Gangetic Delta. Certainly the young after migrating down the river spend much time feeding in estuarine waters (Sujansingani 1957), and some stocks may not even leave river influence (Pillay 1958).

Despite the large amount of circumstantial evidence indicating a causal relationship between food supply and migration outlined above, as well as the strong statements by Heape (1931), Nikol'skii (1963) and others that shortage of food in itself may serve as the immediate stimulus for feeding migrations, there do not seem to be many experimental studies to draw on in support of this view. Certainly, experimental enrichment of food supply has been shown in lakes and streams to increase growth and population density of fish residing there at least temporarily (Barraclough & Robinson 1972, Huntsman 1948, Mills 1969, Mason 1974c). In laboratory systems the importance of prey abundance in setting aggression levels, territory size and emigration rate has been indicated for young coho salmon, (Mason & Chapman 1965). Although Symons (1971) in artificial stream experiments with hatchery-reared Atlantic salmon parr could not demonstrate any immediate effect of food abundance on density or territory size of socially dominant young, he did show higher emigration rates at low than at high food ration and noted that 'changes in the abundance of food may cause a shift in numbers of fish from an area of poor supply to one of better supply.' That food abundance might directly affect short term migration patterns of rainbow trout fry was suggested in field studies and supported by laboratory experiments (Slaney & Northcote 1974). In stream channels provided with abundant food the frequency of aggressive encounter was low, territory size small and after 8 days less than a third of the fry had emigrated (Fig. 13.2). On the other hand when food was scarce the frequency of aggression was high, territory size large and nearly 90% of the fry emigrated from the stream channels.

AVOID UNFAVOURABLE CONDITIONS

While many feeding migrations may be interpreted as a strategy to increase growth, fecundity or survival and perhaps production by optimizing the feeding habitat, 'wintering' migrations, on the other hand, may be seen as a

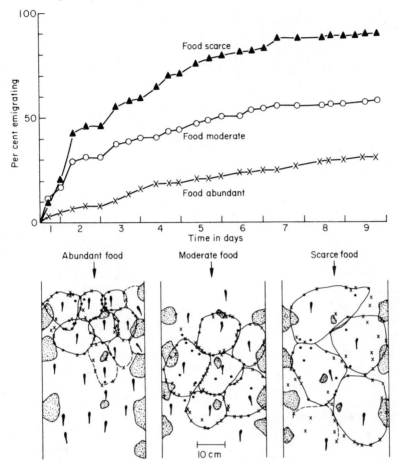

Fig. 13.2 Effect of food abundance on emigration rate (upper) and territory size (lower) of rainbow trout fry in laboratory stream channels; low food=15 mg/m², moderate= 50 mg/m², abundant=250 mg/m², at 10 minute intervals over 9 days. Lower: territory size plotted for fry remaining in stream after 8 days; solid fry show location of 'home station' within territory; •=location of aggressive interactions; x=location of peripheral feeding; stones on stream bottom shaded; broken lines indicate probable territory limits; territories not determined for all fry within study section of stream channels, especially at abundant and moderate food levels. (Adapted from Slaney & Northcote 1974).

waters, even though food resources there are minimal (Mathisen & Berg 1968, Armstrong 1970, 1971, 1974, Moore & Moore 1974, Moore 1975a, b). Also in arctic river systems there may be extensive migrations of fish from shallow tactic developed in seasonally fluctuating environments to reduce high mortality which would result if the population attempted to remain feeding in habitats which periodically become unsuitable.

Thus in subarctic and arctic waters several anadromous species leave the

very cold marine feeding areas and move back into slightly warmer fresh-
streams which freeze up during winter to very localized regions of warmer
ground water inflow (McCart & Craig 1973, Craig & Poulin 1975). Juvenile
salmonids overwintering in temperate streams make 'micro-migrations' into
habitats with suitable cover and protection from freshets, anchor ice and other
unfavourable conditions which reduce survival (Hartman 1965, Hunt 1969,
Bustard & Narver 1975). The migrations of many tropical fishes back to
lakes or to mainstem rivers to avoid being trapped in lateral marshes and
sidechannels (Santos 1973, Lowe-McConnell 1975) minimize mortalities
associated with desiccation of aquatic habitats during the dry season.

   To overwinter successfully in a habitat where survival is possible but feeding opportunities are minimal, fish must build up adequate energy reserves
before leaving their feeding habitat. Summer-run steelhead trout move up
spawning rivers in July and August but overwinter in deep headwater pools
and do not spawn until next spring (Smith 1969). They leave marine feeding
areas with much heavier deposits of body and visceral fat than do winter-run
fish. The anchovy (*Ewgraulis* sp.) in the Sea of Azov make a regular, massive
wintering migration and shortly before doing so 'stuff themselves' in their
feeding habitat to prepare for the migration by a rapid accumulation of fat
(Shul'man 1974).

   Animal migration in general may be regarded as a strategy to deal with
areas subject to marked fluctuations and frequent periods of temporarily
unsuitable conditions (Southwood 1962, Southwood *et al.* 1974, Cohen 1967,
Lack 1968). Such a view would fit well with Nikol'skii's (1963) observation
that migratory behaviour is more highly developed in temperate than in
tropical fishes. Along the Pacific coast of North America the diversity of
migratory species and their wide flexibility in patterns of migration would
seem well adapted to exploit a freshwater habitat subject to extreme variations
in discharge, temperature and other environmental conditions.

OPTIMIZE REPRODUCTIVE SUCCESS

The previous two migratory strategies discussed may optimize abundance,
size and fecundity relationships within a population, as well as minimize
losses resulting from onset of unfavourable conditions. But these benefits
would not necessarily enhance production of a population unless there were
means for ensuring that successful reproduction followed along with optimal
survival of eggs and larvae. This, of course, brings in the third of the three
key strategies involved in most migratory freshwater fish, namely reproductive
homing. The value of such behaviour has long been recognized (Lindsey *et al.*
1959, Thompson 1959, Nikol'skii 1963, Harden Jones 1968). Reproductive
homing promotes the orderly arrival on appropriate spawning habitats of

mates in the correct physiological condition. It also serves to ensure that in a complex of localized, patchily distributed spawning sites (as may occur in the dendritic branches of watersheds, in the many tributaries to large lakes, or along the shoreline of large lakes where only a particular substrate is suitable for spawning) all sites are optimally utilized, not leaving some grossly underused or unused and others heavily oversaturated so that egg survival is greatly reduced.

When the spawning habitat of a species is widely spread as in the case of the several species of salmon and anadromous trout along the west coast of North America, or even throughout a major river system, the species must adapt to a great variety of spawning conditions with particular areas being suitable only at certain times. If a parental stock was successful in utilizing a particular site, then probably its offspring should also find it suitable when they reach maturity. Furthermore, the young fish often require special rearing or feeding habitats and must make very different and sometimes complex migrations to reach them. Nevertheless, there are disadvantages in reproductive homing becoming too precise or fixed in a population as discussed by Northcote (1967) and Harden Jones (1968). The propensity to migrate seems to be flexible within stocks (Skrochowska 1969) and not entirely under innate control even in headwater streams above waterfalls where selective pressures to reduce migratory traits must be high (Northcote *et al.* 1970).

Reproductive homing has been established or strongly suggested in a number of other species apart from salmon and rainbow trout (Dodson & Leggett 1973, Carscadden & Leggett 1975a, O'Connor & Power 1973) and in cutthroat trout has been shown to be specific to particular small tributaries and even redds (LeBar 1971).

Because feeding stops or is much reduced during most spawning migrations, it would seem advantageous to ensure that their timing was adjusted so that adults cover the distance involved and reach the spawning habitat when conditions were most suitable for both successful spawning and subsequent survival of eggs and young. Nikol'skii (1963) gives an example for salmon which apparently demonstrates such adjustment. The various races of several species of Pacific salmon in the Fraser River, particularly sockeye, are generally thought to be so adapted (Northcote 1974). Nevertheless, when unusual conditions prevail on the spawning habitat (e.g. high temperatures) the timing of arrival may be inappropriate and mortality severe in certain tributaries of the Fraser (Williams 1973). Mature adults of some arctic fishes, such as the arctic grayling, arrive at their spawning tributaries just as the spring flood waters are receding (Craig & Poulin 1975). The timing here would seem to be the best compromise to minimize egg mortality from silting or scouring during the freshet period and to maximize growth of the fry which stay to feed in the tributary during the brief summer (July–August).

ENRICH FEEDING HABITATS

Migrants returning from rich feeding areas to spawn in less productive habitats may bring with them not only a large number of eggs but also a large nutrient source. That these nutrients, on decomposition of the fish which die after spawning, could enrich juvenile rearing habitats (Juday *et al.* 1932, Krokhin 1957) was discussed previously in relation to migration and production (Northcote 1967). Subsequently a number of studies have shown biogenic nutrient enrichment by migration to be an important phenomenon not only in fish (Donaldson 1967, Mathisen 1972, Krogius 1973) but also in birds (Weir 1969, Sturges *et al.* 1974, Manny *et al.* 1975).

During peak runs (15–25 million spawners) in the five-year cycle of sockeye salmon to the Iliamna system in Alaska, phosphorus from salmon carcasses may contribute more to the annual budget than that entering from tributaries and atmospheric sources (Donaldson 1967). Much of this nutrient is rapidly bound up by periphyton along the lake shore and passed up the food web via benthic invertebrates. There also may be some enrichment of phytoplankton and zooplankton but not so obviously or directly. When commercial fisheries have drastically reduced the amount of phosphorus transported by adult salmon from marine feeding habitats to those of their offspring in lakes, biogenic fertilization is no longer adequate (Mathisen 1972, Krogius 1973) and the high yields of the early years of the fisheries could not be realized again without artificial fertilization to bring back the previous productivity of the juvenile feeding habitat.

Biogenic nutrient enrichment via fish migration also may occur solely within a freshwater system. In a tributary to Lake Tahoe, Richey *et al.* (1975) have shown that decomposition of large runs of kokanee salmon after spawning significantly increased nutrient concentrations, heterotrophic activity, primary production and periphyton biomass of the stream. As there were several lags in the nutrient release cycle, it seemed quite possible that minerals released from adults in the stream could be important for enrichment of lake feeding areas near the stream mouth utilized by their offspring. Thus large numbers of adult migrants established conditions conducive to good early growth and survival of their young.

In contrast is the observation by Smith (1966) that a substantial loss of organic matter (1·5–5·1 kg/ha) was incurred in a small unproductive Canadian lake as a result of eel emigration. One wonders if the fertility of the Sargasso Sea could be enhanced by decomposition of the eel carcasses from two continents, if indeed they all migrate to spawn there!

ENHANCE COLONIZATION

Perhaps a minor but not insignificant role of migration in freshwater fish is that of facilitating colonization of new habitats as they become available or of recolonizing former ones tempoɹarily made untenable. Undoubtedly migratory characteristics have been advantageous to the spread of ranges which accompanied deglaciation of North America (McPhail & Lindsey 1970, Power *et al.* 1973), Europe and Asia, and therefore it is not surprising that we find many highly migratory species such as the salmonoids dominating the more northern drainages of these areas today. Minor range extensions and crossing of watershed boundaries are still going on. Again it is usually species such as the salmonoids with well developed migratory tendencies which seem to be in the forefront. Pink salmon introduced to northern Russia (Barents Sea area) have strayed in their spawniṅg migrations as far to the west as Scotland or Ireland and eastwards to the Kara Sea, entering the Yenisei River in thousands (W. E. Ricker, pers. comm.).

In north temperate and arctic freshwater systems of North America the past few thousand years has been a period of considerable change in drainage patterns (McPhail & Lindsey 1970) and one in which marked seasonal, annual and longer term fluctuations in environmental conditions have been the rule rather than the exception. The possibility to migrate to richer feeding areas even if only temporarily, together with movement to overwintering refuges and the common but not invariable return to parental spawning habitats would seem a most successful combination during such a period. Portions of watersheds may temporarily be made inaccessible by slides, blockages and other catastrophic events, either natural or as a result of man. Migration, particularly where reproductive homing was not too precise, would seem an obvious means for recolonization (Stott 1967, Hunt & Jones 1974).

In tropical waters migration also may bring about wide dispersal and facilitate full colonization of river and lake systems. Summing up the argument regarding the importance of predation as a causal factor for river spawning migrations of African lake fishes, Fryer (1965) notes that 'migration for spawning is probably not so much for protection of the young . . . but to ensure dispersal over the whole colonizable river course.'

Migration permits species to utilize habitats which are only temporarily available or open to them such as the many backwaters, oxbow lakes and other waters associated with river floodplains. Young of migratory fishes which spawn during monsoon flooding in marginal shallows become dispersed over wide areas of India, occurring year after year in many seasonal and perennial ponds (Qasim & Qayyum 1961).

13.4 Energetics of migration

That migration could be adaptive strategy to improve the energy balance of a species has been considered for some time, particularly in birds (West 1960, Cox 1961, 1968). To individual fish, energy costs of spawning migrations may be severe especially in some salmon populations which move long distances up turbulent rivers. Sockeye salmon in the Fraser River consume about 90 % of their body fat and about 15 % of their body protein during the up-river stage of their spawning migration (Idler & Clemens 1959). Sharp decreases occur in fat content of anadromous fishes migrating up the Volga and Amur rivers (Nikol'skii 1963), although in some fish which usually migrate short distances up lower reaches of rivers the costs appear to be much lower (Perkins & Dahlberg 1971). The energetic costs of migration for shad (*Alosa sapidissima*) in Florida are lower per unit distance than in Connecticut, but the longer distance of migration to the spawning grounds in Florida (300 km compared to 150 km in Connecticut) result in more extensive depletion and higher mortality in Florida stocks. Survival and thus the percentage of repeat spawners show a clear latitudinal cline (0 in the St. John River in Florida to over 70 % in the St. John River in New Brunswick) but there are obvious energetic as well as reproductive adaptations to specific home rivers in the stocks (Glebe & Leggett 1976).

Little attempt has been made to investigate more general ecological aspects of the energetics of fish migration, at least in salmon, probably because of the grand geographic scale over which the process usually ranges (broad expanses of oceans—large river systems). Some consideration has been given to allocation of energy expenditures on migration as it might affect selection for salmon life history patterns (Gadgil & Bossert 1970, Schaffer 1974). The energetic balance during spawning migration in favour of either fecundity or post-spawning survival is discussed by Carscadden and Leggett (1975b).

Where fish migrations occur entirely or largely within small streams, the possibilities seem more tractable for examining whether or not energy gains may result from the process as well as its general significance to the ecosystem. Apparently, the only study so far which expressly addressed itself to this question has been that of Hall (1972) on New Hope Creek, North Carolina. He tested the hypothesis for the fish community of the creek that 'migration and reproduction are coupled to optimize the use of energy resources'. Although difficulties were encountered even on this small system, the results indicated that the upstream migrations of fish (centrarchids and cyprinids being important components) effected a several fold net energy gain for the populations. If anything the methods used underestimated the gain to be derived by migration (C.A.S. Hall, pers. comm.). Upstream movement, being largely for spawning, resulted in the early juvenile stages emerging in the upper, more productive region of the stream at a time to maximize utilization of energy available for feeding. However, the fact that a positive energy balance resulted from migration in this case does not necessarily imply that such behaviour would therefore be selected for in the populations (Hall 1972).

Small (1975) notes that headwaters of streams may be closer to the major nutrient and energy inputs which could account for the upstream migrations in Hall's study. However, in some cases, especially in streams, fish may find that residency, at least during early juvenile phases rather than migration is more efficient energetically, as they utilize food carried to them in drift (Jenkins 1969, Berrie 1972, Mundie 1974, Lotrich 1975), thus reducing the energy requirements of moving to capture prey.

Juvenile sockeye salmon in Babine Lake make regular and extensive vertical diel migrations during most of the growing season, moving twice a day from cold (5–9C) hypolimnial layers to the warm (12–18C) epilimnion (Narver 1970, McDonald 1973). The possibility that such migrations may confer an energetic bonus on the young fish feeding in warm waters and metabolizing in cool waters has been investigated experimentally by R. Biette (G.H. Geen, pers. comm.). At an intermediate food ration and a cyclic temperature regime similar to that which the young fish would encounter during their diel migrations in Babine Lake, growth was indeed more rapid than at constant high or low temperatures, suggesting that in such thermally stratified waters there may well be a positive selective value for vertical migration.

13.5 Summary and conclusions

Perhaps one should not expect to find any single 'unifying principle' or 'explanation' for the variety of migratory patterns evident in freshwater fish populations, let alone those in marine fishes or in other migratory animals. Because of the diversity of species in widely separated taxonomic groups which exhibit well established migrations over a broad range of habitats and regions from the arctic to the tropics, it would seem that this behaviour has evolved independently many times and for different reasons. Within a single species and even within a single population of fish (e.g. the Loon Lake rainbow trout; Northcote 1969a) different patterns of migration may occur simultaneously in cohorts of a single year class. Nevertheless, if the complete migratory cycle is considered as a sequential unit, it may be possible to recognize general adaptive features characteristic of migration in freshwater fish even in those from different ecosystems.

We have seen several stages in the migratory cycle where production may be increased, but the most important of these is probably that associated with feeding migrations (Fig. 13.3). Although feeding migrations may occur at several periods in the life history of an individual, those that occur early in the life cycle have maximum impact on production of the population because both numbers and growth rates are highest then. Not only does migration to a rich feeding habitat increase growth rate directly by making more food available, but numbers are dispersed so that competition that otherwise would

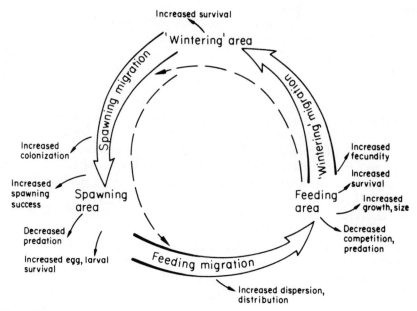

Increased survival

'Wintering' area

Spawning migration

Wintering migration

Increased
colonization

Increased
fecundity

Increased
survival

Increased
spawning
success

Spawning
area

Feeding
area

Increased
growth, size

Decreased
predation

Decreased
competition,
predation

Increased egg, larval
survival

Feeding migration

Increased dispersion,
distribution

Fig. 13.3 The cyclic pattern of freshwater fish migrations emphasizing features which enhance production. Broken lines indicate by-pass loops.

occur on crowded early rearing habitats is reduced. Furthermore, as growth rates are increased, so too are subadult and adult sizes, which in turn may increase both survival (lower predation) and fecundity. While 'wintering' migrations increase survival, growth rates are not usually increased but often lowered. Nevertheless, such migrations may be essential for survival even though they do little to enhance production at that stage. At the end of the spawning migration, the stage is set for finalizing the numbers which will initiate the next year class. Here again several factors operate through various features of the spawning migration which may increase production (Fig. 13.3).

Thus we see that there are many mechanisms for increasing survival, growth, abundance and hence production of freshwater fish which can result from their migratory behaviour. The importance of migrations seems assured as an adaptive feature of major significance in production of freshwater fish, especially in environments subject to sharp temporal fluctuations or to marked spatial patchiness in habitat fertility. Still badly needed though are studies to determine more precisely the degree to which these phases of migration regulate production and the extent to which they can be purposefully manipulated to better exploit the phenomenon in production of species useful to man.

Acknowledgements

I am grateful to a number of colleagues who read the manuscript and gave me their most helpful criticisms and comments. These include G. Fryer, C.A.S. Hall, W.S. Hoar, P.A. Larkin, J.D. Reynolds, W.E. Ricker, D. Scott, and K. Thompson. Others such as G.H. Geen, W.C. Leggett and B.C. Marcy kindly provided relevant information about specific aspects of migration on which they or their students were working. Professor J.G. Pendergrast, University of Waikato, New Zealand was most generous in providing facilities and assistance for preparation of the manuscript. The library staff at the University of Waikato and the Fisheries Research Division at Wellington, New Zealand assisted greatly in obtaining reference material as did Mrs. Ann Nelson, Librarian at the Institute of Animal Resource Ecology, University of British Columbia.

References

AASS P. (1970) The winter migrations of char, *Salvelinus alpinus* L., in the hydroelectric reservoirs Tunhovdfjord and Pålsbufjord, Norway. *Rep. Inst. Freshwat. Res. Drottningholm* **50**, 5–44.

ALEXANDER D.R. & MACCRIMMON H.R. (1974) Production and movement of juvenile trout (*Salmo gairdneri*) in a headwater of Bothwell's Creek, Georgian Bay, Canada. *J. Fish. Res. Bd. Canada* **31**, 117–121.

ANDREWS C.W. & LEAR E. (1956) The biology of artic char (*Salvelinus alpinus* L.) in northern Labrador. *J. Fish. Res. Bd. Canada* **13**, 843–860.

ANDRUSAK H. & NORTHCOTE T.G. (1971) Segregation between adult cutthroat trout (*Salmo clarki*) and Dolly Varden (*Salvelinus malma*) in small coastal British Columbia lakes. *J. Fish. Res. Bd. Canada* **28**, 1259–1268.

ARMSTRONG R.H. (1970) Age, food and migration of Dolly Varden smolts in southeastern Alaska. *J. Fish. Res. Bd. Canada* **27**, 991–1004.

ARMSTRONG R.H. (1971) Age, food and migration of sea-run cutthroat trout, *Salmo clarki*, at Eva Lake, southeastern Alaska. *Trans. Am. Fish. Soc.* **100**, 302–306.

ARMSTRONG R.H. (1974) Migration of anadromous Dolly Varden (*Salvelinus malma*) in southeastern Alaska. *J. Fish. Res. Bd. Canada* **31**, 435–444.

ARNOLD G.P. (1974) Rheotropism in fishes. *Biol. Rev.* **49**, 515–576.

AVERETT R.C. & MACPHEE C. (1971) Distribution and growth of indigenous fluvial and adfluvial cutthroat trout, *Salmo clarki*, St. Joe River, Idaho. *NW. Sci.* **45**, 38–47.

AZUMA M. (1973) Studies on the variability of the landlocked Ayu-fish *Plecoglossus altivelis* T. et S. in Lake Biwa. II. On the segregation of populations and the variations in each population. *Jap. J. Ecol.* **23**, 126–139.

BAGENAL T.B. (1969) The relationship between food supply and fecundity in brown trout (*Salmo trutta*) L. *J. Fish Biol.* **1**, 167–182.

BANKS J.W. (1969) A review of the literature on the upstream migration of adult salmonids *J. Fish Biol.* **1**, 85–136.

BARRACLOUGH W.E. & ROBINSON D. (1972) The fertilization of Great Central Lake. III. Effect on juvenile sockeye salmon. *Fishery Bull. Fish Wildl. Serv. U.S.* **70**, 37–48.

Bayley P.B. (1973) Studies on the migratory characin, *Prochilodus platensis* Holmberg

1889, (Pisces, Characoidei) in the River Pilcomayo, South America. *J. Fish. Biol.* **5**, 25–40.

BECKER C.D. (1973) Food and growth parameters of juvenile chinook salmon, *Oncorhynchus tshawytscha* in central Columbia River. *Fishery Bull. Fish Wildl. Serv. U.S.* **71**, 387–400.

BEKESHEV A.B. & PESERIDI N.E. (1966) Downstream migration of young sturgeons in the Ural River, pp. 49–51. *Biological bases of the fishing industry in Central Asian and Kazakhstan waters.* Nauka, Alma-Ata.

BELYI N.D. (1972) Downstream migration of pike-perch *Lucioperca lucioperca* (L.) and its feeding in the lower reaches of the Dnieper during early stages of development. *Vop. Ikhtiol.* **12**, 513–520.

BENZIE V. (1968) Some ecological aspects of the spawning behaviour and early development of the common whitebait, *Galaxias maculatus attenuatus* (Jenyns). *Proc. N.Z. ecol. Soc.* **15**, 31–39.

Berg L.S. (1934) Vernal and hiemal races among anadromous fishes. Fish. Res. Bd Can., Transl. Ser. No. 206, *J. Fish. Res. Bd. Canada* **16**, 515–537.

BERRIE A.D. (1972) Productivity of the River Thames at Reading. *Symp. zool. Soc. London.* **29**, 69–86.

BJORNN T.C. (1971) Trout and salmon movements in two Idaho streams as related to temperature, food, stream flow, cover and population density. *Trans. Am .Fish. Soc.* **100**, 423–438.

BLACKETT R.F. (1973) Fecundity of resident and anadromous Dolly Varden (*Salvelinus malma*) in southeastern Alaska. *J. Fish. Res. Bd. Canada* **30**, 543–548.

BOGATU D. & STANCIOIU S. (1968) Contribution to the taxomic and ecological study of the barbel (*Barbus barbus* (L.)) from the Danube (near Galati). *Hydrobiologia* **9**, 161–171.

BONETTO A.A., DIONI W. & PIGNALBERI C. (1969) Limnological investigations on biotic communities in the Middle Paranā River valley. *Verh. int. Verein. theor. angew. Limnol.* **17**, 1035–1050.

BRUNEL P. (1973) Les migrations horizontales des poissons, en particulier de la morue: résumé critque de connaissances et d'opinions recentes. *Trav. Pêch. Québec* No. 33, 22 pp.

BRUTON M.N. & ALLANSON B.R. (1974) The growth of *Tilapia mossambica* Peters (Pisces: Cichlidae) in Lake Sibaya, South Africa. *J. Fish Biol.* **6**, 706–715.

BRUTON M.N. & BOLTT R.E. (1975) Aspects of the biology of *Tilapia mossambica* (Pisces: Cichlidae) in a natural freshwater lake (Lake Sibaya, South Africa). *J. Fish Biol.* **7**, 423–445.

BUSTARD D.R. & NARVER D.W. (1975) Aspects of the winter ecology of juvenile coho salmon (*Oncorhynchus kisutch*) and steelhead trout (*Salmo gairdneri*). *J. Fish. Res. Bd. Canada.* **32**, 667–680.

CALA P. (1970) On the ecology of the ide *Idus idus* (L) in the River Kävlingeån, South Sweden. *Rep. Inst. Freshwat. Res. Drottningholm* **50**, 45–99.

Campos H. (1973) Migration of *Galaxius maculatus* (Jenyns) (Galaxiidae, Pisces) in Valdivia Estuary, Chile. *Hydrobiologia* **43**, 301–312.

Carscadden J.E. & Leggett W.C. (1975a) Meristic differences in spawning populations of American shad *Alosa sapidissima*: evidence for homing to tributaries in the St. John River, New Bruswick. *J. Fish. Res. Bd. Canada* **32**, 653–660.

CARSCADDEN J.E. & LEGGETT W.C. (1975b) Life history variations in populations of American shad, *Alosa sapidissima* (Wilson) spawning in tributaries of the St. John River, New Brunswick. *J. Fish Biol.* **7**, 595–609.

CHAPMAN D.W. (1965) Net production of juvenile coho salmon in three Oregon streams. *Trans. Am. Fish. Soc.* **94**, 40–52.

COHEN D. (1967) Optimization of seasonal migratory behaviour. *Am. Nat.* **101**, 5–17.

Cox G.W. (1961) The relation of energy requirements of tropical finches to distribution and migration. *Ecology* 42, 253–266.

Cox G.W. (1968) The role of competition in the evolution of migration. *Evolution* 22, 180–192.

Craig P.C. & Poulin V.A. (1975) Movements and growth of arctic grayling (*Thymallus arcticus*) and juvenile arctic char (*Salvelinus alpinus*) in a small arctic stream, Alaska. *J. Fish. Bd. Canada* 32, 689–697.

Dingle H. (1972) Migration strategies of insects. *Science, (N.Y.)* 175, 1327–1335.

Dodson J.J. & Leggett W.C. (1973) Behavior of adult American shad (*Alosa sapidissima*) homing to the Connecticut River from Long Island Sound. *J. Fish. Res. Bd. Canada* 30, 1847–1860.

Donaldson J.R. (1967) Phosphorus budget of Iliamna Lake, Alaska, as related to the cyclic abundance of sockeye salmon. *Ph.D. thesis, Univ. of Washington, Seattle.*

Drucker B. (1972) Some life history characteristics of coho salmon of the Karluk River system, Kodiak Island, Alaska. *Fishery Bull. Fish. Wildl. Serv. U.S.* 70, 79–94.

Dunford W.E. (1975) Space and food utilization by salmonids in marsh habitats of the Fraser River estuary. *M.Sc. thesis, Dept. Zool. Univ. Brit. Col.*

Dushin A.I. (1966) The migration of the omul (*Coregonus autumnalis*), in Pyasina Bay of the Kara Sea. *Uchen. Zap. mordov. gos. Univ.* 54, 210–214.

Everhart W.H. & Waters C.A. (1965) Life history of the blueback trout (*Salvelinus alpinus* Linnaeus). *Trans. Am. Fish. Soc.* 94, 393–397.

Fedorov A.V., Afonyushkina E.V. & Alfeev K.M. (1966) Contributions to the study of fish migrations in the Upper Don. *Rab Nauch-Issled Rybokhoz Lab. Voronezh Univ.* 3, 34–64.

Ferguson R.G. & Derksen A.J. (1971) Migrations of adult and juvenile walleyes (*Stizostedion vitreum vitreum*) in southern Lake Huron, Lake St. Clair, Lake Erie and connecting waters. *J. Fish. Res. Bd. Canada* 28, 1133–1142.

Fittkau E.J. (1970) Role of caimans in the nutrient regime of mouth-lakes of Amazon affluents (a hypothesis). *Bitropica* 2, 138–142.

Fittkau E.J. (1973) Crocodiles and the nutrient metabolism of Amazonian waters. *Amazoniana* 4, 103–133.

Foerster R.E. (1968) The sockeye salmon. *Bull. Fish. Res. Bd. Canada* 162, 422 pp.

Fryer G. (1965) Predation and its effects on migration and speciation in African fishes: a comment. *Proc. zool. Soc. Lond.* 144, 301–310.

Fryer G. & Iles T.D. (1972) *The cichlid fishes of the great lakes of Africa.* Oliver and Boyd, Edinburgh.

Gadgil M. & Bossert W. H. (1970) Life history consequences of natural selection. *Am. Nat.* 104, 1–24.

Ganapati S.V. (1973) Ecological problems of man-made lakes of south India. *Arch. Hydrobiol.* 71, 363–380.

Geen G.H., Northcote T.G., Hartman G.F. & Lindsey C.C. (1966) Life histories of two species of catostomid fishes in Sixteenmile Lake, British Columbia, with particular reference to inlet stream spawning. *J. Fish. Res. Bd. Canada* 23, 1761–1788.

Gerbilsky N.L. (1958) The question of the migratory impulse in connection with the analysis of intraspecific biological groups. *Acad. Nauk. SSSR, Ihktiol. Komissiia, Trudy Soveshchanii,* No. 8, 142–152.

Glebe B.D. & Leggett W.C. (1976) Weight loss and associated energy expenditure of American shad during freshwater migration. *U.S. Nat. Nav. Fish. Serv. Rept., Proj. AFC* 8, 110 pp.

Godoy M.P. de (1972) Brazilian tagging experiments, fishes' migration, and upper Parana River basin ecosystem. *Recta. bras. Biol.* 32, 473–484.

GRAINGER E.H. (1953) On the age, growth, migration, reproductive potential and feeding habits of artic char (*Salvelinus alpinus*) Frobisher Bay, Baffin Island. *J. Fish. Res. Bd. Canada* **10**, 326–370.

GULLESTAD N. (1973) On the biology of char (*Salmo alpinus* L.) in Svalbard: I. Migratory and non-migratory char in Revvatnet, Spitsbergen. *Nor. Blarinst. Arbok.* **1973**, 125–140.

HALL C.A.S. (1972) Migration and metabolism in a temperate stream ecosystem. *Ecology* **53**, 585–604.

HALL C.A.S. (1975) Models and the decision-making process: the Hudson River power plant case, *Ecosystem analysis and prediction* (Levin S. ed.) pp. 1–16 SIAM Inst. Math. & Soc., Philadelphia.

HARDEN JONES F.R. (1968) *Fish migration*. Edward Arnold, London.

HARDISTY M.W. & POTTER I.C.(1971) The general biology of adult lampreys, *The biology of lampreys* (Hardistry M.W. & Potter I.C. eds.) pp. 127–206. Acad. Press, London.

HARTMAN G.F. (1965) The role of behavior in the ecology and interaction of underyearling coho salmon (*Oncorhynchus kisutch*) and steelhead trout (*Salmo gairdneri*). *J. Fish. Res. Bd. Canada* **22**, 1035–1C81.

HARTMAN G.F., NORTHCOTE T.G. & LINDSEY C.C. (1962) Comparison of inlet and outlet spawning runs of rainbow trout in Loon Lake, British Columbia. *J. Fish. Res. Bd. Canada* **19**, 173–200.

HARTMAN W.L., HEARD W.R. & DRUCKER B. (1967) Migratory behavior of sockeye salmon fry and smolts. *J. Fish. Res. Bd. Canada* **24**, 2069–2099.

HASLER A.D. (1966) *Underwater guideposts*. University of Wisconsin Press, Madison.

HEAPE W. (1931) *Emigration, migration and nomadism*. Heffer, Cambridge.

HEMPEL G. (1973) Productivity of the oceans. *J. Fish. Res. Bd. Canada* **30**, pt. 2, 2184–2189.

HOSS D.E. (1974) Energy requirements of a population of pinfish *Lagodon rhombiodes* Linnaeus) *Ecology* **55**, 848–855.

HOWARD-WILLIAMS C. & LENTON G.M. (1975) The role of the littoral zone in the functioning of a shallow tropical lake ecosystem. *Freshwat. Biol.* **5**, 445–459.

HUNT R.L. (1969) Overwinter survival of wild fingerling brook trout in Lawrence Creek, Wisconsin. *J. Fish. Res. Bd. Canada* **26**, 1473–1483.

HUNT P.C. & JONE; J.W. (1974) A population study of *Barbus barbus* (L.) in the River Severn, England. II. Movements. *J. Fish Biol.* **6**, 269–278.

HUNTSMAN A.G. (1948) Fertility and fertilization of streams. *J. Fish. Res. Bd. Canada* **7**, 248–253.

IDLER D.R. & CLEMENS W.A. (1959) The energy expenditures of Fraser River sockeye salmon during the spawning migration to Chilko and Stuart lakes. *Prog. Rept. int. Pacif. Salm. Fish. Comm.* **6**, 80 pp.

ILES T.D. (1971) Ecological aspects of growth in African cichlid fishes. *J. Cons. perm. int. Explor. Mer* **33**, 363–385.

ISLAM B.N. & TALBOT G.B. (1968) Fluvial migration, spawning and fecundity of Indus River, Hilsa, *Hilsa ilisha*. *Trans. Am. Fish. Soc.* **97**, 350–355.

JACKSON P.B.N., ILES T.D., HARDING D. & FRYER G. (1963) Report on the survey of northern Lake Nyasa. *Govt. Printer, Zomba*, 171 pp.

JENKINS T.M. (1969) Social structure, position choice and micro-distribution of two trout species (*Salmo trutta* and *Salmo gairdneri*) resident in mountain streams. *Anim. Behav. Monogr.* **2**, 55–123.

JOHNSON C.G. (1969) *Migration and dispersal of insects by flight*. Methuen, London.

JOHNSON L. (1972) Keller Lake: Characteristics of a culturally unstressed salmonid community. *J. Fish. Res. Bd. Canada* **29**, 731–740.

JONES S. (1957) On the late winter and early spring migration of the Indian shad, *Hilsa ilisha* (Hamilton), in the Gangetic Delta. *Indian J. Fish.* **4**, 304–314.

JUDAY C., RICH W.H., KEMMERER, G.I. & MANN A. (1932) Limnological studies of Karluk Lake, Alaska, 1926–1930. *Bull. Bur. Fish., Wash.* **47**, 407–436.

KATZ P.L. (1974) A long-term approach to foraging optimization. *Am. Nat.* **108**, 758–782.

KENNEDY W.A. (1953) Growth, maturity, fecundity and mortality in the relatively unexploited whitefish, *Coregonus clupeaformis*, of Great Slave Lake. *J. Fish. Res. Bd. Canada* **10**, 413–441.

KROGIUS F.V. (1973) Population dynamics of growth of young sockeye salmon in Lake Dalnee. *Hydrobiologia* **43**, 45–51.

KROKHIN E.M. (1957) Sources of enrichment of spawning lakes in biogenic elements. Izv. Tikhookean. nauch. issled. Inst. ryb. Khoz., **45**, 29–35. (*Fish. Res. Bd. Canada Transl.* No. 207).

KUNIN M.A. (1974) The size and age characteristics and feeding of the eastern bream (*Abramis brama orientalis*) of the Aral Sea. *Vop. Ikhtiol.* **14**, 260–266.

LeBAR G.W. (1971) Movement and homing of cutthroat trout (*Salmo clarki*) in Clear Bridge creeks, Yellowstone National Park. *Trans. Am. Fish. Soc.* **100**, 41–49

LACK D. (1968) Bird Migration and natural selection. *Oikos* **19**, 1–9

LEGGETT W.C. & WHITNEY R.R. (1972) Water temperature and the migrations of American Shad. *Fishery Bull. Fish. Wildl. Serv. U.S.* **70**, 659–670.

LINDSEY, C.C., NORTHCOTE, T.G. & HARTMAN G.F. (1959) Homing of rainbow trout to inlet and outlet spawning streams at Loon Lake, British Columbia. *J. Fish. Res. Bd. Canada* **16**, 695–719.

LOTRICH V.A. (1975) Summer home range and movements of *Fundulus heteroclitus* (Pisces: Cyprinodontidae) in a tidal creek. *Ecology* **56**, 191–198.

LOWE-MCCONNELL R.H. (1964) The fishes of the Rupununi savanna district of British Guiana. Pt. 1. Groupings of fish species and effects of the seasonal cycles on the fish. *J. Linn. Soc. Zoology* **45**, 103–144.

LOWE-MCCONNELL R.H. (1975) *Fish communities in tropical freshwaters.* Longman Inc., New York.

MAKSIMOV V.A. (1972) Some data on the ecology of the Kamchatka trout (*Salmo mykiss* Walbaum) from the Utkholok River. *Vop. Ikhtiol.* **12**, 759–765.

MANNY B.A., WETZEL R.G. & JOHNSON W.C. (1975) Annual contribution of carbon, nitrogen, and phosphorus by migrant Canada geese to a hardwater lake. *Verh. int. Verein. theor. angew. Limnol.* **19**, 949–951.

MARCY B.C. (1976) Early life history studies of American shad in the Lower Connecticut River, and the effects of the Connecticut Yankee Plant. *The Connecticut River ecological Study: the impact of a nuclear power plant* (D. Merriman and L. Thorpe eds.) *Trans. Amer. Fish. Soc. Monograph* No. 1.

MASON J.C. (1974a) Behavioural ecology of chum salmon fry (*Oncorhynchus keta*) in a small estuary. *J. Fish. Res. Bd. Canada* **31**, 83–92.

MASON J.C. (1974b) Movements of fish populations in Lymn Creek, Vancouver Island: a summary from weir operations during 1971 and 1972, including comments on species life histories. *Tech. Rep. Envir. Can. Fish. Mar. Serv.* No. 483, 35 pp.

MASON J.C. (1974c) A further appraisal of the response to supplemental feeding of juvenile coho, (*Oncorhynchus kisutch*) in an experimental stream. *Tech. Rep. Environ. Can. Fish. Mar. Serv.* N. 470, 26 pp.

MASON J.C. & CHAPMAN D.W. (1965) Significance of early emergence, environmental rearing capacity and behavioral ecology of juvenile coho salmon in stream channels. *J. Fish. Res. Bd. Canada* **22**, 173–190.

MATHISEN O.A. (1972) Biogenic enrichment of sockeye salmon lakes and stock productivity. *Verh. int. Verein. theor. angew. Limnol.* **18**, 1089–1095.

MATHISEN O.A. & BERG M. (1968) Growth rates of the char *Salvelinus alpinus* (L.) in the

Vardnes River, Troms northern Norway. *Rep. Inst. Freshwat. Res. Drottningholm* **48**, 176–186.

McCart P. & Craig P.C. (1973) Life history of two isolated populations of Artic char (*Salvelinus alpinus*) in spring-fed tributaries of the Canning River, Alaska. *J. Fish. Res. Bd. Canada* **30**, 1215–1220.

McDonald J. (1973) Diel vertical movements and feeding habits of underyearling sockeye salmon (*Oncorhynchus nerka*) at Babine Lake, B.C. *Tech. Rep. Fish. Res. Bd. Canada* **378**, 55 pp.

McDowall R.M. (1968a) The application of the terms anadromous and catadromous to the southern hemisphere salmonoid fishes. *Copeia* **1968**, 176–178.

McDowall R.M. (1968b) *Galaxias maculatus* (Jenyns), the New Zealand whitebait. *N.Z. Marine Dept., Fish. Res. Bull* No. 2 (N.S.), 84 pp.

McDowall R.M. (1970) The galaxiid fishes of New Zealand. *Bull. Mus. comp. Zool. Harv.* **139**, 341–432.

McDowall R.M. (1971) The galaxiid fishes of South America. *Zoology. Linn. Soc.* **50**, 33–73.

McPhail J.D. & Lindsey C.C. (1970) Freshwater fishes of northwestern Canada and Alaska. *Bull. Fish. Res. Bd. Canada* **173**, 381 pp.

Meek A. (1916) *The migrations of fish.* Edward Arnold, London.

Mills D. (1969) The survival of juvenile Atlantic salmon and brown trout in some Scottish streams, pp. 217–228. *Symposium on salmon and trout in streams* (Northcote T.G. ed.) MacMillan Lectures in Fisheries, Univ. Brit. Col.

Mills D. (1971) *Salmon and trout: a resource, its ecology, conservation and management.* Oliver & Boyd, Edinburgh.

Moore J.W. (1975a) Reproductive biology of anadromous arctic char, *Salvelinus alpinus* (L.) in the Cumberland Sound area of Baffin Island. *J. Fish Biol.* **7**, 143–151.

Moore J.W. (1975b) Distribution, movements and mortality of anadromous arctic char, *Salvelinus alpinus* L., in the Cumberland Sound area of Baffin Island. *J. Fish Biol.* **7**, 339–348.

Moore J.W. & Moore I.A. (1974) Food and growth of arctic char *Salvelinus alpinus* (L.) in the Cumberland Sound area of Baffin Island. *J. Fish Biol.* **6**, 79–92.

Moroz V.M. (1970) Biological description of the vimba from the lower reaches of the Danube. *Vop. Ikhtiol.* **10**, 29–39.

Morse D.H. (1971) The insectivorous bird as an adaptive strategy. *Annual Rev. Ecol. Systematics* **2**, 177–200.

Mundie J.H. (1974) Optimization of the salmonid nursery stream. *J. Fish. Res. Bd. Canada* **31**, 1827–1837.

Narver D.W. (1970) Diel vertical movements and feeding of underyearling sockeye salmon and the limnetic zooplankton in Babine Lake, British Columbia. *J. Fish. Res. Bd. Canada* **27**, 281–316.

Nikol'skii G.V. (1963) *The ecology of fishes.* Academic Press, N.Y.

Northcote T.G. (1962) Migratory behaviour of juvenile rainbow trout, (*Salmo gairdneri,*) in outlet and inlet streams of Loon Lake, British Columbia. *J. Fish. Res. Bd. Canada* 23, 201–270.

Northcote T.G. (1967) The relation of movements and migrations to production in freshwater fishes, pp. 315–344. *The Biological Basis of Freshwater Fish Production.* (Gerking S.D. ed.) Blackwell Scientific Publications, Oxford.

Northcote T.G. (1969a) Patterns and mechanisms in the lakeward migratory behaviour of juvenile trout, *Symposium on Salmon and Trout in Streams* (Northcote T.G. ed.) pp. 183–203. MacMillan Lectures in Fisheries, Univ. Brit. Col.

Northcote T.G. (1969b) Lakeward migration of young rainbow trout (*Salmo gairdneri*) in the Upper Lardeau River, British Colombia. *J. Fish. Res. Bd. Canada* **26**, 33–45.

NORTHCOTE T.G., WILLISCROFT S.N. & TSUYUKI H. (1970) Meristic and lactate dehydrogenase genotype differences in stream populations of rainbow trout below and above a waterfall. *J. Fish. Res. Bd. Canada* 27, 1987–1995.

NORTHCOTE T.G. (1974) Biology of the lower Fraser River: a review. *Tech. Rep. No. 3, Westwater Res. Centre. Univ. Brit. Col.*, 94 pp.

O'CONNOR J.F. & POWER G. (1973) Homing of brook trout (*Salvelinus fontinalis*) in Matamek Lake, Quebec. *J. Fish. Res. Bd. Canada* 30, 1012–1014.

ÖSTERDAHL L. (1969) The smolt run of a small Swedish river, *Symposium on Salmon and Trout in Streams*. (Northcote T.G. ed.) pp. 205–215. MacMillan Lectures in Fisheries, Univ. Brit. Col.

PERKINS R.J. & DAHLBERG M.D. (1971) Fat cycles and condition factors of Altramaha River shads. *Ecology* 52, 359–362.

PETERSON H.H. (1968) The grayling (*Thymallus thymallus* (L.)) of the Sundsvall Bay area. *Rep. Inst. Freshwat. Res. Drottningholm* 48, 36–56.

PETR T. (1974) Distribution, abundance and food of commercial fish in the Black Volta and the Volta man-made Lake in Ghana during the filling period. (1964–1968) II Characidae. *Hydrobiologia* 45, 303–337.

PETR T. (1975) On some factors associated with the initial high fish catches in new African man-made lakes. *Arch. Hydrobiol.* 75, 32–49.

PETTIT S.W. & WALLACE R.L. (1975) Age, growth and movements of mountain whitefish, *Prosopium williamsoni* (Girard), in the North Fork Clearwater River, Idaho. *Trans. Am. Fish. Soc.* 104, 68–76.

PILLAY T.V.R.. (1958) Biology of the Hilsa, *Hilsa ilisha* (Hamilton) of the River Hooghly. *Indian J. Fish.* 5, 201–257.

PODDUBNY A.G., GORDEEV N.A. & PERMITIN I.E. (1968) Migration pattern of the feeding fish populations in relation to environmental factors, *Biological and hydrological factors of local movements of fish in reservoirs* (Kuzin B.S. ed.) pp. 278–349. USSR Acad. Sci., Inst. Biol. Inland Waters. Trudy No. 16 (19). Transl. Bur. Sport Fish Wildl., U.S. Dept. Int. and Nat. Sci. Found., Wash., D.C., by Amerind Publ. Co. Pvt. Ltd., New Dehli, 1974.

PODLESNYY A.V. (1968) The fundamental difference between migratory and non-migratory teleost fishes. *Problems of Ichthyology*, *Am. Fish. Soc.* 8, 165–168.

POLLARD D.A. (1971) The biology of a landlocked form of the normally catadromous salmoniform fish *Galaxais maculatus* (Jenyns) I. Life cycle and origin. *Aust. J. mar. Freshwat. Res.* 22, 91–123.

POLLARD D.A. (1972) The biology of a landlocked form of the normally catadromous salmoniform fish *Galaxias maculatus* (Jenyns) IV. Nutritional cycle. *Aust. J. mar. Freshwat. Res.* 23, 39–48.

POWER G. (1973) Estimates of age, growth, standing crop and production of salmonids in some north Norwegian rivers and streams. *Rep. Inst. Freshwat. Res. Drottningholm* 53, 78–111.

POWER G., POPE G.F. & COAD B.W. (1973) Postglacial colonization of the Matamek River, Quebec, by fishes. *J. Fish. Res. Bd. Canada* 30, 1586–1589.

QASIM S.Z. & QAYYUM A. (1961) Spawning frequencies and breeding seasons of some freshwater fishes with special reference to those occurring in the plains of northern India. *Indian J. Fish.* 8, 24–43.

RAJYALAKSHMI T. (1973) The population characteristics of the Godavary Hilsa over the years 1963–1967. *Indian J. Fish* 20, 78–94.

RALEIGH R.F. & CHAPMAN D.W. (1971) Genetic control in lakeward migrations of cutthroat trout fry. *Trans. Am .Fish. Soc.* 100, 33–40.

REIMERS P.E. (1973) The length of residence of juvenile fall chinook salmon in Sixes River, Oregon. *Oreg. Fish. Comm., Research Reports* 4, 43 pp.

REYNOLDS J.D. (1971) Biology of the small pelagic fishes in the new Volta Lake in Ghana. II. Schooling and migrations. *Hydrobiologia* **38**, 79–91.

REYNOLDS J.D. (1974) Biology of the small pelagic fishes in the new Volta Lake in Ghana. Part III: Sex and reproduction. *Hydrobiolgia* **45**, 489–508.

RICHEY J.E., PERKINS M.A. & GOLDMAN C.R. (1975) Effects of kokanee salmon (*Oncorhynchus nerka*) decomposition on the ecology of a sub-alpine stream. *J. Fish. Res. Bd. Canada* **32**, 817–820.

RICHKUS W.A. (1975) Migratory behavior and growth of juvenile anadromous alewives, *Alosa pseudoharengus*, in a Rhode Island drainage. *Trans. Am. Fish. Soc.* **104**, 483–493.

RINGSTAD N.R. (1974) Food competition between freshwater sculpins (Genus *Cottus*) and juvenile coho salmon (*Oncorhynchus kisutch*) an experimental and ecological study in a British Columbia coastal stream. *Tech. Rep. Environ. Can., Fish. Mar. Serv.* **457**, 88 pp.

ROBERTS T.R. (1972) Ecology of fishes in the Amazon and Congo basins. *Bull. Mus. comp. Zool. Harv.* **143**, 117–147.

ROYCE W.F., SMITH L.S. & HARTT A.C. (1968) Models of oceanic migrations of Pacific salmon and comments on guidance mechanisms. *Fishery Bull. Fish Wildl. Serv. U.S.* **66**, 441–462.

RUSSELL E.S. (1937) Fish migrations. *Biol. Rev.* **12**, 320–337.

SANTOS U. DE M. (1973) Beobachtungen über Wasserbewegungen, chemische Schichtung und Fischwanderungen in Várzea-Seen am mittleren Solimões (Amazonas). *Oecologia (Berl.)* **13**, 239–246.

SCHAFFER W.M. (1974) Selection for optimal life histories: the effects of age structure. *Ecology* **55**, 291–303.

SCHAFFER W.M. & ELSON P.R. (1975) The adaptive significance of variations in life history among local populations of Atlantic salmon in North America. *Ecology* **56**, 577–590.

SCHOENER T.W. (1971) Theory of feeding strategies. *Annu. Rev. Ecol. Syst.* **2**, 369–404.

SCOTT D.B.C. (1974) The reproductive cycle of *Mormyrus kannume* Forsk. (Osteoglossomorpha Mormyriformes) in Lake Victoria, Uganda. *J. Fish Biol.* **6**, 447–454.

SCOTT W.B. & CROSSMAN E.J. (1973) Freshwater fishes of Canada. *Bull. Fish. Res Bd. Canada* 184, 996 pp.

SHIKHSHABEKOV M.M. (1969) Different forms of vobla, bream and carp in the Arakum waters of Dagestan. *Vop. Ikhtiol.* **9**, 34–38.

SHILOV V.I., KHAZOV Y.K. & BATYCHKOV G.A. (1970) Races of Volga-Caspian sturgeon (*Acipenser güldenstädti* Brandt). *Vop. Ikhtiol.* **10**, 460–466.

SHULMAN G.E. (1974) *Life cycles of fish: physiology and biochemistry.* John Wiley, New York.

SINHA V.R.P., JHINGRAN V.P. & GANAPATI S.V. (1974) A review on spawning of the Indian major carps. *Arch. Hydriobiol.* **73**, 518–536.

SKROCHOWSKA S. (1969) Migrations of the sea-trout (*Salmo trutta* L.) brown trout (*Salmo trutta* M. *fario* L.) and their crosses. I. Problem, methods and results of tagging. *Polskie Archwm. Hydrobiol.* **16**, 125–140.

SLANEY P.A. & NORTHCOTE T.G. (1974) Effects of prey abundance on density and territorial behavior of young rainbow trout (*Salmo gairdneri*) in laboratory stream channels. *J. Fish. Res. Bd. Canada* **31**, 1201–1209.

SMALL J. W. (1975) Energy dynamics of benthic fishes in a small Kentucky stream. *Ecology* **56**, 827–840.

SMITH M.W. (1966) Amount of organic matter lost to a lake by migration of eels. *J. Fish. Res. Bd. Canada* **23**, 1799–1801.

SMITH S.B. (1969) Reproductive isolation in summer and winter races of steelhead trout, *Symposium on Salmon and Trout in Streams* (Northcote T.G. ed.) pp. 21–38. MacMillan Lectures in Fisheries, Univ. Brit. Col.

SOUTHWOOD T.R.E. (1962) Migration of terrestrial arthropods in relation to habitat. *Biol. Rev.* **37**, 171–214.

SOUTHWOOD T.R.E., MAY R.M., HASSELL M.P. & CONWAY G.R. (1974) Ecological strategies and population parameters. *Am. Nat.* **108**, 791–804.

SPRULES W.M. (1952) The arctic char of the west coast of Hudson Bay. *J. Fish. Res. Bd. Canada* **9**, 1–15.

STAUFFER T.M. (1972) Age, growth and downstream migration of juvenile rainbow trout (*Salmo gairdneri*) in a Lake Michigan tributary. *Trans. Am. Fish. Soc.* **101**, 18–28.

STEPHAN B. (1973) Population dynamics and bird migration. *Mitt. Zool. Mus. Berl.* **49**, 175–183.

STOTT B. (1967) The movements and population densities of roach (*Rutilus rutilus* L.) and gudgeon (*Gobio gobio* (L.)) in the River Mole. *J. Anim. Ecol.* **36**, 407–423.

STURGES F.W., HOLMES R.T. & LIKENS G.E. (1974) The role of birds in nutrient cycling in a northern hardwood ecosystem. *Ecology* **55**, 149–155.

SUJANSINGANI K.H. (1957) Growth of the Indian shad, (*Hilsa ilisha* Hamilton), in the tidal stretch of the Hooghly. *Indian J. Fish.* **4**, 315–335.

SYMONS P.E.K. (1970) The possible role of social and territorial behaviour of Atlantic salmon parr in the production of smolts. *Tech. Rept. Fish. Bd. Canada* No. 205, 25 pp.

SYMONS P.E.K. (1971) Behavioral adjustment of population density to available food by juvenile Atlantic salmon. *J. Anim. Ecol.* **40**, 569–587.

TESCH F.-W. (1973) *Der Aal. Biologie und Fischerie*. Paul Parey, Hamburg.

TESCH F.-W. (1975) Orientation in space: Animals, Fishes, *Marine Ecology. Volume 2 Physiological Mechanisms Part 2*. (Kinne O. ed.) pp. 657–707. John Wiley and Sons, London.

THOMPSON W.F. (1959) An approach to population dynamics of the Pacific red salmon. *Trans. Am. Fish. Soc.* **88**, 206–209.

THORPE J.E. (1974) The movements of brown trout, *Salmo trutta* (L.) in Loch Leven, Kinross, Scotland. *J. Fish. Biol.* **6**, 153–180.

USTYUGOV A.F. (1972) The ecological and morphological characteristics of the Siberian cisco (*Coregonus albula sardinella* (Valenciennes)) from the Yenisey basin. *Vop. Ikhtiol.* **12**, 745–759.

VAN SOMEREN V.D. (1963) Freshwater fishery research in East Africa. *Sci. Prog., Lond.* **51** 1–11.

WARD F.J. & LARKIN P.A. (1964) Cyclic dominance in Adams River sockeye salmon. *Prog. Rept. int. Pacif. Salm. Fish. Comm.* **11**, 1–116.

WEIR J. (1969) Importation of nutrients into woodlands by rooks. *Nature, Lond.* **221**, 487–488.

WELCOMME R.L. (1969) The biology and ecology of the fishes of a small tropical stream. *J. Zool. Lond.* **158**, 485–529.

WELCOMME R.L. (1974) A brief review of the floodplain fisheries of Africa. *Afr. J. Trop. Hydrobiol. Fish., Special Issue* **1**, 67–76.

WELCOMME R.L. (1975) The fisheries ecology of African floodplains *FAO-CIFA Tech. Paper* No. 3, 51 pp.

WEST G.C. (1960) Seasonal variation in the energy balance of the tree sparrow in relation to migration. *Auk.* **77**, 306–329.

WILLIAMS I.V. (1973) Investigation of the prespawning mortality of sockeye in Horsefly River and McKinley Creek in 1969. *Prog. Rept. int. Pacif. Salm. Fish. Comm.* **27**, Part 2, 42 pp.

WILLIAMS R. & HARCUP M.F. (1974) The fish populations of an industrial river in South Wales. *J. Fish Biol.* **6**, 395–414.

ZAKHARCHENKO G.M. (1973) Migrations of the grayling (*Thymallus thymallus* (L.)) in the upper reaches of the Pechora. *Vop. Ikhtiol.* **13**, 628–629.

Chapter 14: Social Behaviour as it Influences Fish Production

David L. G. Noakes

14.1 Introduction

Gerking (1967) noted that fish behaviour in relation to production was only beginning to receive badly needed study. He suggested that studies of behaviour would lead to interesting viewpoints on the analysis of fish production. Recent studies of social behaviour and social systems have sketched the outlines of relationships to ecology and fish production. Perhaps we can put these outlines in some perspective, and judge those areas requiring further study and definition.

The theoretical framework for the study of social behaviour is most strongly rooted in studies of birds and mammals (e.g. Altmann & Altmann 1970, Crook 1970), compared to which our understanding of fish social systems is primitive (Barlow 1974a). To understand fish social behaviour, we must assemble information from diverse, often opportunistic, studies. There have been few, if any, planned programmes of study of the behaviour of any one species in the variety of circumstances necessary to give an understanding of its social system. Comparative studies usually emphasize reproductive behaviour, as evidenced by the impressive compilation by Breder and Rosen (1966).

The importance of social behaviour to population dynamics has been recognized and demonstrated in a variety of species. The recent books by Brown (1975) and Wilson (1975) are major contributions summarizing a wealth of studies, and reviewing the 'state of the art' in ethology. Ecological considerations and implications have been incorporated into books such as those by Krebs (1972) and Ricklefs (1973). The importance of social behaviour to production in a wider range of animal species strongly suggests that it has important implications to fish production. General ecological theories, or quantitative models of fish production, usually ignore the social behaviour of the fish, or at best oversimplify its role with the assumption of uniform behaviour for all individuals within a population. Maynard Smith (1974) has proposed several theoretical models in ecology, including one concerning some ecological consequences of territorial behaviour. Barlow (1974b) has given an example to support the prediction of this model. Further testing of these and similar ecological models on fish populations would lead to profitable lines of research.

14.2 Social Systems

THEORETICAL CONSIDERATIONS

Terminology

In this review I will follow commonly accepted terminology and definitions, at least for the more general terms. Production refers to the total tissue elaborated in a given period of time, regardless of the ultimate fate of that tissue (Ivlev 1945, Chapman 1967, Balon 1974). Social behaviour includes all behaviour directly related to actual or potential encounters between individuals within a species. This includes agonistic behaviour and reproductive behaviour in addition to the gregarious and co-operative behaviour we typically think of as being social. Social organization, or the social system of a particular species, is the totality of social relationships among all members of a species (Brown 1975).

General review

Altmann (1974), and others, have pointed out the two complementary, but basically different approaches to the study of social organization and social behaviour. One approach is concerned with proximal causes, i.e. the study of immediate behavioural or motivational causes in individuals within a social group. While this is of considerable interest and importance, it will not be dealt with here, since it has relatively less direct bearing on fish production, than the other approach.

The second approach attempts to investigate the adaptive significance or ecological function of social behaviour. This approach is based on the assumption, supported by a large and rapidly growing body of factual evidence, that the social system of a species is a function of, and an evolutionary response to, the ecological situation of the species (Brown 1975). Crook (1970), Altmann (1974), and Geist (1974) give excellent reviews of such data for birds, primates, and ungulates, respectively.

I will not attempt any such exhaustive reviews or compilation of social systems and/or their likely (or possible) ecological correlates among fishes. Field studies of social behaviour of fishes are relatively scarce, and few meaningful generalizations are available (Barlow 1974a, 1976). My intention is to point out particular areas of immediate interest or importance and to suggest promising areas for future studies. The recent proposal by Balon (1975), for example, to categorize fish according to reproductive guilds, indicates the utility of more general consideration and assimilation of behavioural data into ichthyological theory.

I will give particular attention to spacing and agonistic behaviour, and schooling behaviour in the ecology of fishes. These aspects of social behaviour

will be considered in relation to feeding and/or growth, reproduction, survival
and harvest of fish. Both laboratory and field studies will be discussed to
construct as complete a picture as possible.

The study of social behaviour in relation to ecology has recently become
active for both freshwater and marine fishes. For example, there are a number
of general surveys of behaviour in marine fish communities (e.g. Hobson 1974,
Itzkowitz 1974a, Reese 1975, Sale 1975), as well as more detailed studies for
individual species (e.g. Fishelson 1970a, Keenleyside 1972a, Myrberg 1972,
Potts 1973, Losey 1974). Comparable survey studies exist for freshwater
communities (e.g. Fryer & Iles 1972, Emery 1973, Barlow 1974a, Hoar
1976), as well as detailed studies of individual species (e.g. Keenleyside 1967,
1972b, Baylis 1974).

SPACING SYSTEMS AND AGONISTIC BEHAVIOUR

Dominance-subordinance relations

A number of phenomena, with diverse implications, such as enhancement
and/or inhibition of growth as a result of undercrowding or overcrowding,
are apparently more or less direct consequences of population density. They
do not appear to fit the definition of social behaviour, and are considered by
Backiel and LeCren elsewhere in this book. Fish movements and migrations
are not considered in this chapter, although they may in some cases involve
social responses by fish (e.g. Nordeng 1971, Solomon 1973), since the
phenomenon of migration is considered at length in the chapter by Northcote.

The occurrence and consequences of dominance-subordinance interactions
are commonly observed, particularly in the confines of laboratory tanks or
aquaculture facilities, in species such as anabantids (Forselius 1957, Frey &
Miller 1972), centrarchids (Allee *et al.* 1948, Greenberg 1974, Beitinger &
Magnuson 1975), cichlids (Barlow 1974a), cyprinodontids (Magnuson 1962,
Ewing 1975), eleotrids (Yamagishi *et al.* 1974), hiodontids (Fernet & Smith
1976), ictalurids (Todd 1971, Konikoff & Lewis 1974), poeciliids (Braddock
1949), and salmonids (Brown 1946, Fabricius 1953, Fabricius & Gustafson
1954, Kawanabe 1958, Kawanabe *et al.* 1957, Keenleyside & Yamamoto
1962, Hartman 1963, 1965, Yamagishi 1962, 1964, Mason & Chapman 1965,
Edmundson *et al.* 1968, Everest & Chapman 1972). Such interactions are
readily observed, and the inference of the existence of a dominance heirarchy
is sometimes drawn. There are shortcomings, some quite serious, to such
an approximation, however. We must have individual identification of the
fish within a group, and evidence that dominance is independent of location
within the study area, before reasonably concluding that a dominance
heirarchy is present. The latter point may be critical, as this dominance may
grade into site-dependent dominance, and/or territoriality (Brown 1975).

It is also possible to have dominance interactions without necessarily having a stable dominance hierarchy in a population.

We must likewise consider the effects of artificial crowding or enclosure, which often increase these dominance interactions (Schein & Hafez 1969, Hinde 1970, Martin 1975). This is not to say that dominance heirarchies cannot, or do not exist under natural conditions, but to recommend caution when extrapolating from a particular set of circumstances to the general conditions for a species.

The importance of dominance-subordinance relations has been reported in a number of salmonoid species under field conditions (Onodera 1967, Chapman & Bjornn 1969). Newman (1956) reported loose heirarchies in small, mixed populations of rainbow trout, *Salmo gairdneri*, and brook trout, *Salvelinus fontinalis*, in streams. Staples (1975a) recorded in New Zealand a marked size heirarchy among male upland bullies, *Philypnodon breviceps*, with the larger males responsible for breeding and more successful in defending eggs. Jenkins (1969) described local heirarchies in populations of brown trout, *Salmo trutta*, in California streams. He concluded that the resident population would accommodate to altered density by forming relatively stable social structures around preferred areas. He assumed that situations of single fish defending territories would occur only in sparse populations, and he did not consider this proof of control of numbers by social structure.

Size (weight), but not sex of brown trout is important in determining success in agonistic encounters (Jenkins 1969). Prior residence has a marked positive effect on dominance. This latter observation is supported by the studies of Miller (1957) on cutthroat trout, *Salmo clarki*, of Chapman (1972) on coho salmon, *Oncorhynchus kisutch*, of Mason et al. (1967) on brook trout and Mason (1969, 1975) on coho salmon, which show greatly increased movement and reduced survival of fish recently introduced into a stream with a resident population. Introduced individuals are apparently forced to move, feed much less, and consequently suffer much higher mortality rates than resident fish (Stringer & Hoar 1955, Newman 1956, Miller 1958, Saunders & Smith 1962, Mason & Chapman 1965, Chapman 1966, Backiel & LcCren 1967). The mechanism for these effects has generally been concluded to be dominance of the resident fish, whether territorial or not, over the introduced fish. The relative stability and site permanence of many species of freshwater fish in streams has, of course, been known for some time (Gerking 1959). It is worth noting in this context, the unusual relationship between size, sex, and prior residence and dominance in goldeye (*Hiodon alosoides*). Smaller fish, females, and intruders, are dominant over larger fish, males, and residents (Fernet & Smith 1976). They interpreted these apparently contradictory results as reflecting differential growth rates, and hence food demands, in this species.

Observations of fish under confined conditions supplement and reinforce these conclusions. Dominance heirarchies have frequently been observed, with important consequences for both dominant and subordinant fish. Yamagishi (1962) and Yamagishi *et al.* (1974) found that in laboratory populations of rainbow trout, and an eleotrid, *Odontobutis obscurus*, the dominant fish were generally larger in size and grew more rapidly. Brown (1946), Fabricius (1953), and Barlow *et al.* (1975) showed that among salmonids and cichlids those fish higher in a dominance heirarchy have better access to food, eat more, and grow more rapidly. Dominant bluegills, *Lepomis macrochirus*, exclude subordinate fish from areas of preferred temperatures (Beitinger & Magnuson 1975). However, dominance heirarchies do not necessarily form in all such cases (Kalleberg 1958, Backiel & LeCren 1967). Rainbow trout, in a confined laboratory tank, will show territorial and/or dominance behaviour, and differences in operant behaviour to obtain food, but if sufficient food is provided, all fish eat and small individuals are not deprived to the point of losing condition (Landless 1976). Magnuson (1962), in a carefully controlled study of the medaka, *Oryzias latipes*, showed that aggressive behaviour of dominant individuals was largely responsible for the growth depensation under conditions of restricted food and/or space. However, in the zebra danio, *Brachydanio rerio*, growth depensation with restricted food is not a consequence of agonistic behaviour (Eaton & Farley 1974).

Barlow (1973) and Barlow *et al.* (1975) have reported some important correlates of colour polymorphism in a Central American cichlid, *Cichlasoma citrinellum*. The less common yellow or red (Xanthic) morphs dominate otherwise equal fish of the more common colour (grey). This higher dominance status, and hence, greater advantages in feeding and other competitive interactions, are consequences of the colour itself. There are also important differences in the response of young of this species to parents of the different colour morphs (Noakes & Barlow 1973b). Others (Ferno & Sjolander 1976, Weber & Weber 1976) have tested for possible imprinting preferences on selection of mates in cichlids, but only with species in which the colour morphs occur as a domestic variety, and not in natural populations. Their results are equivocal, but suggest that there is an effect of imprinting on mate selection. This and other effects, where the polymorphism occurs naturally, could have significant effects on production. For example, Sage and Selander (1975) have demonstrated trophic radiation in a polymorphic cichlid (*Cichlasoma* sp.) from Mexico. A similar situation appears to exist with the limnetic and benthic forms of stickleback, *Gasterosteus aculeatus*, in a lake in British Columbia (Larson 1976). These studies clearly indicate that the relationship between polymorphism and fish production deserves serious study.

Under controlled stream conditions, brown trout which are better fed become sexually mature earlier, and produce more smaller eggs than do less well fed individuals (Bagenal 1969, but see Balon 1963, for differing conclusions from similar studies on other species). Gall (1975) reported a positive correlation between egg volume and egg number in domestic rainbow trout. Reduced rations cause a reduction in fecundity in rainbow trout (Scott 1962), guppies, *Poecilia reticulata* (Hester 1964), brook trout (Vladykov 1956, Wydoski & Cooper 1966), and winter flounders, *Pseudopleuronectes americanus* (Tyler & Dunn 1976). The mechanism, whether follicular atresia or decreased recruitment of oocytes, varies with the species, but the consequence is the same. Success in feeding seems to depend on relative dominance, so we may conclude that dominance would affect not only survival and growth, but also reproduction. Under the crowded conditions of aquaculture facilities, dominance-related differences in feeding and growth are often of considerable significance (Brett 1974, Weatherly 1976), and, in extreme cases, there may be mortality of subordinate fish. Such correlates and consequences of dominance heirarchies are well known among domestic birds and mammals, and measures are taken to minimize their occurrence and/or impact (Guhl 1969). Under more natural conditions, where fish are less crowded, and subordinates have at least the opportunity for escape or emigration, such severe effects of dominance are less common. Nonetheless, dominance relations may be of considerable importance under natural conditions, and may affect one or more aspects of production process.

Whether this, or other aspects of social behaviour cause stunting in populations is a matter of considerable interest. Several species are known to produce stunted populations, apparently in response to differences in food supply or predators (Nikol'skii 1969). Social control of growth has been demonstrated in marine anemone fish (Allen 1972). Fishelson (1970b) and Robertson (1972) have reported on the control of sexual differentiation and maturation in other marine fishes (Anthiidae and Labridae) by social behaviour. Adult males occur either with a harem of females, or as one of a few active, territorial fish. If the male is removed, the most dominant female assumes the male's status and behaviour, and soon transforms into a functioning, reproductive male. The details of the mechanisms for these effects are not known, but they are the most striking immediate effects of social behaviour on production.

Smith and Tyler (1975) suggested that in monospecies communities, such as farm ponds, intraspecific competition will limit the size of individuals, but that in a complex community, such as the marine coral reef, this competition would limit the number of individuals, rather than their size. They base this on the unproven assumption, that space, but not food, is limiting in the corral reef situation. Sale (1975), similarly proposed that occurrence of a fish of a particular species on a coral reef is largely a matter of chance, depending on initial colonization, since space is limiting. Closer study of the behaviour of fish in these circumstances could lead to useful insights into such general ecological relationships.

Territoriality

Commonly used definitions of territoriality include such phrases as defended, or exclusive, area. However, we should exercise caution in ascribing conse- quences to, or even inferring the existence of territoriality, without direct confirmatory observations. Similar consequences could result in most cases from simple agonistic encounters, or dominance heirarchies. Territoriality is a specific term, whatever its precise definition, and should not be used as a catch-all (Brown 1975).

Territories continuously grade into, and perhaps are inseparable from dominance heirarchies. In fact, the major criterion separating the two is often the site-fixation of dominance in territoriality. Here we may conveniently avoid some of the difficulties of defining territorial behaviour by considering it as site-dependent (and site-fixed) dominance. Both field and laboratory studies have indicated the general occurrence and importance of such behaviour among fishes.

Staples (1975b), for example, found that in the New Zealand bully, adult males were territorial during the breeding season. Males defended the few, preferred spawning sites, and spawned polygamously with females. Only the larger males were able to defend territories successfully, and reproduce. Territorial behaviour during the breeding season occurs quite commonly among such freshwater species as etheostomids, centrarchids, cichlids, cyprinodontids, and at least some cyprinids (e.g. Winn 1958, Keenleyside 1971, Barlow 1974a, Itzkowitz 1974b, Smith & Murphy 1974, Smith 1976). A variety of reproductive strategies may be involved with this territorial behaviour, including monogamous pair bonds, sequential polygyny, or simultaneous polygyny, but certain features are common. Defense of a spawning site, at least for the duration of the spawning act, is typical. If the number or distribution of spawning sites are limited, territorial behaviour could restrict breeding by non-territorial fish, and enhance the breeding and subsequent survival of young of the former. If territorial defense extends to the period when young are in the territory, their survival would certainly be enhanced. In such species, failure of the parent to constantly defend the young results in immediate losses to predators (Noakes & Barlow 1973a). The considerable differences in detail of territorial systems may be related to the overall production process (Maynard Smith 1974). However, since these details are so poorly studied for fish, we can only draw attention to them as promising areas for further study.

Fish may exhibit territoriality not associated with reproductive behaviour. For example, juveniles or adults outside the breeding season may defend territories related to feedings. LeCren (1965) suggested that reduction in growth rate with increased population density of young brown trout may be due, in part, to the increased time spent in defense of territories, thereby distracting them from feeding activity. Above a critical density (9 fry m^{-2})

some fry did not secure territories and starved. The relatively constant production of brown trout per unit area in English streams is due to the aggressive territorial behaviour of the fish soon after emergence from the gravel (Backiel & LeCren 1967). Such territorial behaviour is proposed to act as a population controlling (spacing) mechanism before any trophic demands are made on the environment. This would be an important effect, since it would be of major importance in mortality of young, and most production is accomplished by young fish (Mahon 1976). Maynard Smith (1974) has discussed the ecological consequences of territories, and proposed models of three patterns of territories which may have consequences for population regulation. These models make specific predictions which would be worthwhile to test on fish populations. Keenleyside (1962), and Keenleyside & Yamamoto (1962) reported that most of the available habitat of non-breeding Atlantic salmon, *Salmo salar*, was taken up by a mosaic of territories. Others (e.g. Elson 1942, Hoar 1951, Kalleberg 1958, Hartman 1963, 1965) have made similar observations on salmonids, leading to the suggestion by Jenkins (1969) that such a situation results from a combination of uniform body size (and, therefore, aggressiveness), uniform substrate, and individual position choice (and permanence) by the fish. These territories seem obviously related to feeding behaviour, and some have suggested that feeding is the prime factor producing such a social system. However, other advantages have been demonstrated for territorial fish, so we should not jump to immediate conclusions, obvious though they may seem. For example, territorial juvenile Atlantic salmon are less subject to predation than are non-territorial individuals (Symons 1974). Also, the factors influencing position choice by salmonids are varied, including light intensity, overhead cover, water depth, current velocity, substrate colour, and bottom cover (McCrimmon 1954, Saunders & Smith 1962, Gibson & Keenleyside 1966, Kwain & MacCrimmon 1969, Bustard & Narver 1975, Gibson & Power 1975).

Territoriality, whether related to reproduction or to feeding, is only one possible behavioural strategy. Reproductive strategy generally characterizes a species (Balon 1975), whereas feeding strategy depends more on proximate factors, such as type and abundance of food items, and their dispersion pattern in space and time. For example, within a given section of stream, trout (*Salmo* sp.) and chub (*Siphatales* sp.) employ different feeding strategies, and exhibit different social systems (Jenkins 1969). Limnetic and benthic forms of sticklebacks, in the same lake, show significant differences, based on social behaviour and feeding ability (Larson 1976). The limnetic form is gregarious, nonaggressive, and feeds more successfully on zooplankton, while the benthic form is solitary, aggressive, and feeds better on macrobenthos.

The common relationship between territories and food supply (Pianka 1974, Brown 1975), suggests there may be a basic principle involved. Juvenile Atlantic salmon desert their territories more often if food supply is reduced

(Symons 1971). Territorial structure in juvenile rainbow trout can be altered by manipulating food availability (Slaney & Northcote 1974). Extreme crowding (relative to population densities of unconfined fish) usually inhibits territoriality and dominance interactions, presumably by exceeding the maximum group size possible for individual recognition (e.g. Mundie 1974). Manipulation of such factors as space, and population density to optimize production is commonly practiced with other domestic animals (Guhl 1969), often with considerable success. The relationship between food supply, social system, and production has great promise for practical, as well as theoretical, consequences. We could reduce territorial or dominance effects for intensively managed populations, and so minimize stunting and mortality.

A recent study in my laboratory (Cole 1976) illustrated striking effects of fish density and water flow on social behaviour of young rainbow trout. High water flow rate encouraged station maintenance and territorial behaviour, especially at low fish density. Increased crowding of these fish suppressed territorial behaviour, and dominance heirarchies formed. Fish in less turbulent water flow tended to form dominance heirarchies, with higher levels of agonistic encounters, at all densities. This illustrates the plasticity of social organization, the necessity for studying it under a variety of circumstances, and the potential for manipulating it through physical or biological factors.

SCHOOLING BEHAVIOUR

Since schooling can occur in such a great number of different species under a wide variety of circumstances it probably has no single function, or even the same major function, in all cases. In fact, attempts at such a generalization seem to hamper the investigation of this phenomenon. There is considerable evidence which clearly implies significant positive effects of schooling on feeding, survival, and possibly, reproduction. Undoubtedly there are a number of possible adaptive features of schooling, and species may have to be considered as individual cases to interpret these in some circumstances. Williams (1964), Shaw (1970), and Radakov (1973) have given particularly good reviews of the subject and consideration of its possible evolution.

Probably no other type of social organization is more commonly associated with fish than is schooling behaviour. Not surprisingly, it has attracted considerable attention (e.g. Breder 1951, 1954, 1959, 1965, 1967, Keenleyside 1955, Hunter 1966, Shaw 1970, Radakov 1973). Many important marine and freshwater fisheries are based on the exploitation of dense schools of fish. The correlates of this behaviour with physical and biological factors and its significance in relation to production are likely of a general nature. Many freshwater species school at some time in their life, for example the young with their parents or shortly after leaving the nest (e.g. cichlids, Noakes & Barlow 1973b; centrarchids, Robbins & MacCrimmon 1974).

Despite the study it has received, schooling behaviour is by no means well-understood. Radakov (1973) has given a good survey of the topic, particularly the work of Soviet authors. As he points out, schooling behaviour is the subject of only a small proportion of studies on fish behaviour, and these in turn are a minority of all studies on fish biology. Investigators have tended to study specific problems or individual features of schooling with little effort to characterize the phenomenon as a whole (Radakov 1973). In particular, there has been a dichtomy between those concerned with the mechanisms of schooling and others concerned with the ecological context or significance of this behaviour. Both aspects are important, and must be brought together to formulate a comprehensive synthesis of the behaviour. A consideration of schooling in relation to production will not be such a synthesis, but may suggest new questions, or viewpoints for analyzing the phenomenon.

Schooling as a defensive strategy

Populations of guppies from areas without fish predators do not show schooling behaviour; those from areas with fish predators school normally. These differences have significant effects on survival, and are genetically determined (Haskins *et al.* 1961, Seghers 1974, Liley & Seghers 1975).

Many species school facultatively. In these, schooling is readily induced by fearful stimuli, such as sudden disturbances, or appearance of a potential predator. The defensive value of this response for the individuals has been considered by several authors, using qualitative and quantitative models (Brock & Riffenburgh 1960, Williams 1964, Breder 1967, Hamilton 1971, Vine, 1971, Dill 1974, Treisman 1975). These models generally predict a high selective advantage of schooling in reducing mortality by predation. Ginetz and Larkin (1976) have provided field data to show increased survival of young sockeye salmon, *Oncorhynchus nerka*, which school in the presence of rainbow trout as predators. Radakov (1973) cites several references which similarly illustrate a higher survival of fish in schools, compared to isolated individuals.

Schooling and reproduction

Enhanced reproduction has also been suggested as a possible beneficial consequence of schooling, but details of spawning behaviour are known for only a few such species (Magnuson & Prescott 1966). Reproduction by fish within a school, whether or not they form brief pair bonds, is likely to be of secondary importance to other functions of schooling. If the species schools most of the time, spawning behaviour would only occupy a brief period, and so would be an adjunct to schooling, for whatever other reason. The carp (*Cyprinus carpio*), for example, gathers in numbers (schools?) only at spawning time (McCrimmon 1968). If in fact such congregations are schools, then repro-duction would be the primary function, and would be of considerable significance. At present, sufficient data are lacking about these aggregations to speculate on their importance.

Schooling and feeding behaviour

Feeding behaviour and schooling are closely related. The significance of gregarious feeding behaviour has been postulated (e.g. Brown & Orians 1970, Cody 1971, Pianka 1974, Thompson *et al.* 1974) and has been demon-strated in some species of birds (Murton *et al.* 1971, Krebs *et al.* 1972). There are suggestions in the literature of similar advantages for fish (Keenleyside 1955, Ivlev 1961, Ommanney 1964, Radakov 1973). In general, the effect is one of enhanced foraging efficiency of animals in the group, whether by increasing the chance of finding food, or by decreasing the risk of not finding food. The importance of feeding strategy has been stressed in several studies (Beukema 1968, Glass, 1971, Werner & Hall 1974, Lett *et al.* 1975).

Changes in social organization in relation to feeding behaviour (or vice versa) have seldom been reported due to the lack of attention to this question. For instance, Barlow (1974c) described an interesting adaptation in the social behaviour of a Hawaiian surgeonfish called the manini, *Acanthurus triostegus*. This species uses schooling behaviour as a protective device to reach the bottom to feed when the bottom is held by territory-holding food competitors. Individuals in such schools have a much lower probability of being attacked by a territorial fish than if they were to attempt to feed on the bottom as solitary individuals, outside a school. However, in the same species, schooling in other situations, where no interspecific feeding competition exists, is an anti-predator behaviour.

Schooling and harvest

A recent paper by Clark (1974) elaborates an important hypothesis concerning possible effects of schooling on the exploitation of fish. He proposes a situa-tion of critical depensation, in which a schooling species may enter an irreversible path to extinction. Species normally subject to heavy predation, which have evolved the habit of forming large compact schools as a defensive mechanism are cited as especially liable to such a fate. Such a negative consequence of schooling would be of major importance, not just as a theoretical concept but as a basis for management policy. Such a simple, discrete model, which makes such clear predictions, should be the aim of those attempting to relate social behaviour to production.

SOCIAL FACILITATION

Other effects of social groupings are lumped under this term. These do not include such things as undercrowding or overcrowding, as mentioned earlier, but do include enhanced or decreased learning, feeding, etc., from animals being in close proximity. Speed of learning to traverse a maze, to avoid a net or to operate a self-feeder are increased by the presence of conspecifics (Hale 1956, Hunter & Wisby 1964, Adron *et al.* 1973). The mere presence of a conspecific markedly alters the spatial pattern of swimming in goldfish (Timms 1975), but one may question if this is necessarily a part of social behaviour. Of course, some species show obvious social responses even to an artificial dummy or to a mirror image (Brown & Noakes 1974). Social enhancement of feeding has been reported in several species (Adron *et al.* 1973, Radakov 1973). I have commonly observed, as I am sure most fish keepers have, that fish kept in groups will feed more readily, and adjust to changes in diet more quickly, than will isolated individuals. This enhancement of feeding, known as social facilitation, is common in domestic animals (Scott 1969), and in fact in other species which typically live in groups. The positive effects of social groupings on foraging have already been considered.

14.3 Some prospects

While these are by no means the only, or necessarily the most likely, I feel the following types are the most interesting and exciting prospects in behavioural studies of fish.

BEHAVIOURAL GENETICS

This area of research has received increasing attention (Hirsch 1967, Ehrman & Parsons 1976), leading to some of the most interesting breakthroughs in our understanding of behavioural mechanisms (e.g. Bentley 1976). Unfortunately, fish have rarely been the subject of such studies (Clark, Aronson & Gordon 1954, Franck 1970).

Salmonids have relatively high heritability and high phenotypic variance for a number of characteristics, so that selection, similar to that used to produce strains of other domestic animals, could be profitably applied, and should yield fish with desired traits (Gall 1975, Gjedrem 1976). Behavioural differences in salmonids could likely be affected by this artificial selection. Successfully stocking salmonid hybrids in natural waters or in ponds (Flick & Webster 1976, Ihssen 1976), requires a better understanding of the genetic basis of behaviour in these fish.

Some studies on salmonids, comparing hatchery and wild stocks, strongly suggest significant behavioural differences which are genetically determined.

Vincent (1960) reported faster growth, less fright behaviour, less resistance to accumulated metabolites, less resistance to higher temperature, and a surface response (fish moved toward the surface in a rearing trough or tall aquarium) in a domestic stock or brook trout compared to a wild stock of the species. Wild fish also had greater stamina, as measured by swimming to exhaustion, and better survival in natural (pond or stream) conditions. Moyle (1969) confirmed that domestic brook trout fry tend to be closer to the surface and wild ones near the bottom, although his results were more variable than Vincent's (1960). This difference appeared to be related to the greater activity of domestic fish which in turn led to more agonistic encounters.

Fenderson and Carpenter (1971) concluded that wild Atlantic salmon were most aggressive in aquaria at low density and hatchery fish were most aggressive at intermediate or high density. Since social interactions interferred with feeding, hatchery fish grew less well as density increased. More recently, Holzberg and Schroder (1975) have proposed a genetic basis for differences in aggressiveness between normal and irradiated convict cichlids, *Cichlasoma nigrofasciatum*.

BEHAVIOURAL ENERGETICS

Behavioural energetics has received sufficient attention to illustrate its potential (Wolf *et al.* 1972, Pianka 1974). Energetics will be the most useful common denominator for relating social behaviour to production, and brings this chapter full circle. Intensive laboratory studies have provided a wealth of data on metabolic expenditures, conversion, and nutrient requirements (Beamish & Dickie 1967, Beamish 1974, Webb, 1975, Halver 1976), and these studies have led to models of varying complexity and sophistication (e.g. Kerr 1971). We now need for such models and for more general considerations of fish production, good data on the energetic costs of behaviour, including social behaviour. With such data, we could assess the relative advantages of various behavioural strategies, and perhaps gain a better understanding of the ecological significance, and possible evolution of social behaviour in fish. Carline and Hall (1973) and Feldmeth and Jenkins (1973) have shown that such data can be obtained under field conditions. The acquisition and incorporation of such data into broader considerations of fish production should be of considerable significance.

14.4 Conclusions

Fish production necessarily stresses data obtained from field studies, but much information, useful to those interested in the subject, can be derived from observations under carefully controlled laboratory conditions. For a number of reasons, both practical and historical, behavioural observations of fish in the field have been relatively limited in scope. Such field observations for brief periods of time have typically been made on diurnal species, living in shallow water. On the other hand, laboratory studies provide a good body of information on reproductive behaviour, dominance heirarchies, schooling behaviour, and, increasingly, development of behaviour all of which are most likely to have significant effects on fish production. Practical limitations in the laboratory usually restrict studies to species which mature at a small size, under reasonable conditions of temperature and photoperiod. Also, the considerable body of information and experience from the aquarium hobby shares a similar bias. Families such as the Poeciliidae, Cichlidae, and Cyprinodontidae thus are disproportionately represented in these studies (Breder & Rosen 1966).

Conclusions from these studies, in relation to production, must be drawn with some caution, since our knowledge of the ecological context, if any, of the behaviour observed in such laboratory situations, is limited (e.g. Martin 1975). In addition, we know that behaviour can be very plastic, depending on proximate factors to a great extent for its expression. A better knowledge of how these factors affect behaviour would be very useful in allowing us to extrapolate from controlled studies to the complexity of the field.

Few behavioural studies, even those framed in an ecological context, have been specifically directed to the question of the relationship between social behaviour and production. The social behaviour and organization of young salmonids in streams is probably the best studied (Hoar 1976), and illustrates the need for detailed study of other species. Since the bulk of production is by young fish, this period of life should obviously be carefully considered. This emphasis lends itself well to the current trend towards studying the development of behaviour in fish.

We can best understand social systems as adaptive features of species in relation to their ecology. As the interface between the organism and its environment, behaviour must combine flexibility and stereotypy in its patterns (Barlow 1968). Certain types of behaviour, for example reproductive patterns, are often highly stereotyped and rigidly pre-set, if the context of this behaviour is either highly predictable, or inflexible. On the other hand, feeding behaviour may be much more adaptable within an individual, depending on its development, and/or immediate circumstances. Social behaviour, since it involves more than one individual, will of necessity be inherently complex. Our concern, in this case, is with the relationships among interacting organisms, and this is the level at which we will find the explanation for the social phenomena we observe. The movement patterns of individual fish, the rate at which they interact with each other, their allocation of time to various activities, whether feeding, agonistic or reproductive, are the kinds of measures we need in order to understand the social organization. As with other areas of science, a knowledge of the factors regulating these relationships will allow us to predict, and ultimately manipulate the relationships, in this context, to enhance fish production.

Acknowledgements

I thank Eugene Balon for his insightful comments on this manuscript, and the National Research Council of Canada for their financial support, through grant No. A6981.

References

ADRON J.W., GRANT P.T. & COWLEY C.B. (1973) A system for the quantitative study of the learning capacity of rainbow trout and its application to the study of food preferences and behaviour. *J. Fish Biol.* **5**, 625–636.

ALLEE W.C., GREENBERG B., ROSENTHAL G.M. & FRANK P. (1948) Some effects of social organization on growth in the green sunfish, *Lepomis cyanellus. J. Exp. Zool.* **108**, 1–20.

ALLEN G.P. (1972) *The Anemone Fishes, Their Classification and Biology.* T.F.H. Publications.

ALTMANN S.A. (1974) Baboons, space, time and energy. *Amer. Zool.* **14**, 221–248.

ALTMANN S.A. & ALTMANN J. (1970) *Baboon Ecology: African Field Research.* University Chicago Press.

BACKEIL T. & LeCREN E.D. (1967) Some density relationships for fish population parameters. *The Biological Basis of Freshwater Fish Production* (S.D. Gerking ed.) Blackwell Scientific Publications, Oxford.

BAGENAL T.B. (1969) The relationship between food supply and fecundity in brown trout, *Salmo trutta* L. *J. Fish Biol.* **1**, 167–182.

BALON E.K. (1963) Altersstruktur der Populationen und Wachstumgesetzmässigkeiten der Donaubrachsen (*Abramis brama, A. sapa, A. ballerus*). *Bul. Inst. Chem. Tech.* (Prague), **7**, 459–542.

BALON E.K. (1974) Fish production of a tropical ecosystem. *Lake Kariba: A Man-Made Tropical Ecosystem in Central Africa* (E.K. Balon & A.G. Coche ed.) W. Junk, Amsterdam.

BALON E.K. (1975) Reproductive guilds in fishes: A proposal and definition. *J. Fish. Res. Bd. Canada* **32**, 821–864.

BARLOW G.W. (1968) Ethological units of behavior. *The Central Nervous System and Fish Behaviour* (D. Ingle, ed.) Univ. Chicago Press, Chicago, Ill.

BARLOW G.W. (1973) Competition between color morphs of the polychromatic Midas cichlid *Cichlasoma citrinellum. Science* **179**, 806–807.

BARLOW G.W. (1974a) Contrasts in social behavior between Central American cichlid fishes and coral reef surgeon fishes. *Amer. Zool.* **14**, 9–13.

BARLOW G.W. (1974b) Hexagonal territories. *Anim. Behav.* **22**, 876–878.

BARLOW G.W. (1974c) Extraspecific imposition of social grouping among surgeonfishes (Pisces: Acanthuridae). *J. Zool.* **174**, 333–340.

BARLOW G.W. (1976) The Midas cichlid in Nicaragua. *Investigations of the Ichthyofauna Nicaraguan Lakes* (T.B. Thorson ed.) University of Nebraska Press, Lincoln.

BARLOW G.W., BAUER D.H. & McKAYE K.R. (1975) A comparison of feeding, spacing, and aggression in color morphs of the Midas cichlid. I. Food continuously present *Behaviour* **54**, 72–96.

BAYLIS J.R. (1974) The behavior and ecology of *Herotilapia multispinosa* (Teleostei, Cichlidae). *Z. Tierpsychol.* **34**, 115–146.

BEAMISH F.W.H. (1974) Apparent specific dynamic action of largemouth bass, *Micropterus salmoides. J. Fish. Res. Bd. Canada* **31**, 1763–1769.

BEAMISH F.W.H. & DICKIE L.M. (1967) Metabolism and biological production in fish. *The Biological Basis of Freshwater Fish Production* (S.D. Gerking ed.) Blackwell Scientific Publications, Oxford.

BEITINGER T.L. & MAGNUSON J.J. (1975) Influence of social rank and size on thermo-selection behaviour of bluegill (*Lepomis macrochirus*). *J. Fish. Res. Bd. Canada* **32** 2133–2136.

BENTLEY D.R. (1976) Genetic analysis of the nervous system. *Simpler Networks and Behaviour* (J.C. Fentress ed.) Sinauer Associates Inc, Boston.

BEUKEMA J.J. (1968) Predation by the three-spined stickleback (*Gasterosteus aculeatus* L.): The influence of hunger and experience. *Behaviour* **31**, 1–126.

BRADDOCK J.C. (1949) The effect of prior residence upon dominance in the fish *Platypoecilus maculatus. Physiol. Zoöl.* **22**, 161–169.

BREDER C.M. JR. (1951) Studies on the structure of the fish school. *Bull. Amer. Mus. Nat. Hist.* **98**, 1–28.

BREDER C.M. JR (1954) Equations descriptive of fish schools and other animal aggregations. *Ecology* **35**, 361–370.

BREDER C.M. JR (1959) Studies on social groupings in fishes. *Bull. Amer. Mus. Nat. Hist.* **117**, 393–482.

BREDER C.M. JR (1965) Vortices and fish schools. *Zoologica* **50**, 97–114.

BREDER C.M. JR (1967) On the survival value of fish schools. *Zoologica* **52**, 25–40.

BREDER C.M. JR & ROSEN D.E. (1966) *Modes of Reproduction in Fishes.* Natural History Press, New York.

BRETT J.R. (1974) Marine fish aquaculture in Canada. *Aquaculture in Canada* (H.R. MacCrimmon, J.E. Stewart & J.R. Brett ed.) *Fish Res. Bd. Canada Bull.* **188**.

BROCK E.V. & RIFFENBURGH R.H. (1960) Fish schooling: a possible factor in reducing predation. *J. Cons.* **25**, 307–317.

BROWN D.M.B. & NOAKES D.L.G. (1974) Habituation and recovery of aggressive display in paradise fish (*Macropodus opercularis* (L.)). *Behav. Biol.* **10**, 519–525.

BROWN J.L. (1975) *The Evolution of Behavior.* W.W. Norton, New York.

BROWN J.L. & ORIANS G. (1970) Spacing patterns in mobile animals. *Rev. Ecol. Syst.* **1**, 239–262.

BROWN M.E. (1946) The growth of brown trout (*Salmo trutta* L.) I. Factors influencing the growth of trout fry. *J. Exp. Biol.* **22**, 118–129.

BUSTARD D.R. & NARVER D.W. (1975) Preferences of juvenile coho salmon (*Oncorhynchus kisutch*) and cutthroat trout (*Salmo clarki*) relative to simulated alternations of winter habitat. *J. Fish. Res. Bd. Canada* **32**, 681–687.

CARLINE R.F. & HALL J.D. (1973) Evaluation of a method for estimating food consumption rates of fish. *J. Fish. Res. Bd. Canada* **30**, 623–629

CHAPMAN D.W. (1962) Aggressive behaviour in juvenile coho salmon as a cause of emigration. *J. Fish. Res. Bd. Canada* **19**, 1047–1080.

CHAPMAN D.W. (1966) Food and space as regulators of salmonid populations in streams. *Amer. Nat.* **100**, 345–357.

CHAPMAN D.W. (1967) Production in fish populations. *The Biological Basis of Freshwater Fish Production* (Gerking S.D. ed.). Blackwell Scientific Publications, Oxford.

CHAPMAN D.W. & BJORNN T.C. (1969) Distribution of salmonids in streams, with special reference to food and feeding. *Symposium on Salmon and Trout in Streams* (T.G. Northcote T.G. ed.) University of British Columbia, Vancouver.

CLARK C.W. (1974) Possible effects of schooling on the dynamics of exploited fish populations. *J. Cons.* **36**, 7–14.

CLARK E., ARONSON L.R. & GORDON M. (1954) Mating behavior patterns in two sympatric species of xiphophorin fishes: their inheritance and significance in sexual isolation. *Bull. Amer. Mus. Nat. Hist.* **103**, 135–226.

CODY M.L. (1971) Finch flocks in the Mojave Desert. *Theoret. Pop. Biol.* **2**, 142–158.

COLE K.S. (1976) Social behavior and social organization of young rainbow trout, *Salmo gairdneri*, of hatchery origin. M.Sc. thesis, Zoology Department, University of Guelph, Guelph, Ontario.

CROOK J.H. (1970) *Social Behavior in Birds and Mammals*. Academic Press, New York.

DILL L.M. (1974) The escape responses of the zebra danio (*Brachydanio rerio*). II. The effect of experience. *Anim. Behav.* **22**, 723–730.

EATON R.C. & FARLEY R.D. (1974) Growth and the reduction of depensation of zebrafish, *Brachydanio rerio*, reared in the laboratory. *Copeia* **1974**, 204–209.

EDMUNDSON E., EVEREST F.H. & CHAPMAN D.W. (1968) Permanence of station in juvenile chinook salmon and steelhead trout. *J. Fish. Res. Bd. Canada* **25**, 1453–1464.

EHRMAN L. & PARSONS P.A. (1976) *The Genetics of Behavior*. Sinauer Associates, Boston.

ELSON R.F. (1942) Behaviour and survival of planted Atlantic salmon fingerlings. *Trans. N. Amer. Wildl. Conf.* **7**, 202–211.

EMERY A. (1973) Preliminary comparisons of day and night habits of freshwater fish in Ontario lakes. *J. Fish. Res. Bd. Canada* **30**, 761–774.

EVEREST F.H. & CHAPMAN D.W. (1972) Habitat selection and spatial interaction of juvenile chinook salmon and steelhead trout in two Idaho streams. *J. Fish. Res. Bd. Canada* **29**, 91–100.

EWING A.W. (1975) Studies on the behaviour of cyprinodontid fish. II. The evolution of aggressive behaviour in Old World revulins. *Behaviour* **52**, 172–195.

FABRICIUS E. (1953) Aquarium observations on the spawning behaviour of the char, *Salmo alpinus*, L. *Rep. Inst. Freshwat. Res. Drottningholm* **35**, 14–48.

FABRICIUS E. & GUSTAFSON K.-J. (1954) Further aquarium observations on the spawning behaviour of the char, *Salmo alpinus* L. *Rep. Inst. Freshwat. Res. Drottningholm* **35**, 58–104.

FELDMETH C.R. & JENKINS T.M. Jr. (1973) An estimate of energy expenditure by rainbow trout (*Salmo gairdneri*) in a small mountain stream. *J. Fish. Res. Bd. Canada* **30**, 1755–1759.

FENDERSON O.C. & CARPENTER M.R. (1971) Effects of crowding on the behaviour of juvenile hatchery and wild landlocked Atlantic salmon (*Salmo salar* L.). *Anim. Behav.* **19**, 439–447.

FERNET D.A. & SMITH R.J.F. (1976) Agonistic behaviour of captive goldeye (*Hiodon alosoides*). *J. Fish. Res. Bd. Canada* **33**, 695–702.

FERNO A. & SJOLANDER S. (1976) Influence of previous experience on the mate selection of two colour morphs of the convict cichlid, *Cichlasoma nigrofasciatum* (Pisces, Cichlidae). *Behav. Proc.* **1**, 3–14.

FISHELSON L. (1970a) Behaviour and ecology of a population of *Abudefduf saxatilis* (Pomacentridae, Teleostei) at Eilat (Red Sea). *Anim. Behav.* **18**, 225–237.

FISHELSON L. (1970b) Protogynous sex reversal in the fish *Anthias squamipinnis* (Teleostei, Anthiidae) regulated by the presence or absence of a male fish. *Nature* **227**, 90–91.

FLICK W.A. & WEBSTER D.A. (1976) Production of wild, domestic, and interstrain hybrids of brook trout (*Salvelinus fontinalis*) in natural ponds. *J. Fish. Res. Bd. Canada* **33**, 1525–1539.

FRANCK D. (1970) Verhaltengenetische Untersuchungen an Artbastarden der Gattung *Xiphophorus*. (Pisces). *Z. Tierpsychol.* **27**, 1–34.

FORSELIUS S. (1957) Studies of anabantid fishes. I–III. *Zool. Bidrag, Uppsala*, **32**, 93–598.

FREY D.F. & MILLER R.J. (1972) The establishment of dominance relationships in the blue gourami, *Trichogaster trichopterus* (Pallus). *Behaviour* **42**, 8–62.

FRYER G. & ILES T.D. (1972) *The Cichlid Fishes of the Great Lakes of Africa*. Oliver & Boyd, London.

GALL G.A.E. (1975) Genetics of reproduction in domesticated rainbow trout. *J. Anim. Sci.* **40**, 19–28.

GEIST V. (1974) On the relationship of social evolution and ecology in ungulates. *Amer. Zool.* **14**, 205–220.

GERKING S.D. (1959) The restricted movement of fish populations. *Biol. Rev.* **34**, 221–242.

GERKING S.D. (1967) *The Biological Basis of Freshwater Fish Production*. Blackwell Scientific Publications, Oxford.

GIBSON R.J. & KEENLEYSIDE M.H.A. (1966) Responses to light of young Atlantic salmon (*Salmo salar*) and brook trout (*Salvelinus fontinalis*). *J. Fish. Res. Bd. Canada* **23**, 1007–1024.

GIBSON R.J. & POWER J. (1975) Selection by brook trout (*Salvelinus fontinalis*) and juvenile Atlantic salmon (*Salmo salar*) of shade related to water depth. *J. Fish. Res. Bd. Canada* **32**, 1652–1656.

GINETZ R.M. & LARKIN P.A. (1976) Factors affecting rainbow trout (*Salmo gairdneri*) predation on migrant fry of sockeye salmon (*Oncorhynchus nerka*). *J. Fish. Res. Bd. Canada* **33**, 19–24.

GJEDREM T. (1976) Possibilities for genetic improvements in salmonids. *J. Fish. Res. Bd. Canada* **33**, 1094–1099.

GLASS N.W. (1971) Computer analysis of predation energetics in the largemouth bass. *Systems Analysis and Simulation in Ecology*, (B. Patten, ed.). Academic Press, New York.

GREENBERG B. (1947) Some relations between territory, social heirarchy and leadership in the green sunfish (*Lepomis cyanellus*). *Physiol. Zool.* **20**, 269–299.

GUHL A.M. (1969) The social environment and behaviour. *The Behaviour of Domestic Animals*, (E.S.S. Hafz, ed.). Williams & Wilkins, Baltimore.

HALE E.B. (1956) Effects of forebrain lesions on the aggressive behaviour of green sunfish *Lepomis cyanellus*. *Physiol. Zoöl.* **29**, 107–127.

HALVER J.E. (1976) Formulating practical diets for fish. *J. Fish. Res. Bd. Canada* **33**, 1032–1039.

HAMILTON W.D. (1971) Geometry for the selfish herd. *J. Theoret. Biol.* **31**, 295–311.

HARTMAN G.F. (1963) Observations on behaviour of juvenile brown trout in a stream aquarium during winter and spring. *J. Fish. Res. Bd. Canada* **20**, 769–787.

HARTMAN G.F. (1965) The role of behaviour in the ecology and interaction of underyearling coho salmon (*Oncorhynchus kisutch*) and steelhead trout (*Salmo gairdneri*). *J. Fish. Res. Bd. Canada* **22**, 1035–1081.

HASKINS C.P., HASKINS E.F., MCLAUGHLIN J.J.A. & HEWITT R.E. (1961) Polymorphism and population structure in *Lebistes reticulatus*. *Vertebrate Speciation*, (W.F. Blair, ed). University of Texas Press, Austin.

HESTER F.J. (1964) Effect of food supply on fecundity in the female guppy, *Lebistes reticulatus* (Peters). *J. Fish. Res. Bd. Canada* **21**, 757–764.

HINDE R.A. (1970) *Animal Behaviour. A Synthesis of Ethology and Comparative Psychology*. McGraw-Hill, New York.

HIRSCH J. (1967) *Behavior–Genetic Analysis*. McGraw-Hill. N.Y.

HOAR W.S. (1951) The behaviour of chum, pink, and coho salmon in relation to their seaward migration. *J. Fish. Res. Bd. Canada* **8**, 241–263.

HOAR W.S. (1976) Smolt transformation: Evolution, behavior and physiology. *J. Fish Res. Bd. Canada* 33, 1233–1252.

HOBSON E.S. (1974) Feeding relationships of teleostean fishes on coral reefs in Kona, Hawaii. *Fish. Bull. U.S. Nat. Mar. Fish. Serv.*, NOAA 72, 915–1031.

HOLZBERG S. & SCHRODER J.H. (1975) The inheritance of aggressiveness in the convict cichlid fish, *Cichlasoma nigrofasciatum* (Pisces: Cichlidae). *Anim. Behav.* 23, 625–631.

HUNTER J.R. (1966) Procedure for analysis of schooling behaviour. *J. Fish. Res. Bd. Canada* 23, 457–562.

HUNTER J.R. & WISBY W.J. (1964) Net avoidance behaviour of carp and other species of fish. *J. Fish. Res. Bd. Canada* 21, 613–633.

IHSSEN P. (1976) Selective breeding and hybridization in fisheries management. *J. Fish. Res. Bd. Canada* 33, 316–321.

ITZKOWITZ M. (1974a) A behavioural reconnaissance of some Jamaican reef fishes. *J. Zool. Linn. Soc.* 55, 87–118.

ITZKOWITZ M. (1974b) The effects of other fish on the reproductive behaviour of the male *Cyprinodon variegatus* (Pisces, Cyprinodontidae). *Behaviour* 48, 1–22.

IVLEV V.S. (1945) The biological productivity of waters. *Usp. sovrem. biol.* 19, 98–120 (Russian) (1966) *J. Fish. Res. Bd. Canada* 23, 1727–1759. (English).

IVLEV V.S. (1961) *Experimental Ecology of the Feeding of Fishes.* Yale University Press, New Haven.

JENKINS T.M. Jr. (1969) Social structure, position choice and microdistribution of two trout species (*Salmo trutta* and *Salmo gairdneri*) resident in mountain streams. *Anim. Behav. Monogr.* 2, 57–123.

KALLEBERG H. (1958) Observations in a stream tank of territoriality and competition in juvenile salmon and trout (*Salmo salar* L. and *S. trutta.*). *Rep. Inst. Freshwat. Res. Drottningholm* 39, 55–98.

KAWANABE H. (1958) On the significance of the social structure for the mode of density effect in a salmon-like fish, 'Ayu', *Plecoglossus altivelis* Temminck et Schlegel. *Mem. Coll. Sci., Kyoto Univ.* B25, 171–180.

KAWANABE H., MORI S. & MIZUNO N. (1957) Modes of utilizing the river pools by a salmon-like fish, *Plecoglossus* or 'Ayu', in relation to its population density. *Jap. J. Ecol.* 7, 22–26. (English summary.)

KEENLEYSIDE M.H.A. (1955) Some aspects of the schooling behaviour of fish. *Behaviour* 8, 183–248.

KEENLEYSIDE M.H.A. (1962) Skindiving observations of Atlantic salmon and brook trout in the Miramichi River, New Brunswick. *J. Fish. Res. Bd. Canada.* 19, 625–634.

KEENLEYSIDE M.H.A. (1967) Behavior of male sunfishes (genus *Lepomis*) towards females of three species. *Evolution* 21, 688–695.

KEENLEYSIDE M.H.A. (1971) Aggressive behavior of male longear sunfish (*Lepomis megalotis*). *Z. Tierpsychol.* 28, 227–240.

KEENLEYSIDE M.H.A. (1972a) The behaviour of *Abudefduf zonatus* (Pisces, Pomacentridae) at Heron Island, Great Barrier Reef. *Anim. Behav.* 20, 763–774.

KEENLEYSIDE M.H.A. (1972b) Intraspecific intrusions into nests of spawning longear sunfish (Pisces: Centrarchidae). *Copeia* 1972, 272–278.

KEENLEYSIDE M.H.A. & YAMAMOTO F.T. (1962) Territorial behaviour of juvenile Atlantic salmon (*Salmo salar* L.). *Behaviour* 19, 138–169.

KERR S.R. (1971) A simulation model of lake trout growth. *J. Fish. Res. Bd. Canada* 28, 815–819.

KONIKOFF M. & LEWIS W.M. (1974) Variation in weight of cage-reared channel catfish. *Prog. Fish-Cult.* 36, 138–144.

KREBS C.J. (1972) *Ecology. The Experimental Analysis of Distribution and Abundance.* Harper & Row, New York.

Krebs J.R., MacRoberts M.H. & Cullen J.M. (1972) Flocking and feeding in the great tit, *Parus major*—an experimental study. Ibis **114**, 507–530.

Kwain W. & MacCrimmon H.R. (1969) Age and vision as factors in bottom colour selection by rainbow trout, *Salmo gairdneri. J. Fish. Res. Bd. Canada* **26**, 687–693.

Landless P.J. (1976) Demand-feeding behaviour of rainbow trout. *Aquaculture* **7**, 11–25.

Larson G.L. (1976) Social behavior and feeding ability of two phenotypes of *Gasterosteus aculeatus* in relation to their spatial and trophic segregation in a temperate lake. *Can. J. Zool.* **54**, 107–121.

LeCren E.D. (1965) Some factors regulating the size of populations of freshwater fish. *Mitt. Verein. theor. angew. Limnol.* **13**, 88–105.

Lett P.F., Beamish F.W.H. & Farmer G.J. (1975) System simulation of the predatory activities of sea lampreys (*Petromyzon marinus*) on lake trout (*Salvelinus namaycush*). *J. Fish. Res. Bd. Canada* **32**, 623–631.

Liley N.R. & Seghers B.H. (1975) Factors affecting the morphology and behaviour of guppies in Trinidad. *Function and Evolution in Behaviour*, (G. Baerends, C. Beer & A. Manning, eds.). Clarendon Press, Oxford.

Losey G.S. (1974) *Aspidontus taeniatus*: Effects of increased abundance on cleaning symbiosis with notes on pelagic dispersion and *A. filamentosum* (Pisces, Bleniidae). *Z. Tierpsychol.* **34**, 430–435.

Magnuson J.J. (1962) An analysis of aggressive behavior, growth, and competition for food and space in medaka (*Oryzias latipes* (Pisces, Cyprinodontidae)). *Can. J. Zool.* **40**, 313–363.

Magnuson J.J. & Prescott J.H. (1966) Courtship, locomotion, feeding, and miscellaneous behaviour of Pacific bonito (*Sarda chiliensis*). *Anim. Behav.* **14**, 54–67.

Mahon R. (1976) A second look at bluegill production in Wyland Lake, Indiana. *Env. Biol. Fish.* **1**, 85–86.

Martin R.G. (1975) Sexual and aggressive behavior, density and social structure in a natural population of mosquitofish, *Gambusia holbrooki. Copeia* 1975, 445–454.

Mason J.C. (1969) Hypoxial stress prior to emergence and competition among coho salmon fry. *J. Fish. Res. Bd. Canada* **26**, 63–91.

Mason J.C. (1975) Seaward movement of juvenile fishes, including lunar periodicity in the movement of coho salmon (*Oncorhynchus kisutch*) fry. *J. Fish. Res. Bd. Canada* **32**, 2542–2547.

Mason J., Brynildson O.M. & Degurse P.E. (1967) Comparative survival of wild and domestic strains of brook trout in streams. *Trans. Amer. Fish. Soc.* **96**, 313–319.

Mason J.C. & Chapman D.W. (1965) Significance of early emergence, environmental rearing capacity, and behavioral ecology of juveniles coho salmon in stream channels. *J. Fish. Res. Bd. Canada* **22**, 173–190.

Maynard Smith J. (1974) *Models in Ecology*. Cambridge University Press, Cambridge.

McCrimmon H.R. (1954) Stream studies on planted Atlantic salmon. *J. Fish. Res. Bd. Canada* **11**, 362–403.

McCrimmon H.R. (1968) Carp in Canada. *Fish. Res. Bd. Canada. Bull.* 165.

Miller R.B. (1957) Pemanence and size of home territory in stream-dwelling cutthroat trout. *J. Fish. Res. Bd. Canada* **14**, 687–691.

Miller R.B. (1958) The role of competition in the mortality of hatchery trout. *J. Fish. Res. Bd. Canada* **15**, 27–45.

Moyle P.B. (1969) Comparative behavior of young brook trout of domestic and wild origin. *Prog. Fish-Cult.* **31**, 51–56.

Mundie J.H. (1974) Optimization of the salmonid nursery stream. *J. Fish. Res. Bd. Canada* **31**, 1827–1837.

Murton R.K., Isaacson A.J. & Westwood N.J. (1971) The significance of gregarious feeding behaviour and adrenal stress in a population of wood-pigeons, *Columba palumbus. J. Zool.* **165**, 53–84.

MYRBERG A.A. Jr. (1972) Ethology of the bicolour damselfish, *Eupomacentrus partitus* (Pisces, Pomacentridae): A comparative analysis of laboratory and field behaviour. *Anim. Behav. Monogr.* **5**, 197–283.

NEWMAN M.A. (1956) Social behavior and intraspecific competition in two trout species. *Physiol. Zoöl.* **29**, 64–81.

NIKOL'SKII G.V. (1969) *Theory of Fish Population Dynamics as the Biological Background for Rational Exploitation and Management of Fishery Resources.* Oliver & Boyd, London.

NOAKES D.L.G. & BARLOW G.W. (1973a) Cross-fostering and parent-offspring responses in *Cichlasoma citrinellum* (Pisces, Cichlidae). *Z. Tierpsychol.* **33**, 147–152.

NOAKES D.L.G. & BARLOW G.W. (1973b) Ontogeny of parent-contacting in young *Cichlasoma citrinellum* (Pisces, Cichlidae). *Behaviour* **46**, 221–255.

NORDENG H. (1971) Is the local orientation of anadromous fishes determined by pheromones? *Nature* **233**, 411–413.

OMMANNEY F.D. (1964) *The Fishes.* Time-Life International, New York.

ONODERA K. (1967) Some aspects of behaviour influencing production. *The Biological Basis of Freshwater Fish Production*, (S.L. Gerking, ed.). Blackwell Scientific Publications, Oxford.

PIANKA E.R. (1974) *Evolutionary Ecology.* Harper & Row, New York.

POTTS G.W. (1973) The ethology of *Labroides dimidiatus* (Pisces, Labridae) on Aldabra Atoll. *Anim. Behav.* **21**, 250–291.

RADAKOV D.V. (1973) *Schooling in the Ecology of Fish.* Halsted Press, New York.

REESE E.S. (1975) A comparative field study of the social behavioral and related ecology of reef fishes of the family Chaetodontidae. *Z. Tierpsychol.* **37**, 37–61.

RICKLEFS R.E. (1973) *Ecology.* Chiron Press, Portland, Oregon.

ROBBINS W.H. & MACCRIMMON H.R. (1974) *The Blackbass in America and Overseas.* Biomanagement and Research Enterprises, Ontario.

ROBERTSON D.R. (1972) Social control of sex reversal in a coral-reef fish. *Science* **177**, 1007–1009.

SAGE R.D. & SELANDER R.K. (1975) Trophic radiation through polymorphism in cichlid fishes. *Proc. Nat. Acad. Sci. USA* **72**, 4669–4673.

SALE P.F. (1975) Patterns of use of space in a guild of territorial reef fishes. *Mar. Biol.* **29**, 89–97.

SAUNDERS J.W. & SMITH M.W. (1962) Physical alterations of stream habitat to improve brook trout production *Trans. Amer. Fish. Soc.* **91**, 185–188.

SCHEIN M.W. & HAFEZ E.S.E. (1969) The physical environment and behaviour. *The Behaviour of Domestic Animals*, (E.S.E. Hafez, ed.). Williams & Wilkins, Baltimore.

SCOTT D.P. (1962) Effect of food quantity on fecundity of rainbow trout *Salmo gairdneri. J. Fish. Res. Bd. Canada* **19**, 715–731.

SCOTT J.P. (1969) Introduction to animal behavior. *The Behaviour of Domestic Animals*, 2nd edition, (E.S.E. Hafez, ed.). Williams & Wilkins, Baltimore.

SEGHERS B.H. (1974) Schooling behavior in the guppy (*Poecilia reticulata*): An evolutionary response to predation. *Evolution* **28**, 486–489.

SHAW E. (1970) Schooling in fishes: critique and review. *Development and Evolution of Behavior*, (L.R. Aronson, E. Tobach, D.S. Lehrman & J.S. Rosenblatt, eds.). W.H. Freeman, San Francisco.

SLANEY P.A. & NORTHCOTE T.G. (1974) Effects of prey abundance on density and territorial behavior on young rainbow trout (*Salmo gairdneri*)in laboratory streamchannels. *J. Fish. Res. Bd. Canada* **31**, 1201–1209.

SMITH C.L. & TYLER J.C. (1975) Succession and stability in fish communities of dome-shaped patch reefs in the West Indies. *Amer. Mus. Nov.* **2572**, 1–18.

SMITH R.J.F. (1976) Seasonal loss of alarm substance cells in North American cyprinoid fishes and its relation to abrasive spawning behaviour. *Can. J. Zool.* **54**, 1172–1182.

SMITH R.J.F. & MURPHY B.D. (1974) Functional morphology of the dorsal pad in fathead minnows (*Pimephales promelas* Rafinesque). *Trans. Amer. Fish. Soc.* **103**, 65–72.

SOLOMON D.J. (1973) Evidence for pheromone-influenced homing by migratory Atlantic salmon, *Salmo salar* (L.). *Nature* **244**, 231–232.

STAPLES D.J. (1975a) Production biology of the upland bully *Phylipnodon breviceps* Stokell in a small New Zealand lake. I. Life history, food, feeding and activity rhythms. *J. Fish Biol.* **7**, 1–24.

STAPLES D.J. (1975b) Production biology of the upland bully *Philypnodon breviceps* Stokell in a small New Zealand lake. II. Production dynamics. *J. Fish Biol.* **7**, 25–45.

STRINGER G.E. & HOAR W.S. (1955) Aggressive behavior of underyearling Kamloops trout. *Can. J. Zool.* **33**, 148–160.

SYMONS P.E.K. (1971) Behavioural adjustment of population density to available food by juvenile Atlantic salmon. *J. Anim. Ecol.* **40**, 564–587.

SYMONS P.E.K. (1974) Territorial behavior of juvenile Atlantic salmon reduced predation by brook trout. *Can. J. Zool.* **52**, 677–679.

THOMPSON W.A., VERTINSKY I. & KREBS J.R. (1974) The survival value of flocking in birds: A simulation model. *J. Anim. Ecol.* **43**, 785–820.

TIMMS A.M. (1975) Intraspecific communication in goldfish. *J. Fish Biol.* **7**, 377–389.

TODD J. (1971) The chemical language of fishes. *Sci. Amer.* **224**, 98–108.

TREISMAN M. (1975) Predation and the evolution of gregariousness. I. Models for concealment and evasion. *Anim. Behav.* **23**, 779–800.

TYLER A.V. & DUNN, R.S. (1976) Ration, growth, and measures of somatic and organ condition in relation to meal frequency in winter flounders, *Pseudopleuronectes americanus*, with hypotheses regarding population homeostasis. *J. Fish. Res. Bd. Canada* **33**, 63–75.

VINCENT R.E. (1960) Some influences of domestication upon three stocks of brook trout (*Salvelinus fontinalis* Mitchill). *Trans. Amer. Fish. Soc.* **89**, 35–52.

VINE I. (1971) Risk of visual detection and pursuit by a predator and the selective advantage of flocking behaviour. *J. Theoret. Biol.* **30**, 405–420.

VLADYKOV V.D. (1956) Fecundity of wild speckled trout (*Salvelinus fontinalis*) in Quebec lakes. *J. Fish. Res. Bd. Canada* **13**, 799–841.

WEATHERLY A.H. (1976) Factors affecting maximization of fish growth. *J. Fish. Res. Bd. Canada* **33**, 1046–1058.

WEBB P.W. (1975) Hydrodynamics and energetics of fish propulsion. *Fish. Res. Bd. Canada Bull.* 190.

WEBER P.G. & WEBER S.P. (1976) The effect of female colour, size, dominance and early experience upon mate selection in male convict cichlids, *Cichlasoma niqrofasciatum* Gunter (Pisces, Cichlidae). *Behaviour* **54**, 116–135.

WERNER E.E. & HALL D.J. (1974) Optimal foraging and the size selection of prey by the bluegill sunfish (*Lepomis macrochirus*). *Ecology* **55**, 1042–1052.

WILLIAMS G.C. (1964) Measurement of consociation among fishes and comments on the evolution of schooling. *Publs. Mus. Mich. St. Univ. Biol. Ser.* **2**, 351–383.

WILSON E.O. (1975) *Sociobiology. The New Synthesis.* Harvard University Press, Cambridge, Mass.

WINN H.E. (1958) Comparative reproductive behavior and ecology of fourteen species of darters (Pisces – Percidae). *Ecol. Monogr.* **28**, 155–191.

WOLF L.A., HAINSWORTH F.R. & STILES F.G. (1972) Energetics of foraging: rate and efficiency of nectar extraction by hummingbirds. *Science* **176**, 1351–1352.

WYDOSKI R. & COOPER E.E. (1966) Maturation and fecundity of brook trout from infertile streams. *J. Fish. Res. Bd. Canada* **23**, 623–649.

YAMAGISHI H. (1962) Growth relations in some small experimental populations of rainbow trout fry, *Salmo gairdneri* Richardson with special reference to social relations among individuals. *Jap. J. Ecol.* **12**, 43–53.

YAMAGISHI H. (1964) An experimental study on the effect of aggressiveness to the variability of growth in the juvenile rainbow trout, *Salmo gairdneri* Richardson. *Jap. J. Ecol.* **14**, 228–232.

YAMAGISHI H., MARUYAMA T. & MASHIKA K. (1974) Social relation in a small experimental population of *Odontobutis ogscurus* (Temminck et Schlegel) as related to individual growth and food intake. *Oecologia* **17**, 187–202.

THE CONTRIBUTION OF FISH
PRODUCTION TO HUMAN
NUTRITION AND WELL-BEING

The first three sections of the book are theoretical in nature, albeit with practical overtones. The last section is aimed directly at the harvest. We have chosen to look closely at the contribution of freshwaters, because the freshwater yield has usually been underestimated in relation to the total aquatic harvest used for human food, and because fish culture in fresh and estuarine waters and the yield in closely managed large, inland bodies of water have prospects of greater potential growth than does the catch from the oceans. Ponds and reservoirs have the added advantage of being located near the consumer whereas marine products depend on more complicated processing and transportation arrangements.

The first two chapters are devoted to management. The evidence of eroding fish stocks of favoured species is clear and argues strongly for the need to manage the world's fish resources, whether marine, brackish or freshwater. The current plea for the 'law of the sea' is based in large part on over-exploitation of stocks, once viewed as inexhaustible. Many nations are now demanding the control of fishing 200 miles from their coastline, because they have experienced a noticeable decline in landings of traditionally favoured species. Some of the methods of capture used by ocean ships have been adapted on a smaller scale to large inland lakes and reservoirs, which have historically been proven to be vulnerable to intensive fishing pressure. One of the two chapters attempts to predict what the potential harvest might be from large lakes and reservoirs, such as the huge bodies of water recently created in Africa. The other reviews some scientific aspects of fish management and presents an argument that some compromises are often required between the calculated maximum sustained yield under ideal conditions and the social needs of the local people; such as employment, economic return, competition between sport and commercial fishing, and the like.

The same argument – the need for management to control the yield – is used to justify aquaculture in its broad perspective and pond fish culture in particular. Management is not required in pond culture to protect the stock from overexploitation as it is in large water bodies, but all life history stages from egg to harvest are intensively managed in order to lessen natural mortality, such as disease, competition and predation, in order to ensure high yield. Many biologists believe that the contribution freshwater fish will make toward meeting the virtually certain demand for more protein lies in expanding pond culture. As stated in the Introduction, Dr. S. J. Holt, in the first edition of this book, predicted that a 15-fold expansion in freshwater fish production, mainly from pond culture, would barely meet the rising need for fish protein in the next 25 years. One of the two chapters on this question suggests a variety of ways to increase production in intensively managed situations, concentrating heavily on survival of the early life history stages. The other chapter exemplifies the present status of highly developed techniques in rearing adult fish for the market. The example is drawn from experience in Israel, but the principles are the same wherever fish are raised in ponds.

Lest false hopes are raised by the possibility of raising large amounts of fish in artificial cultivation, I hasten to add that there is as yet no centralized effort to put such an expanded scheme into practice. The international cooperation and capital outlay required for a thrust of this magnitude has not yet appeared on the horizon, although the need has been clearly demonstrated. Along with expansion of physical facilities, the technical people must be trained to match them. At the present time our supply of competent manpower in this area is extremely small in comparison to the projected requirements.

The last chapter in the book on the global freshwater fish harvest is the first comprehensive review of its kind. This harvest is quite respectable when placed alongside that part of the ocean's yield used for human food, much more so than is usually conceded. In place of the cod, herring, tuna and plaice, the less familiar names of carp, tilapia, trout and catfish are discussed. China, where pond culture originated, is still the world's greatest producer of pond-reared fish, and this enterprise progressively declines as one moves westward.

The historical viewpoint that this chapter provides gives us an appreciation of the havoc that pollution has caused in our waterways. Inland Europe which once relied on lakes and streams to furnish fish for the table saw the resource dwindle away to a small fraction of its potential during and since the Industrial Revolution. Total rehabilitation of these waters is too much to expect, but a partial abatement will restore the food producing value of these waters, if and when it may be needed in the future.

Chapter 15: Fishing, Fish and Food Production

J. A. Gulland

15.1 Introduction

Man has had a long interest in fish production, first as a source of food and later as sport. His immediate scientific concern has, therefore, been to determine the factors affecting his catch. On the other side of the same coin, he has been keenly concerned about the impact of fishing on the total fish production, and hence on the part of this production that can be harvested. Mathematical models, of varying degrees of complexity, have been derived to describe these interactions. More recently it has become obvious that fishing is not the only impact that man has on fish stocks, particularly in inland and coastal areas. All forms of water management and flood control, as well as effects of pollution also need to be taken into account. In addition, it is clear that even within the fishery sector scientists, as well as administrators, must take account of factors other than the gross catch. For example, the size or species composition of the catch, the economic return, or the employment opportunities may influence decisions regarding management. While the present chapter concentrates on the mathematical analysis of the interaction of fishing and fish populations, mention needs to be made of these other factors in order to give a complete picture of the practical as well as the scientific framework in which these studies are set.

15.2 The fishable stock

The stock, or population, as used here generally refers only to the fishable part of the stock, and specifically excludes the larvae and small fish which usually differ from the larger fish in distribution, behaviour or feeding, etc. For instance, the younger stages of some species of tilapia in the large African lakes are found close along the shore, and only the larger fish occur in the main fishing areas. The process of recruitment to the fishable stock sometimes coincides with maturity, but may not be so simple, especially when the fisheries are carried on by a wide range of techniques, capable of catching between them fish of most sizes, wherever they may be found.

After recruitment the fish may be liable to capture with suitable gear,

387

but the gear actually in use may only capture a limited size range of the fish. A simple type of gear selectivity is that of trawls, which, to a reasonable approximation, retain all fish above a certain size, but release those below that size. Gill nets are more frequently used in freshwater than are trawls and have a rather more complex selection pattern. Fish smaller than the optimum size pass through the meshes to a greater or less extent, and larger fish are not effectively retained. There is no range of sizes (or ages) over which the mortality caused by meshes of a given size is constant. Most gears, like the gill net, are more or less selective towards certain sizes, and also species, of fish. This effect of fishing mortality is difficult to analyse since it varies continuously with age. Usually it is much more convenient, and also more reliable within an acceptable approximation, to consider that fishing mortality changes abruptly, being zero on the pre-recruits below a certain age and constant from a given age upwards. Where this is clearly unacceptable (e.g. in a gill net fishery) fishing mortality may be treated as being constant over a certain size range, and zero for both bigger and smaller fish (Beverton 1959).

15.3 Production model from the sigmoid growth curve

The simplest analyses of fish populations consider them as single entities that can be described by a single parameter, e.g. the biomass, B, of fish of a fishable size, and subject to simple rules of population growth. Thus Graham (1939) considered that the typical form of population growth was the sigmoid curve—slow at first at low population levels, then rapid, and finally flattening out at the maximum stable population. This curve can be transformed to give the absolute rate of increase of the population, $\dfrac{dB}{dt}$ at any population level, and this has a maximum at some moderate population size. The curves describing the growth of a population in time from a very low level up to the maximum may also be assumed to describe satisfactorily the potential rate of increase—recruitment plus growth less natural mortality (plus the net effect of migration, if any)—at any intermediate population level if kept at that level by fishing. Then if the catch is equal to the potential increase, known as the sustainable yield, the stock will remain unchanged. At large stocks (i.e. with light fishing) the losses due to natural mortality are not much less than the gains from growth and recruitment, and the sustainable yield is small. At small stocks (i.e. when fishing is very intense) the growth and recruitment are large when related to the biomass of the stock, but in absolute terms they are small, and so too is the sustainable yield. Thus, the maximum sustainable yield occurs at moderate stocks, with a moderate amount of fishing.

This sigmoid curve theory has been put on a more quantitative basis by Schaefer (1954) using the expression

$$\frac{dB}{dt} = a(B_M - B)B \tag{1}$$

where

B_M = maximum population abundance,
a = constant

and the maximum sustainable yield is $\dfrac{aB_M^2}{4}$ at a population abundance of $\dfrac{B_M}{2}$.

A more general formulation for the rate of natural increase has been given by Pella and Tomlinson (1969) in the form

$$\frac{dB}{dt} = HB^m - KB \tag{2}$$

where H, K, are constants
of which the special case $m = 2$ results in equation (1). They also give, in the computer programme GENPROD, an objective method of obtaining the best fit (i.e. the fit that gives the least deviation between observed and predicted catches) for a given series of data of catch and fishing effort.

This approach of analyzing the observed changes in stock with increasing fishing effort, as measured by catch per unit effort, gives in quantitative terms a comprehensive explanation of the population dynamics of exploited fish stocks. It is not easy, however, to improve the results obtained in this manner to take into account additional biological information, e.g. growth, predation, etc.

15.4 Analytic models

The more useful approach is to express the stock and the catch explicitly in terms of the various vital parameters, particularly the growth and mortality rates of the individual fish, so that the effect of changes in any parameter can be studied.
Thus, denoting

t_r = age at which fish recruit to the fishery;
$N_t dt$ = number of fish between ages t and $t + dt$ alive at a given time;
$R_t dt$ = number of fish recruiting to the fishery between times $t - t_r + dt$ and $t - t_r$ earlier,

then

$$N_t dt = R_t S_t dt$$

where

S_t = proportion of fish surviving from recruitment to age t,

then

$$N = \int_{t_r}^{t_\lambda} R_t S_t dt \tag{3}$$

where N = total numbers of fish in the fishable stock,
and t_λ = maximum age of fish in the stock.

Also, if

W_t = average age of fish of age t

$$B = \int_{t_r}^{t_\lambda} R_t S_t W_t dt \tag{4}$$

= total biomass of fish in the fishable stock.

If the fish do not mature until some age t_M which is greater than t_r, then B', the weight of mature fish in the stock, is

$$B' = \int_{t_M}^{t_\lambda} R_t S_t W_t d_t \tag{5}$$

and this expression can be useful in studying the relation between parent stock and subsequent recruitment, since in many fish individual egg production is proportional to the weight of the fish, i.e. total egg production is proportional to qB'_M, where q is the proportion of females in the adult stock.

The catch rate, Y, is then given by

$$Y = \int_{t_r}^{t_\lambda} F_t R_t S_t W_t dt \tag{6}$$

where

F_t = fishing mortality coefficient for fish of age t.

The value of F_t will depend (1) on the total amount of fishing, or fishing effort, f, (2) on the pattern with which this is distributed between different ages of fish in accordance with the selectivity of the gear, and (3) on the tactics of the fishermen and the distribution of the fish. In the simplest case, e.g. of a trawl fished without differential attention to places with different ages of fish, we may have

$F_t = 0$, when $t < t_c$

$F_t = F = $ constant, when $t \geqslant t_c$

t_c = age at first capture, i.e. the age of the youngest fish taken by the fishing gear.

By integrating equation (6) over different ages at the same instant in time, the total yield in unit time (say a year) from all year classes can be obtained; alternatively, by integrating over time for the same year class, the yield from that year class during its whole life is observed. Under constant conditions of recruitment, fishing, etc., the results of these two integrations will be the same.

The integration process requires first noting that

$$S_t = R \exp\left(-M(t_c - t_r) - (F + M)(t - t_c)\right)$$

where M=natural mortality, for $t > t_c$
and substituting an appropriate function for the weight of the individual fish, W_i. The first models of this type used growth curves that made integration simple, for instance exponential (Ricker 1975), or the von Bertalanffy (Beverton & Holt 1957). The latter gives the expression for the yield

$$Y = FRW_x \exp\left(-M(t_c - t_r)\right) \sum_{n=0}^{3} \frac{U_n \exp\left(-nK(t_c - t_0)\right)}{F + M + nK} \tag{7}$$

where W_x, K, and t_0 are constants in the Bertalanffy growth curve, which gives the weight at age t as $W_t = W_\infty(1 - e^{-K(t-t_0)})^3$, and U_n are constants $U_0 = 1$, $U_1 = -3$, $U_2 = 3$, $U_3 = -1$

The integration can also be achieved by dividing the lifespan into sufficiently small intervals, so that there are small changes in numbers or weight within each period. The calculations then only involve lengthy but simple arithmetic, which can be made easy by using computers. This method allows any empirically observed growth curve to be used, and also for seasonal changes in growth or mortality rate to be taken into account.

Typical yield curves showing the catch taken by the fishermen as a function of the amount of fishing, as obtained from equation (7) are shown in Fig. 15.1, when recruitment, growth and natural mortality rates are assumed to be constant. Similar calculations can be made of the total production, i.e. the gross additions to the fishable stock through growth and the addition of young recruits. A typical curve of production is shown in Fig. 15.2, which also shows the corresponding yield curve. The figure suggests that while the catch, at least initially, increases with increasing fishing, though perhaps falling off at high rates of fishing, the total production is greatest with no fishing, and steadily decreases with increasing fishing.

From these equations some estimates of the efficiency of the fishery in harvesting the fish production can be made. This efficiency can be defined in

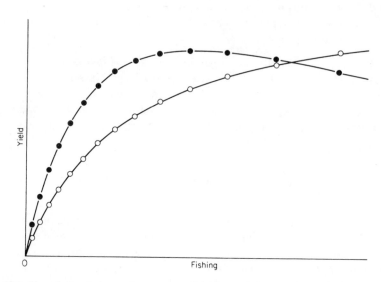

Fig. 15.1. The relation between the amount of fishing and the yield in weight for fast-growing fish with relatively low natural mortality (open circles) and for slow-growing fish with high natural mortality (full circles).

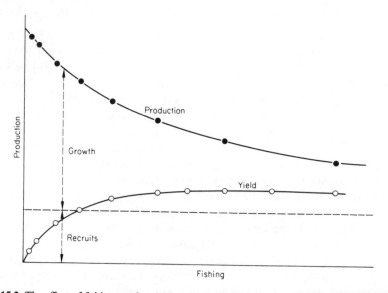

Fig. 15.2. The effect of fishing on the yield and on total production of fish, distinguishing the contribution to the biomass of a fish in the exploited phase, from the recruits and from growth within the exploited phase.

various ways: the catch may be expressed as a percentage of (1) the actual production of fish flesh, (2) the potential production of fish flesh, (3) food consumed by the fish, or (4) the total production of fish food. These will all give different results, not only in absolute terms, but also in the way that the 'efficiency' changes with different rates of fishing.

Thus, in terms of the potential production of fish flesh, the efficiency will follow the same curve as the curve of total catch having, at least when the size of fish capture is small, a maximum at some moderate fishing rate and then decreasing. However, increasing fishing reduces the actual production, so that the efficiency of the catch expressed as a proportion of the production will increase steadily with increasing fishing.

This contrast between the two expressions of 'efficiency' is still greater when considering the catch of fish as a proportion of the food consumed or available to the fish. While the food consumed cannot be expressed precisely in quantitative terms, the qualitative changes are obvious. At low rates of fishing, the fish are relatively large, old and slow-growing, so that a larger proportion of the food consumed is used for maintenance, and less for growth; at higher rates of fishing the fish are younger and smaller, and hence using more of the food consumed for growth. Thus, as the amount of fishing increases, and the fish become fewer and on the average younger, the amount of food consumed will decrease, and at a faster rate than the decrease in production of fish biomass. Therefore, the greatest efficiency of utilization for growth of that part of the food supply that is actually consumed by the fish will be achieved at a high rate of fishing, when the population is composed mainly of small fast-growing fish (Paloheimo & Dickie 1965). These changes within a single species also tend to occur among several species, where the larger species which grow to a greater age are usually more affected by fishing than the smaller species.

However, the food consumed by the fish is rarely a fixed proportion of the total food production. With a reduced fish stock, the proportion of the available food which is consumed will decrease. This lowered efficiency of grazing by a reduced stock of small fast-growing fish is more than sufficient to balance the greater efficiency with which these same fish utilize the food they do consume.

15.5 Density-dependent changes

The advantage of the analytic expressions for the catch, such as the equation (7) over the simpler equations such as (1), is not that one is more accurate or precise than the other in their simplest forms, but that the analytic form enables additional information to be incorporated into the equation to provide progressively more accurate and realistic estimates. This is particularly true when considering changes in growth, natural mortality or recruitment due to changes in the stock abundance, which are themselves being caused by changes in fishing mortality. The probable qualitative effects of these density-dependent changes are easy to assess; increased abundance will generally increase the natural mortality (e.g. by increased spread of parasites or disease), reduce the growth rate, but may increase the number of recruits. The changes in growth and mortality, therefore, tend to counteract the changes in abundance, and to flatten out the curves of Figs 15.1 and 15.2. Best yields are likely to be taken at rather greater levels of fishing, and with a rather greater size at first capture, than is suggested by the constant parameter model. Changes in the number of recruits, however, will tend to reinforce the changes in abundance; thus they will exaggerate the peaks in the yield curves, and the best yields will be taken with smaller amounts of fishing and larger sizes at first capture than is suggested by the constant parameter model. The density-dependent effects on recruitment are, therefore, likely to be more important in determining the productivity of a fishery than the effects on mortality or growth.

Such density-dependent effects on recruitment will be most critical when, because of their behaviour, all or virtually all the adult population can be caught before spawning, e.g. when the fish move from a lake or the sea into a river, and can all be caught by suitably blocking a river. Thus, for example, on the Zambia-Congo border the adult Mpumbu (*Labeo altivelis*) move from Lake Mweru up to the Luapula River to spawn; intensive and unrestricted fishing at the mouth of the river at the time of the spawning migration has reduced the stock and consequently the recruitment and subsequent catches to a negligible level (Mortimer 1965). More often, however, fishing will cause a moderate reduction in the spawning stock. This reduction can be estimated from equation (5); from which, assuming that the egg production is proportional to the weight of the mature stock, the number of eggs produced, E, is given by

$$E = b \int_{t_M}^{t_\lambda} R_t S_t W_t \, dt \qquad (8)$$

where $b = $ constant.

Substituting appropriately into equation (8) the egg production can be shown to be proportional to the number of recruits from which the adults originate, the constant of proportionality between them being determined by the intensity and pattern of fishing. This is indicated by the straight lines in Fig. 15.3, for different values of fishing mortality, assuming a fixed value of the age of first capture. Also in Fig. 15.3 is shown the relation between the adult stock and the subsequent recruitment. Note that when representing the adults produced from a given recruitment, Fig. 15.3 differs from normal

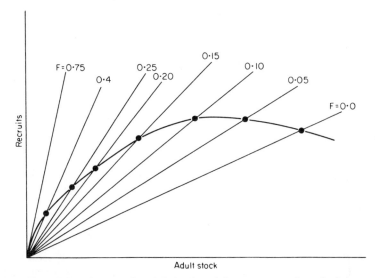

Fig. 15.3. The relations between the adult stock and the average number of subsequent recruits and the average resultant adult stage at various fishing mortailties (straight lines).

usage in that the vertical axis is the independent variable, recruits, and the horizontal axis the dependent variable.

The figure can be used to indicate the equilibrium values of stock and recruitment for a given amount of fishing, which will be when the stock/recruit curve and the appropriate line intersect. These equilibrium values can be substituted into equation (7) (or similar equations giving the yield as a function of the number of recruits and the fishing pattern), to give a yield curve that takes account of the indirect effects of fishing on recruitment. This is shown in Fig. 15.4 (broken curve), which also shows the corresponding curve (full line) of the yield per recruit (or of yield if recruitment were constant). It will be seen that for the stock/recruit relation assumed here, the yield curve has a more pronounced maximum and decreases much more seriously at high levels of fishing than does the curve of yield per recruit.

The practical problem in using this approach is to determine the form of the stock/recruit curve. This has attracted much attention, particularly in respect of marine fish (Ricker 1954, Cushing, 1977, Parrish 1973), but for most species, especially those with high fecundity, the form of the relation is not well established. It is probably best known for Pacific salmon, for some stocks of which the stock-recruit curve has a clear maximum at some intermediate stock level (Larkin, 1977). The studies of the stock/recruit relation is made more difficult because there is not one unique relation for a given stock. Environmental changes can affect the form of the stock/recruitment curve. For example, in Fig 15.5 the solid line may represent the relation under

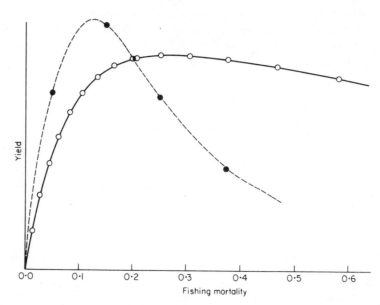

Fig. 15.4. The effect of fishing on the yield per recruit (full line) and total yield (dotted line) if the stock/recruitment curve has the form of Fig. 15.3.

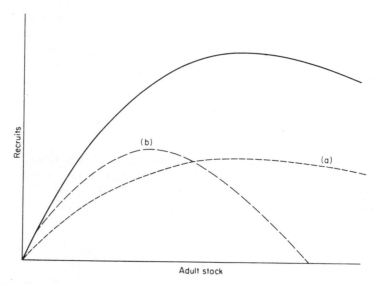

Fig. 15.5. The relation between adult stock and recruitment under good conditions, and two possible relations under poor conditions.

certain 'good' conditions (without specifying what is involved in terms of hydrographic or biological factors to make conditions 'good'). Under 'poor' conditions the resultant recruitment may be half what it would be under good conditions as shown by the dashed line (a). Alternatively, if the declining survival rate (number of recruits per unit stock) at higher stock levels is due to shortage of food and competition between the young fish, then when conditions are poor this competition is likely to become significant at lower stock levels. At even lower stock levels the difference between good and bad conditions may not be great, but the differential effects of good or bad conditions increases with increasing stock. Thus, under poor conditions the greatest recruitment will be less, and will occur at a lower level of adult stock than under good conditions, as shown by curve (b).

The two types of curves have different implications as to how the fishery should be managed to maintain optimum production. Curve (a) presents few problems, in that there is an optimum abundance of the adult stock, which is the same under all conditions. For curve (b), however, the optimum stock under poor conditions is less than under good conditions, which, if these could be forecast, would imply the need for less fishing when future conditions for spawning are likely to be good. Even if conditions cannot be forecast, there may be an important practical decision between maintaining a large spawning stock to take advantage of good conditions when they occur, with the risk of possibly very small recruitment under poor conditions, and maintaining a rather smaller stock which may produce moderate recruitment under good or bad conditions.

15.6 Other human intervention

The classical analyses of fish population dynamics, outlined very briefly in the preceding section, treat fishing as the only independent variable in the system. Fishing affects directly only the value of the mortality rate over a restricted range of sizes (or ages). In addition it may have indirect effects through changing the abundance of the population, especially the adults. Man can alter the population, and his catch, either by altering the value of the fishing mortality through changes in the amount of fishing, or the ranges of sizes to which it is applied, through changes in the selectivity of the fishing gear or the seasons or areas in which fishing is done. The rest of the system is treated as being essentially constant. This is far from being a realistic assumption.

There can be very big natural changes in the environment—seasonal changes, year-to-year fluctuations and longer term trends—many of which can seriously affect fish production. Also many human activities, other than fishing, can have an influence. These range from the accidental or thoughtless impact of landbased activities on water-systems (e.g. by discharge of waste material), through alterations of the water-system for purposes other than fishing (e.g. irrigation, hydro-electric works), to deliberate measures to increase fish production. These last may include stocking with preferred species, fertilization and engineering works to adjust the water regime to one favourable to fisheries, or at least the preferred fish species.

Many of the natural fluctuations, especially regular seasonal changes, tend to average out over a period, and calculation of what occurs under average conditions—for instance, the incorporation of single, average values for the natural parameters of growth, mortality and recruitment—will be sufficient for many purposes. However, this is not necessarily always the case. Figure 15.5 shows that in relation to recruitment the optimum spawning stock under average conditions may be quite different from the optimum for individual years, and may not even give the best average recruitment over a period of years. A better spawning stock may be one yielding recruits at slightly less than the optimum under 'average' conditions, but doing very much better under extreme conditions. Similarly the fishing mortality rate that gives the best result as an average over a period of years may not be the one that is best under constant average conditions.

All these factors can act on one or more of the parameters of the stocks of individual species, and to the extent that the effects may be different for different species, they can have a big effect on the balance between species. A noticeable instance has been the discharge of nutrients into cold oligotrophic lakes (Loftus & Regier 1972). Given the Anglo-Saxon worship of the salmon family for food and sport, this has generally been considered a bad practice, though the total production of fish (and the possible catch) may increase considerably (see particularly Fig. 1 of Colby *et al.* 1972). The whole subject of pollution, especially the discharge of toxic material is attracting an enormous volume of studies and a flood of scientific and semi-scientific papers. Although these are strictly outside the scope of the present review, the various aspects of pollution need to be borne in mind in any study of fish production.

The more interesting activities for the present purpose are those that alter the water system. Previous chapters have shown how fish production depends on the flow of material through the food web, and on a complex of behavioural processes. Disturbance of these at any point can have drastic effects on the production. The best known example is, of course, the migration of salmon upstream to spawn, and the consequent need for fish-passes round dams to allow the passage. Failure to do so—or the blockage of rivers to fish from other causes—caused the collapse, sometimes only temporary, of important salmon runs, the best known being the Hell's Gate slide on the Fraser River which, following railway construction work in 1913, blocked the passage of salmon to their spawning grounds for several decades.

Another important type of movement is the lateral dispersion of many tropical fish onto the flood-plains beyond the limits of the dry season river banks at times of seasonal floods. Breeding often takes place on these flood-plains, which is also the area where much growth occurs. Thus, during each flood season there is an outburst of fish production. The best fishing often occurs as the water falls and the fish can be trapped in isolated pockets of water, or intercepted on their way back to the main river. In Africa these flood-plains support some of the biggest fisheries in the continent (Welcomme 1975), even though they have on the whole attracted less attention than the better known fisheries of the large lakes, where most of the major research centres have been.

Simulation studies (Welcomme & Hagborg, in press) have shown the pattern of the spread of water onto and off the flood-plain, and the extent of flooding can be critical in determining the volume of fish production. Thus, anything that changes these patterns, such as dam construction up-stream, flood-control or irrigation works can seriously affect fish production, usually adversely. Often fishing on flood-plains is mainly carried out at the subsistence level, and therefore the catches do not figure in official statistics. The flood-plains may then be looked on as so much waste land, to be re-claimed for round-the-year agriculture. The resultant works may well result in a net loss in food production at least in quality if not in volume.

The effects of water-control measures are usually better recognized by policy makers when they can be expected to be beneficial, for example, in bringing into existence new bodies of water behind power or irrigation dams. The economic and social justifications for projects of this type, involving bodies of water as large as all but a handful of the natural freshwater lakes, now almost always explicitly recognize the need to take the effect on fisheries into account to an increasing extent, including the effects downstream of the dam, even as far as the sea. The severe fall in the Egyptian catch of sardine in the eastern Mediterranean is a well-known example of the widespread impact of changes in the water regime (see Ryder, Chapter 16).

The opposite extreme of fisheries benefiting (or losing) incidentally from the construction of large dams for hydro-electric power is the construction of ponds specifically for intensive aquaculture. This arm of fisheries is discussed by Hepher and Bardach (Chapters 17 & 18), but the grey area between inten-sive aquaculture and natural fish stocks disturbed only by fishing is of high practical importance. In many bodies of water, larger but less productive than intensive aquaculture ponds, but smaller and more productive than lakes like Kariba and Volta, fisheries can be a significant proportion of the whole economic activity of the region. Relatively minor adjustments to the water regime, e.g. allowing enough flooding onto the marginal land at a time when such flooding is important to successful spawning, can greatly increase fish production and catches. More direct action to increase fish production, e.g. by stocking of desired species, can also be effective at low cost, and leads imperceptibly into intensive aquaculture (see Chapter 18).

15.7 Objectives in fishery management

The ideas of fish population dynamics, outlined very briefly in the first part of this chapter, evolved as they did, and as early as they did, for good practical reasons in advance of the development of most other quantitative analyses of natural populations. The compelling reasons were the concern of fishermen with declining catch rates (and sometimes also declining total catch) in areas (mostly marine) where fishing had become intense. At first the analyses were simple, e.g. the parabolic curve of catch as a function of the amount of fishing given by equation (1). This has a clear maximum and the immediate practical conclusions were correspondingly simple. 'Overfishing', with the effort above the maximum, was clearly bad (less fish at greater cost), and the objective of wise fishery management should be to hold the fishery at the peak of the curve, thus harvesting the Maximum Sustainable Yield (MSY).

At this simple level the concepts of the scientists and of the administrator responsible for national fishery policy meshed together very well, and the scientific results were immediately relevant to the task of pursuing MSY. A yield curve could be produced and the location of the maximum on it pointed out. If this happy relationship is to continue and scientific research into fish production is to be of practical use in determining fishery policies, both scientists and administrators must be aware of a higher order of complexity that must be grappled with as our knowledge of fish production develops.

One important aspect, touched upon in the previous section, is the fact that fisheries, especially freshwater fisheries do not operate in a social, economic or physical vacuum. A whole range of non-fishery activities can affect (positively or negatively) fish production and potential fish harvest. The fishery scientist needs to be able to point out as early as possible the likely magnitude of these effects. In some cases a modest loss in fisheries must be accepted, but very often small modifications (particularly when planning dams or flood-control systems) can result in eliminating much of this damage, or even give some positive benefit to fisheries.

When the stocks concerned are only affected by fishing, there still can be a wide range of interests other than maximizing the catch of a particular species. Fishing provides employment, as well as food, and usually the economic return (either gross value, or net value in excess of costs) of the catch is at least as important as the actual weight. Even if feeding the local population is the primary concern, this may be better done by a catch that includes highly valued species, some of which can be sold or exchanged for other staple food than from a larger catch of low valued fish. Costs are also important; many yield curves (cf. Fig. 15.1) are flat-topped. Since total costs are usually roughly proportional to the amount of fishing, moving some way to the left of the point of maximum yield can allow a substantial reduction of costs with little loss in total catch. This position would clearly have much more economic attractions than the MSY.

In sports fisheries, which dominate freshwater interests in the richer countries, costs are largely irrelevant, but species composition and size composition among each species become more important. Some fishermen would like to have the pleasure of catching at least one fish whenever they go out, while others are more interested in the possibility of catching a prime fish on rare occasions. The question of an alternative and simply defined objective to substitute for MSY has been commonly debated (cf. Roedel 1975), but the general conclusion has been that the search will be fruitless, and that a more nebulous term such as 'optimum sustained yield' has to be used in framing general objectives of management and conservation. In each particular situation the actual decisions will have to be taken, balancing the need for food, employment, economic returns, sports fishing, etc. This means that the scientist must be able to determine what may happen, as a result of different action—unrestricted development of fishing, controls on types of gear, limits on catches or number of fishermen, etc.—on many characteristics of the fish stock, particularly the abundance and production of the major species, and the catches taken. This in turn calls for a better and wider understanding of the general process of fish production.

Acknowledgements

My thanks are due to Robin Welcomme and my other colleagues in the Department of Fisheries for stimulating discussions and exchange of ideas incorporated in this paper, and comments on early drafts.

References

BEVERTON R.J.H. (1959) Report on the state of the Lake Victoria's fisheries. Fisheries Laboratory, Lowestoft. Report **44** (mimeo).

BEVERTON R.J.H. & HOLT S.J. (1957) On the dynamics of exploited fish populations. *Fishery Invest. London, Ser. 2.* **19**, 1–533.

COLBY P.J., SPANGLER G.R., HURLEY D.A., & McCROMBIE A.M. (1972) Effects of eutrophication on salmonid communities in oligotrophic lakes. *J. Fish. Res. Bd. Canada* **29** (6), 975–983.

CUSHING D.H. (1977) The problems of stock and recruitment. *Fish Population Dynamics* (Gulland, J.A. ed.), Wiley, New York. pp. 116–133.

GRAHAM M. (1939) The sigmoid curve and the overfishing problem. *Rapp. Proc.-Verb. Cons. Int. Explor. Mer.* **110** (2), 15–20.

LARKIN P.A. (1977) Pacific salmon. *Fish Population Dynamics* (Gulland, J.A. ed.), Wiley, New York. pp. 156–186.

LOFTUS K.H. & REGIER H.A. (eds) (1972) Proceedings of the 1971 Symposium on Salmonid Communities in Oligotrophic Lakes. *J. Fish. Res. Bd. Canada* **29**, 611–986.

MORTIMER M.A.E. (1965) *The fish and fisheries of Zambia.* Ministry of Lands and Natural Resources, Lusaka.

PALOHEIMO J.E. & DICKIE L.M. (1965) Food and growth of fishes. 1. A growth curve derived from experimental data. *J. Fish. Res. Bd. Canada* **22** (2), 521–542.

PARRISH B.B. (ed.) (1973) Symposium on Stock and Recruitment. *Rapp. Proc.-Verb. Cons. Int. Explor. Mer.* 164.

RICKER W.E. (1954) Stock and recruitment. *J. Fish. Res. Bd. Canada* **11** (5), 521–542.

RICKER W.E. (1975) Computation and interpretation of biological statistics of fish populations. *Bull. Fish. Res. Bd. Canada* **191**, 382 pp.

ROEDEL P. (ed.) (1975) Optimum sustainable yield as a concept in fisheries management. *Spec. Publ. Amer. Fish. Soc.* **9**, 89 pp.

SCHAEFER M.B. (1954) Some aspects of the dynamics of populations important to the management of marine fisheries. *Inter-Amer. Trop. Tuna Commn, Bull.* **1** (2), 26–56.

WELCOMME R.L. (1975) The fisheries ecology of African floodplains. *CIFA Tech. Pap.* (3), 51 pp.

WELCOMME R.L. & HAGBORG D. (in press) Towards a model of a floodplain fish population and its fishery. *CIFA/75/Inf.* 9, August 1975, 30 pp.

Chapter 16: Fish Yield Assessment of Large Lakes and Reservoirs—A Prelude to Management*

R. A. Ryder

16.1 Introduction

Successful management of an aquatic ecosystem requires fundamental information about both the state properties and dynamic interactions among the various components of the system—a condition rarely attained on large lakes and reservoirs. Management has suffered despite sometimes generous funding and the provision of sufficient staff with the appropriate expertise, because the effects of future perturbations upon the complex dynamics of large lake systems are uncertain. Watershed development, in particular, is subject to insidious secondary effects in the form of cultural eutrophication and exploitation of biotic resources. These perturbations tend to degrade aquatic systems differently, and often more rapidly than the rate at which management plans can be formulated.

The fisheries management position relies upon the traditional demographic analyses which determine fecundity, recruitment, mortality, age and growth, as well as other reductionist single-stock methods. While possibly appropriate for the management of a farm pond or a trout stream, these procedures are hardly adequate for a body of water with the immensity of Lake Michigan, for example, (57,850 km^2; Hutchinson 1957) or Lake Tanganyika (32,893 km^2; Barbour & Brown 1974). Political jurisdiction is also a complicating factor. Fisheries management of the 114 species (Ryder 1972) occurring in Lake Michigan is handled by 1 federal and 4 state government agencies co-ordinated by a fishery commission; in the case of the 233 fish species of Lake Tanganyika (Poll 1953, 1956) by the governments of four countries aided by one or more international agencies. Also, well-intended management is hindered by increasing population pressures which create greater demand on the aquatic resource than it can conveniently support. Finally, the single-species population approach is impractical for application to fish communities in large water bodies.

In the developing countries, the immediate need for food production in he form of fish flesh is paramount, especially in the face of burgeoning

* Contribution No. 76–8 of the Ontario Ministry of Natural Resources, Fish and Wildlife Research Branch, Thunder Bay, Ontario.

populations. This immediacy precludes the possibility of implementing long-term management plans dependent on the detailed demography of fish populations. Even in the more developed nations, this methodology has met with infrequent success because of rapidly changing conditions within the system or because of over-exploitation and other cultural stresses (e.g. SCOL Symposium; Loftus & Regier eds. 1972).

Large aquatic ecosystems, which are basically intractable via the traditional single-species route, may be successfully managed by treating groups of species, although a degree of both precision and resolution is sacrificed for the sake of timeliness in this instance (Henderson *et al.* 1973). Timely, order of magnitude estimates, refined through feed-back information such as might be obtained from an extant fishery, will be more useful than would a precise demographic analysis which cannot hope to keep abreast of rapid changes. One of the primary requisites for a successful fisheries manager on a large aquatic ecosystem, therefore, is to strike a utilitarian compromise between precision of yield estimates and timeliness in their application. Of foremost importance in this regard is the derivation of yield (that portion of fish production harvested by man) estimates for large lakes and reservoirs prior to the implementation of any management scheme. Once first approximation yield estimates have been derived for the total fish community inhabiting a large body of water, then the first step (and often most important) of management may be taken—provisional apportionment of the ichthyomass to the various fisheries by one of many available methods over an appropriate time frame. This is usually implemented by quota systems, gear regulations, size limits, time limits for exploitation or any of the multiple restrictive measures often utilized to ensure the perpetuation of a fishery.

16.2 Basic Concepts

The proposed *modus operandi* of preliminary fish yield assessment in large lakes and reservoirs is at the community level rather than at the traditional species or population levels. Some fundamental principles must be remembered in adopting this approach. Although ecosystems and communities are open systems (von Bertalanffy 1968), they do, nevertheless, conform to laws derived from and applied to closed systems such as the First Law of Thermodynamics. The essence of an open ecological system lies in the continual inflow and outflow of energy and materials which, if balanced, result in the familiar steady-state condition (see Gallucci 1973). This condition is subject to cybernetic controls, such as positive or negative feedback, which allows the system to remain within certain upper and lower homeostatic bounds. The First Law then, which is essentially the conservation of energy and matter, is fundamental to the processes of a biotic community or ecosystem. While it is usually not necessary to apply the First Law specifically to a management prognosis for an aquatic system, it is, nonetheless, basic to the interactions within the system. Solar energy interacting with mineral nutrients is required for the basic organic synthesis (which fixes solar energy in reduced carbon compounds) in the autotrophs that supports all organic production. As both energy and nutrients are basic to fish production, we should be less concerned with energy-limited or nutrient-limited systems per se in favour of concentrating our attention on limitation of the energy required to transport nutrients within the ecosystem, as suggested by Mann (1974). Availability of, and interaction between energy and nutrients, therefore, might best be considered as the primary constraints on a fish production system.

16.3 Secondary Constraints

There are two major secondary constraints on freshwater production systems which limit the availability of energy and nutrients or modify the supply regime in various ways. These limitations, namely, climatic and edaphic conditions, are so inextricably inter-related that their separate effects are usually difficult to identify. The edaphic condition, however, may be considered as the principal factor relating to nutrient availability. The simultaneous and interacting effects of both the climatic and edaphic conditions are even more important in this respect, as the former influences the latter condition and together they determine nutrient availability, flux and transport.

Climate quite obviously governs the length of the growing season, by influencing the duration and intensity of light reaching the earth's surface, which in turn is essential for photosynthesis. Wind patterns determine the degree to which autochthonous nutrients are recycled, and precipitation influences the degree of nutrient leaching and ultimately, nutrient transport to the water body. Long periods of ice and snow cover in temperate and polar zones temporarily prevent the transport of allochthonous and aeolian materials and inhibit the degree of sunlight penetration to photosynthesizing organisms beneath the frozen lake surface. In general, climate is a basic constraint on energy availability but since energy is required to circulate nutrients, therefore it indirectly, as well as directly, affects the nutrient regime.

Edaphic conditions determine the capacity of the surrounding watershed to supply nutrients and trace elements in sufficient quantities and appropriate proportions for organic synthesis. Judged on an edaphic basis alone the most productive lacustrine zones of the world generally were previously submerged. Years of submergence provides an accumulation of organic matter which in turn enriches contemporary waters. Pleistocene submergence following withdrawal of the Wisconsin ice sheet in North America (Ryder 1964) and the post-Würm submergence period of northern Europe (Toivonen 1972) have created these two rich fish production regions of the North-temperate zone. Former inland seas have deposited a legacy of nutrients that are currently utilized by the biota of contemporary lakes in many parts of the world and tend to be even richer than extinct Pleistocene lakes (Hutchinson 1957).

Relatively infertile zones occur on the acid-intrusive rocks of the Precambrian. Immense areas of North America and Eurasia are pock-marked with myriads of lakes of low fertility lying on Precambrian formations. These zones are generally subjected to unfavourable climatic conditions which result in short growing seasons. Precambrian regions also occur in the tropics and are less productive there than lakes situated on the relatively richer lateritic soils. The latter, however, are subject to rapid leaching and nutrient loss where the prevailing terrestrial plant communities have not been retained because of burning or poor agricultural practices.

16.4 Tertiary Constraint

The morphology of a lake or reservoir system is also a major constraint. Area and volume can be considered the two primary parameters here, but others such as mean depth or shore development may be of greater utility in estimating the capability of a lake to produce fish.

Let us imagine a hypothetical and idealized family of lakes in which the area varies but all other physical and external conditions (e.g. climate) are identical for each lake in the set. Under such conditions each lake may be expected to produce fish in direct proportion to its respective area (e.g. Rounsefell 1946). For the same set of lakes in which area is held constant but for which depth is variable, fish production will vary inversely with depth. Mean depth has been declared by Rawson (1955) to be the most important morphometric feature of a lake while other authors have stressed the depth of the mixing zone (Hargrave 1973). However, depth alone has many interactive and often subtle consequences restricting production in an aquatic ecosystem. It is obvious that great depths limit the effects of the mixing action of the wind and hence restrict the reentry of substrate nutrients into the production system, as well as constrain the effects of solar radiation to the upper strata. These effects are most extreme in meromictic lakes where the concentrated salts below the chemocline provide an additional density resistance to overturn. Lake Tanganyika, therefore, because of its great depth, is likely to be less productive than it would if it were sufficiently shallow to preclude the possibility of the existence of an energy-nutrient sink in its monomolimnion. *Ceteris paribus*, shallow lakes will be more productive of fishes than deep lakes (Ryder *et al.* 1974).

16.5 Other Constraints

Flushing rate is important since it regulates both the degree and regime of nutrient loading (Vollenweider 1969). Flushing rate defines the retention time of any unit volume of water within a lake and is determined by the size and shape of the lake basin as well as by the inflowing and outflowing streams. Flushing rate or one of its correlates (e.g. storage ratio, water level fluctuation; Jenkins 1967) may be one of the major distinguishing features between large natural lakes, and reservoirs.

Shoreline development, another morphometric feature, is measured by the ratio of shoreline length to the circumference of a circle enclosing the same area as that of the lake

$$\frac{s}{2\sqrt{\Pi\, a}}$$

where s = shoreline length and a = area of circle (Welch 1948). A shoreline development factor substantially greater than unity, represents a higher proportion of ecotone (in this case, land-water interface) which for fishes may provide greater habitat diversity (MacArthur 1972) and consequently a greater opportunity for productivity. Interface zones provide areas for breeding, reproduction, nursery grounds and feeding and may generally be more productive than other, relatively homogeneous zones. This is due to the structural heterogeneity that they provide between land and water and also their ability to concentrate nutrients, plants, and invertebrates, and ultimately, fish food. Ecotones are also the first zones to receive nutrients from their primary sources at the mud-water, and the land-water or the air-water interfaces. Surprisingly, in some lakes, aeolian sources of nutrients may supply the major input (Schindler & Nighswander 1970). Horizontal ecotones exist between the open limnetic zones of lakes, which are relatively nutrient-impoverished and the shallow and richer littoral zone. Nutrients regenerated within the water column itself, and recycled there, ultimately must originate from other sources.

Inorganic turbidity, most often caused from suspended particles of colloidal nature, prevents solar radiation from penetrating all but the surface waters and thereby effectively reduces the trophogenic zone. Hence photosynthesis and primary production are restricted to a relatively shallow zone just below the air-water interface. Nominally rich lakes may be rendered relatively unproductive by turbidity because photosynthesis is decreased drastically. Inorganic turbidity enters the lake water either from the bottom or terrestrial soils and hence may be of either autochthonous or allochthonous origin.

Climatic effects vary with latitude and altitude to such an extent that they should be considered as useful, measurable entities in themselves. Lakes at either a high altitude or high latitude are generally less productive than their counterparts at lower latitudes and altitudes, other things being equal. It is important therefore to recognize and quantify these differences when observed and relate them to the ambient climatic regimen.

16.6 Ecological Considerations

Community *production, biomass* or *yield*, depend on fundamental ecological considerations. Two basic premises that have been formulated for community systems are: (1) a community exists as an interacting array of species and consequently, interlocking species' niches; as such, the community resists *decomposition* (inverse of the number of interactions of subsystem components) to varying degrees (see Simon 1969) and may be treated as a single, integral unit. (2) Since the community is a single unit it must exhibit certain recognizable characteristics as does an organism or a population. *Emergent properties* (Kerr 1974a) of a community are not always clearly defined but usually represent external characteristics of a particular community just as morphological and behavioural characteristics identify the species. Two properties of a fish community most likely to be measured prior to yield estimation are standing crop or ichthyomass, and production rates or the inverse, community mortality (Ryder & Henderson 1975).

Interactions among various members of a community maintain its integrity through resistance to decomposibility, thus insuring its continuing identity. In this manner a *system conservatism* is established such that varying combinations of external stresses *in toto* produce similar patterns of community reactions which are themselves predictable (Ryder *et al.* 1974). Large lakes may undergo substantial perturbations over a long period of time resulting from causes such as exploitation by fisheries, cultural stresses, or species changes due to the introduction of non-indigenes, or the initiation of extinction processes (Loftus & Regier 1972, MacArthur 1972). Despite these internal changes, certain properties of the system may remain constant and these system conservatisms are usually best expressed in terms of community biomass or yield. Hence, yield levels on a community basis from a large lake or reservoir stressed in a multiplicity of ways over long periods of time might be expected to remain essentially constant with relatively small variance. During this time, species composition and mean sizes could conceivably fluctuate drastically and perhaps proceed unidirectionally to extinction for some species (Regier *et al.* 1969, Regier & Hartman 1973). Fisheries managers faced with the day by day dilemma of predicting fish yield for exceptionally large lakes, or for large numbers of smaller ones, would be well advised to approach their problem from the fish community or subsystem basis.

16.7 Community and subsystem approaches to yield estimation

MORPHOEDAPHIC INDEX

Concept

The morphoedaphic index (*MEI*) is an empirical model, initially derived from yield measurements for two sets of north-temperate lakes (Ryder 1965). The index is comprised of two basic and easily-measured parameters of a lake system, namely, total dissolved solids (*TDS*), a correlate of nutrient levels (edaphic factor), and mean depth ($\bar{z}$), a morphometric factor (i.e.

$MEI = \sqrt{\dfrac{TDS}{\bar{z}}}$ Since the latter parameter, may be to a large degree directly

related to energy dissipation, and ultimately heat dilution, we have therefore, an expression of the two fundamentals implicit in the First Law; i.e.—inter-convertibility and conservation of energy and matter. While this succinct statement does indeed oversimplify all of the implications of the *MEI* (Ryder *et al.* 1974), it does, nonetheless, emphasize the major fundamental effects reflected by the index. Explicitly, TDS measured at selected surface stations of a lake over an appropriate time frame will provide seasonal and spatial sampling equity and integrates the time and space variability of this parameter into one meaningful value (Ryder 1964).

Mean depth as a single, simple morphometric determination, expresses a dilution ratio for energy and nutrients and a partial unidirectional sink for these two variables. Mean depth also is related in various ways to the extent of littoral zone, shore development, flushing rate, mixing depth and other expressions of morphometry. The *MEI*, then combines many of the primary, secondary and tertiary constraints on organic production into one general index which is positively correlated with fish yield for large lakes and reservoirs (Ryder 1965, Jenkins 1967, Ryder *et al.* 1974).

One major consideration not adequately expressed by the index, however, is the latitudinal climatic effect. Consequently, the simplified metric formulation of the *MEI* as expressed for north-temperate lakes is $Y \sim 0.966 \sqrt{X}$, where *Y* is annual yield to a fishery and $X = MEI$ (Ryder *et al.* 1974) or more simply $Y \sim \sqrt{X}$ (Fig. 16.1; A). An equivalent generalized formulation for tropical lakes or south-temperate reservoirs would be approximated by $Y \sim 10 \sqrt{X}$ as expressed by the Ryder (1965) model (Fig. 16.1; D). All regression curves of Fig. 16.1 regardless of whether they are patterned after the curvilinear model of Jenkins (1967) or the linear model of Ryder (1965), possess a common characteristic—i.e. similar slopes of the ascending limbs. This would seem to indicate that all ascending limbs may be represented by the generalized model $Y = k \sqrt{X}$ where *k* is a constant that represents an appropriate suite of climatic effects.

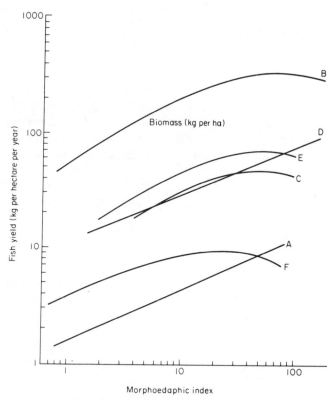

Fig. 16.1. Yield (A, C, D, E, F) and biomass (B) curves for four regional lake sets:
A, yield curve for north-temperate lakes of North America (Ryder 1965); B, biomass
curve for temperate and south-temperate reservoirs of the U.S.A. (Jenkins 1967); C,
yield curve for Jenkins' reservoirs derived by Regier *et al.* (1971); D, yield curve for
tropical African lakes using the model of Ryder (1965), and E, Jenkins (1967); F, yield
curve of Finnish north-temperate lakes (based on data from J. Toivonen) using Jenkins'
(1967) model (after Ryder *et al.* 1974).

Methodology

The *MEI* is usually applied when a first approximation of fish yield is being
made where specific data are lacking for precise estimates. Prior to application
a sufficient limnological assessment will be made in order to place the
candidate lake in the appropriate lake set. Consequently, it will be necessary
to ascertain if specific *a priori* criteria are satisfied, i.e. the lake or reservoir
is subjected to a *relatively* homogeneous set of external environmental conditions (compared to other lakes in the set). These conditions might include
a characteristic and reasonably constant proportionality of ions, a flushing

regime roughly proportional to the lake volume, inorganic turbidity levels on about the same order of magnitude for all lakes in the set, and moderately intensive to intensive fishing effort on most abundant species over a period of years. The latter condition can be objectively quantified using the method of Henderson and Welcomme (1974). Once the criteria have been fulfilled it is a simple matter to apply the index multiplied by an appropriate constant. The *MEI* may then be substituted into the corresponding equation (depending on climate) and solved for total fish yield. Alternately, the appropriate family of lakes (Fig. 16.1) may be used for a graphic solution once the *MEI* has been calculated.

In cases where the regression for a family of lakes at any given latitude has not been previously described, it may have both utilitarian and heuristic value to do so. Preferably, a series of ten or more lakes are selected that appear to be homogeneous with respect to most external abiotic factors, especially climate, and a regression is calculated between fish yield and the *MEI*. Essentially, this linear regression should be made to comply with a slope of about 0·45 using a $\log_{10}$ transformation, since the '*Y*-intercept'* is the value that will anchor the regression line. While it may seem unduly subjective to make this line comply with a 0·45 slope, it is nevertheless a valid process as most of the slopes for global families of lakes have been bracketed by the north-temperate systems (slope = 0·45; Ryder 1965) and the tropical series (slope = 0·45; Regier *et al.* 1971). Once the curve has been described for a new set of climatically homogeneous lakes, the relationship can be tested with new inputs. It is especially useful in such cases to select lakes with extremes in TDS or mean depth or extreme ratios of the two for inclusion in the series. This procedure will insure that future use of the regression line will likely include all new entries from that particular climatic zone without need for extrapolation.

PARTICLE-SIZE HYPOTHESIS

Concept

The particle size hypothesis of Sheldon *et al.* (1972) introduces an important concept that holds much promise as a practical tool for first approximation yield estimation. The hypothesis proposes that approximately equal concentrations of material occur at all particle sizes, when size intervals are segregated on an order of magnitude basis. There will be a slight decrease in mass per unit volume of water, however, with increasing size of organism.

---

* As the regression has been $\log_{10}$-transformed there is, of course, no Y-intercept. Rather it is the lower asymptotic level that is being estimated as it would cross the Y axis were it plotted on arithmetic co-ordinates.

Consequently, for the equatorial Pacific, about equal concentrations (mg l^{-1}) occurred of zooplankton, micronekton and tuna whose relative sizes were 10^3, 10^4 and $10^5 \mu$, respectively. Essentially, the particle-size hypothesis has been successfully applied to organisms ranging in size from bacteria (1μ) to whales ($10^6 \mu$) which covers the total spectrum of marine organisms (Sheldon *et al.* 1972). The authors explained a portion of this phenomenon on the basis that food-chain relationships in the ocean contain relatively large predators that feed on substantially smaller prey. Consequently, the standing stock hierarchy as previously outlined could only be maintained where rate of particle production varies inversely as particle size. Hence, if rates of production vary by an order of magnitude for each particle size, and trophic transfer efficiency is about 10%, then it follows that standing stocks must be approximately equivalent (Sheldon *et al.* 1972).

Theoretical support for the particle-size hypothesis has been provided by Kerr (1974b) who has developed a model explaining the observations of Sheldon *et al.* (1972). He assumed that a simple relationship between the sizes of prey and predator exists in the pelagic zone and that metabolic growth rates satisfy similar functions of body (particle) size if averaged over entire trophic levels. The model was successfully applied to other community types as well and suggested that trophic processes were complementary to competitive displacement since particle density is a linear function of the logarithm of particle size.

Methodology

In order to apply the particle–size hypothesis to an actual community yield determination we must first solve for the unknowns in the Gulland (1970) yield equation as applied to fish communities, i.e. $Y = kMB$, where Y is estimated total fish yield on a long-term basis; k is a constant (such that $0.3 < k < 0.5$, or $= 0.4$ typically); M is a community mortality coefficient based on individual species mortality coefficients weighted on the basis of their proportionality within the total harvest; and B is fish community standing crop expressed as biomass prior to fishing.

To solve for Y, we must possess data that will allow us to estimate both M and B. Two examples of this methodology are available in the literature. One estimates yield for a U.S. reservoir (Regier *et al.* 1971) and the second for a tropical reservoir (Ryder & Henderson 1975). This methodology can also be applied to natural lakes. As a practical example of the method of estimating M and B, we will utilize the data (Table 16.1) from Ryder and Henderson (1975) on the Nasser Reservoir, Egypt. In this instance B was obtained from an independent study by Entz *et al.* (1971) who determined that 10^4 metric tons of zooplankton was the average biomass of a particular particle diameter in the Nasser Reservoir. The assumptions were made that

Table 16.1. Data used in the determination of long-term potential yield for the Nasser Reservoir based on the Gulland (1970) model. Mortality coefficients were approximated as in Regier *et al.* (1971). Table from Ryder & Henderson (1975).

Species	Approximate mortality coefficient[1] M'	Relative Biomass[2] B'	$M' \times B'$
Sarotherodon niloticus[3]	0·25	0·56	0·1400
Alestes sp.	0·40	0·18	0·0720
Labeo sp.	0·25	0·09	0·0225
Lates niloticus	0·10	0·07	0·0070
Bagrus bayad	0·15	0·04	0·0060
Clariidae	0·30	0·01	0·0030
Schilbeidae	0·60	0·01	0·0060
Other	0·50	0·04	0·0200
			$\Sigma M'B' = 0\cdot2765$

[1] Based on known longevity of fish following recruitment to exploitable size.
[2] Based on 1972 percentage composition in the commercial fishery.
[3] The most abundant tilapia species in the reservoir.

suspended particles within the size interval of zooplankton contained living material and that in fresh water there are likely no marked differences between volume or mass in living organisms for each unit volume of water. The 10^4 tons (B) of zooplankters then, was directly equated to the standing crop of fishes which on average had a particle diameter about 2 orders of magnitude greater than zooplankters. No further corrections were made for particle size as the obvious generality of this technique does not warrant further refinement.

The solution of a community mortality coefficient for the Gulland equation required the computation of individual species mortality coefficients from the time the fish entered the fishery, weighted by the proportionality of each species in average catches of the commercial fishery (Table 16.1). In this case individual species mortality coefficients were computed (M'), multiplied by their relative ichthyomass contribution to the fishery (B') and summed to provide a weighted community mortality coefficient, $M'B'$, or more simply, M. The previously determined M and B may now be substituted into the Gulland (1970) yield equation, $Y = 0\cdot4(0\cdot2765)10^4$; therefore $Y = 11,060$ and represents an annual predicted yield of about 10^3 metric tons or 26 kg ha^{-1}yr^{-1} for the Nasser Reservoir (Ryder & Henderson 1975). As this yield closely coincided with other yield estimates based on discrete data sets, it was accepted as a useful corroborative technique of low (order of magnitude) precision but potentially useful because of the relative ease in obtaining data to satisfy its requirements and the possibility it offers for timeliness in application.

GILLNET MONITORING

Concept

Standardized gillnets have long been a valuable sampling tool in routine fisheries surveys where a representative sample of the standing stock of fishes is required (e.g. Moyle 1950). Unfortunately, selectivity by this type of enmeshing gear complicates any possible deductions to be made between estimated standing stocks (biomass) and actual values (e.g. see Hamley 1975 for a discursive presentation on gillnet selectivity). Nevertheless, this complication may be obviated in certain instances where a proportional relationship is to be established between monitoring gillnet catches and an extant commercial fishery using similar gear. Ideally, the sampling nets are designed to simulate those used in the commercial fishery. It is postulated that catch per unit of effort (*CUE*) of the sampling gear is related to standing stocks of fishes, often termed abundance or availability (Hile & Buettner 1959). Similarly, for the commercial fishery, *CUE* should be related to the standing stocks and, in a moderately intensive to highly intensive fishery, to the yield as well. It is easy, therefore, to make the transfer from *CUE* in the monitoring series to *CUE* in the commercial fishery in order to deduce yield from the commercial fishery. Since we are primarily interested in predicting yield rather than standing crop in this instance, gillnet selectivity does not bias the results because the population sampled in the monitoring series is essentially the same as that taken by the commercial fishery. This technique is particularly applicable to large lakes and reservoirs (e.g. Laurentian Great Lakes, African Rift Lakes and reservoirs) where the fishing is sufficiently intensive to evoke a response reaction in terms of density-dependent responses, such as growth and mortality, (Ware 1975), and therefore be truly representative of the vulnerable portion of the fish community. Aside from being an estimator of standing crop and yield, this technique allows the short-term prognosis of a fishery prior to the occurrence of actual fishing. Where a fishery follows seasonal trends, this technique may allow for the fishery and the market for the produce to be activated simultaneously, with obvious economic benefits to both.

Methodology

On the Volta Reservoir, Ghana, gillnet monitoring employed standardized series of gillnets as the principal assessment gear (Henderson *et al.* 1973). This large, 8300 km² (Obeng 1973) and diverse reservoir was divided into eight zones, and each zone was fished systematically in both time and space by the sampling gear. Concurrently, yield estimates were derived by various methods from a subsampling of the catches (Bazigos 1971). Since 80% of the total yield was taken by gillnets which the sampling gear simulated, it was expected that a constant proportionality in catch would exist between the samples and the total catch. Over the course of one year, the actual catches kg ha^{-1} closely followed the trends in *CUE* (kg 100m^{-2}) established from the sampling nets (Fig. 16.2). Predictions were made for the fishery on

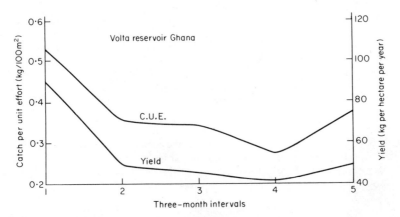

Fig. 16.2. Time series plots of catch per unit of effort for monitoring gillnets and estimated harvest by the commercial fishery at Volta Reservoir, Ghana (after Henderson *et al.* 1973).

an annual basis and also on a short-term, quarterly basis. For example, seasonal variations in the fishery resulted from the effects of rainy or dry seasons, annual floods, etc. Once an absolute value of yield for one season was established, a predicted value for any other season could be estimated on the basis of past annual trends (Fig. 16.2). When the relationship between predicted catch as measured by sampling gear and actual catch as taken by the fishery has been well established, a regression can be used to estimate future total catches from catches made by the sampling gear. For example, on the Volta Reservoir, the relationship between yield (*Y*) in kg ha^{-1}yr^{-1} and standing crop (*X*) as expressed by *CUE* in kg 100m^{-2} was: $Y = 204X - 20$ (Fig. 16.3, Henderson *et al.* 1973). Various levels of sampling effort should be exerted in order to determine the minimum intensity required to obtain satisfactory predictive results.

A SYSTEMS APPROACH

Concept

Following the lead of von Bertalanffy (1968) we may aggregate the yield problem at one or two hierarchal levels above that of the fish community

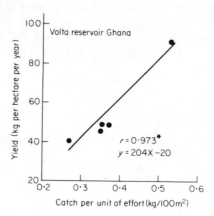

Fig. 16.3. Relation between computed harvest by the commercial fishery and catch per unit of effort by gillnets used to monitor standing crops of fishes for the Volta Reservoir, Ghana. Each point represents a mean value of a 3-month period. The regression line was fitted by the least squares method (after Henderson *et al.* 1973).

per se and relate the loss of biological production in a particular lake or reservoir to the augmentation of production in the contiguous watershed. Since Forbes (1887) first described a lake as a microcosm, it has been increasingly evident that water bodies are inextricably influenced by interconnected aquatic systems as well as the lake basin and the surrounding terrain (e.g. Hasler 1975). Because strong interactions are intrinsic to the system, we can reasonably expect that a significant change in one part of the system might effect a departure from steady-state in another part of the overall system. This effect will usually be resilient and compensatory in nature. Of necessity, because we are dealing with open systems, the causal effect in a second subsystem will not often be precisely equivalent to the initiating effect in the primary subsystem but rather will be proportional to it.

In brief then, we are measuring a time-series of emergent behaviours of one subsystem (e.g. production, yield) and equating them to a time-series in a contiguous portion of a second subsystem. Because of time-lags between the onset of stress in the first subsystem and its expression in the form of an emergent response in the second subsystem, concomitant data are not necessarily appropriate.

Methodology

Application of this technique for yield estimation has been rare in the fisheries literature. However, data are often readily available for timely, order-of-magnitude estimates that will serve as confirmatory evidence for yield determinations made using other techniques.

An application of this method has been described by Ryder and Henderson (1975) for the coupled subsystems of the Nassar Reservoir and the Mediterranean Sea. These two water bodies are connected by about 900 km of the Nile River. Despite the great distance separating them, both the Nasser Reservoir and the Mediterranean Sea may be considered as subsystem components of one of the world's largest and most diverse aquatic systems—the Nile River watershed. The creation of the High Dam at Aswan resulted in some dramatic changes in at least three separate components of the subsystem, namely, the Nasser Reservoir, the Nile Delta Lakes and the Mediterranean Sea (Ryder & Henderson 1975).

Following the completion of the High Dam, the water level of Lake Nasser rose to the highest stage in its history, which commenced with the construction of a dam in 1902 (Greener 1962) and continued with two subsequent additions to the top of this dam in 1907 and 1929. The last modification was effected more than thirty years prior to the initiation of work on the High Dam in 1960, and the system at that time might be assumed to have reached a new steady-state. During its first fifty-odd years of existence, this relatively stable (in 1960) reservoir has been transformed from a lotic environment populated principally with riverine fish species to a largely lacustrine environment in which lake species were predominant. *Ceteris paribus*, there would be no reason to expect the new lentic community to be any more productive (on a per unit area basis) than the former lotic one except for the augmentation of nutrients following dam construction. The stockpiling of nutrients behind the dam resulted in a diminution of nutrients in the downstream portion of the system. An observed decline in fish yield of the Nile Delta Lakes was noted by El-Sedfy and Libosvárský (1974), and it was inferred that this might be attributed to the construction of the High Dam. The alterations in river flow and consequently nutrient supply, probably accounted for the yield losses of the Nile Delta Lakes subsystem. Additionally, it was predicted by El-Zarka and Koura (1965) that construction of the High Dam might cause noticeable changes in the fishery of the Eastern Mediterranean through withdrawal of nutrients by the biotic component of Lake Nasser. These nutrients would normally be deposited on the Nile Delta or be swept into the Mediterranean in the form of either dissolved or suspended solids. The documented decline of the fishery in the Nile Delta lakes might, therefore, be deduced to be due to reduced nutrient levels in the lower Nile resulting from High Dam construction and a corresponding decline in yield could be predicted for the Mediterranean due to a similar attenuation in nutrient supply.

In this instance attention was focused on two plant-eating fish species on the assumption that changes in nutrient concentrations and regime would be felt first at the primary consumer level. *Sarotherodon niloticus* (Tilapia) in the Nasser Reservoir and *Sardinella* sp. in the eastern Mediterranean are both diatom feeders and coincidentally constituted 48% of the total fish harvest for both the Nasser Reservoir and the Mediterranean. Tilapia yields might be expected to increase with the rise in water levels and corresponding nutrient increases while *Sardinella* yields could be predicted to decline through nutrient losses to the lower Nile and the eastern Mediterranean. These effects might conceivably be inversely related and proportional, but not necessarily equivalent or coincidental because of the inevitable time-lags between nutrient alteration and the effect on the fish populations. In actual fact, Tilapia increased both its biomass and yield rapidly following first flooding (Ryder & Henderson 1975) perhaps because of the stimulation of diatom growth through increase of nutrients. The sardinellas on the other hand suffered drastic yield declines between 1962 and 1964 which was almost certainly due to their dependence on the high biomass of diatoms carried downstream by the annual floods in the Delta region of the Nile.

The actual yield decline (all species) for the Egyptian fishery in the eastern Mediterranean between 1962 and 1969 was about 26 thousand metric tons (Fig. 16.4) and may be roughly equated to a future yield for the Nasser Reservoir of a similar amount if fished to its full potential. More importantly, the recorded yield loss to the Mediterranean allowed us to place an order of magnitude estimate on the potential yield of the Nasser Reservoir which was

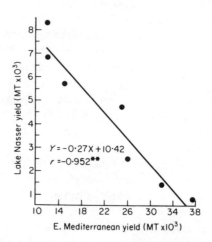

$$Y = -0.27X + 10.42$$
$$r = -0.952**$$

Fig. 16.4. Regression of the Nasser Reservoir annual yield on the annual yield from the E. Mediterranean four years earlier. It is postulated that the construction of the High Dam near Aswan has witheld nutrients from the Mediterranean thereby creating a reduction in standing stocks. These nutrients, however, may have contributed in part to the increase in ichthyomass of the Nasser Reservoir (after Ryder & Henderson 1975).

not dissimilar from other estimates made from discrete data sets (Ryder & Henderson 1975). This technique then, provides an alternate method of relatively low precision and resolution, that will allow rapid assessment of yield potential on the basis of data that may be readily available and quickly analysed.

ESTIMATING ASYMPTOTIC YIELD LEVELS IN RESERVOIRS

Concept

This technique depends on the quantification of substantially rapid changes in abiotic variables following stream impoundment as well as the subsequent return to steady-state and the eventual stabilization of the system. It follows, therefore, that newly created reservoirs are best suited for analysis by this method. This technique may occasionally be applied to natural lake systems with a sufficient departure from steady-state in the normal regime of abiotic variables. Suitable natural lakes may be those in which fish are introduced for the first time or those recovering from pollution or other stresses.

Reservoirs in North America and Africa, immediately following impoundment, show a change in unison of several abiotic variables at a predictable rate (Henderson *et al.* 1973, Rawson 1958). The concentrations of nutrients and other ions, turbidity levels, and certain other edaphic features increase with the dissolution of ionic compounds from the inundated soils or from the decomposition products of drowned vegetation. Nutrient concentrations reach a peak and then descend to a lower asymptotic level approaching steady-state conditions. Sets of morphological variables, usually partial correlates, also change in unison following impoundment. Area, volume, mean depth and shoreline length change at predetermined rates barring any macrosystem emergent inconsistencies (Kerr 1974a). These morphological changes, as opposed to edaphic changes, are generally uni-directional, rising to an upper asymptotic level somewhat early in the sequence of events.

Interestingly, several biotic and socio-economic parameters follow the above sequence of events (Henderson *et al.* 1973). Such factors as standing crop, yield, number of fishermen, number of fishing boats, effort and *CUE* may be proportional to one or more abiotic variables over time. Usually there is a substantial time lag between cause and effect resulting in similarly shaped curves being offset on the time axis. The upper and lower asymptotes may be calculated for either abiotic or biotic variables using the Gompertz curve technique (Croxton *et al.* 1967). Normally the asymptote would be calculated for the most complete data set of any single variable and then related to other variables with an appropriate time lag. This method, then, tends to be entirely mathematical in nature and does not allow for any unpredicted inconsistencies in the biotic sector that may change parameter proportionality.

Methodology

The Gompertz curve technique (Croxton *et al.* 1967) used for yield estimation was first applied in Ghana, after the large Volta Reservoir had reached its peak in water levels in 1968 following impoundment in 1964. The fish yields reached a peak and then declined as expected, and the lower asymptotic level of fish yield was predicted (Fig. 16.5), employing the Gompertz curve

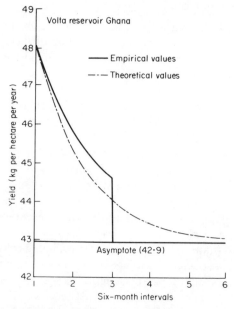

Fig. 16.5. Asymptotic level of community fish yield as determined from the Gompertz curve for the Volta Reservoir, Ghana, upon return to steady-state (after Bazigos 1971, Henderson *et al.* 1973).

(Bazigos 1971, Henderson *et al.* 1973). Yield levels from the Volta Reservoir in recent years have remained very close to 43 kg ha^{-1} yr^{-1}, the predicted value.

16.8 Future development of yield estimation techniques

If successful management of the fisheries of large lakes and reservoirs is ever to be achieved, we must develop new whole systems models that will relate to yield assessment. Those methods of fish yield estimation at the community or subsystem levels currently in vogue require modification or refinement so that fisheries yield values may be determined with greater resolution and higher precision. The approach to the development of new models requires appreciable innovation. Inferences can often be drawn from poorly defined or discontinuous data if the analytical approach is free of traditional constraints. The future of our global fisheries will depend on the formulation of models sufficiently precise to allow fisheries managers to apply adequate controls in ample time to preserve the integral structures of the production systems.

Acknowledgements

Many of the concepts and methodology in this chapter have been developed with the help or encouragement of various colleagues, especially G. P. Bazigos, H. F. Henderson, S. R. Kerr, A. H. Lawrie, K. H. Loftus and H. A. Regier. S. D. Gerking, S. R. Kerr, A. H. Lawrie and K. H. Loftus reviewed drafts at various stages.

I am grateful to J. C. Stevenson, editor, and the Fisheries Research Board of Canada for permission to publish the five figures.

References

BARBOUR C.D. & BROWN J.H. (1974) Fish species diversity in lakes. *Am. Nat.* **108**(962), 473–489.

BAZIGOS G.P. (1971) Yield indices in inland fisheries with special reference to Volta Lake. *UN/FAO Stat. Stud.* **F10**:SF/GHA/10: 1–25.

CROXTON F.E., COWDEN D.J. & KLEIN S. (1967) *Applied General Statistics.* Prentice-Hall Inc., Englewood Cliffs, N.J.

EL-SEDFY H.M. & LIBOSVÁRSKÝ J. (1974) Some effects of Assuan [sic] High Dam on water and fishes of Lake Borullus, *A.R.E. Zool. Listy*, **23**(*1*), 61–70.

EL-ZARKA S. & KOURA R. (1965) Seasonal fluctuations in production of the principal edible fish in the Mediterranean Sea off the United Arab Republic. *Proc. Gen. Fish. Coun. Medit.* **8**, 227–259.

ENTZ B.A.G., DEWITT J.W., MASSOUD A. & KHALLAF E.S. (1971) Lake Nasser, United Arab Republic. *Evaluation of Fisheries Resources in African Fresh Waters* (H.A. Regier, ed.), *Afr. J. Trop. Hydrobiol. Fish.* **1**(1), 69–83.

FORBES S.T. (1887) The lake as a microcosm. *Bull. Peoria Sci. Ass.*, 1887, 77–87.

GALLUCCI V.F. (1973) On the principles of thermodynamics in ecology. *Ann. Rev. Ecol. Syst.* **4**, 329–357.

GREENER L. (1962) *High Dam Over Nubia.* Cassel, London.

GULLAND J.A. (1970) The fish resources of the oceans. *FAO Fish Tech. Pap.* **97**, 425 pp.

HAMLEY J.M. (1975) Review of gillnet selectivity. *J. Fish. Res. Bd. Canada* **32**, 1943–1969.

HARGRAVE B.T. (1973) Coupling carbon flow through pelagic and benthic communities. *J. Fish. Res. Bd. Canada* **30**, 1317–1326.

HASLER A.D. (ed.) (1975) *Coupling of Land and Water Systems.* Springer-Verlag, New York.

HENDERSON H.F., RYDER R.A. & KUDHONGANIA A.W. (1973) Assessing fishery potentials of lakes and reservoirs. *J. Fish. Res. Bd. Canada* **30**, 2000–2009.

HENDERSON H.F. & WELCOMME R.L. (1974) The relationship of yield to morphoedaphic index and numbers of fishermen in African inland fisheries. *UN/FAO CIFA Occ. Pap.* No. 1, 19 pp.

HILE R. & BUETTNER H.J. (1959) Fluctuations in the commercial fisheries of Saginaw Bay 1885–1956. *U.S. Fish Wildl. Serv. Res. Rept.* **51**, 38 pp.

HUTCHINSON G.E. (1957) *A Treatise on Limnology. Vol. I.* John Wiley and Sons, New York.

JENKINS R.M. (1967) The influence of some environmental factors on standing crop and harvest of fishes in U.S. reservoirs. *Proc. Reservoir Fish. Symp. Southern Div. Am. Fish. Soc.*: 298–321.

KERR S.R. (1974a) Structural analysis of aquatic communities. *Proc. 1st Int. Congr. Ecol. The Hague.* pp. 69–74.

KERR S.R. (1974b) Theory of size distribution in ecological communities. *J. Fish. Res. Bd. Canada* **31**, 1859–1862.

LOFTUS K.H. & REGIER H.A. (1972) Introduction to the proceedings of the 1971 symposium on salmonid communities in oligotrophic lakes. *J. Fish. Res. Bd. Canada* **29**, 613–616.

MACARTHUR R.H. (1972) *Geographical Ecology.* Harper and Row, New York.

MANN K.H. (1974) Comparison of freshwater and marine systems: the direct and indirect effects of solar energy on primary and secondary production. *Proc. 1st Int. Congr. Ecol., The Hague.* pp. 168–173.

MOYLE J.B. (1950) Gillnets for sampling fish populations in Minnesota waters. *Trans Am. Fish. Soc.* **79**, 195–204.

OBENG L.E. (1973) Volta Lake: Physical and biological aspects. *Man-made Lakes: Their Problems and Environmental Effects.* (Ackermann, White and Worthington, eds.) Am. Geophys. Union, Washington, D.C.

POLL M. (1953) Exploration hydrobiologique du Lac Tanganika (1946–1947): Poissons non-Cichlidae. *Inst. Roy. Sci. Nat. Belg.*, **111**(5A), 251 pp.

POLL M. (1956) Exploration hydrobiologique du Lac Tanganika (1946–1947): Poissons Cichlidae. *Inst. Roy. Sci. Nat. Belg.* **111**(5B), 619 pp.

RAWSON D.S. (1955) Morphometry as a dominant factor in the productivity of large lakes. *Verh. Int. Ver. Limnol.* **12**, 164–175.

RAWSON D.S. (1958) Indices to lake productivity and their significance in predicting conditions in reservoirs and lakes with disturbed water levels. H.R. MacMillan Lect. in *Fish. Univ. Brit. Col.* pp. 27–42.

REGIER H.A., APPLEGATE V.C. & RYDER R.A. (1969) The ecology and management of the walleye in western Lake Erie. *Great Lakes Fish. Comm. Tech. Rep.* No. 15, 101 pp.

REGIER H.A., CORDONE A.J. & RYDER R.A. (1971) Total fish landings from fresh waters as a function of limnological variables, with special reference to lakes of east-central Africa. *FAO Fish Stock Assessment Working Paper* No. 3. 13 pp.

REGIER H.A. & HARTMAN W.L. (1973) Lake Erie's fish community: 150 years of cultural stresses. *Science* **180**, 1248–1255.

ROUNSEFELL G.A. (1946) Fish production in lakes as a guide for estimating production in proposed reservoirs. *Copeia* **1**, 29–40.

RYDER R.A. (1964) Chemical characteristics of Ontario lakes as related to glacial history. *Trans. Am. Fish. Soc* **93**, 260–268.

RYDER R.A. (1965) A method for estimating the potential fish production of north-temperate lakes. *Trans. Am. Fish. Soc.* **94**(3), 214–218.

RYDER R.A. (1972) The limnology and fishes of oligotrophic glacial lakes in North America (about 1800 A.D.). *J. Fish. Res. Bd. Canada* **29**, 617–628.

RYDER R.A., KERR S.R., LOFTUS K.H. & REGIER H.A. (1974) The morphoedaphic index, a fish yield estimator—review and evaluation. *J. Fish. Res. Bd. Canada* **31**, 663–688.

RYDER R.A. & HENDERSON H.F. (1975) Estimates of potential fish yield for the Nasser Reservoir, Arab Republic of Egypt. *J. Fish. Res. Bd. Canada* **32**, 2137–2151.

SCHINDLER D.W. & NIGHSWANDER J.E. (1970) Nutrient supply and primary production in Clear Lake, Eastern Ontario. *J. Fish. Res. Bd. Canada* **27**, 2009–2036.

SHELDON R.W., PRAKASH A. & SUTCLIFFE W.H. Jr. (1972) The size distribution of particles in the ocean. *Limnol. and Oceanogr.* **17**(3), 327–340.

SIMON H.A. (1969) *The Sciences of the Artificial.* M.I.T. Press, Cambridge, Mass.

TOIVONEN J. (1972) The fish fauna and limnology of large oligotrophic glacial lakes in Europe (about 1800 A.D.). *J. Fish. Res. Bd. Canada* **29**, 629–637.

VON BERTALANFFY L. (1968) *General System Theory.* George Braziller, New York.

VOLLENWEIDER R.A. (1969) Möglichkeiten und Grenzen elementarer Modelle der Stoffbilanz von Seen. *Arch. Hydrobiol.* **66**, 1–36.

WARE D.M. (1975) Relation between egg size, growth, and natural mortality of larval fish. *J. Fish. Res. Bd. Canada* **32**, 2503–2512.

WELCH P.S. (1948) *Limnological Methods.* Blakiston Co., Toronto, Ont.

Chapter 17: The Growing Science of Aquaculture

John E. Bardach

17.1 Introduction

The emphasis in the title to this chapter should be on the word 'growing'. Even though aquaculture is centuries old, improvements in its practices have proceeded slowly by empirical, trial-and-error methods until recent years. The scientific approach to controlling and enhancing the biomass production from managed waters is of relatively short standing. The application of fisheries science and limnology to aquaculture has been late in arriving, because these sciences themselves had to develop specific technical and conceptual tools in population biology, fish physiology, the analysis of nutrient dynamics in water, and the like. It is this growth, the growth of the scientific approach to aquaculture, which may eventually develop into an applied scientific discipline, to which I address myself in this chapter.

In most kinds of aquaculture the organisms are under man's control until the time of harvest, and management practices are aimed at cycling materials and energy at accelerated rates in artificial settings. Increases in production of one or several species are thus obtained over that which would ensue under unmanaged or natural conditions. Some fish-rearing containments can safely be called ecosystems in a biological sense (ponds) while some represent highly artificial production facilities (silos, raceways). In the latter, like an industrial chicken farm, the infrastructure is simply a holding device for the passage of imported food through the animals as protein-synthesizing machines. Since the species so reared are, by and large, carnivores, the energy costs are high in terms of calories supplied both for feed and for system maintenance (Pimentel *et al.* 1975).

Aquaculture directed at increasing human food supplies produces animal protein which competes with that produced from mammalian or avian livestock. As a result, it is interesting to compare food production on the land with that in the water. Terrestrial pasture, for example, is capable of supporting either food grains or soy beans for direct human consumption or several species of domestic herbivores. By comparison, we do not have such a choice when harvesting food from water. Whether it is derived from ocean fishing aquaculture, the food is animal protein.

However, searches are in progress for new staple plant crops since the calorie shortfall threatening humanity in the next few decades appears to be somewhat more serious than the protein one (Chancellor & Goss 1976). With reference to this search, the U.S. National Academy of Sciences has made the following suggestion: The grain-bearing eelgrass *Zostera marina* should be investigated for its potential starch production in underwater cultivation (NAS 1975). Perhaps of wider pertinence are the investigations of algologists and sanitary engineers with unicellular algae as animal feed components and, eventually, as food additives for man (Waslien *et al.* 1976). While yields per unit surface and the protein content of some of these algae are high and their amino acid profiles favourable, e.g., *Spirulina maxima*, the cost of processing them is still prohibitive. Generally speaking, it is not likely that food plants competitive with those of land-based agronomy and useable as staple primary calorigenic food for mankind will be grown in rivers, ponds, lakes or oceans.

Much aquaculture is now done in settings such as estuaries and reclaimed swamps that do not compete for land with agricultural endeavours. In areas where a choice between aquaculture and agriculture must be made, economic factors often influence the choice. For example, marketability primarily controls the development of catfish farming in the Southern U.S. In other areas the factors are ecologic, like in the trout farming district of the Snake River Valley with an abundant water supply and optimal temperature (Bardach *et al.* 1972). More aquaculture will be done on marginal land in the future (Pantulu 1976, Ryther 1975, Pillay 1976) as costs of agricultural land rise with population increases and staple food needs. Yet cultural preferences with respect to food choice often take precedence over economic, ecological, or even nutritional factors when a choice of this kind has to be made.

Yields obtainable per unit of effort are higher for certain aquaculture species than for terrestrial animal crops, and they may also be cheaper to produce when assessed from various economic vantage points than are stock husbanded on land, especially when we include constraints of total energy inputs such as cost equivalents of fuel, machinery, and the like. In fact it is not well known that fish are better converters of metabolizable energy than birds or mammals. A comparison by Smith (1976) based on whole animal, direct calorimetry bears out this fact:

Animal	g protein/Mcal*
cow	2
pig (hog)	6
chicken	15
milk (grazing)	16
milk (fed)	20
fish (trout)	30–40

* 1×10^6 calories in terms of metabolizable energy

As biological science and engineering techniques pertaining to aquaculture become further perfected, this aspect of aquatic animal husbandry will make it compete successfully with the growing of land animals as well as with commercial fisheries, as it has already done in Taiwan (Shang 1973).

It is incongruous that the need for animal protein in the human diet is most pronounced in the tropics, but that aquaculture producing a high biomass per unit of input (usually not including labour) traditionally is practiced more intensively in the temperate and subtropical zones. Recent research activities indicate, however, not only a great potential in tropical, fresh and brackish water aquaculture but also offer the possibility of culturing indigenous species (Ling 1973). Chaudhuri *et al.* (1975), for instance, report an average harvest of over 8 tons/ha/yr of Indian carps, mainly *Catla catla*, *Cirrhinus mrigala* and *Labeo rohita*, reared in experimental polyculture.

I will select in this chapter certain tricks of the aquaculturist that are designed to increase production with emphasis on the scientific approach. From the early life history phase, the facilitation of spawning will be examined and some aspects of enhancing larval survival. From the growout phase, I will mention ocean ranching and its freshwater counterpart. I will also touch on polyculture and joint land-water nusbandry. Selective breeding will be mentioned after early life history production and growout. The chapter will end with a prognosis for tropical fresh and brackish water aquaculture and with some considerations of its economics.

17.1 Manipulation of reproduction

Control over reproduction is essential to success in aquaculture. The primary goals of such control may differ as do, of course, strategies and tactics to achieve control. For instance, in the freshwater culture phase of ocean-ranched Pacific salmon (*Oncorhynchus* sp.), one may wish to time as exactly as possible the return of certain species or strains to the home stream. This can be accomplished by selecting progenitors from a first or last wave of ripe adults returning to a natural or imprinted home stream. An alternate course may be followed if the objective is to extend the timespan of spawning readiness of a stock (Donaldson 1970). In this case breeding from both first and last arrivals of such cohorts fulfilled the requirement in a run of chinook salmon (*Oncorhynchus tshawytscha*) established in a small stream of Union Lake on the University of Washington campus. Since gonad maturation within the lifespan of an individual is genetically determined, the California Department of Fish and Game, for instance, employed selective breeding to increase the number of rainbow trout (*Salmo gairdneri*) spawning at two years of age from 53 to 98% in three generations (Leitritz 1960).

Light and temperature.

Seasonal spawning is controlled by the external environment, such as light and temperature, and also by the interaction of these factors with various hormones, although as yet the interaction is not completely understood

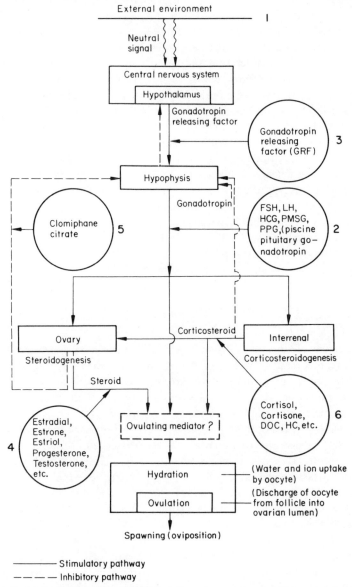

Fig. 17.1. Endocrine pathways and relations influencing ovarian development and ovulation in fishes. (After Kuo & Nash 1975.)

(Fig. 17.1). Numbers on this scheme from Kuo and Nash (1975) pinpoint way stations numbered [1] to [6] in the negative feedback pathways where human interference can or could be attempted to force ovulation for the sake of convenience and/or the extension of the spawning season. In some cases, e.g., tilapia, temperature in the natural environment provides full control ([1] Fig. 17.1) over breeding (Pruginin *et al.* 1975).

Since gonad maturation occurs in the upper ranges of a species' temperature range and is often also governed by day length, spawning seasons can be extended by temperature and light control. Kuo *et al.* (1973) extended the spawning season of grey mullet (*Mugil cephalus*) in Hawaii from one or two months to nearly five months within a seasonal cycle ([1] in Fig. 17.1), by regulating light and temperature in combination with hormone injections ([2] Fig. 17.1).

In the tropics, light and temperature vary less drastically throughout the year than in the temperate zone, and longer spawning seasons prevail. The subtle light influence of lunar cycles comes into play in the tropics on the eye-hypothalamus-pituitary axis governing the priming of the gonads. These have been demonstrated to determine spawning almost to the hour in the Indo-Pacific marine shore species *Polydactylus sexfilis* (May 1976); other species are suspected to behave similarly ([1] Fig. 17.1). Since lunar cycles influence migration waves of freshwater fishes in monsoon regions (Sao-Leang & Dom-Saveun 1955), moon-phase affect the spawning of tropical freshwater species during their several month-long spawning seasons.

Hormones

The practice of injecting pituitary extract and pituitary hormone is widely used in the forced spawning of fishes. Injections are often limited to the females in order to facilitate hydration of the eggs, ovulation and oviposition, because in most species males remain in reproductive condition longer than females. Nevertheless, in some fishes, e.g., the grass carp (*Ctenopharyngodon idellus*) both sexes generally receive hormone treatment (Chaudhuri 1976). Evidence is accumulating for the existence of only one gonadotrophic hormone in fishes as opposed to two in mammals, with the former resembling the mammalian LH factor (Hoar 1969). Nevertheless, pronounced species or higher-group-specificity of the hormone seems to be emerging from the research. Specificity is expressed by the recalcitrance of some species to yield to treatment with hormones other than those of their congeners. By and large though, fish gonadotropins are effective, albeit with dosage differences, across generic, familial and order lines. Mammalian hormones (e.g. human chorionic gonadotropin) are also employed either alone or in conjunction with piscine hormones (Liao *et al.* 1971) to achieve spawning success. Fish pituitary gonadotropin has been purified in the last decade from salmon, carp (*Cyprinus carpio*) and mullet. Much smaller dosages are required to elicit ovulation with purified gonadotropin than with the more commonly used less refined extraction methods. Yet hypophysation is still frequently practiced with unpurified preparations, and intraperitonal injection of macerated fresh pituitaries suspended in isotonic saline solution is also common. For dosages, timing of the repeated injections, and other techniques regarding the manipulations of the spawning facilitation (see [2] in Fig. 17.1), the reader is referred to Bardach *et al.* (1972) and Chaudhuri (1976).

In fishes as in other vertebrates, the release of pituitary gonadotropin is under the control of the hypothalamus, and various other hormones also affect the reproductive state. The reproductive state of the goldfish (*Carassius auratus*) has been influenced by a gonadotropin releasing factor or factors ([4] in Fig. 17.1) and with mammalian fertility inducing drugs, both of which appear to stimulate the hypothalamus (Lam *et al.* 1975). Ovulation can be induced with a synthetic joint luteinizing hormone and releasing hormone (LH-RH) and also with a medicinal chemical, clomiphene citrate, which has been shown to mainly, but not unequivocally, block the effects of estrogen on the pituitary (Pandey *et al.* 1973), ([3] in Fig. 17.1). Moreover, by using this latter material, ovulation has been induced in goldfish at 13–14C, a temperature at which the species does not ovulate in nature (Pandey & Hoar 1972).

At least one species of fish, *Clarias lazera*, an African airbreathing catfish, has been induced to spawn with injections of desoxycortiscosterone acetate, DOCA (Micha 1975). This action ([6] in Fig. 17.1) can probably be explained by the similarity of this hormone to the gonadel steroids. Jalabert (1976) bears this out by stating that there seems to be a '2-way control triggering maturation, a direct gonadotropin route and an indirect corticotropic route'. He also demonstrated that epinephrine acts on ovulation *in vitro*. Summarizing the remarks on spawning facilitation, it is likely that gonadotropins will remain the main tool, but that other hormones and hormone analogues will be used increasingly together with environmental modifications as the total endocrine mechanism of fishes becomes better undertsood (Fontaine 1976).

17.3 From spawning to the early juvenile stage

Blaxter (1969) classifies the eggs of fishes into several types. Some are laid singly and no parental care is given to them. These are further subdivided into buoyant, produced mostly by marine fishes, nonbuoyant, produced mostly by freshwater fishes, and nonbuoyant buried, also produced by freshwater fishes, especially the salmonids. While species which lay nonbuoyant, attached or buried eggs are among those which have been best adapted to aquaculture in the past, the development of controlled spawning now offers great opportunities to procure offspring from species with very numerous but delicate single, small, pelagic eggs. Most unresolved and serious larval rearing problems are generally restricted to these species, e.g. mullet, the Chinese and Indian carps (*Ctenopharyngodon idellus, Hyptothal-michthys molitrix, Cirrhinus molitorella, C. mrigala, Mylopharyngodon piceus, Catla catla* and *Labeo rohita*), and tropical catfishes, e.g. *Clarias lazera.*

Some kind of parental care is exercised over the eggs of other species. To this category belong several important freshwater species of tilapia either already adapted or now undergoing adaptation for aquaculture. Tilapia protect their eggs either by mouthbreeding (*Sarotherodon*) or nest building (*Tilapia*) and a high percentage of larvae survive. Survival is especially high when the species is protected from predation, as it is or has been in most variants of tilapia culture. This leads to stunting, and several approaches have been tried, all more or less successful, to counteract the poor growth. The most important of these methods are: stocking predators (e.g. *Lates* spp.), and rearing nest building *Tilapia* species in cages without contact with the substrate. Additionally, the faster growing males have been selected for rearing by the early sexing of fish before they reach maturity, as well as rearing hybrids of one sex. Several kinds of crosses tend to produce only male hybrids, for example that of *Sarotherodon mossambicus* females with *Sarotherodon hornorum* males and *S. niloticus* females and *S. aureus* males. Ruwet *et al.* (1975) treat this subject in considerable detail while Pruginin *et al.* (1975) emphasize the difficulties in the consistent production of 100% male offspring in such crosses, among which the maintenance of pure parental stock is prominent.

Small larvae are difficult to rear because they need small sized food organisms at a high density. Shirota (1970) describes the mouth sizes of 40 species of fish larvae and makes a number of observations pertinent to aquaculture of the larvae with a small mouth size. These feed on phytoplankton, protozoa and nauplii in contrast to those that feed on medium to large-sized copepods. The latter, incidentally, are generally among faster growing carnivores of the species measured. It is not only the size of the mouth, but also the ability to open it which determines the size of larval foods and that also differs substantially among different species. Mouth sizes and total lengths of larvae at the very onset of feeding, though, are only weakly correlated (Shirota 1970).

The mouth size of salmon at the onset of feeding was not included in Shirota's review. They start their feeding on relatively large particles since their eggs have a diameter of several millimeters and their larvae are thus large. In fact, not only the size but also the specific weight of the eggs, both adaptations related to egg burial in streams, are the prime reasons which led to great success with the culture of trout and salmon.

Some species have far more delicate larvae than others. Grey mullet, for instance, should not be handled, that is, changed from one container to another during the first 40 days of their lives (Nash, pers. communication), while others of comparable adult feeding habits, such as the milkfish, *Chanos chanos*, are very hardy. They can be handled a few days after they begin feeding.

Food size and density are of prime concern in larval rearing. Braum (this book) has investigated the capability of larvae in sweeping a certain volume of water of its food organisms. This information is of interest to aquaculturists since they must supply organisms of a proper size and of sufficient numbers to satisfy larval needs. As Braum indicates, vision is of primary importance in many species. Certain other species develop gill rakers very early and filter rather than take individual bites. Atlantic menhaden (*Brevoortia tyrannus*) were shown to feed in this mode after reaching 32 mm in length (June *et al*. 1971) and milkfish larvae feed on *Chlorella* when they are 10 to 13 mm total length, presumably filtering the algal cells (Pantastico, personal communication).

Virtually nothing is known about the chemosensory equipment of fish larvae in their early stages, though it is likely that chemical clues would be important under certain conditions. Diel changes in activity also play a role as do angle and mode of illumination. All investigators agree that a certain particle density in the close vicinity of a larva is of paramount importance. The larvae of grey mullet, for instance, grow adequately if they are surrounded at the onset of feeding by at least 3 individual rotifers (*Brachionus plicatilis*) per millilitre. The rotifers, in turn, require a density of at least 45 *Chlorella* cells per millimetre. Small-egged species, such as the mullet, rabbitfish (*Siganidae*), and threadfin (*Polydactylus sexfilis*) are best reared in cultures of algae and rotifers which should have reached the high densities indicated above before the fish larvae are introduced (Nash & Koningsberger, in press). In addition to being necessary for the sustenance of the rotifers which form the earliest larval food, *Chlorella* reduces the ammonium content in larval rearing containers and appears to have antibiotic properties beneficial to larval survival (Brick 1974, May & Hashimoto, personal communication). As the larvae grow, *Artemia* (brine shrimp) eggs and nauplii and finally *Artemia* adults replace the smaller rotifers as rations; oyster trochophores are also used. The number of fish larvae per litre also have to be monitored carefully; numbers may range from five for rabbitfish (*Siganus*) up to 50 for plaice (*Pleuronectes*) (Nash & Koningsberger, in press).

For the rearing of delicate larvae, such as grey mullet or silver carp (*Hypothalmichthys molitrix*), experience has indicated that large containers are preferable to small ones and that 5,000 litres is a minimum, but that containers can profitably reach far larger sizes. A safe, even conservative, stocking rate in such containers is 10 mullet larvae per litre. Round containers with sloping bottoms are preferable to square or rectangular ones and black, green or blue inside colouration of the container with strong illumination on an automatic daylight cycle to account for diel habits of the larvae facilitate food finding. Persoone and Sorgeloos (1975) and Sorgeloos and Persoone (1976) suggest detailed, mainly experimental, technologies for gentle stirring, water exchange, rearing and introduction of food, etc.

If the cultured species has medium sized eggs, such as the common carp whose eggs attach to aquatic plants, or cichlid eggs that are protected by one or both parents, the larvae are reared in brood ponds. There the establishment of mixed plankton cultures that would contain bite-sized morsels for larvae of various sizes follows methods now well established and described in several books, most notably those of Huet (1970) and Bardach *et al.* (1972).

The subject of sanitation and disease is extremely important to aquaculture. The denser the animals, the greater the possibility of disease and parasite transmission. Under culture conditions, fish are in greater proximity to one another for longer periods of time than in nature. Parasites and diseases are not restricted to larvae or juveniles, but epizootics are often more severe in the early culture stages. The subject is beyond the scope of this chapter, but a few methods of combating disease problems will be mentioned. Selective breeding of trout and carp has been relatively successful against various bacterial diseases (Snieszko 1970, Schaeperclaus 1954). Immunological procedures against viral, bacterial and even protozoan diseases have been developed for several salmonid species (Table 17.1 in Anderson 1974, Bullock & Wolf 1975). Various chemical baths have been perfected against a number of ectoparasites, and antibiotics have been mixed with prepared rations to combat bacterial infections (Reichenbach-Klinke 1966).

This brief discussion reveals that substantial gaps exist in the biological knowledge needed to rear larval fishes. For instance, larval behaviour is far less well studied than is the behaviour of juveniles or adults, and the information on the sensory equipment of larvae is very spotty indeed. Systematic studies of these various aspects of the physiological ecology of fish larvae would greatly contribute to the aquaculturists' ability to increase the survival rate. It is also evident that these studies are of a long term nature and that it will take longer still to translate the results into practical procedures.

17.4 Growout

A glance at a generalized schema of an aquatic foodweb (Fig. 17.2) suggests to an aquaculturist a spectrum of approaches to increase production. One may isolate the species one rears from the biota with which it shares its

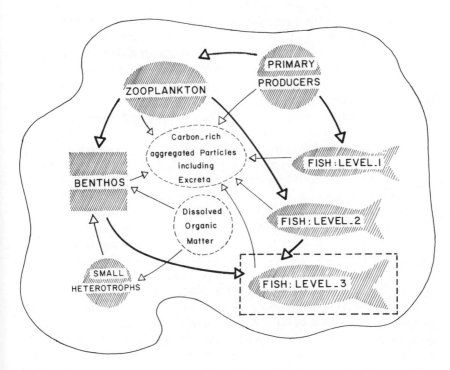

Fig. 17.2. Schema (simplified) of interactions among various important trophic components in fish farming; fish: level 3 is often reared on artificial food only.

niche in nature (Fish—level 3). This is the obvious practice in rearing carnivores at the top of a food pyramid (trout, yellowtail *Seriola*) where the fish farmer's prime concern is the supply of food. As the stock is reared at high densities, he will also be confronted with problems of sanitation and accumulating waste products. Such culture in ponds, cages, artificial raceways, etc. is usually highly capital intensive, relying on rapid turnover, dependable markets and a high technical capability. At the other end of the spectrum is fish ranching, involving culture only to a certain stage at which the fish farmer ceases either to feed the fish or to care for them altogether. Yet the ranching of fish in all its variants does rely on hatcheries and/or the intensive care of larvae and/or juveniles in case the fish in question cannot be spawned. The milkfish is a good example of this latter type; the larvae are routinely gathered from the wild.

A successful variant of ranching has been developed by rearing milkfish to the harvesting stage in large bamboo and nylon net enclosures in the freshwater lake Laguna del Bay near Manila in the Philippines. The water is highly eutrophic due to intensive land use and the enclosures are placed so as to make maximal use of the concentration of plankton by the wind. The milkfish, which are generally considered to be predominantly herbivorous, thrive on zooplankton, with a greatly improved growth rate over that resulting from their usual fare of bluegreen and green algae.

Still other comparable practices exist that employ containment without either feeding or fertilizing the water for greater productivity. One such aquacultural practice is cage culture of carp with yields of up to 50 kg m^{-2} of water surface in sewage streams of a bottom type that enables growth of tubificid and similar worms (Bardach *et al.* 1972). Another involves the marine genus *Siganus* (rabbitfish) which feeds on various seagrasses (e.g. *Enhalus*) and macro-algae (e.g. *Enteromorpha*). They have been spawned artificially with considerable success (Soh & Lam 1973, Popper *et al.* 1976) and appear to lend themselves to a mode of ranching in large enclosures on reef flats where they can graze (McVey, personal communication).

The best known mode of fish ranching is, of course, the rearing of anadromous salmonid fishes to the onset of the oceanic stages in their life cycles. Needless to say, this kind of aquaculture is species- and region-limited but it is, at the same time, also an animal husbandry practice with a very high rate of gain; in fact annual yields expressed by weight of returning fish compared to weight of juveniles released results in a gain that is several million-fold (Calaprice 1976). Under certain circumstances and with landlocked strains of salmon, no ocean is needed for this feat. A large lake, such as Lake Michigan, will do (Tody & Tanner 1966). The introduction and with it the ranching of various species and strains of salmonids has been so successful the world over that an imaginative, albeit high risk, scheme is being undertaken under the aegis of the Rockefeller Foundation of establishing salmon runs in Southern Chile (Joyner 1975). It is anticipated that the salmon, once hatched, will return to the waters with the scent of which they were imprinted and that they will feed during their ocean foraging on antarctic krill (*Euphausia superba*) which form the richest plankton pastures in existence.

In the centre of spectrum of practices suggested by the trophic dynamics of piscine consumers in aquatic ecosystems lies the main range of fish pond management in fresh or brackish waters. Depending on a variety of circumstances, such as availability of capital and cost of containment as well as socio-cultural factors the fish culturist will manage the pond or small lake by fertilization or by fertilization and supplementary feeding of the species he chooses to grow. That he may have to add chemical elements and calories continuously or intermittently is also clear if one remembers that biomass is being removed, i.e. fish are harvested and consumed offsite. Especially if he wants to use pond ecology to his advantage, he is likely to practice polyculture, or at a minimum the rearing of two compatible species, such as carp and tilapia, shrimp and milkfish, etc. Polyculture need not be restricted to the rearing together of aquatic species only; fish—duck, fish—geese, and fish—pig culture has been practiced most successfully in Asia and Eastern Europe. Savings, from the vantage point of pond fertilization, and gains in fish yields are clearly obtainable (Ho 1961, Renner & Sarodnik 1973, Woynarovich 1976).

Thus the two main reiterative inputs into containment-aquaculture, aside from labour, are fertilizer and feed; in the case of the carnivore grown in isolation, it is feed only. The latter has been researched thoroughly in the case of salmonid nutrition (e.g. Halver 1972). Salmonid rearing requires feeding a diet that is high in animal protein, especially if fast-growing salmon are reared, such as those bred selectively by Donaldson (Donaldson & Menasveta 1961), and this is a luxury that can be afforded only by high technology nations. Continued success will depend upon how long forage fishes, to be processed into fish pellets, can be harvested cheaply in large quantities. As the world's population increases further and as fossil fuels become depleted, the cost of harvesting these fishes, the mainstay of fish meal in animal feed, is becoming more expensive. We will increasingly use these forage fish for human nutrition directly, in one form or another, and trout may then well be fed predominantly with enriched soy or protein rations derived from single-cell organism cultures.

Successful pond management which involves both fertilizer and feed depends on availability and cost of the nutrients. Obviously, local conditions and species available for culture give fish farmers all over the world many opportunities to enhance pond productivity at minimal costs. For instance, fertilization with inorganic fertilizers was fairly widely practiced in Philippine milkfish ponds until recently. The rise in the price of fossil fuels in 1973 and with it the three-fold or higher rise in price of inorganic fertilizer, such as urea, has led to a reversal of the intensifying trend in fish farming that could be noted earlier (Rabanal & Shang 1976). The result, of course, is a reduction of the animal protein availability in a number of nations, e.g. India, where high yield polyculture of fishes is in the process of development (Chaudhuri *et al.* 1975).

The use of natural plant as well as animal or human waste-derived fertilizers is being fostered (chapter 18, Bardach 1976, Ryther *et al.* 1972). A number of obstacles to this practice should be mentioned, however. First distributing the fertilizer or establishing fishponds where manure or other usable wastes occur in abundance are practical problems to be taken into account. Since aquaculture is now undergoing vigorous development in Africa, where cost constraints of fertilizer are critical, the broad array of organic fertilizers in use on that continent are listed (Table 17.1).

A great variety of supplemental feeds are used for keeping fishes at higher densities than the water body can support naturally. Local availability largely

Table 17.1. Percent composition of various organic fertilizers for use in fish culture in Africa. (From Miller 1975.)

Organic fertilizer	N	P	K	Ca	Protein	Carbohydrates
Grass hay	1·12	0·20	1·20	—	—	—
Peanut hay	1·62	0·13	1·25	—	10·1	38·5
Cassava leaves	6·80	0·31	—	—	—	7·5
Cotton seed cake	7·02	2·50	1·60	0·30	41·1	26·4
Peanut meal	6·96	0·54	1·15	—	43·5	31·3
Soybean oil cake	7·07	0·59	1·90	—	44·2	29·0
Palm oil cake	3·07	1·10	0·50	0·30	19·2	46·5
Meat scrap	8·21	5·14	—	—	51·0	3·5
Horse manure	0·49	0·26	0·48	—	—	—
Chicken manure	1·31	0·40	0·53	—	—	—
Sheep manure	—	—	—	—	5·0	15·4
Pig manure	—	—	—	—	—	—
Blood meal	12·90	0·22	0·31	0·05	81·0	1·5
Brewers grain waste	2·90	1·60	0·20	0·40	18·3	45·9
Cow stomach contents	—	—	—	—	11·8	—
Beer yeast extract (dry)	—	—	—	—	50–75·0	—
Distillers grain waste	—	—	—	—	22·0	37·0

determines what is fed and, as an example, a number of local fish foods are given for Africa (Table 17.2). In tropical aquaculture, such as in Africa, feeding is usually supplemental and the choice and intensity of use is cost-determined. Where the greatest constraint is land and water, as in Israel, more manufactured feeds are used for supplemental feeding. These feeds tend to be low in animal protein, but with proper attention to the complete-ness of diet. In Yugoslavia such carp pellets even contain some medicating chemicals (Ghittino 1972).

Recent progress reports from several tropical regions strongly suggest that it is economically and ecologically sound to grow in polyculture fishes of various feeding types and, perhaps, even to include some invertebrates in these ranks (Fig. 17.2). Polyculture of lower and mid-trophic level fishes is hampered by a two-pronged substantial deficiency in basic knowledge. One concerns the limnology of tropical fishponds, where, unfortunately, local variations in soil types require many local and regional studies before the effectiveness of pond fertilization practices can be enhanced. The other deals with the nutritional physiology of tropical fishes that are used in pond culture. Given the trend to polyculture, ecological management is of great importance with fertilization as well as stocking and cropping rates being the main tools to optimize pond production. A certain amount of extraneous feeding is usually also practiced. However, as increasing production is sought, additional feeding will rely increasingly on compounded rather than on natural feeds (chapter 18, Tal 1974). Consequently, information will

Table. 17. 2. Composition of various feedstuffs for use in fish culture in Africa. (From Miller 1975.)

Feed stuffs	% Carbohy-drates	% Fats	% Total Protein	% Fibre
Baobab press cake	76·7	0·8	2·2	6·8
Beer waste	46·4	7·8	22·8	18·8
Cabbage leaves	4·8	0·1	1·7	1·2
Cassava flour dry	83·2	0·5	1·6	1·7
Cassava leaves	14·3	1·0	7·0	4·0
Cassava tubers	34·6	0·2	1·2	1·1
Cocoa hulls	57·5	0·8	8·7	23·7
Coffee hulls	33·5	7·2	12·2	39·0
Corn cooked	79·2	4·8	8·0	1·9
Corn bran	64·4	8·6	12·2	2·8
Corn flour	71·5	3·8	9·3	1·9
Corn grain	81·3	4·6	10·3	2·3
Corn leaves and stalks dry	46·6	1·6	5·9	30·9
Cotton seed cake	38·5	7·4	47·3	9·6
Cotton seed	29·6	18·8	22·8	24·1
Cow stomach dried	37·6	1·9	16·7	28·2
Cow stomach fresh	36·2	1·0	11·6	37·8
Kale	6·1	0·8	3·5	1·6
Lettuce	3·7	0·2	1·2	0·6
Millet	81·0	2·8	9·0	3·0
Mill sweepings	58·0	14·0	12·5	7·5
Napier grass	1·0	0·2	2·6	1·1
Palm nut press cake	53·0	8·9	19·9	14·0
Peanut press cake	27·3	7·6	53·5	6·2
Peanut shells ground	46·3	1·0	4·0	46·7
Plaintain banana whole	79·2	1·8	6·5	5·3
Potatoes	19·7	0·1	2·1	0·9
Pumpkin	4·7	0·1	1·0	0·8
Rice	77·7	2·2	7·4	0·4
Rice bran	56·9	3·8	0·7	22·6
Sorghum	81·0	2·8	9·0	3·0
Soybeans ground	31·4	15·7	33·7	5·5
Spinach	4·5	0·2	2·1	0·8
Sugar cane fibre	55·4	0·6	1·3	40·0
Sweet potatoes	27·5	0·2	1·6	1·0
Wheat bran	59·7	3·8	4·5	14·5
Yam	25·6	0·1	1·5	0·9
Blood fresh	36·2	1·0	11·6	0·0
Blood meal	—	1·0	76·6	0·0
Smoked, salted fish waste (local)	—	—	35·8	—

have to be sought on the virtually unknown dietary needs of cichlids and tropical catfishes as well as, perhaps, atherinids, the mullets, the Indian carps, and the milkfish in Asia which promise to be the next mainstays of fresh and brackishwater fish culture.

Larvae and juveniles even of herbivorous fishes prefer animal protein in their diets (Ghittino 1972), but there is little or no indication of the time and mode of change of the diet. Endogenous enzymes act synergistically with the enzymes of their prey in the stomachless digestive tract of carp (Schaeperclaus 1961), and it is likely that other fishes may gain similarly from the digestive enzymes of their prey. Further knowledge of this matter would clearly influence feeding strategies. Several species of fishes have been shown to have gut bacteria, and speculations have been made about facilitation in the breakdown of certain food components by this means (Stickney & Shumway 1974).

Admixtures to feed to steroids, some of them hormones, and of thyroid hormones have enhanced growth rates of experimental salmonids (Matty 1975). He closes a lecture review of this subject as follows:

'Is research into hormonal growth additives to fish diets worthwhile and timely? In the case of salmonids, I think Yes, but much screening will have to be made for the most appropriate compound and for the compound having the least toxic effect and for the compound most quickly degraded. In the case of other fish, the correct formulation of diet in respect to carbohydrate, fat, protein and minerals has I think a much higher priority. Lastly, we must remember that there are growth promoters other than hormones (cyproheptadine BPN and methalibure) that may be candidates for enhancing the value of fish foods.'

Attention is also being given to the biochemical role of nutrients such as interrelationships of dietary minerals, the status of essential fatty acids and the like (Cowey 1976). It is also to be noted that different species and various life history stages of single species differ in their complements of carbohydrate metabolizing enzymes including amylase (Shaklee *et al.* 1974).

17.5 Fish breeding

Purdom (1974) stresses that fish lend themselves to application of conventional approaches of animal breeding, amply tested with domestic animals. These include selection and inbreeding with the use of heterosis. The reproductive peculiarities of fishes also enable the breeder to explore advantages of interspecific or intergeneric hybridization. In addition it is possible, with fish, to induce gynogenesis and polyploidy. Finally, in concert with endocrinological measures, one may even exert some control over the sex of the offspring.

The large number of offspring that fish produce, compared to that of birds and mammals, is ideal for mass selection. The characters of importance to the aquaculturist—growth rate, early or late maturity and the like—are usually polygenically determined, fall into a normal distribution and are strongly influenced by environmental factors. Genetic improvement by mass selection therefore requires good controls reared under rigidly uniform environmental conditions. Controls for comparison with the experimental genetic stock are based either on simultaneously rearing unselected cohorts or on selecting for the character in question in both positive and negative directions (e.g. fast or slow growth). A careful analysis of genotype-environment interactions for growth rate has been performed by Moav *et al.* (1975) in comparing the Chinese and European races of the common carp.

The most spectacular advances in plant and poultry breeding have been achieved through the use of inbred lines and the ensuing F1 heterosis. Little advantage has been taken of this technique in fishes because few species have been bred rigorously, but fish lend themselves well to this practice, especially because they have high fecundities. Carp is the only species which has been inbred for a long time in many different countries and under varying conditions of culture. Thus various races differ conspicuously in many characters, a reflection of what the farmers desired them to be. For instance, the Chinese race of common carp has poorer growth rate, higher resistance to crowding, higher viability and higher fertility than does the European carp. The European carp, in contrast, grows faster but is more sensitive to stress and is less tolerant to crowding, reflecting the fact that European carp culturists bestowed substantial care on their fish. Crossbreeds between the races show heterosis for some traits (Wolfarth *et al.* 1975).

Hybrid vigour is one of the desired results of crossing species or genera; the other, production of a monosex F1 generation, was mentioned above (Pruginin *et al.* 1975). Some positive results have been attained in the quest for such hybrid superiority, but it has also been observed that 'the majority of fish hybrids are less fit than their parents and that the process of hybridization is somewhat speculative' (Purdom 1974). Perhaps the path to follow in practical fish genetics is to attempt to obtain interspecies hybrid vigour after the advantages of heterosis have been explored in both parents of the interspecific or intergeneric crosses.

Fish can be made to produce gynogenetic offspring by means of inactivating spermatozoa by irradiation and inducing retention of the polar body in the egg by cold shock so as to produce a batch of diploid progeny (Purdom 1969). This process is equivalent to several generations of inbreeding and Purdom (1974) estimates that 'one generation of gynogenesis is equivalent to 10–15 generations of conventional sib mating'.

Salmonids are evolutionarily tetraploid (Ohno *et al.* 1968) compared with closely related families of fishes. They also underwent an important part of their evolution during the Pleistocene (Schreck & Behnke 1971, Behnke, personal communication), a turbulent portion of the earth's history with regard to drainage patterns in the Northern Hemisphere. Variation in environment together with polyploidy may well have contributed to the present great variety of strains of trout and salmon. In fact polyploidy seems to be a starting point for genetic engineering with fishes. The induction of polyploidy and gynogenesis follow similar paths, except that the former does not require the inactivation of sperm. Triploidy was established in plaice (*Pleuronectes platessa*) with the advantage of the triploid condition being that the fish grow faster than the normal genotype and do not mature sexually (Purdom 1972).

Sex control is at least possible since a number of fishes show sex reversal (Smith 1967, Reinboth 1962), and some degree of control over sexuality has also been obtained by hormone treatment (Yamamoto 1969). Control of the sex of the offspring has been achieved for aquarium fishes (Purdom 1974), and sex control is a strong possibility in species important to aquaculture now or in the future. The results would, again, shortcut the establishment of inbred lines.

Finally one ought to mention that electrophoretic markers can be used to monitor strains. Two or more strains can be kept in the same ponds, thus enabling continuous production of desirable F1 hybrids (Moav *et al.* 1976). It is anticipated also that fish bred for certain characteristics may be released into a wild population with the intent of influencing the gene complement and hence the phenotype of the latter (Moav 1976).

17.6 Conclusions and prognosis

The preceding glimpses about aquaculture indicate that the practice now has an uneven scientific base. If it were enhanced, fish farmers would certainly be able to grow more per unit area. It is estimated that the application of presently known 'best' practices could raise the annual supply of farmed fish (and shellfish) by a factor of two or three (Pillay 1973). It is, therefore, germane to examine the forces that are constraining and those that are favourable to aquaculture. Economics will form an important part of this appraisal.

Various aquaculture enterprises have been subjected to studies of production economics. They range from small family, to large-scale enterprises. Given the availability of capital, certainly one of the most frequent constraints, the returns that are realized in aquaculture range from adequate, when viewed in competition with other extractive industries, to very high. For instance, depending on quality of management, several channel catfish (*Ictalurus punctatus*) culture establishments in the USA showed, in 1970 and

1971, between 23 and 55% return on capital (before taxes), the range reflecting variance in management skills (Bardach *et al.* 1972). The equivalent figure for home industry rearing of *Pangasius*, a Southeast Asian catfish, in Cambodia was 37% (Bardach *et al.* 1972). Milkfish culture in the Philippines, again depending on management and use of certain natural advantages for which the farmer need not pay (copious, run-off produced plankton as fish food), spread from 20% to 48% of return on investment and 65% to 86% on operating costs (Rabanal & Shang 1976); much higher returns on operating costs are realized in other areas (Pillay 1975).

There are other free or cheap inputs which a fish farmer may seize upon to enhance his earning picture, such as heated wastewater effluents, sewage, high volume springs of an optimum temperature as prevail in the Snake River Valley trout rearing district, to mention only a few. Such amenities are the exception, however, and commercial polyculture at various levels of intensity in an economy like that of Israel recently showed net profits below 10% (Hepher, personal communication).

Whether or not such amenities are at his disposal, an aquaculturist's expenses vary greatly with the type and intensity of operation. Feed costs usually account for more than 50% in rearing carnivores, while in tropical polyculture seed, fertilizer, labour, marketing, and interest are the major cost items, with the latter two being highly prominent (Rabanal & Shang 1976). Intensification of the culture procedures, usually dictated by high land values and/or pressure on available freshwater, means higher fixed expenses and with them a tendency to mechanize with a saving in labour (Tal 1974). Mechanization need not always result in an enhancement of profit, however.

By no means all constraints of aquaculture emerge from business balance sheets. Skills are also often lacking among those who are rearing the fish. These deficiencies are notable in several parts of the tropics; not even Southeast Asia is free of them even though aquaculture there has a long tradition. Even traditional cultural patterns can slow down successful introduction of aquaculture into regions that seem suited for it and would benefit from an upgrading of the animal protein portion of the people's diet. Farmers may, in fact, take to aquaculture more easily than do fishermen, especially under conditions of limited literacy. It seems to be the similarity in production methods in land and water animal husbandry that attracts the farmers (Bardach 1959).

Pollution and aquaculture is a complex subject, and even though what is organic pollution to some is a boon to aquacultural production for others, there is clear evidence from various parts of the globe that brackish water aquaculture installations have been affected by siltation, organic agricultural chemicals, industrial effluents and oil (Andren 1975, Shang 1973). Where pollution abatement is enforced, such enforcement may lead to ecologically sound recycling of aquacultural wastewater. In Hawaii, for instance, such high value is placed on the pristine condition of shorewaters into which the effluents of an aquafarm would be voided that a greater percentage of the effort on that farm would go into the production of vegetables with pumped pond irrigation water than on the rearing of Malaysian prawns (*Macrobrachium rosenbergii*). Here the price fetched for the truck garden produce from the land is such that the land crop is, in fact, subsidized by the aquatic one.

Another conundrum in the development of aquaculture is that of new species. Should research be expended on the domestication of a few species (see section on Fish Breeding) or should research in aquaculture go towards finding and testing 'better' species? The argument is very much alive; Bailey (1975) in an address to the American Fisheries Society comments that '... fishery experts ... often introduce an untested intruder ... successful, useful or popular in the manager's parochial experience but of unpredictable response in alien ecosystems ...'. In spite of this admonition, the search goes on for wild species in various parts of the globe that seem 'culturable' and have local culinary acceptance (George 1975, Sivalingam 1975, Ruwet *et al.* 1975). Serious efforts should be made to avoid the ecological backlashes of introductions. But, are not the grass carp, the common carp, the largemouth black bass (*Micropterus salmoides*), the rainbow trout or *Sarotherodon mossambicus* already the equivalent of the chicken, the duck, the pig and cattle? That is, have these species not been spread so far and wide and are they not so generally accepted that selective breeding or the more sophisticated measures of applied genetics mentioned earlier should be applied mainly to them with higher pay-off potential than if 'new' species were chosen for these treatments? The answer, I believe, lies in plying the trade on both sides of the street, at least for awhile. The hybridization potential of cichlids is great (Lovshin & da Silva 1975), to mention only one promising family, and the genetic variability that may emerge under cultivation of milkfish, mullet and the Indian carp is unknown because their reproduction is still not controlled. Thus neglecting the search for new species would mean the potential loss of locally or regionally well adapted species.

Dissemination of present management skills in science-based aquaculture through extension services, coupled with easier credit in many developing nations could at least double aquacultural output in the world. There is no question that there will be more buyers of aquacultural produce. Further advances in scientific disciplines applicable to aquaculture would greatly enhance the yields of aquaculture. That would be true especially if the emphasis went to the culture of feeders on algae, macrophytes, plankton and small bottom organisms rather than to that of predaceous carnivores. To achieve a several-fold increase in aquaculture yield by the year 2000, many shifts to new and unconventional practices must be employed. Waste water must be used as much as possible (Ryther *et al.* 1972), fish breeding must be improved, aquaculture must become part and parcel in planning for integrated land and water use (Pantulu 1976), and ecologically sound renewable-energy-intensive aquaculture must be fostered (Bardach 1968). Whether or not this will be the case depends as much on scientific advances as on international and local organizations, on training and, perhaps most importantly, on a changing world view of fish as food both in the technologically advanced and ineloping nations.

References

ANDERSON D.P. (1974) Fish immunology, Vol. 4. *Diseases of Fishes* (S.F. Snieszko and H.R. Axelrod, eds.). TFH Publications Inc. Ltd., Hong Kong.

ANDRÉN L.E. (1975) Pollution and degradation of environment affecting aquaculture in Africa. *FAO/CIFA Symp. Aquac. Africa, Accra Ghana. WM/G2165, CIFA/75/SR12, FAO Rome,* 16 pp.

BAILEY R.M. (1975) Comments by the President. *Newsletter Amer. Fish. Soc.* **19**(96), 3/4. (Sept./Oct.).

BARDACH J.E. (1959) Report on fisheries in Cambodia to USOM Cambodia. Publications Division, Committee for the Investigation of the Lower Mekong Basin, ESCAP, Bangkok, Thailand, 55 pp., mimeo.

BARDACH J.E. (1968) Aquaculture. *Science* **161**(3846), 1098–1106.

BARDACH J.E. (1976) Aquaculture revisited. *Jour. Fish. Res. Bd. Canada* **33** 4(2), 880–887.

BARDACH J.E., RYTHER J.H. & McLARNEY W.O. (1972) *Aquaculture: The Farming and husbandry of Freshwater and Marine Organisms.* Wiley Inter-Science, J. Wiley & Sons, Inc., N.Y., London, Sydney, Toronto.

BLAXTER J.H.S. (1969) Development: eggs and larvae. Ch. 4, Vol. 111, *Fish Physiology* (W.S. Hoar and D.J. Randall, eds.), pp. 178–252. Academic Press, N.Y., London.

BRICK R.W. (1974) Effects of water quality, antibiotics, phytoplankton and food on survival and development of larvae of *Scylla serrata* (Crustacea, Portunidae). *Aquaculture* **3**, 231–244.

BULLOCK G.L. & WOLF K. (1975) Recent advances in the diagnosis and detection of some infectious diseases of fishes. *Proc. Third U.S.–Japan Meeting on Aquaculture at Tokyo, Japan,* Oct. 15–16, 1974, A. Furukawa and W.N. Shaw, panel chairmen, Spec. Publ. Fishery Agcy. Jap. Govt. & Jap. Sea Reg. Fish. Res. Lab. Nagata, Japan, pp. 99–104.

CALAPRICE I.R. (1976) Mariculture—Ecological and Genetic Aspects of Production. *Jour. Fish. Res. Bd. Canada* **33**, 4(2), 1068–1987.

CHANCELLOR W.J. and GOSS J.R. (1976) Balancing energy and food production, 1975–2000. *Science* **192**(4236), 213–218.

CHAUDHURI H. (1976) The use of hormones in induced spawning of carps. *Jour. Fish. Res. Bd. Canada* **33**, (42), 940–947.

CHAUDHURI H., CHAKRABARTY R.D., SEN P.R., RAO N.G.S. & JENA S. (1975) A new high in fish production in India with record yields by composite fish culture in freshwater ponds. *Aquaculture* **6**(4), 343–356.

COWEY C.B. (1976) The use of synthetic diets and biochemical criteria in the assessment of nutrient requirements of fish. *Jour. Fish. Res. Bd. Canada* **33** 4(2), 1040–1045.

DONALDSON L.R. (1970) Selective breeding of salmonid fishes. *Marine culture* (W.J. McNeil ed.), pp. 65–74. Oregon State Univ. Press. Corvallis.

DONALDSON L.R. & MENASVETA D. (1961) Selective breeding of chinook salmon. *Trans. Amer. Fish. Soc.* **90**(2), 160–164.

FONTAINE M. (1976) Hormones and the control of reproduction in aquaculture. *Jour. Fish. Res. Bd. Canada* **33**, 4(2), 922–939.

GEORGE T.T. (1975) Introduction and transplantation of cultivatable species into Africa. *FAO/CIFA Symp. Aquac. Africa, Accra Ghana.* *WM/GL*1967, *CIFA/75/SR*7, *FAO Rome*, 25 pp.

GHITTINO P. (1972) *Fish Nutrition* (J.E. Halver, ed.), pp. 539–650. Academic Press, N.Y. and London.

HALVER J.E. (ed.) (1972) *Fish Nutrition.* Academic Press, N.Y. and London.

HO R. (1961) Mixed farming and multiple cropping in Malaya. *Proc. Symp. on Lane Use and Mineral Deposits in Hong Kong, Southern China and Southeast Asia*, pp. 88–104.

HOAR W.S. (1969) Reproduction. *Fish Physiology*, Vol. III (W.S. Hoar and D.J. Randall, eds.), pp. 1–59. Academic Press, N.Y. and London

HUET M. (1970) *Traité de Pisciculture*, 4th ed. (Ch. de Wyngaert, Brussels, ed.).

JALABERT B. (1976) *In vitro* maturation and ovulation in trout, pike and goldfish. *Jour. Fish. Res. Bd. Canada* **33**, 4(2), 974–988.

JOYNER T. (1975) Farming the ocean range. *Pacific Northwest Sea* **8**(4), 12–15. Oceanogr. Comm. of Washington, Seattle.

JUNE F.C. & CARLSON F.T. (1971) Food of young Atlantic menhaden *Brevoortia tyramus*, in relation to metamorphosis. *Fish. Bull.* **68**(3), 493–512.

KUO C.M. & NASH C.E. (1975) Recent program on the control of ovarian development and induced spawning of the grey mullet (*Mugil cephalus* L.). *Aquaculture* **5**, 19–29.

KUO C.M., NASH C.E. & SHEHADEH Z.H. (1973) The effects of temperature and photoperiod on ovarian development in captive grey mullet (*Mugil cephalus* L.). *Aquaculture* **3**, 25–43.

LAM T.J., PANDEY S. & HOAR W.S. (1975) Induction of ovulation in goldfish by synthetic luteinizing hormone-releasing hormone (LH-RH). *Can. Jour. Zool.* **53** (8), 1189–1192.

LEITRITZ E. (1960) Trout and salmon culture (hatchery methods). *Cal. Dept. Fish & Game Fishery Bull.* 107.

LIAO I.C., LU Y.I., HUANG T.L. & LIN M.C. (1971) Experiments on induced breeding of the grey mullet *Mugil cephalus* (Linnaeus). *Fisheries series, Chinese–American Joint Committee on Rural Reconstruction* **11**, 1–29.

LING S.W. (1973) Status, potential and development of coastal aquaculture in the countries bordering the South China Sea. *S. China Sea Fisheries Development and Coordinating Program, FAO Rome, SCS/Dev/73/5*, 51 pp.

LOVSHIN L.L. & DA SILVA A.B. (1975) Culture of monosex and hybrid Tilapias. *FAO/CIFA Symp. Aquac. Africa, Accra Ghana.* *WM/G*2056, *CIFA/75/SR*9. *FAO Rome*, 16 pp.

MATTY A.J. (1975) Endocrine control of growth and protein metabolism in aquaculture. Abstracts of papers 13th Pac. Sci. Congr., Record of Proceedings, Vol. 1, p. 58.

MAY R.C. (1976) Studies on the culture of the threadfin *Polydactylus sexfilis* in Hawaii. *World Aquaculture Conference Kyoto, Japan.* *FIR:AQ/CONF/76/E5/5pp FAO Rome.*

MICHA J.C. (1975) Synthèse des essais de reproduction d'alevinage et de production chez un silure Africain: *Clarias lazera. FAO/CIFA Symp. Aquac. Africa, Accra Ghana.* *WM/G*1379, *CIFA/75/SE*5, *FAO Rome*, 23 pp.

MILLER J.W. (1975) Fertilization and feeding practices in warm water pond fish culture in Africa. *FAO/CIFA Symp. Aquac. Africa, Accra Ghana.* *WM/G*1325, *CIFA/75/SR*4, *FAO Rome*, 29 pp.

MOAV R. (1976) Genetic improvements in aquaculture industry, *World Aquaculture Conference, Kyoto, Japan.* *FIR/AQ/Conf/76/R*9, *WM/G*3661. *FAO Rome*, 32 pp.

MOAV R., BRODY T., WOLFARTH G. & HULATA G. (1976) A proposal for the continuous production of F1 hybrids between the European and Chinese races of the common carp in traditional fish farms of Southeast Asia. *World Aquaculture Conference, Kyoto, Japan.* *FIR/AQ/Conf/76/E*78, *W/G*3680, *FAO Rome.*

MOAV R., HULATA G. & WOLFARTH G. (1975) Genetic differences between the Chinese and European races of common carp. I. Analysis of genotype-environment interactions for growth rate. *Heredity* **34**(3), 323–340.

NASH C.E. and KONINGSBERGER R.M. (in press) Artificial propagation. *The Grey Mullet* (C. Orin, ed.). Int. Biol. Progr. Publ. London.

NATIONAL ACADEMY OF SCIENCES (1975) *Underexploited Tropical Plants with Promising Economic Value.* Ad Hoc Panel Advis. Comm. on Technology Innovation Board on Science and Technology for International Development, Commission on International Relations. Washington, D.C. ix + 188 pp.

OHNO S., WOLF U. & ATKIN N.B. (1968) Evolution from fish to mammals by gene duplication. *Hereditas* **59**, 169–187.

PANDEY S. & HOAR W.S. (1972) Induction of ovulation in goldfish by clomiphene citrate. *Can. Jour. Zool.* **50**, 1679–1680.

PANDEY S., STACEY N. & HOAR W.S. (1973) Mode of action of clomiphene citrate in inducing ovulation of goldfish. *Can. Jour. Zool.* **51**, 1315–1316.

PANTULU V.R. (1976) Role of aquaculture in water-resources development: A case study of the Lower Mekong Basin project. *World Aquaculture Conference, Kyoto, Japan FIR:AR/Conf/76/E20, WM/G3036, FAO Rome*, 9 pp.

PERSOONE G. & SORGELOOS P. (1975) Technological improvements for the cultivation of invertebrates as food for fishes and crustaceans. I. Devices and Methods. *Aquaculture* **6**(3), 275–289.

PILLAY T.V.R. (1973) The role of aquaculture in fishery development and management. *Jour. Fish. Res. Bd. Canada* **30**(12/2), 2202–2217.

PILLAY T.V.R. (1975) Editorial in FAO Aquaculture Bulletin 7(1–2):1. *FAI Rome.*

PILLAY T.V.R. (1976) The state of aquaculture, 1975. *World Aquaculture Conference. Kyoto, Japan. FIR: AQ/Conf/76/R36, WM/G3650, FAO Rome*, 13 pp.

PIMENTEL D., DRITSCHILO W., KRUMMEL J. & KUTZMAN J. (1975) Energy and land constraints in food protein production. *Science* **190**, 754–761.

POPPER D., MAY R.C. & LICHATOWICH T. (1976) An experiment in rearing larval *Siganus vermiculatus* (Valenciennes) and some observations on its spawning cycle. *Aquaculture* **7**, 281–290.

PRUGININ Y., ROTHBARD S., WOLFARTH G., HALEVY A. & HULATA G. (1975) All male broods of *Tilapia nilotica* and *T. aurea* hybrids. *Aquaculture* **6**, 11–21.

PURDOM C.E. (1969) Radiation-induced gynogenesis and androgenesis in fish. *Heredity* **24**, 431–444.

PURDOM C.E. (1972) Induced polyploids in plaice (*Pleuronectes platessa*) and its hybrid with the flounder (*Platichthys flesus*). *Heredity* **29**, 11–24.

PURDOM C.E. (1974) Breeding the domestic fish. *Fish Farming in Europe Conference Report;* Oyen Intl. Bus. Communicns. Ltd., London, England, 61–68.

RABANAL H.R. & SHANG Y.C. (1976) The economics of various management techniques for pond culture of finfish. *World Aquaculture Conference, Kyoto, Japan. FIR: AQ/Conf/76/R22, WM/G3351, FAO Rome*, 22 pp.

REICHENBACH-KLINKE H. (1966) *Krankheiten und Schaedigungen der Fische.* Gustav Fischer Verlag, Stuttgart, Germany.

REINBOTH R. (1962) Morphologische und Funktionelle Zweigeschlechtlichkeit bei Marinen Teleostiern (Serranidae, Sparidae, Centracanthidae, Labridae). *Zool. Jb. Physiol. Bd.* **69**, S. 405–480.

RENNER E. & SARODNIK W. (1973) Jahresbericht 1972 über die Produktion des Wirtschaftszweiges Binnenfischerei der Deutschen Demokratischen Republik *Zeitsch f. d. Binnenfischerei der DDR* **20**, 134–153.

RUWET J.Cl., VOS J., HANON L. & MICHA J.C. (1975) Biologie et élevage des Tilapia. *FAO/CIFA Symp. Aquac. Africa. Accra, Ghana. WM/G2016, CIFA/75/SR10*, 32 pp.

RYTHER J. (1975) Mariculture. *Oceanus* **18**(2), 10–22.

RYTHER J., DUNSTAN W. TENORE K. & HUGUENIN J. (1972) Controlled eutrophication-increasing food production from the sea by recycling human wastes. *BioScience* **22**, 144.

SAO-LEANG & DOM-SAVEUN (1955) Aperçu général sur la migration et la réproduction des poissons d'eau douce du Cambodge. *Proc. Indo Pac. Fish Council* (5th meeting, 1954) (2/3): 138 142 (IPFC/FAO), Bangkok, Thailand.

SCHAEPERCLAUS W. (1954) *Fischkrankheiten.* Akademie Verlag, Berlin.

SCHAEPERCLAUS W. (1961) *Lehrbuch der Teichwirtschaft.* Paul Parey, Hamburg and Berlin, 2nd Ed.

SCHRECK C.B. & BEHNKE R.J. (1971) Trouts of the Upper Keon River Basin, California with reference to systematics and evolution of western North American *Salmo. Jour. Fish. Res. Bd. Canada* **28**, 987–998.

SHAKLEE J.B., CHAMPION M.J. & WHITT G.S. (1974) Development genetics of teleosts: a biochemical analysis of lake chubsucker ontomogy. *Dev. Biol.* **38**, 356–382.

SHANG Y.C. (1973) Comparison of the economic potential of aquaculture, land animal husbandry and ocean fisheries: the case of Taiwan. *Aquaculture* **2**, 187–195.

SHIROTA A. (1970) Studies on the mouth size of fish larvae. *Bull. Jap. Soc. Scient. Fisheries* **36**(4), 353–368.

SIVALINGAM S. (1975) The biology of cultivable brackish water and marine finfish in Africa. *FAO/CIFA Symp. Aquac. Africa, Accra, Ghana. WM/G1310, CIFA/75/SR1, FAO, Rome* 8 pp.

SMITH C.L. (1967) Contribution to a theory of hermaphroditism. *Jnl. of Theoretical Biology* **17**(1), 76–90.

SMITH R.R. (1976) Nutrient utilization by fish. Paper 45, *Proceedings of the First International Symposium on Feed Composition, Animal Nutrient Requirements, and Computerization of Diets.* Logan, Utah, 11–16 July, 1976.

SNIESZKO S.F. (ed.) (1970) A symposium on diseases of fishes and shellfishes. *Am. Fish. Soc. Spec. Publ.* #5. Wash. D.C. 526 pp.

SOH C.L. & LAM T.J. (1973) Induced breeding and early development of the rabbitfish *Siganus oramin* (Schneider). *Proc. Symp. Biol. Res.+ Nat. Dev.* May 5–7, 1972, University of Malaya, Kuala Lumpur. pp. 49–56.

SORGELOOS P. & PERSOONE G. (1976) Technological improvements for the cultivation of invertebrates as food for fishes and crustaceans II. Hatching and culturing of the brine shrimp *Artemia saline* L. *Aquaculture* **6**(4), 303–317.

STICKNEY R.R. & SHUMWAY S.E. (1974) Occurrence of cellulase activity in the stomachs of fishes. *Jour. Fish Biol.* **6**, 779–790.

TAL S. (1974) Preliminary observations on new methods of fish culture in Israel. Bamidgeh **26**(3), 51–56.

TODY W.H. & TANNER H.A. (1966) Coho salmon for the Great Lakes. *Fish. Man. Rep.* #1. Mich. Dept. Conser. Lansing, Mich.

WASLIEN C., KOK B., MYARS T. & OSWALD W.J. (1976) Photosynthetic single cell protein. (In press.) *Scrimshaw* (N. and M. Milner, eds.). NSF/MIT Protein Resources Study.

WOLFARTH G., MOAV R. & HULATA G. (1975) Genetic differences between the Chinese and European races of the common carp. II. Multicharacter variation—a response to the diverse methods of fish cultivation in Europe and China. *Heredity* **34**(3), 341–350.

WOYNAROVICH E. (1976) The feasibility of combining animal husbandry with fish farming, with special reference to duck and pig production. *World Aquaculture Conference, Kyoto, Japan. FIR: AQ/Conf/76/R6, WM/G3363, FAO, Rome,* 11 pp.

YAMAMOTO T. (1969) Sex differentiation. *Fish Physiology* (W.S. Hoar and D.J. Randal, eds.). Vol. 3, 117–175. Academic Press, N.Y. and London.

Chapter 18: Ecological Aspects of Warm-Water Fishpond Management

B. Hepher

18.1 Introduction

In spite of the fact that the sea, which covers more than three-fifths of the world's surface, provides the principal fraction of the fish consumed in the world, fish culture is considered today as one of the most promising sources of fish for the future. This is not surprising in view of the increasing gap between the total marine catch and the world demand for fish and fishery products. According to FAO statistics (FAO 1974), the world total catch of fish, which had been constantly increasing since 1948, levelled off in the seventies, and contrary to some optimistic forecasts, did not rise above 70 million tons per year. Craig (1972) estimates the world demand for fish in 1980 to be 100·6 million tons. This leaves about 30 million tons to be supplied from inland waters or aquaculture.

Today fish culture produces only a very small fraction of the total catch. Bardach *et al.* (1972) estimates the total world annual production of fish through aquaculture at about 4 million tons including freshwater, brackish water and marine species. This amount, which is considered by some to be an overestimation, is less than 6 % of today's total catch. However, the optimistic view on the future role of fish culture in the supply of fish gets support when the production potential per unit area is considered. Fishing in natural waters produces much lower yields than can be achieved in ponds. The present catch in the oceans (area of about $3·6 \times 10^5$ km²) amounts to less than 2 kg ha^{-1} yr^{-1}. Lakes, in general, are much more productive than the oceans. Henderson and Welcomme (1974) studied the potential catch in African lakes. Most of the 31 lakes analyzed by these authors, which are approaching or have reached their maximum level of exploitation, yielded 10 to 200 kg ha^{-1} yr^{-1}. Lake Kinneret, Israel, yields about 130 kg ha^{-1} yr^{-1}. These yields are still much lower than those obtained even in fishponds which are not intensively managed. Yields of up to 400 kg ha^{-1} yr^{-1} are obtained from such ponds in Europe without any supplementary feed added (Mann 1961).

One is amazed, however, by the wide variation in yields obtained from fishponds under different management methods. It is sufficient to compare the yields mentioned above for the extensively managed ponds in Europe with those obtained from intensive running water fish ponds in Japan. The yield obtained from these small ponds (each 10–40 m²), in which the water changes quite rapidly (Kawamoto 1957, Brown 1969, Tamura 1961), when extrapolated to one hectare, is about 2,000 tons ha⁻¹ yr⁻¹. No doubt differences in yield are partly due to differences in climatic conditions and water temperatures; however, pond management is a major consideration. This large difference in pond yield of over 4,000 fold indicates that a considerable increase in fish production can be achieved in existing fish ponds by improvement and intensification of the management methods.

Assuming that the chemical composition of the water (salinity, pH, etc.) is suitable for fish culture, intensification of fish culture can be achieved by controlling four main ecological factors: (a) water temperature, (b) supply of adequate food, (c) supply of oxygen and (d) removal of metabolites. Controlling these factors makes it possible to increase the stocking rate without reducing the growth rate of the individual fish and thus obtain a higher yield. Such a control has been achieved to a considerable extent in the running water systems, such as in trout raceway culture, and in the Japanese running water carp culture. In these cases the water current provides a favourable constant temperature, supplies the necessary oxygen and removes the metabolites produced by the fish. However, under these conditions not much natural food develops, and practically all the food is supplied artificially. It has to be, therefore, of high quality, nutritionally balanced, and is usually quite expensive.

In contrast to this, only a partial control over the four above mentioned factors can be achieved in stagnant water fish ponds. Daily and seasonal changes in light and temperature become significant and metabolites accummulate in the water. There is, however, a production of natural food, which can support a certain fish population. Increasing this population in order to attain higher yields is possible by one or a combination of the following methods: (a) increasing the utilization of the existing natural food resources by culturing in the same pond several different fish species concurrently, each having a different trophic niche (polyculture); (b) increasing the natural food production in the pond by chemical fertilizers or organic manures; (c) adding supplementary feed.

The increase in the stocking rate and in yield achieved by these methods is limited because of the decrease in oxygen concentration and the increase in metabolite accumulation which are associated with an increase in fish density. The yield per unit area is, therefore, lower in stagnant water fishponds than in intensively stocked, flowing water systems.

The intensification of fish culture is thus very often associated with the use of greater amounts of supplementary feed of high nutritive value, thereby increasing costs of fish production. The degree of intensification should, therefore, be critically examined with regard to the economics involved.

This can be done from two different points of view: profitability for the fish farmer and efficiency of protein conversion. Tal and Hepher (1967) pointed out that with the increasing costs for building ponds and for labour to operate them, both proportional to the pond area, higher yields per unit area are required to cover these costs. This can only be achieved by intensification and the use of supplementary feed, even though this increases the cost per unit weight of fish yield. It is only when labour and building of the pond is inexpensive, or not taken into account, as is the case in many old ponds in Europe, which are inherited from one generation to another, that management without supplementary feed is possible. When, however, the difference between the price of the fish and the price of its feed is large, as in the case of the trout, it is possible to go to the other extreme and use very costly feed.

The more intensive the fish culture, the less natural food protein is utilized by the fish; therefore, more supplementary protein is required for the production of the same quantity of fish. The protein contained in 1 kg of trout is about 150 g. The amount of protein in the feed used to obtain it is about 600 g (40 % of protein in the feed × 1·5 kg feed required to produce 1 kg of fish). Most of this dietary protein is of animal origin. As long as the price of 1 kg of trout is higher than that of the protein in the feed, it may be profitable to the individual farmer, but from the point of view of world protein supply it is doubtless a waste of valuable material.

The best practice of fish culture should aim to achieve the highest possible yields per unit area with as little supplementary feed as possible. This can be reached by a better utilization of the natural food produced in the pond which will lead to a more rational supplementary feeding. To achieve that, a better knowledge of the ecological principles involved is necessary. The present work aims at clarifying some of these concepts.

18.2 Quantitative food requirement

The relationship between the yield of carp (*Cyprinus carpio*) and the stocking density in ponds with favourable growth conditions is linear (Fig. 18.1). These favourable conditions were achieved under different management methods. Some of these ponds were stocked with only a few fish per unit area. In others, fertilizers and/or feed were added and the stocking rate increased accordingly. No doubt a similar relationship can be found for other warm-water pond fishes. From this it can be concluded that when food is not a limiting factor, it is the number of fish per unit area that determines the yield, rather than the growth rate of the individual fish. When enough food is available, growth reaches a maximum physiological limit. Any further increase in yield can then be obtained only by increasing the number of fish

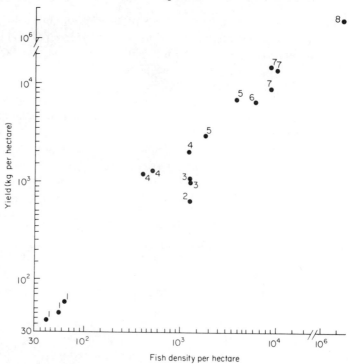

Fig. 18.1. The relationship between yield of carp (kg/ha/year) and fish density (fish/ha) in ponds with favourable growth conditions. For comparison a year was considered as 250 growing days and the data were calculated accordingly.
1. Low stock rate, no feeding in Japan (Nakamura *et al.* 1954). **2.** No fertilization, no feeding (Hepher 1962). **3.** Fertilized ponds, no feeding (Hepher 1962). **4.** Feeding with cereals (Yashouv 1969). **5.** Feeding with protein-rich pellets (Yashouv 1969). **6.** Feeding with protein-rich pellets with a demand feeder (Hepher, unpublished). **7.** Feeding with protein-rich pellets, aerated ponds (Sarig & Marek 1974). **8.** Flowing water system in Japan (Kawamoto 1957).

per unit area. Fertilization and feeding increase the amount of available food. A proper management of fish ponds should, therefore, aim at stocking the maximum possible number of fish per unit area so that all available food will be utilized and each fish will have enough food to attain its maximum physiological growth rate. In order to achieve this, two basic parameters should be known: the food requirement of the individual fish and the amount of natural food present in the pond.

If considered merely from a calorific point of view, the metabolizable energy of the food, i.e. the energy assimilated by the fish after losses in feces and urine, is utilized by the fish for two purposes: the maintenance of the body and its essential physiological processes, and for growth. The amounts of energy required for both purposes are a function of the fish's weight: the larger the fish, the more energy required for maintenance, and the greater its capacity to grow. The food requirement of the fish can thus be expressed by the following equation:

$$Rp = aM(W) + b \frac{\Delta w}{\Delta t} (W) \qquad (1)$$

where R is the food requirement of the individual fish; $p =$ the correction factor for conversion of ration consumed to ration assimilated (metabolizable energy); $M(W) =$ energy required for maintenance, a function of W; $\Delta w/\Delta t$ $(W) =$ weight change per unit time, a function of W; $W =$ the weight of the fish; a and b are the coefficients of energy conversion for maintenance and growth, i.e. the amount of metabolizable energy (in calories) required to produce one calorie of free energy for these purposes. These are equal to the reciprocal of the coefficients of utilization of the metabolizable energy for maintenance and growth.

The components of equation (1) are affected by a great number of factors which can be divided into three groups:

(1) Those related to the fish: besides the weight of the fish, which appears in the equation itself, factors such as activity, physiological state (diseases, state of gonad development, etc.), genetic characteristics, etc. may affect both metabolism and growth.

(2) Those related to the environment: water temperature; water composition (dissolved oxygen concentration, the concentration of minerals, pH, etc.); the presence in water of toxic substances, such as metabolites, which have recently been given special attention.

(3) Those related to the amount and composition of the food.

When so many factors affect the food requirement, it is obvious that the requirement will vary with individual fish and environmental conditions. No attempt will be made here to solve the food requirement equation for different conditions. From the equation itself it is apparent that the function of fish size (weight) on the food requirement is of major importance and, therefore, affects stocking rate and yield in ponds. While most of the environmental conditions cannot be controlled by the fish farmer, he can control the fish population in the pond. The average weight of the fish and its rate of change are the major yardsticks which the fish farmer has for the conditions in his ponds. The relationship between maintenance, growth, total food requirement and the weight of the individual fish is, therefore, worth further discussion.

REQUIREMENT FOR MAINTENANCE

Winberg (1956, 1961) has added an important contribution to the study of metabolic rate of the fish by reviewing a considerable amount of data on this subject. He found a very good correlation between the average standard metabolism and the weight of the fish. When the standard metabolism, Qs, is expressed in cal/day (taking the oxycalorific value to be 4·8 cal per ml O_2), at a temperature of 20C, and the weight of the fish (W) is expressed in grams, his equation takes the form:

$$Qs = 35·4 \ W^{0·8} \tag{2}$$

The regression coefficient of 0·8, which determines the slope of the line, has since been confirmed by many (Fry 1957, Paloheimo & Dickie 1966 and others). The specific metabolic rate (the y intercept in the equation) is the energy expense for maintenance by a fish of 1 g.

In his paper, Winberg suggests that the energy expense for maintenance in natural conditions, when the fish gets enough food to maintain growth and is active in the search for this food, may be taken as twice the standard metabolism, or:

$$aM = 2Qs = 71 \ W^{0·8} \tag{3}$$

Several studies support Winberg's assumption. Paloheimo and Dickie (1966) state that while the 'weight exponent' does not change with feeding and activity, the specific metabolic rate, i.e. the metabolic rate of a 1 g fish, will differ with the amount of food. On a maintenance diet the metabolic rate is close to standard metabolism, while at maximum or *ad libitum* feeding rates it approaches active metabolism. Kausch (1968), who studied the metabolic rate of fed and active carp, found it to be 1·9 times higher than that of the standard metabolism (Kausch 1968, Table 15b). Winberg's corrected equation also agrees satisfactorily with calculation of the calorific value of the feed consumed by the fish. Marek (1966) states that according to the practice of Israeli fish farmers, carp of 500 g each, stocked very densely in a storage pond (thus excluding natural food), can be held without appreciable change in weight at a temperature of 25C if fed sorghum at a rate of 1% of their body weight per day. Calculation will show that this will provide about 13·5 kcal/day per 500 g fish of metabolizable energy as compared to 10·9 derived from equation (3). Schäperclaus (1966) conducted similar experiments in aquaria using carp of approximately 300 g. The daily energy requirement for maintenance in his experiments amounted to 6·3 kcal day^{-1} as compared to 7·2 kcal day^{-1} according to equation (3). An even better agreement is found in data given by Kevern (1966). For maintenance of carp of an average weight of 137 g, 3·7 kcal day^{-1} is required; the requirement as derived from equation (3) is 3·8.

It seems, therefore, that for all practical purposes the corrected Winberg's equation (3) may serve as a basis for calculating the relationship between the energy requirement for maintenance and the weight of fish in natural pond conditions.

Winberg (1961) quotes studies carried out on freshwater fish in the tropics, where metabolic rates were much lower than those found in experiments carried out in the temperate zone at identical temperatures. In all these cases, the rate of metabolism of carp in the tropical zone at 25–30C was approximately that found in the temperate zone at 20C. Winberg explains these differences by the adaptation of the fish to the temperature. This was also discussed by Fry (1958) who measured respiratory rates over a wide range of temperatures. Frequently, a double sigmoid curve is found with a central interval, over which the rate is little affected by increasing temperature. This adaptation effect has been also noted by Kanungo and Prosser (1959). It seems, therefore, that one should distinguish between changes in the metabolic rate caused by rapid changes in temperature as opposed to slow changes where adaptation may be acquired. It may well be that for temperature adapted eurythermal warmwater fish, the effect of temperature on the requirement for maintenance at the range of 20–30C is small. The temperature aspect, therefore, will be ignored in our further discussion.

REQUIREMENT FOR GROWTH

The growth of fish has been extensively discussed by many authors. For our purpose, however, there is a special significance to the works of Parker and Larkin (1959), and Paloheimo and Dickie (1965, 1966), who discuss the growth-weight relationships. These are given by the equation:

$$\frac{dw}{dt} = kw^x \tag{4}$$

where w = body weight,
or in its linear form:

$$\log \frac{dw}{dt} = \log k + x \log w$$

This relationship holds true only when there is enough food to produce the maximum physiological growth rate. When a deficit in food occurs, the straight line of the log equation is inflected. With the increase in weight of the fish, more food is required for maintenance and for attaining the maximum physiological growth capacity; a condition that is not always met in natural conditions. Natural populations of fish, therefore, do not always conform to the equation. Growth is usually measured over long periods and, therefore, can hardly be expected to give the accurate parameters for this equation. Much better information can be obtained from fish ponds or aquaria, where enough food is available to sustain maximum growth rate, and frequent weighings of the entire population, or a considerable part of it, are carried out.

Lühr (1967) reports results of experiments with carp carried out in aquaria at a constant water flow and a more or less constant temperature. The fish were fed a protein-rich diet. The data given in the growth curves (Figs. 2, 4, 6 in his paper) were analysed by comparing the growth rate for any period ($\frac{Wt - Wo}{t}$) to the average weight of fish at the same period ($\frac{Wt + Wo}{2}$). This analysis resulted in the following regression:

$$\frac{dw}{dt} = 0\cdot090 \ W^{0\cdot667} \ (r = 0\cdot79)$$

where $\frac{dw}{dt}$ is measured in g day^{-1} and W grams = correlation coefficient

Yashouv (1969) describes the growth of carp in a pond where they have been stocked at a very low rate (500 per ha). The frequent sample weighings were analysed by us in the same way as before, giving the regression:

$$\frac{dw}{dt} = 0\cdot20 \ W^{0\cdot65} \ (r = 0\cdot924)$$

An analysis of much more comprehensive data on carp growth gathered during 15 years in a 24 pond system (see Fig. 18.3) resulted in a regression similar to the previous two:

$$\frac{dw}{dt} = 0\cdot179 \ W^{0\cdot66} \ (r = 0\cdot88)$$

Although all these cases were related to carp, it is interesting to note that the regression coefficients were not very different from those found by Parker and Larkin (1959) for steelhead trout (*Salmo gairdneri*) and chinook salmon (*Oncorhynchus kisutch*). Most of the regression coefficients found by them were under 0·8 and generally between 0·83 and 0·50. It may well be that when food is not a limiting factor, the regression coefficient of the growth–weight ratio is about 0·66 and is determined by an intrinsic physiological character, while the specific growth rate (*Y* intercept) is determined by the genetic characteristics of the fish and by the environment.

Three main conclusions may be drawn from the foregoing discussion on the food requirement of the fish:

(a) Since both the requirements for maintenance and the growth capacity increase with an increase in fish weight, the total food requirement also increases with the increase in the weight of the fish.

(b) Since both the requirement for maintenance and the requirement for maximum growth increase at a lower rate than the increase in the weight of the fish, the relative food requirement, i.e. the requirement per unit weight, will decrease with the increase in the weight of the fish.

(c)   Since the requirement for maintenance increases with the increase in weight at a greater rate than the requirement for growth (a weight exponent of $W^{0.8}$ against $W^{0.66}$), an increasing part of the food consumed will be utilized for maintenance. The food consumed to produce a unit of weight gain will therefore increase with the increase in the weight of the fish.

As will be seen later, these points have an effect on the yield and the method of management.

18.3 Relationships between food, growth and yield

It is apparent that as long as the environmental factors do not become limiting, the optimal stocking rate for achieving the highest possible yield will depend not only on the food requirement of each fish, but also on the amount of available food. If no supplementary feed is added, yield will depend on natural food only, and when added, on the quantity and quality of the feed. Because the purpose of the added feed is to supplement the natural food, it is always important to determine the amount of natural food available for the fish population. Unfortunately, this is very difficult and no satisfactory method has been developed yet. In many studies, the estimated standing crop of natural food was taken as a measurement of the available food. However, since the standing crop is only the result of an equilibrium between the production of the natural food on the one hand and its removal from the pond (including consumption by fish) on the other, it is obvious that the standing crop cannot serve as true estimate of the amount of natural food available. This estimation has to be done, therefore, by indirect methods. Such a method might involve estimating standing crop of the fish population which can be supported at its maximum growth capacity.

Because the maximum physiological growth capacity depends on the weight of the fish, and this changes with its growth, we will limit ourselves to the above-mentioned food-fish population relationship for a given fish weight within a very short time period. This will also exclude variations in the amount of natural food with time, which no doubt occur in the pond.

When fish density is low, the total amount of food, when divided among the number of fish, will be higher than the individual requirement of the fish. The full physiological growth capacity is attained. As long as the available natural food in the pond is sufficient to provide each fish with its entire food requirement, this growth rate will be maintained irrespective of the fish density. However, over a certain density (the 'critical standing crop'), the amount of food per fish becomes less than that required to maintain full growth. Growth will decrease with an increase in density of fish (Fig. 18.2)

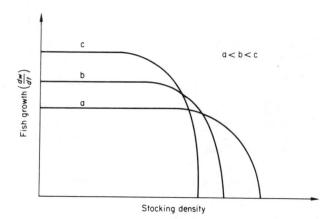

Fig. 18.2. Schematic presentation of the relationship between the stocking density and short interval fish growth at different fish weights (a weighs less than b which in turns weighs less than c).

until the amount of food per fish will be sufficient for maintenance only and will cease entirely (the 'carrying capacity').

If, at the 'critical standing crop', food of sufficient quality is added, the maximum growth rate will be retained at higher densities up to a point where some limiting factor in the food will show up or until a build-up of unfavourable environmental conditions, such as lack of oxygen or accumulation of metabolites, will inhibit growth.

Fish of a larger average individual weight have a greater maximum growth potential than smaller fish, provided enough food is available. However, since the food requirement is greater, in order to maintain growth and maintenance, the amount of the available food in the pond will be sufficient for a lesser number of larger fish than smaller ones. The 'critical standing crop' and the 'carrying capacity' will, therefore, be at a lower density.

It is clear that at a given density, smaller fish can attain their maximum physiological growth capacity. When they increase in size, the 'critical standing crop' is reached and their growth will be less than maximum and eventually will decrease. Growth will cease entirely when the 'carrying capacity' is reached. With lower density of fish in the pond, these points will be reached at a higher individual weight. Feeding of fish will also increase the weight in which these points will be reached. This is illustrated by the results from 24 ponds in Israel (each of 0·1 ha) during a period of 15 years in which four different treatments and fish densities were used (Fig. 18.3).

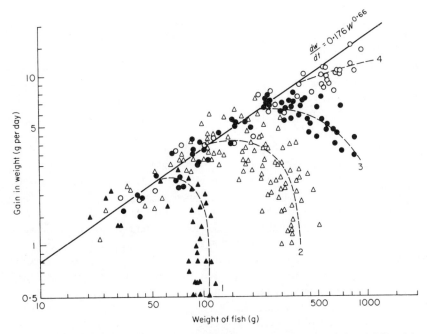

Fig. 18.3. Relationship between short interval fish growth rate (1–2 weeks) and fish weight, at different treatments at the Fish Culture Station, Dor (for nature of treatments see text).

(1) No fertilization and no supplemental feeding at a density of 1,200 fish/ha;

(2) Chemical fertilization but no supplemental feeding, at a density of 1,200 fish ha^{-1};

(3) Chemical fertilization and feeding with cereal grains (sorghum) at a density of 2,000 fish ha^{-1};

(4) Chemical fertilization and feeding with protein-rich pellets (25 % crude protein with no added vitamins) at a density of 6,000 fish ha^{-1}.

It can be seen that at small fish weights the growth rate is equal in all treatments, irrespective of stocking rate or treatment. The line describing this growth rate in relation to fish body weight is the maximum growth capacity. It is only when lack of food develops that the growth rate deviates from this line (at the 'critical standing crop'). With more natural food (by fertilization) and more supplemental food, the 'critical standing crop' is raised. The regression of the line describing the relation between maximal growth rate (expressed in g day^{-1}) and fish weight (in grams) at prevailing environmental conditions at Dor was found to be:

$$\frac{dw}{dt} = 0.179 \ W^{0.66} \ (r = 0.88) \tag{5}$$

Under given environmental and management conditions the 'critical standing crop' and 'carrying capacity' are dependent both on the density and the weight of the fish. It is justifiable, therefore, to express these points by the product of the last two variables, i.e. by the standing crop, which is the biomass of fish in the pond kg ha^{-1} at the given moment. The standing crop at which the growth deviates from the maximum growth line is the critical standing crop and the standing crop at which growth ceases, the 'carrying capacity' of the pond.

It is obvious that one of the main factors affecting the critical standing crop and the carrying capacity is the productivity of the pond, which may be increased artificially by fertilization or supplemental feeding. Naturally the food should contain all nutrients necessary for growth. A lack of any essential nutrient will cause a reduction of growth. These required nutrients may be different for different fish species and different life stages of any one species. The essential nutrient needed in the fry stage is not necessarily required by the adults. The 'critical standing crop' and the 'carrying capacity' may, therefore, be different for different feeds and at different life stages. From Fig. 18.3 and the information on the fish density one can calculate the critical standing crops and carrying capacities of the different treatments for the ponds in which the experiments were carried out. This calculation has shown the following:

Treatment	Critical standing crop kg ha^{-1}	Carrying capacity kg ha^{-1}
No fertilization, no feeding	65	130
Fertilized but no feeding	140	480
Fertilized and fed cereals	550	2,500 (estimated)
Fertilized and fed protein-rich pellets	2,400	

These values for the carrying capacity are very similar to those found by Yashouv (1959) in the same ponds.

Because the relative food requirement of the fish (unit food per unit fish weight) decreases with increasing fish size, a given amount of food in the pond should suffice for a larger biomass of larger fish than of smaller ones. One may expect, therefore, that the 'critical standing crop' and the 'carrying capacity' will also be higher for the larger fish. However, practical experience has shown that there is almost no difference in this respect. Moreover, some experimental evidence, that should be studied further, even indicates the reverse situation, i.e. a higher critical standing crop for smaller fish than for larger. This can probably be explained by the fact that since a unit fish weight contains more small fish than bigger ones, they may be able to hunt and find individual food items more easily. A similar situation is discussed by Ivlev (1961). According to him the relation between the amount of food eaten by a fish (*r*) and the concentration of its food in the feeding area (*p*) is proportional to the difference between this actual ration and the maximal ration which can be consumed by the fish (*R*). If the concentration of the food is given but the number of fish (*N*) vary, as in our case, the share of each individual fish will amount to *p/N*, and the amount of food per fish can be expressed according to Ivlev by the equation:

$$r = R\,(1 - e - kp/N)$$

where *k* is the coefficient of proportionality. From this equation it is clear that the relation between the increase in the number of fish and the decrease in the amount of food per fish is not linear. The nearer the amount of food available per fish gets to the maximal ration, the greater is the deviation from linearity. An increase in the number of fish does not bring a proportional decrease in the amount of food per fish, which means, that the whole population gets more food out of its feeding area. This behaviour compensates or even overcompensates for the effect of fish size on the critical standing crop and the carrying capacity.

An experiment on feeding carp fry in nursing ponds at densities of 15,000 and 30,000 fry ha^{-1}, and three diets, has shown that the effect of density on growth was more pronounced than that of the supplementary feed even when fed with protein-rich diet. The fry, which were stocked at identical average weight, reached the following average final weights (in grams) after 48 days:

Feed	15,000 carp fry per ha	30,000 carp fry per ha
Sorghum	27·7 g	17·0 g
Pellets containing 25% crude protein	34·5 g	24·7 g
Pellets containing 30% crude protein and vitamins	42·7 g	22·6 g

Significant statistical differences were found only between the three diets in the lower density and between the two densities, but not between the diets at the higher density. This indicates that the fry are dependent on natural food at the higher density. The added feed is utilized but it does not entirely replace the natural food. It seems that a nutritional factor is missing in this fry stage that is not required in the larger fish. The critical standing crop and the carrying capacity will, therefore, be lower than that of larger fish fed the same diet, but somewhat higher than that for natural food only.

Weight-growth data have been plotted for fry fed on cereal grains (Sorghum) and weighing less than 25 g (Fig. 18.4). While the slope of the log-log line remains similar to the other cases ($W^{0.644}$), the critical standing crop is much lower and is about 200 kg ha^{-1} as compared with 550 kg ha^{-1} in the larger fish.

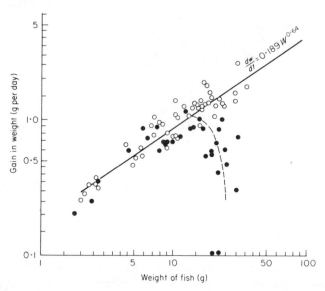

Fig. 18.4. Relationship between the short interval fish growth (1–2 weeks) and fish weight in carp nursing pond (below 50 gm) at different density ranges. Open circles—5,000–10,000 fish ha^{-1}; solid circles—11,000–15,000 fish/ha.

The yield per unit area is a product of the average gain per fish and the number of fish per unit area. This is presented schematically (Fig. 18.5) showing the change in yield with an increase in fish density when measured during short intervals for a given fish weight. When the growth rate is not affected by the stocking rate, i.e. below the critical standing crop, the yield increases with the increase in density in a linear proportion. Above the critical standing crop, as long as the growth rate decreases at a smaller rate than the increase in density, there is still an increase in yield, though sublinear. When, however, the growth rate decreases at a higher rate than the increase in stocking rate, the yield drops and reaches zero at the point of carrying capacity.

It is clear that for a short period the highest yield is attained at a certain fish density which is between the critical standing crop and the carrying capacity. Since these points move toward a lower density with the increase in the weight of the fish, so will the density of maximum yield. This can also

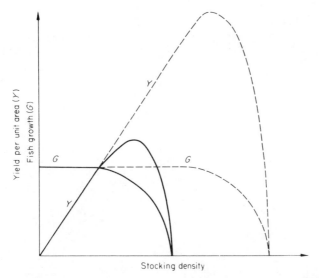

Fig. 18.5. Schematic presentation of the relationships between the stocking density, the short interval growth rate and the short interval yield per unit area, with (broken line) and without (solid line) supplementary feeding.

be expressed in terms of standing crop (Fig. 18.6). There is an increase in yield with the increase in standing crop up to a maximum which is between the critical standing crop and the carrying capacity. From this point, the yield drops with the increase in standing crop until it reaches zero at the point of carrying capacity. Since, however, the relative growth rate (i.e. the growth rate per unit weight) is smaller as the fish becomes larger, the yield for a given standing crop will be higher when this consists of smaller fish than when it consists of larger fish. The overall yield of a unit area is the area under the curve and this, of course, will be larger when the peak in yield is attained at a higher standing crop.

18.4 Application of the ecological principles to management

From the foregoing discussion a number of conclusions can be drawn which have a direct application to the management of fish ponds.

THE CONTROL OF POPULATION DENSITY

Population control is the most important factor affecting pond yield and the utilization of supplement feed. Highest yields will be attained at a certain density of fish in the pond. If the density changes, it will affect the yield.

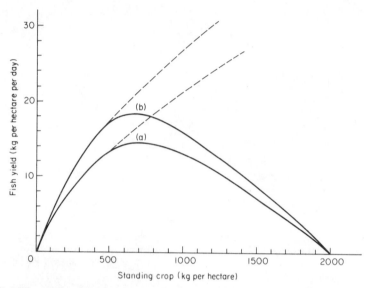

Fig. 18.6. The relationship between standing stock and yield per unit area for fish receiving sorghum supplementary feed in ponds stocked with two fish densities: (a) 2,000 ha^{-1}; (b) 4,000 ha^{-1}. Solid lines give calculated values based on actual average growth rates (Fig. 18.3). Broken line gives extrapolated possible growth when food is not a limiting factor according to equation (5).

When the density is too low, the food resources are not fully utilized and there is loss of yield. When the density is too high, the carrying capacity may be reached and the yield will then drop to zero. At this point the entire food resource will be utilized for maintenance. The fish density is especially important when the fish are large and stocked at lower numbers than small fish. Each fish has a greater relative effect on the standing crop. It is also important when the natural productivity of the pond is poor, since then the range between the critical standing crop and the carrying capacity is small.

The number of fish in the pond can change unintentionally and without the knowledge of the fish farmer. Diseases or deoxygenation may reduce the number while nucontrolled spawning in the pond may increase it. Both may lead to loss in yield and waste of feed. Short growing seasons can reduce this hazard. After each season the pond should be drained completely, all the fish should be harvested, and a new growing period started. If some fish are left in the pond, they will spawn readily in the next season and reduce the possibility of getting a good yield. It is obvious, therefore, that undrainable ponds cannot be used for controlled commercial fish culture. This is an important point when considering enclosing areas in a lake or a lagoon for fish culture. If the control of the population is not achieved, there is small chance of success in these enclosures. The culture of fish which cannot spawn freely in the pond, such as the Chinese carps (e.g. silver carp—*Hypophthalmichthys molitrix*; grass carp—*Ctenopharyngodon idellus*; and others) have no doubt an advantage in this respect.

The highest instantaneous yield will be received at a certain standing crop of fish. However, stocking this quantity of fish is impractical since, with their growth, the yield will immediately decrease. There are, therefore, two possible approaches:

(a) Frequent thinning out of the fish population, thus keeping the standing crop constant at or near the maximum yield value.

(b) Allowing for a certain amount of growth and stocking at a lower standing crop. The fish are harvested for marketing at a time when the standing crop will be at or somewhat higher than that of maximum yield. It is clear that in this way some yield potential is lost.

The first approach will give the highest possible yield but it calls for careful planning and management, so that there will be allowance in other ponds for culturing the smaller-than-market-size fish which were taken out of the pond. The expense of the extra work involved in this operation should also be considered.

THE SIZE OF THE FISH CULTURED

Below the point of maximum yield, for a given stocking rate of fish in the pond, the bigger the fish, the higher their growth rate and the higher the yield per hectare (see Fig. 18.6). However, the carrying capacity limits growth and, therefore, the fish culturist must choose between culturing fewer fish per hectare or smaller fish. Culturing the smaller fish will result in higher yields but then, when the carrying capacity is reached, the fish may be too small for marketing.

Walter (1934) has divided fish ponds into three productivity categories. He suggests using the pond with poorer productivity for nursing fry and the better one for fattening. This is logical because the relatively smaller size of the required fry enables stocking at a higher density, thus providing a sufficient number of fry for the fattening ponds.

Fattening fish in the less productive pond will require a considerable decrease in stocking rate in order to reach marketable size and this will decrease the yield per hectare. The importance of the supplementary feed becomes obvious in this respect. By using supplementary feed, the carrying capacity is shifted to higher levels. This enables culturing a greater number of marketable fish and thus considerably increases the yield.

SUPPLEMENTARY FEEDING

Supplementary feed is not required when the standing crop is below the critical point. Each fish then gets its required food to sustain maximum growth rate and the supplementary feed will have no effect. In natural waters, populations of fish are usually in a state of equilibrium with their trophic base and the amount of food per fish permits growth. Increasing the food resources of these water bodies by fertilization or by supplemental feeding without increasing the number of fish so as to exceed the critical standing crop will have only a limited effect until a new equilibrium between a larger fish population and the natural food develops. This indeed seems to be the result of most of the attempts to fertilize lakes (Ball 1948, Hasler & Einsele 1948), sea lochs (Gross et al. 1946) or sea bays (Buljan 1960). In most of these cases a considerable increase was observed in the lower links of the food chain after quite a short period, but the effect on fish was small and delayed, since there were not enough fish to utilize this increase in the trophic base.

When deficits in various essential nutrients develop, the supplementary feed should cover these deficits as they develop, so that the total food, natural and added, becomes nutritionally balanced. The critical standing crop may be different for each of these individual essential nutrients. While the natural food may still supply the requirements in vitamins it may be short in protein or in energy, as indeed is often the case with carp in ponds. There is no justification in using nutritionally balanced, and therefore expensive, feed immediately beyond the critical point, since many of the nutrients are not yet required and will be wasted. A considerable increase in yield can be achieved by using simple feeds like cereals or a relatively low protein supplemental feed. It is worthwhile, therefore, to increase the standing crop to a level which will require the use of these feeds and thus obtain high yields, rather than keep the standing crop low and obtain a lower yield without feeding.

On the other hand, if the feed will not satisfy the nutritional requirement of the fish, the feed conversion coefficient will increase rapidly, making fish culture unprofitable. One should distinguish between:

(a)   apparent feed conversion coefficient

$$= \frac{\text{supplementary feed utilized by the fish}}{\text{gain in fish yield}}$$

(b)   true feed conversion coefficient

$$= \frac{\text{supplementary feed utilized by the fish}}{\text{gain in fish yield} - \text{gain attributed to natural food}}$$

As long as the feed balances the natural food, there is a moderate increase in both feed conversion coefficients. This is caused by two factors:

(a)   With the increase in the weight of the individual fish, there is an ncrease in the ratio between the requirement for maintenance and the requirement for growth, increasing the true conversion coefficient.

(b)   With the increase in standing crop, the total food requirement increases; however, the amount of natural food does not increase accordingly. At best it remains constant, and more often it decreases during the growing season. The supplementary feed has, therefore, to replace this part of the natural food, increasing the apparent feed conversion coefficient.

Both of these factors cause an increase in the amount of food required to produce a unit growth. When the feed does not meet the requirements of the fish, a critical point is created and the growth rate drops, resulting in rapidly increasing the true conversion coefficient (Fig. 18.7).

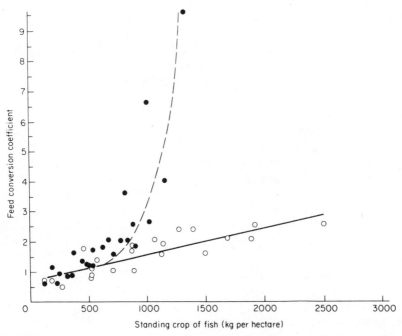

Fig. 18.7. The supplementary feed conversion coefficient at different standing crops of fish in ponds (stocking density—2,000 fish ha^{-1}). Solid circles—fish fed on sorghum; open circles—fish fed on protein rich pellets. Each point is an average of 3 replications. From unpublished data of an experiment at the Fish & Aquaculture Research Station at Dor, Israel.

This emphasizes the importance of the natural food in the pond. If more natural food is in the pond, more nutrients are recycled and less have to be supplemented by the feed. The feed can thus be of a lower quality and less expensive. The critical standing crop of this supplementary feed is higher and the sharp increase in feed conversion coefficient occurs, if at all, at such a size that the fish should be harvested and marketed.

There are two major ways to increase the availability of the natural food in the pond both of which can be used simultaneously:

(a)   A better utilization of the existing natural food resources in the pond by polyculture of a number of suitable fish species in the same pond. Each species has its own trophic niche, so that the total utilization of the natural food in the pond is greater.

(b)   Increasing the natural food productivity of the pond by chemical fertilization or organic manuring.

Yashouv (1959) states that the carrying capacity of a monoculture of carp in Israel was about 460 kg ha^{-1} in fertilized ponds, but without supplemental feeding. A mixed culture of carp and tilapia reared in these ponds had a carrying capacity of 600 kg ha^{-1}. A further increase can be achieved by adding more fish species, such as silver carp (*Hypophthalmichthys molitrix*), which have different feeding habits than the previous two.

The effect of chemical fertilization on the critical standing crop and the carrying capacity can be seen in Fig. 18.3. However, the effect of the fertilization on natural food production is limited. Hepher (1962) showed that beyond a certain level of fertilization light becomes a limiting factor and the production of algae (which is the first link in the food chain) per unit area levels off. Schroeder (1974) has shown that this can be overcome by a 'short cut' in the food chain using organic wastes (fluid cow manure). This manure which is dispersed over the entire pond area, acts in two pathways. The mineralized part acts through the autotrophic pathway, similar to the action of the chemical fertilizer. The dissolved and particulated organic matter acts through the hetrotrophic pathways supplying food for the zooplankton and the bottom fauna in the pond. This has proven to be an important means of increasing the natural food production in the pond.

In conclusion, it can be stated that the increase of yield in fish ponds does not have to follow the intensive but also expensive methods used in running water systems. Much can still be gained by better management of the existing fish ponds. Management based on ecological principles, taking into account the growth capacity of the fish, the proper utilization of the food resources, and ways to increase these resources, will succeed in increasing fish yields from ponds. The increased profits to the fish farmer and greater contribution to the economy of the expanding population of the world will benefit all concerned.

Acknowledgements

Thanks are due to Prof. K. Reich of the Hebrew University of Jerusalem and Dr. John E. Bardach of the Hawaii Institute of Marine Biology (University of Hawaii at Manoa) for their valuable and helpful remarks on this paper.

References

BALL C.R. (1948) Fertilization of natural lakes in Michigan. *Trans. Amer. Fish. Soc.* **78**, 145–155.

BARDACH J.E., RYTHER J.H. & MCLARNEY W.O. (1972) Aquaculture, the farming and husbandry of freshwater and marine organisms. *Wiley-Interscience*, N.Y.

BROWN E.E. (1969) The freshwater cultured fish industry of Japan. *Univ. of Georgia, College of Agric. Exp. Sta. Research Reports* **41**, 57 pp.

BULJAN M. (1960) Some results of fertilization experiments carried out in Yugoslav marine bays. *General Fish. Counc. Mediterr. (FAO), Tech. Pap.* **6**(33), 237–243.

CRAIG M.V. (1972) Fish farming. *Res. Rep., Long Range Planning Serv., Stanford Res. Inst.*, Menlo Park, Calif. 20 pp.

FAO (1974) Yearbook of fishery statistics, Vol. 36, catches and landings, 1973. *FAO*, Rome.

FRY F.E. (1957) The aquatic respiration of fish. *The Physiology of Fishes*, Vol. 1. Metabolism (M.E. Brown, ed.), pp. 1–63. Academic Press, N.Y.

FRY F.E.J. (1958) Temperature compensation. *Ann. Rev. Physiol.* **20**, 207–224.

GROSS F., RAYMONT J.E.G., NUTMAN S.R. & GOULD D.T. (1946) Application of fertilizers to an open sea loch. *Nature, Lond.* **158**, 187.

HASLER A.D. & EINSELE W.G. (1948) Fertilization for increasing productivity of natural inland waters. *Trans. Amer. Wildl. Conf.* **13**, 527–551.

HENDERSON H.F. & WELCOMME R.L. (1974) The relationship of yield to Morpho-Edaphics Index and number of fishermen in African inland fisheries. *CIFA/FAO Occas. Pap.* **(1)**, 19 pp.

HEPHER B. (1962a) Primary production in fishponds and its application to fertilization experiments. *Limnol. Oceanogr.* **7**(2), 131–136.

HEPHER B. (1962b) Ten years of research in fish ponds fertilization in Israel. I. The effect of fertilization on fish yield. *Bamidgeh* **14**(2), 29–38.

IVLEV V.S. (1961) *Experimental Ecology of the Feeding of Fishes.* Yale Univ. Press, New Haven.

KANUNGO M.S. & PROSSER C.L. (1959) Physiological and biochemical adaptation of goldfish to cold and warm temperatures. I. Standard and active oxygen consumptions of cold and warm acclimated goldfish at various temperatures. *J. Cell. Comp. Physiol.* **54**, 259.

KAWAMOTO N.Y. (1957) Production during intensive carp culture in Japan. *Prog. Fish-Cult.* **19**(1), 26–31.

KAUSCH H. (1968) Der Einfluss der Spontanaktivität auf die Stoffwechselrate junger Karpfen (*Cyprinus carpio* L.) im Hunger und bei Fütterung. *Arch. Hydrobiol. Suppl.* **33**(3/4), 263–330.

KEVERN N.R. (1966) Feeding rate of carp estimated by a radioisotopic method. *Trans. Amer. Fish Soc.* **95**(4), 363–371.

LÜHR B. (1967) Die Fütterung von Karpfen bei Intensivhaltung. *Vortraqsveranstaltung über neue Methoden der Fischzüchtung und -haltung.* Bundesforschungsanstalt für Fischerei und Max Planck Institut für Kulturpflanzen züchtung in Hamburg, 47–58.

MANN H. (1961) Fish cultivation in Europe. *Fish as Food*, Vol. 1. Production, Biochemistry and Microbiology (G. Borgstrom, ed.), pp. 77–102. Academic Press, N.Y. and London.

MAREK N. (1966) The effect of carp size on the food coefficient. *Bamidgeh* **18**(1), 14–25.

NAKAMURA K. *et al.* (1954) Fish production in seven farm ponds in Shioda Plain, Nagano Perfecture, with reference to natural limnological environment and artificial treatment. *Bull. Freshwat. Fish Res. Lab.* Tokyo **3**(1), 27–79.

PALOHEIMO J.E. & DICKIE L.M. (1965) Food and growth of fishes. I. A growth curve derived from experimental data. *J. Fish. Res. Bd. Canada* **22**(2), 521–542.

PALOHEIMO J.E. & DICKIE L.M. (1966) Food and growth of fishes. II. Effects of food and temperature on the relation between metabolism and body weight. *J. Fish. Res. Bd. Canada* **23**(6), 869–908.

PALOHEIMO J.E. & DICKIE L.M. (1966) Food and growth of fishes. III. Relations among food, body size and growth efficiency. *J. Fish. Res. Bd. Canada* **23**(8), 1209–1248.

PARKER R.R. & LARKIN P.A. (1959) A concept of growth in fishes. *J. Fish. Res. Bd. Canada* **16**, 721–745.

SARIG S. & MAREK N. (1974) Results of intensive and semi-intensive fish breeding techniques in Israel in 1971–1973. *Bamidgeh* **26**(2), 28–48.

SCHÄPERCLAUS W. (1966) Weitere Untersuchungen über Grösse und Bedeutung des Naturnahrungsteils an der Gesamtnahrung der Karpfen bei Fütterung mit Getreidenkörnern in Abwachteichen. *Z. Fisch.* N.F. **14**(1/2), 71–99.

SCHROEDER G.L. (1974) Use of fluid cowshed manure in fishponds. *Bamidgeh* **26**(3), 84–96.

TAL S. & HEPHER B. (1967) Economic aspects of fish feeding in the Near East, *FAO Fish. Rep.* **44**(3), 285–290.

TAMURA T. (1961) Carp cultivation in Japan. *Fish as Food*, Vol. 1, Production, Biochemistry and Microbiology (G. Borgstrom, ed.), pp. 103–120. Academic Press, N.Y. and London.

WALTER E. (1934) Grundlagen der allgemeinen fischereilichen Produktionslehre, einschliesslich ihrer Anwendung auf die Fütterung. *Handb. Binnenfisch. Mitteleurop.* **4**(5), 481–662.

WINBERG G.G. (1956) Rate of metabolism and food requirements of fishes. *Nauchnye Trudy Belorusskovo Gosudarstvennoyo Universiteta imeni V.I. Lenina*, Minsk, 253 pp. *Fish. Res. Bd. Canada Transl. Ser.* **194**, 1960.

WINBERG G.G. (1961) New information on metabolic rate in fishes. *Vop. Ikhtiol.* **1**(1), 157–165. *Fish. Res. Bd. Canada Transl. Ser.* 362, 1961.

YASHOUV A. (1959) Studies in the productivity of fish ponds. I. Carrying capacity. *Pro. Gen. Fish. Counc. Medit.* **5**, 409–419.

YASHOUV A. (1969) The fish pond as an experimental model for study of interactions within and among fish populations. *Verh. int. Ver. Limnol.* **17**, 582–593.

Chapter 19: The Contribution of Freshwater Fish to Human Food

Georg Borgstrom

19.1 Introduction

Shellmounds around the globe, almost in every continent, bear witness to man's early dependence on either freshwater or marine molluscs for survival, subsequently to be supplemented by fish from lakes, rivers and marine coastal areas as catching devices were developed. Bones of fish in subsequent layers reveal which species were used and the sequence of their importance.

China appears to be the first country that reached the massive human numbers requiring expanded management of natural freshwaters. This took place one to two millennia B.C. and gradually fish cultivation emerged which unquestionably China pioneered (Drews 1961).

Medieval Europe depended heavily on the many lakes from the postglacial period for securing their food. Dried pike was such a dependable source of nutrition that it even took the place of money in Sweden and Central Europe. Monasteries became the centres for carp production. Both the dominant rural population as well as town dwellers sought supplementation from the rich herring catches along the ocean coasts. From the start of the 16th century on, lengthy forays were made across the Atlantic, and during these voyages the rich banks of Newfoundland were initially exploited. Up to this day the Grand Banks remain an indispensable part of European food. Then in the 18th century came the large-scale drainage programmes that paved the way for expanded agriculture in Central Europe, but jeopardized the standing of freshwater fish as an important source of animal protein. Serious opposition and complaints came from all those who were accustomed to eating cheap food from the fish of freshwaters and cultivated ponds.

Industrialization continued the large-scale destruction of the freshwater biotope. Land grabbing became critical, but far more decisive was the growing pollution of natural waters—finally knocking out freshwaters as a source of food.

469

19.2 Role in global household

How important is freshwater fish in the world household? Judged on the over-all scene, it is certainly not as significant as one would be inclined to imagine. To a large extent, due to the adverse effects of industrialization just mentioned, drastic decline has been registered in catches from freshwaters in most industrialized countries such as Japan, the U.S.S.R., western Europe, and the United States. Approximately one-sixth (14·9%) of the total global aquatic catch comes from freshwaters, i.e. around 10·2 mmt (million metric tons) (av. 1973-74) (Table 19.1). When related to the total aquatic catch

Table 19.1. Aquatic harvests, 1965–1974, in million metric tons

	1965	1966	1967	1968	1969	1970	1971	1972	1973	1974
Freshwater catch	7·9	7·9	7·8	8·0	8·3	9·2	9·7	9·1	10·2	10·2
Shellfish + misc.	4·2	4·3	4·5	5·0	5·6	6·0	5·9	6·2	6·4	5·3
Marine fish catch	40·6	44·2	47·1	50·0	48·7	*54·8*	54·6	49·5	49·5	53·3
—human food	24·2	25·4	25·6	26·0	26·2	28·2	29·1	29·1	*31·0*	*32·8*
—animal feed	16·4	18·8	21·5	24·0	22·5	*26·6*	25·5	20·4	18·5	20·5
Total	52·7	56·4	59·4	63·0	62·6	70·0	*70·2*	65·6	66·1	68·8
Total human food	36·3	37·6	37·9	39·0	40·1	43·4	44·7	45·2	47·6	*48·3*
Human population (million)	3,285									3,905

Increase, 1965–1974

Total human food	13·0 million metric tons 35·8%
Population	620 million 18·8%

Peak values italicized.
Source: FAO Fisheries Yearbook.

moving into human consumption it accounts for 20·6%. Three-fourths of this take (6·83 mmt) is attributed to Asia (Table 19.2). This corresponds to only little less than one-third of the ocean fish catch used as human food (Table 19.1). In addition there exists a widespread freshwater subsistence

Table 19.2. Freshwater catch as to continents (ave. 1973/74) in million metric tons

Asia	6·83	Europe	0·24
Africa	1·34	Latin America	0·14
USSR	0·47	North America	0·09
		Oceania	0·02

Computed by author on basis of FAO Fisheries statistics.

fishery, for which catch records are scanty. Its volume has been estimated at approximately six million tons. Furthermore, sportfishing plays a considerable role in North America, USSR, Europe and Japan, in the order they are listed here. Taking all these sources into account, it can be estimated that freshwater fish presumably provides human food in a quantity not far below the registered ocean catch of fish as such. Nonetheless, freshwater fish, as accounted for in commercial statistics, is really significant only in 4 to 6 countries, accounting for 68% to 73% of the catch in this category (Table 19.3).

Table 19.3. World freshwater catch in 1973 and 1974 (registered) in million metric tons

Country	Weight	Accumulated weight	%	Accumulated percentage
China	4·57		50·3	
India	0·77	5·34	8·5	58·8
USSR	0·47	5·81	5·2	64·0
Indonesia	0·40	6·21	4·4	68·4
Bangladesh	0·21	6·42	2·3	70·7
Nigeria	0·20	6·62	2·2	72·9
Thailand	0·17	6·79	1·9	74·8
Uganda	0·17	6·96	1·9	76·7
Tanzania	0·15	7·11	1·6	78·3
Burma	0·13	7·24	1·4	79·7
Chad	0·11	7·35	1·2	80·1
Japan	0·11	7·46	1·2	82·1
Zaire	0·11	7·57	1·2	83·3
Colombia	0·09	7·66	0·9	84·2
North Vietnam	0·09	7·75	0·9	85·1
South Vietnam	0·09	7·84	0·9	86·0
Philippines	0·08	7·92	0·87	86·87
Brazil	0·07	7·99	0·77	87·64
Egypt	0·07	8·06	0·77	88·41
USA	0·05	8·11	0·55	88·96
World	9·13			
	(10·22)*			

* Within parenthesis comprises anadromous fish captured in freshwaters.

Fish cultivation yields far more per acre than does freshwater fisheries, relatively speaking, but is forced into controlled, largely artificial waters. This may best be illustrated with an example from East Germany. The 'natural' waters yield about the same amount of fish in volume as the fish ponds, but the latter cover merely one-eleventh of the area of the natural waters. Consequently, pond yields are, as a rule, higher per acreage unit.

The global annual yield of fish obtained through cultivation, in this strict sense of the term, amounts to about 700,000 metric tons with average yields around 110 to 170 kg ha^{-1} in temperate regions and twice that figure in the tropics.

More than half of this cultivated fish is produced in China, which has retained its global leadership in this field for all of recorded history. The fish pond acreage of the Soviet Union is almost twice as large, but due to its northern location, the total yield is only one-half that of China's. About one-third of all cultivated fish on earth is harvested from lagoons along the ocean coasts in Indonesia, the Philippines, and in the brackish Po delta of Italy and other places not covered in this survey.

Fish cultivation is of crucial importance not only to China but in large parts of Southeast Asia and Central Africa. In Hong Kong, for instance, it is much cheaper to raise fish in ponds than to invest in long-distance fishing, and in China fish-raising is an indispensable source of first-rate protein. Without this resource many millions would get an insufficient amount of protein in a country where the transportation of fish, be it by air, by rail, or by road, is still too complicated and costly. In the future a mounting number of millions will bless the raising of fish as the key to their survival.

Fish farms and ponds sited in the vicinity of potential consumers offer an obvious way to avoid transport difficulties. They also may constitute an attractive cash crop. Suitable terrain and adequate water supplies are obvious prime prerequisites. Ponds may utilize lands of otherwise low value.

19.3 Sportsfishing versus commercial fishing

In the United States considerable friction prevails between sportsfishing and commercial catching, since sportsfishing in the United States nowadays provides the country with far more fish from inland waters than do commercial enterprises. Thus, in reviewing the role of freshwater fish, both sources must be taken into account. A clear distinction cannot be drawn between purely recreational fisheries and fishing in order to supplement daily food. The management of these freshwater sources is hampered by this two-pronged fishing pressure, frequently at loggerheads. In good years both overextend themselves, resulting in serious decline in the stock. A telling example of such drops is the catch of walleye (*Stizostedion v. vitreum*) in Green Bay (Wisconsin), down to 22,700 kg yr^{-1} after exceeding 455,600 kg in a good year. The whitefish (*Coregonus clupeaformis*) catch in Lake Michigan has almost collapsed due to such excessive catches after reaching 0·6 million kg in 1966.

Recreational fishing has also become something of a people's movement in the USSR with a special monthly magazine (Rybolovstvo) reviewing what is happening in this vast lake-rich empire (one-sixth of the globe's land area). Even in wintertime sportsfishing through the ice is quite popular. Few data are available as to the size of these catches, but unquestionably fish constitutes a significant supplement to the diet along many rivers and around several lakes.

19.4 Fish management

On a world scale man has seriously jeopardized freshwater commercial fish yields, requiring large-scale fish management. Such endeavours are pursued with increasing fervour and are rapidly becoming a matter of key importance as hundreds of millions of additional people demand their share of food. Protective measures thus are imperative, but so far the record of the industrialized world is rather miserable. Very few clear advances can be registered. Man has by and large been fighting a rear-guard battle in this field of endeavour. The industrialized world has reconciled itself to this very fact and most countries are inclined to believe that these steps primarily favour sportfishing and serve to keep the fish biologists busy and happy.

19.5 Ponds and fish cultivation

China was the first empire in history that felt the damaging effects of population congestion upon water resources. It is undoubtedly no mere coincidence that China actually became the birthplace of fish cultivation. China also initiated large-scale utilization of natural waters for food production and developed methods for the hatching of fish eggs and the raising of fish fry as well as the collection of young fish during spawning seasons for local ponds. Through all these measures, the foundation was laid for the monopoly China still enjoys in the exportation of fingerlings and young fish. She supplies, via Singapore, the entire Asiatic region with fish fry both for stocking in natural waters and in fish ponds.

The Chinese probably started this practice as early as 2000 B.C. A voluminous textbook on this subject, by an author named Fan-Li, was issued in the year 850 B.C. They also expanded this practice into the rice paddies. Sometimes references are given to a much later starting point, attributed to a man called Tao-chu-kung, living in the fifth century B.C. The practice could have evolved independently in distant points of the country.

The pond area of China is today 2·94 million hectares, of which approximately one-third are marine. The average productivity is reported as 862 kg ha^{-1} (Saburenkov 1962).

Artificial cultivation of fish in freshwater ponds or rotation of fish with crops of rice or soybeans has been practiced for many centuries in Asian countries. It was independently developed by the Romans who left the art behind in England, later to reach midsections of the European continent via the monasteries.

Pond culture was introduced into central Europe during the fourteenth century, and later into the Balkans (Mann 1961). In the European fish ponds, the principal species was the common carp (*Cyprinus carpio*); in Southeast Asia and Japan many other species are reared (Drews 1961).

In the Far East, pond fish are raised in relatively small bodies of water; boglands, swamps, and ravines are utilized, as well as the waters of irrigation canals and reservoirs. In many areas (China, Japan, Java, and others), fish are raised on flooded rice paddies. Annual yields of 2000 kg or more per hectare are not infrequent.

Israel started fish farming before World War II with the carp introduced from Yugoslavia; mullet (*Mugil* sp.) were brought in 15 years ago. The average annual production is at present 6,000 metric tons. The availability of dependable water resources is hampering further expansion and increased salt content of the pond water is becoming a menace.

There are more than 2 million fish ponds on United States farms, chiefly in the southern states, largely built in the last 30 years. They have been started with multipurpose aims in view: (1) as a reliable source of water for stock and irrigation, (2) for water conservation, (3) for recreation—and not least, (4) for fish raising. Initially, the farm pond evolved as a method of supplementing the daily food. Each hectare of such water could be made to yield 170–500 kg of pan-size fish each year. When well managed, they support 450–650 kg ha^{-1}. Fertilizing is required at regular intervals. After several convulsive starts in the 1950's, channel catfish (*Ictalurus punctatus*) farming has evolved into a significant commercial operation in fourteen southern states. Some 7 million kg are raised with an estimated area of 36,500 ha. Charging a small fee for pond-side fishing is a profitable sideline. Other fish species are now being tested.

Rice-fish farming, patterned after Southeast Asia, is reported from the rural areas of the southern states of the United States with favourable results. The principal species is the bigmouth buffalo, *Ictiobus cyprinella*. This fish is preferred to carp by the public. Yields of more than 1120 kg/ha have been obtained in heavy delta soils; 560 kg ha^{-1} constitutes a good average when heavy stands of stubble have been left over from the preceding rice crop.

Freshwater crawfish grow in swampy streams and impoundments in the Mississippi Basin. Only in Louisiana do they support fisheries. The densest accumulation occurs along the Atchafalaya River where 900,000 kg are caught in good seasons. Crawfishing is also a popular sport, believed to render a still greater poundage. Cultivating crawfish is rewarding but complicated. They have been reared in ponds in France since 1880 and started in U.S. in the 1950's. It has expanded into Southeast Texas. It is the expert opinion that a considerable potential market exists. Crawfish and rice have been grown in recent years in rotation on 1100 ha in Louisiana and Texas. The yields of each crop are mutually improved—sometimes as much as tripled.

Rainbow trout (*Salmo gairdneri*) is raised commercially across the northern belt of states where temperature conditions are favourable. This culture reaches its greatest yield in ponds fed by springs with large volumes of water. The rapid turnover of water washes away metabolic waste products and allows fantastic (150,000 kg ha^{-1} yr^{-1}) amounts of fish to be reared in small areas of water. Japan and Denmark hold first (17·4 thousand metric tons) and second (13·1 tmt) rank, respectively, as sources of pond-reared trout out of a world total of 48·2 tmt. There is no breakdown by species, but rainbow trout probably dominate. The frozen products of these two countries chiefly serve a limited luxury market.

19.6 Carp and tilapia

Two species have played a key role in the history of fish cultivation, namely tilapia and carp. In some regions ide (*Idus* spp.) is raised, but not as commonly. Several species and subspecies of the carp are used, of which the Chinese have bred the majority. Carp cultivation is presumed to have emerged originally as a sideline to silkworm production. The cocoons and other waste from the raising of these worms were used as feed. The Chinese not only taught the cultivation of carp to the Japanese but spread the techniques over Southeast Asia. It is frequently said that these practices never reached India, but this does not appear to be true, see further p. 480. Large-scale cultivation of carp was common in Europe in the Middle Ages, primarily in the monasteries, which were on the whole centres of advanced food production. The first European carp ponds appear to have been located in Poland (Mann 1961).

The Soviet planners, despite the promotion of deep-sea fishing on a large scale, count increasingly on carp as a locally available cheap fish. The raising of carp is flourishing all the way from central Europe deep into Siberia. As in animal husbandry on land, a constant battle is raging against diseases, in particular 'belly-burn', a common bacterial disease. It is difficult to keep under control at the northern climatic boundary for the carp, and it often gets the upper hand in the ponds. Under these strenuous conditions carp are far more susceptible than otherwise, despite the use of a considerable arsenal of antibiotics and other chemical treatment.

Mixed cultivation of carp and tilapia with greater production is reported from Uganda. Yet only 500 tons was raised in 1965 (Stoneman 1966).

Tilapia may be said to be one of the few new domesticated animals man has acquired in modern times. This fish and its prolific nature reportedly was discovered during World War II in Africa. Biologists noticed its rapid growth and its abundance in some African lakes. It was later established that tilapia were well suited for mass cultivation. There are thousands of tilapia ponds in the Republic of Congo, in the whole of East Africa, on Madagascar (with more than 90,000 ponds reported in 1970), in India, Ceylon, and Indonesia, and in the Central American republics.

Tilapia renders valuable protein food for millions of people; how many depends on how the estimate is made, but probably in the range of ten million. The carp without question provides protein for ten times that many.

19.7 Various tilapia species

More than one hundred million African silver carp (*Sarotherodon mossambicus*) now live in the ocean and river waters of China's southern and central provinces. They are allegedly the offspring of twenty fry presented in 1951 to a Chinese fishery delegation visiting Hanoi (North Vietnam). Tilapia is popularly known as 'Vietnamese fish' and is a native to the Mozambique Channel off Africa. A prolific breeder, the female spawns up to 20,000 fry a year. Within twelve months, in either salt or freshwater, these grow to a weight of 250 grams, at which time they are ready to eat.

Fish-breeders in the south of China rear tilapia in considerable quantities, despite the rather low temperatures of Chinese waters. In order to accomplish this, several aquatic institutes bred tilapias conditioned to these lower temperatures. Large-scale propagation is carried on quite successfully in the Fukien province.

Large-scale projects were initiated in Zaire (Belgian Congo) back in 1944. Indigenous tilapia species (mainly *Sarotherodon macochir* and *T. melanopleura*) turned out to be good pond fish. Yields up to 8·6 tons of fish per hectare were claimed. By 1953 several thousand ponds, some exceeding 80 ha were in operation.

Uganda followed in 1954 with slightly different species (*S. niloticus, T. zillii* and *S. leucostictus*). *Tilapia zillii* is a true herbivore but receives supplementary feed in the form of grass clippings and leaves. The other two species are plankton- and detritus-feeders and normally receive no extra feed except indirectly by means of fertilization. By 1956 some 5000 ponds were in operation, some of which were poorly dug. Well managed ponds render 650 kg ha^{-1} a year but several have yielded poorly or been abandoned. In the sixties considerable research and innovation further improved productivity (Stoneman 1966).

19.8 Other species

Freshwater shrimp production has made major advances in recent years and is practiced on a large scale in China and Japan. In special hatcheries the young ones (two million eggs from each female) are released into tanks of special construction for feeding and growth. The Japanese goal is an annual production of 2,000 metric tons.

Japan has no less than 760 eel (*Anguilla* sp.) farms, a further development of the cultivation of this fish by the Moors in Albufera's wet rice fields near Valencia in Spain. The Japanese also raise salmon (*Oncorhynchus* sp.), trout (*Salmo* sp.), ayu (*Plecoglossus altivelis*) and carp.

19.9 Fish culture in rice paddies

Fish rearing in rice paddy fields dates back to emperor Wu of the Wey Dynasty (Third Century A.D.), and was commonly practiced in pre-war China. During the last twenty years cooperatives and communes have been encouraged to cultivate fish in paddy fields (in the south claimed to cover several 100,000 ha). This method is claimed to be beneficial. Fish dig the bottom of the fields, aerate the soil and eliminate harmful mosquitoes and other insects. The fish fertilize the soil with their faeces.

This method has, however, in most parts of the world entered into a new, almost critical phase, due to the urgent need to maximize rice yields, which makes spraying against insects and fungi almost indispensable. This excludes fish-raising because of the toxicity of the chemicals used. Such joint cultivation has been practiced in Japan for centuries, but it is now reportedly vanishing for this very reason. India has resorted to deep small ponds in a corner of the paddy field in order to avoid the poisonous sprays.

Nevertheless, treatises and popular reviews elaborate on the enormous potentialities of rice fields for fish-raising and on the advisability of introducing these age-old Chinese and Japanese procedures to all rice-growing areas. Millions of tons of fish are anticipated, but it is not always recognized that only one-third of the world's total rice fields are paddies sown with wetland rice—and it is only in such rice fields that fish can be cultivated.

19.10 Dual carp and duck ponds

In the 1960's agricultural and fishery co-operatives in Hungary started stocking their duck ponds with carp. The practice, which began as an experiment about a decade ago, is being carried out on a growing scale. The carp and the duck are carefully tended until they develop into flourishing colonies, then harvested and sent to market. The ponds are drained, and rice, maize and other crops are planted in the exposed bottom. The resulting harvests are, even on poor soils, 15–20% higher than for similar crops grown elsewhere.

After each harvest the ponds are re-flooded and the process is repeated, generally on a three to five-year cycle. Thus, a continuous chain is established and the same plot of normally unproductive land is made to yield fish, poultry, and crops. The ponds vary in size from a few hectares to several hundred hectares and average 1·3 metres in depth. Artificial islands in the middle ensure havens where the ducks can feed from dispensing machines, set in the water.

19.11 Reservoirs

A great deal has been written not only about how man taps and pollutes natural waters but also how he creates new waters by such means as storage dams for hydroelectric plants and for irrigation. In particular, Soviet fishing journals have carried numerous visionary articles about the large quantities of fish that would result from exploiting fully the more than five million hectares (12·5 million acres) of huge reservoirs of various types largely created since the end of World War II. Some of them have almost been looked upon as high-producing fish ponds. Their depth, however, is the first major obstacle to a complete utilization of these water volumes. To introduce life to all levels of these manmade lakes has proven to be quite difficult. In addition, most estimates of potential yields disregard the fact that the productivity of the reservoirs in most cases is highly limited due to the cold environments in many northern locations. Only during a few weeks annually do light and temperatures allow a reasonable growth of the primary producers in the latitudes of Siberia. The United States has experienced on a minor scale these same difficulties. In arid and semi-arid areas, reservoir levels fluctuate considerably, in turn affecting fish production and introducing seasonality.

19.12 The marauders of fish cultivation

The ingenious combination of rice cultivation and fish raising has been practiced by the Chinese and Southeast Asians for centuries. By and large this system functions quite well, and it has been successfully introduced in the rice paddies of Arkansas. But the difficulties are rarely mentioned in these writings. Many rice paddies in Southeast Asia and India are infested by both landcrabs (*Coenobita* spp. or *Birguo* spp.) and seacrabs, which operate in quite a devastating way, especially during the young stages of the rice plants. The crabs themselves drill into the walls of the dams and cause rapid deterioration and serious damage. In Burma the crab menace is so serious that the replanting of rice has to be postponed until the marauders depart for the sea after spawning at the end of July. Efforts to eradicate these pests have so far been futile despite energetic attempts.

Birds often become a hazard to ponds and dams. Birds sometimes also constitute a menace to fish populations in lakes and rivers. But it becomes much more serious in the case of pond cultivation, where the fish stock is much denser. Such stock frequently has to be protected by nets above the water level. The economic losses may reach considerable proportions, and bird eradication becomes necessary as a protection. The higher man pushes yields, i.e. the more lavish tables he spreads in nature, the more serious these attacks become.

Fish parasites are frequently encountered as cumbersome in pond cultivation, inducing mortality or imposing health hazards on prospective consumers. Tilapia seems to be less susceptible to such infestations.

19.13 Asia

Asia remains in the lead with three-quarters of the total registered catch recorded for the whole world. Three countries (China, India, and Indonesia) account for 84·5 % of this total (Table 19.4).

Table 19.4. Freshwater fish catch (registered, ave. 1973/74) in the countries of Asia in relation to the total catch of 6·83 mmt.

Country	Million metric tons	% of Asian portion	Accumulated percentage
China	4·57	66·9	—
India	0·77	11·3	78·2
Indonesia	0·40	5·9	84·1
Bangladesh	0·21	3·1	87·2
Thailand	0·17	2·5	89·7
Burma	0·13	1·9	91·6
Japan	0·11	1·6	93·2
North Vietnam	0·09	1·3	94·5
South Vietnam	0·09	1·3	95·8
Philippines	0·08	1·2	97·0
Pakistan	0·03	0·4	97·4
Sri Lanka	0·01	0·1	97·5
Others	0·17	2·5	100

FAO Fisheries Yearbook.

19.14 China

Only in China does freshwater fishing occupy a prominent position nowadays accounting for 60 % of its national aquatic harvests, and half the globe's registered take from such waters.

Inland landings

1953	1·5 mmt	(45·6% of total)
1965	3·5 mmt	(61·5%)
1970	4·2 mmt	(60%)
1975	4·8 mmt	(60%)

The total of inland freshwater area in China (excluding paddy fields) has been estimated at 18–20 million hectares (Solecki 1966), of which one-third could be used for fish culture (Saburenkov 1962). This encompasses 1,600 rivers and 70–80 lakes covering some 0·6 to 14,000 hectares. Another 1·5 million tons of fish is reportedly obtained from paddy fields, totaling 18 to 24 million hectares. Furthermore there are some 80 reservoirs ranging in size from one to 230,000 hectares (Denisov 1961).

Four rivers exceed 2,000 km in length, namely the Yangtze, the Hwang, the Pearl, and the Amur. China counts more than 500 rivers with basins in excess of 100 square kilometres.

In the freshwater areas of China more than 500 species of fish are counted, of which 30 to 40 are commercially important. All inland water resources belong to the government since the agricultural reforms of 1950–1952.

As far back as historical records go, fishing has played a major role in the economic life of China. This was recognized in the 12th century B.C., when Chiang Tsu-ya developed fishing and salt industries during the reign of Wen Wang. The first pisciculturist is believed to have been Tao Chu-kung, who lived in the fifth century B.C. (Yen 1910). Fish culture in paddy fields dates back to the third century A.D. The fishing laws in old China were very strict and the fishing industry enjoyed protection (Radcliff 1926).

Freshwater fish yield has played a leading role in providing protein to the Chinese people, and still is three times larger than the marine catch. Fishes of a kind which in Europe are scarcely deemed worthy to be caught, and then for sport rather than food, are utilized in China and form the base of an extensive commerce as the staple food of large sections of society. Due to decades of war, the fishing industry was in considerable disarray when the Mao regime was installed in 1949, but has subsequently enjoyed high priority.

19.15 India

India has an abundance of freshwaters, good fish stocks, a large number of fishermen and has for centuries practiced fish cultivation. Availability of water is, however, not dependable even in West Bengal and neighbouring states. Many rivers have been fished since far back in time, but commercial fishing is mostly concentrated in the river mouths which constitute the prime resource. Many impoundments, largely built in the last 20 years, constitute additional assets. Countless tanks and ponds have been created through the ages. Lakes, swamps and lagoons exist in various sections of the country. The total area of cultivatable waters has officially been estimated to be 0·5 million hectares. Ponds (half a million in number) and tanks cover 0·24 million hectares (Sen 1975). In contrast to China and USSR, reservoirs have only occasionally been the site for the development of fisheries, largely on a research basis. Lagoons exist in many places along the lengthy coastline. They are mostly brackish or marine and many are considered as having great potentiality. Few have been developed for commercial production.

Four carp species have been cultivated through the centuries by employing indigenous methods. They are commonly eaten, chiefly in the northeastern states. These fish spawn in the rivers, primarily the Ganges, and an elaborate system for collecting, transporting and rearing fry and fingerlings has evolved.

Waterweeds, both floating and submerged, have greatly hampered fish cultivation. Many of the ponds, tanks and other bodies became run down due to absentee ownership and extensive silting, paving the way for the choking weeds. Many tanks and ponds require redesigning to allow effective fish production.

Inland fish production is low as a consequence—100,000 mmt year^{-1} With aid from FAO, however, efforts are under way to treble or quadruple this production and restructure many old ponds and tanks. A modernized system combined with hatcheries was initiated in 1957 to supplant collecting from rivers. Chinese carp was added to the list of producing fish. Cross breeding resulted in new and better strains. Fry transportation has been restructured, but it has not been possible to fill demands, and 10–15,000 fishermen are still engaged in traditional harvesting from natural waters.

Tilapias were introduced in the Madras region around 1952 and has spread throughout the south. At one research station it has yielded 5,780 kg/ha. In 1947 mirror carp was brought in and has stimulated fish production in several areas. Artificial feeding, employing rice bran and peanut cake, is generally practiced because natural food in the water only supports limited production.

Fish cultivation in rice paddies has been hampered by insecticides and herbicides used to achieve higher rice yields. As mentioned before, this is solved nowadays by creating a deep pond in a corner of the rice field.

19.16 Indonesia

One-third of the aquatic catch of Indonesia comes from internal waters (400 thousand metric tons in 1973/74) divided in the following manner:

lakes and rivers, 73%; brackish waters, 12%; cultivation ponds, 10%; rice paddies, 5%.

One-sixth of the population is involved in one kind of fish farming or another (Schaible 1967). Since 1968 the trend has reached a plateau. The average yield from freshwater cultivation is a modest 200–400 kg ha^{-1}. Carp, gourami, and tilapia are the dominating species. Due primarily to continued deforestation, the marshlands, reservoirs, and ponds are silting up. Insecticides infiltrate many waters and are most sharply felt in the rice paddies. Furthermore, there is a manpower shortage and lack of fingerlings for stocking.

19.17 Africa

Freshwater fish, presumably with a tradition far back in history, has played a significant role in the feeding of this continent. In relative terms Africa may be said to top Asia in this regard (see further Table 19.5). The freshwater fisheries of the African continent are concentrated chiefly in three areas: (1) Lake Victoria, Lake Nyasa and adjacent areas; (2) the Nile; and (3) the Congo basins. As to countries, Nigeria is the leader with 0·26 mmt. Chad and Zaire together represent 0·23 mmt. Tanzania and Uganda together account for 0·18 mmt. The Victoria waters are rich in fish, but the major food catch of the lakes consists of various tilapias. Since time immemorial, fish have been a great asset to the dense population living around these shores (Hickling 1961).

Nigeria has the largest African catch from freshwater sources extracted from its many internal rivers and Lake Chad. Many brackish lagoons make a further contribution. There is a strong drive to expand marine fisheries, yet freshwater fish play a major regional role. The Nigerian sector of Lake Chad produced around 15,000 tons in 1973. It is considered possible to treble that catch without any danger of over-exploitation.

Fish farming is strongly supported both in ponds and reservoirs and several experimental farms are underpinning this drive (Olatunbosun et al. 1972). Community ponds in public hands engaged in commercial fish farming are reported as quite successful. Regional freshwater operations, nevertheless, fail to keep up with the rapidly growing population. There is currently a critical overtaxing of available game meat on which three-fourths of Nigeria's population depends. Against this background, the great emphasis placed on expanding ocean fisheries in recent development programmes is understandable. Nigeria has been depending on imported dried ocean fish from Norway, Canada, and other countries for many years. Recent bans on such imports has increased the urgency of a more coherent policy

Table 19.5. Freshwater catch of African countries as percentage of total aquatic catch, based on an average of 1973 and 1974

African country	Total catch in thousand metric tons	Freshwater catch	%
Nigeria	678·4	203·5	30·0
Uganda	167·5	167·5	100
Tanzania	167·7	144·7	86·3
Zaire	123·9	111·3	89·8
Chad	110·0	110·0	100
Mali	90·0	90·0	100
Mauritania	73·0	23·3	31·9
Egypt	94·9	66·8	70·4
Cameroon	71·6	50·0	69·8
Madagascar	61·6	45·0	73·1
Ghana	209·5	42·8	20·4
Malawi	41·4	41·4	100
Zambia	36·9	36·9	100
Kenya	29·2	25·0	85·6
Dahomey	32·9	22·5	68·4
Niger	15·7	15·7	100
Burundi	9·8	9·8	100
Congo	20·0	7·5	37·5
Ivory Coast	62·9	7·5	11·9
Liberia	23·0	4·0	17·4
Cent. Afr. Rep.	3·5	3·5	100
Upper Volta	3·5	3·5	100
Togo	11·2	3·0	26·8
Rhodesia	2·0	2·0	100
Botswana	1·2	1·2	100
Rwanda	1·0	1·0	100
Gambia	6·0	0·8	13·3
Gabon	4·0	0·4	10

FAO Fisheries statistics.

of developing in a more rational manner the lacustrine and riverine fish wealth.

On the basis of Nigerian data (FAO & Olatunbosun *et al.* 1972) fish account for almost as much animal protein as all meats and 2·7 times the bushmeat input—5·5 times that of milk. Freshwater fish is 41% of fish consumption. The fish acreage* is 90% of the tilled land of which the freshwaters stand for 73·5%, which gives a good idea of the significance of this source of protein.

* Calculated as the tilled acreage required to raise an equivalent amount of animal protein under present state of agricultural techniques (Borgstrom 1962).

Lake Victoria is considered the most heavily exploited of the great lakes of Central Africa, serving Uganda and Tanzania, even to the point of risking overfishing. This is reflected in the efforts to regulate mesh sizes and improve marketing facilities. A more equitable geographical spread of the fishing effort is needed on most of the large lakes of Central Africa in order not to allow the rapid growing numbers of consumers to overwhelm basic productivity. Early fishing was determined by the subsistence requirements of the tribes living around the lake near the shore. Fish was plentiful and dominated by *Sarotherodon esculentus*. Daily requirements could be procured with little effort using indigenous methods, but the warm climate and lack of preservation prevented marketing. Nets were only introduced in 1898–1909 (Graham 1929).

A flourishing gill net fishery developed in the 1910's in Kavirondo Gulf, presumably the richest fishing ground. Drying the fish and transporting them by a newly opened railroad opened a large market. In other parts of the lake, Asian traders likewise developed a trade in dried fish. A further upturn in total production occurred in the late 1920's. Greatly expanded markets for this dried fish were created when many of the local people moved to towns and plantations but remained attached to their traditional eating habits. Alternative meat sources were presumably more expensive. The demand became so great that illegal mesh sizes were used to increase the catch. This led to crucial stock depletion (Garrod 1961). The fallacy has been perpetrated that biological overfishing rarely can take place. This might be true of fecund temperate species, but it is a real danger with the mouth-brooding tropical tilapias which, in contrast, produce very few eggs.

In Uganda an increasing proportion of income has been spent on fish (Crutchfield 1958). But most of the revenue ensuing from fish consumption ends up with the middleman who becomes the chief beneficiary while the fisherman even in expanding markets is shortchanged.

19.18 USSR

Since 1958 USSR is second ranking in world aquatic catch, but holds third place globally in the size of freshwater catches. Expanding irrigation, together with the effects of the rapid industrialization (pollution, ascending figures for water usage, hydroelectric plants), has drastically affected freshwater fishing.

Despite a series of almost desperate measures, large inland waters such as the Caspian Sea, the Black Sea, the Sea of Azov, and numerous other larger and smaller lakes, among them the often praised Lake Baikal in far-off Siberia, show steadily decreasing catches. In the Caspian Sea they are down to two-thirds of those of the thirties, and a large portion of the fish caught are less suitable as direct human food and go into the manufacture of animal feed.

The growth of the cities and the industrialization process follow the European and North American pattern exactly. The waters serve as recipients for mounting volumes of waste of various kinds and are simultaneously tapped at an accelerating pace to fill the water needs of population centres, of irrigation, and of industrial production. Besides, large hydroelectric plants, several with huge dams in key areas already short of water, interfere with water distribution and induce increased evaporation losses. The pollution crisis has hit the Soviet Union much faster than in Europe and North America, mainly due to the fact that its water reserves are in relative terms infinitely more scarce. They become even more scarce the further one moves eastward into the land mass of this giant. Soviet scientists made an ardent appeal for drastic measures to save Lake Baikal, with its unique flora and fauna. Among the latter is the omul (*Coregonus autumnalis*), indigenous to the lake which is in jeopardy, despite repeated attempts to save it. Artificial waters (irrigation canals, impoundments, dams, etc.) offer some compensation. Official support to maintain village ponds and special fish-raising areas in cities constitute a major endeavour to hold the freshwater front. Fishing from natural fishing waters is greatly supported by stocking with fingerlings and food organisms. Proportionately, the greatest expansion is taking place in the large Siberian rivers with the mouth of the Ob being the largest source. This vast body of water is already delivering more fish than the entire Arctic Sea waterfront.

19.19 United States

The Indian population on the Great Plains has been estimated to have been not more than 150,000 when white settlers moved into the area. Their catch has been estimated as 6·8 million kilograms. Game, waterfowl and wildlife dominated as the source of animal proteins, but fishing was essential along waterways and around lakes. Both in the northeast, the northwest, the Great Lakes as well as in the south, ingenious devices and contraptions were used to capture fish. Literally two hundred different kinds of fish were abundant. No wonder fish became the mainstay in the diet of early settlers. The Great Lakes became the centre of flourishing commercial fisheries serving the many urban centres that came into being. Catches started to decline in the 19th century resulting in broad scale efforts in propagation and stocking towards the latter part of that century, reaching a peak in 1899.

Great Lakes fisheries is nowadays under strict regulation, Lake Michigan and Lake Erie contributing most to the total catch and Lake Ontario very little. About half of the fish taken in the Great Lakes today come from Lake Erie and three-fourths of this from the Canadian side. Most of this (95%) goes to large cities. Statistics are easily misleading. Both in 1897 and 1968 116 million lbs of fish were taken from the lakes. This could indicate a considerable stability over a period of some 70 years. But in 1897 the biggest single catch was that of lake herring, *Leucichthys* spp., or cisco. In 1968, the perch (*Perca flavescens*), earlier not considered a particularly good fish, had increased, but most of the catch was now chiefly smelt (*Osmerus mordax*), carp, channel catfish and alewife (*Alosa pseudoharengus*). Only in Lake Superior was herring still a major catch (Cable 1971).

The Great Lakes fishery resources, once called the world's largest and most valuable, have suffered one abuse after the other. Lake Ontario was once the spawning ground of the Atlantic salmon (*Salmo salar*). Milldams, forest cutting and waste disposal ended this in 1890, when the salmon no longer could reproduce. Sturgeon (*Acipenser* sp.) was intentionally eradicated. Overfishing and pollution brought down the lake herring catch to a few million kg from catches which were 30 to 50 times larger in the mid-1920's. The lamprey (*Petromyzon marinus*) and the alewife (*Alosa pseudoharengus*) moved in from the Atlantic through the canals, upsetting the species balance. Alewives, valueless as human food, constituted four-fifths of the Lake Michigan catch in 1966. Various pollutants reached critical levels and accentuated the destructive trends. Sewage, fertilizers, industrial waste, made a dangerous mix. Mercury started appearing in fish in hazardous concentrations, particularly in piscivorous fish, those most desirable as human food. DDT and other pesticides evolved as another danger to aquatic life and human consumers. Finally PCB's (polychlorinated biphenyls) appeared as a true monster. Most of these dangerous chemicals or their breakdown products accumulate in the fish and reach levels resulting in withdrawal from the market. Sportsfishermen are occasionally not allowed to keep their fish for consumption. Estimates currently place the catches of sport fishing as larger than the commercial catches (Shapiro 1971).

The Great Lakes provide the United States and Canada with about 60,000 tons of fish a year, but in comparison to the immensity of these waters, yield per unit area is low. The water surface of lakes and rivers open to commercial fishing is about 31,000 square kilometers, i.e. 20% of the United States part of the Great Lakes. Yet the Mississippi basin landings equal or in some years exceed those of the Great Lakes. Nine-tenths of those catches are dominated by carp, buffalo (*Ictiobus* spp.), catfish (*Ictalurus* spp.), bullheads (*Ameiurus* spp.), and sheepshead (*Aplodinotus grunniens*).

Many natural and artificial lakes as well as farm ponds dot the basin of the mighty Mississippi. Several dams and impoundments have been built to control flooding and provide hydroelectricity. The upper Mississippi has been transformed from a freeflowing river to a series of slack water pools creating new habitats for productive commercial and sport fisheries. The same can be said for the main tributary of the Missouri.

Official catch statistics for the Mississippi basin are not representative of the actual catch from the shores of lakes and streams. Not even all catches marketed get into statistics. Catches vary considerably from year to year. This partly explains the upsurge in fish farming, particularly in Arkansas, Louisiana, and Mississippi. This is done in impoundments or artificial ponds. Channel catfish is the prime species.

19.20 Latin America

Latin American inland fisheries are important locally as a food resource in the lake regions in Bolivia, Argentina, Brazil, Venezuela, Peru and Paraguay and of Central America. Lake Titicaca has, right through history, sustained people around it with fish protein and does so still today, despite modest operations. The Amazon River, the world's largest freshwater river, contains many species, among them the 'pirarucu' (*Arapaima gigas*). This gigantic fish, which weighs up to 90 kg and is found only in the Amazon, is dried and marketed as a substitute for cod (*Gadus morhua*). In most of Latin America, however, freshwater fisheries are operated by fishermen owning minor fishing craft and a few hooks and lines. Lack of capital blocks the introduction of modern equipment.

19.21 Europe

Freshwater fishing in Europe—totaling 0·75 mmt—has continued its decline and only plays a substantial role around the lakes of Scandinavia, Finland in particular. Sportsfishing is still active in waters protected by conservation regulations. Norway, Denmark, and the United Kingdom have led the efforts to raise trout in artificial ponds. Eastern Europe still registers catches from many rivers and lakes. Spain pursues cultivation frequently based on impoundments created by the Arabs in earlier times.

19.22 The devastating effects of pollution

Lakes, rivers, and other freshwaters have poorly withstood the blessings of human civilization. Almost everywhere man has leaned upon these waters not only as the great life givers, but also as cleansers. They have served as recipients for a major part of human waste both directly and indirectly, wastes which frequently became excessive wherever people concentrated in cities and densely populated areas.

Europe felt the consequences of this large-scale pollution of nature much earlier than did North America. The waters became overburdened with organic matter; the oxygen content became depleted, accumulations in the bottom started fermenting, malodorous substances surfaced, and diseases spread. The initially favourable effect of the added nutritive substances on fish and other living organisms soon had the reverse influence, inducing death by suffocation. Then modern industry emerged, adding immensely to the pollution load, frequently aggravated by toxic matter.

In many regions fish management as a kind of repair and maintenance service had to be organized, later followed by water protection measures. Both North America and Europe are now in the midst of this second stage.

Extensive management programmes have been only partially successful. The Great Lakes have not returned to the position of a vital reserve of freshwater fish they once had. This would require far cleaner water. Several major European rivers have in recent decades become far too polluted to sustain fish life. These cases are, however, merely a few innumerable others in North America, Europe, and the USSR. Many instances have also been reported in recent years from China, India, Indonesia, and other major countries, where waters close to steel mills, fertilizer plants, oil refineries, mining enterprises and other industrial plants have affected freshwater fish in a detrimental manner and thereby affected the nutritional base of food-short countries.

One more dimension needs to be added, namely the spread of bacterial diseases and parasites via sewage, another effect of human congestion that undermines sustenance.

In many regards Japan is the most heavily affected, since industrial pollution has eliminated most of the inland waters in all heavily populated areas and has spread out into the coastal marine areas.

From the point of view of food, water pollution meant less to eat. The fish yields diminished drastically, in many instances to zero. Fish species of such high food value as salmon, trout, sturgeon, and walleye yielded to such less palatable species as whitefish, pike, loach (*Cobitis* sp.), and others. Lakes, rivers, and other freshwaters gradually lost their importance as food resources, and in spite of desperate and costly efforts to save the stock. These efforts were, however, principally in vain, as the pollution mounted with the growth of the numbers of humans to feed.

For decades a lively debate has gone on among fish experts regarding the value of artificial stocking of fish. It was once looked upon as a rather self-evident cultivation measure comparable to the seeding of fields. This notion overlooked the fact that the procreative capability of fish, the number of eggs laid by each female, is so enormous that the small contribution man can make is miniscule. Nature works with an extremely high waste percentage. The Western World has, furthermore, neglected a series of other factors, which the Chinese have so far felt very little, but which might become serious in that country's present industrialization drive. It is of little avail to plant young fish in waters which are so strongly polluted that the oxygen supply is too low to enable the fish to survive. Such steps become even more futile if the waters are poisoned through pollutants. Fish conservation is useless when the optimum ecological requirements are not met, let alone the basic conditions for simple survival. To this might be added that adverse biological factors other than man's chemicals and wastes might induce drastic cuts in the stock of some fish. To pursue a man-led implantation policy then becomes foolish obstinacy and is doomed under the strict ecological framework prevailing.

The blue revolution (food from aquaculture) is more realistic in the well-to-do world that can pay the price in capital, engineering, feed and energy. In other words, this world is more likely to pay the high cost of such foods. It is even visualized that fish cultivation might become big industry in the future with ranches with automated control of feeding, harvesting and processing. Natural waters, if not polluted, are visualized to be invaded by cage culture of fishes. Culture is also likely to be employed in inevitable massive recycling of sewage.

Several developing countries in S.E. Asia, tropical Africa and Central America, extend cultivation of freshwater fish as a way of mitigating protein shortage. They have the advantage that temperature is not a limiting factor and pollution is not as imminent. Projections into the future even talk about a blue revolution as a counterpart to the green one. It is claimed that it is more profitable in Taiwan to invest in aquaculture than either in ocean fishing or pig production (Shepherd 1975). Yet, in giving yield figures the added acreage involved in raising supplementary feed is frequently not taken into account. This explains why it may very well be more profitable to pond-produce one ton of carp as in parts of Asia than raise 2 to 3 tons through added fertilizers and feed as in Israel.

19.23 Recycling systems

Reservoirs built for the conversion of urban waste and sewage into food fish exist in several Soviet and East German cities, but the cold climate raises serious obstacles to large scale production of this kind in northern latitudes. A century before William the Conqueror landed on the English coast (1066), the city of Angkor stood in all its splendour as the capital and centre of the Khmer empire in Southeast Asia. The impressive marble palaces, which still stand as ruins in the virgin forest of Cambodia, were surrounded by a system of fish ponds. Most waste was utilized and converted into food fish. Costly transportation was superfluous, and the protein needs of the population were met. The calories were largely produced in the surrounding rice paddies. This ideal recycling can still serve as an example to follow.

We can foresee the day when the large sewage treatment plants that serve the rapidly growing cities will become huge centres for food production. Sewage constitutes a much too valuable raw material to be discarded, dumped, or burned; still less acceptable is diluting it with water by using lakes and rivers as recipients. Most likely, sewage will in the future be fed into huge algal factories. The algae may be utilized as animal feed, but a simpler way is to produce protein by cultivating yeast and fungi, employing the algae as substrate. Both yeast and algae, however, may also serve fish-raising directly in combination with duck raising.

This is an old Chinese method which is currently being successfully employed in the U.S.S.R. and East Germany. Fish cultivation in combination with pig raising, practiced on some United States farms, is also of Chinese origin. The pig sheds are built so that they hang over the rim of the pond; the droppings fall directly into the pond, fertilizing the water. It is important that not too many pigs are kept for each pond, otherwise the water easily becomes overloaded, causing severe imbalance. In Thailand poultry pens are constructed in a similar way, and the nutrient-rich droppings go directly into the cycle and yield new food.

This kind of direct conversion may require considerable sanitary precautions when practiced on a very large scale. It is essential that such future recycling centres of our spaceship Earth do not become focuses for the propagation and spread of epidemics. Much study is needed to avert this calamity.

References

BORGSTROM G. (1962) Fish in world nutrition. *Fish as Food* Vol. II. (G. Borgstrom, ed.), pp. 267–376. Academic Press, New York.

CABLE L.E. (1971) Inland fisheries. *Our Changing Fisheries* (S. Shapiro, ed.), pp. 324–358. U.S. Govt. Printing Office, Washington, D.C.

CRUTCHFIELD J. (1958) A report on the structure of fish marketing in Uganda. Mimeo, FAO, Rome.

DENISOV L. (1961) Orudia technika lovaryby na vnutrennikh vodoyamakh Kitaisko Narodnoi Republika. *Rybn, Khoz.* 6, 77–79.

DICKINSON J.C. III (1974) Fisheries of Lake Izabel, Guatemala, *Geogr. Rev.* 64(3), 385–409.

DREWS R.A. (1961) Raising of fish for food in Southeast Asia. *Fish as Food*, Vol. I. (G. Borgstrom, ed.), pp. 121–143. Academic Press, New York.

GARROD D.J. (1961) The history of the fishing industry of Lake Victoria, East Africa, in relation to the expansion of marketing facilities. *E. African Agric. & Forestry J.* 27, 95–99.

GRAHAM M. (1929) *The Victoria Nyanza and its Fisheries.* Crown Agents, London.

HICKLING C.F. (1961) *Tropical Inland Fisheries*. Longmans, London.

MANN H. (1961) Fish cultivation in Europe. *Fish as Food*, Vol. I. (G. Borgstrom, ed.), pp. 77–102, Academic Press, New York.

OLATUNBOSUN D., IDUSOGIE E.O. & OLAYIDE S.O. (1972) The role of fish and animal products in Nigerian agricultural development and nutrition. *J. Ecol. Food and Nutrition* **1**, 235–243.

RADCLIFF W. (1926) *Fishing From The Earliest Time*. 2nd ed. John Murray, London.

SABURENKOV N.N. (1962) Rybolovstvo v Kitaisksi Narodnoi Republike. *Rybn. Khoziastvo* **9**, 85–89.

SCHAIBLE Ph.J. (1967) Feedlots under water. *Proc. Distillers Feed Research Council* **22**, 19–31.

SEN S. (1975) *Reaping the Green Revolution*. Orbis, New York.

SHAPIRO S. (ed.) (1971) *Our Changing Fisheries*. U.S. Govt. Printing Office, Washington, D.C.

SHEPHERD C.J. (1975) General review of the present state of the biological and economic development of the fish farming industry. *J. Roy. Soc. Arts* **45**, 807–809.

SOLECKI J.J. (1966) *Economic Aspects of the Fishing Industry in Mainland China*. Office of Naval Research, U.S. Dept. of the Navy and University of British Columbia.

STONEMAN J. (1966) Development of fish farming in Uganda. *E. Afr. Agric. & Forestry J.* **31**, 441–444.

SURROCA C. (1971) La pêche sur la Comoé (Côte d'Ivoire): L'emprise des immigrants ghanéens. *Cahiers d'Outre Mer* **24**(93) 75–93.

YEN W.W. (1910) The fisheries of China. *4th Intern. Fisheries Congress*. U.S. Govt. Printing Office, Washington, D.C.

Species Index

492

Author Index

The figures *in italics* refer to the pages on which the complete references are cited.

499

Subject Index